Detrital Mineral U/Pb Age Dating and Geochemistry of Magmatic Products in Basin Sequences: State of the Art and Progress

Detrital Mineral U/Pb Age Dating and Geochemistry of Magmatic Products in Basin Sequences: State of the Art and Progress

Editors

Wilfried Winkler
Albrecht von Quadt

MDPI • Basel • Beijing • Wuhan • Barcelona • Belgrade • Manchester • Tokyo • Cluj • Tianjin

Editors
Wilfried Winkler
Department of Earth Sciences
ETH Zurich
Zurich
Switzerland

Albrecht von Quadt
Department of Earth Sciences
ETH-Zurich
Zurich
Switzerland

Editorial Office
MDPI
St. Alban-Anlage 66
4052 Basel, Switzerland

This is a reprint of articles from the Special Issue published online in the open access journal *Minerals* (ISSN 2075-163X) (available at: www.mdpi.com/journal/minerals/special_issues/DMDGMP).

For citation purposes, cite each article independently as indicated on the article page online and as indicated below:

LastName, A.A.; LastName, B.B.; LastName, C.C. Article Title. *Journal Name* **Year**, *Volume Number*, Page Range.

ISBN 978-3-0365-4954-5 (Hbk)
ISBN 978-3-0365-4953-8 (PDF)

Contents

About the Editors

Wilfried Winkler

Wilfried Winkler is Professor emeritus at the Department of Earth Sciences of ETH Zurich. He graduated in 1977 at University of Fribourg/Switzerland, where he subsequently obtained a Ph.D. in geology in 1981. From 1981 to 1988, he was a postdoctoral fellow and lecturer at both Fribourg and Basel Universities. From 1988 to 2013, he was a senior researcher and professor at the Geological Institute of ETH Zurich. His research interests concern orogenic processes, the relationships between sedimentation, climate and tectonics, i.e., basin analysis with a focus on provenance and age of detrital grains (e.g., zircons). Whilst his earlier research was decided on Alpine, Carpathian and Pyrenean flysch and mélange formations, since 1991 he co-leaded several Master's, Ph.D., and Postdoctoral projects in the Andes of Ecuador, Colombia, and Peru, in Egypt and Morocco, Iran, Myanmar, and Mongolia.

Albrecht von Quadt

Albrecht von Quadt is Senior Scientist (retired) at the Department of Earth Sciences of ETH Zurich. He graduated in 1980 at University of Goettingen/Germany; in 1981 he started a Ph.D. project at ETH "Geochronological and isotope-chemistry investigation of the Scheelite deposit of Felbertal (Austria)" and obtained the Ph.D. in 1985. From 1986 to 1988, he was a postdoctoral fellow at ETH Zurich involved with projects working on the origin, geochemistry and age dating (U-Pb zircon, Sm-Nd minerals) of eclogitic rocks (Bohemian Massif, Alps). From 1989 to 2020 he was a senior researcher, lecture, and leader of the Geochronological Laboratory at the Institute of Crystallography, Institute of Mineralogy and Petrology and Institute of Isotope Geochemistry and Mineral Resources of ETH Zurich. His research interests concern magmatic processes, the life-time of magmatic processes, the relationships between magmatism and time of mineralization, new concepts of analytical approaches to benefit the quality of the clean lab. At the beginning of his career his research was decided on Alpine, Bohemian Massif, since 1996 he co-leaded several Master's, Ph.D., and postdoctoral projects in the ABTS of Eastern Europe, Andes of Argentina (e.g., Bajo de la Alumbrera), North America (Bingham), OK Tedi (Papua New Guinea), Batu Hijau, (Indonesia), in Myanmar, Mongolia, Namibia, and Botswana.

Preface to "Detrital Mineral U/Pb Age Dating and Geochemistry of Magmatic Products in Basin Sequences: State of the Art and Progress"

Our motivation to edit this Special Issue came from our conviction that various methods of dating detrital minerals have set a milestone in the study of basins, their formation and development as controlled by tectonic processes. Detrital grains in sandstones reflect the rocks that were on the surface of the basin margins. Direct dating of the age of grains by various methods reveal the age of the rocks and their exhumation. In addition, geochemical signatures provide information about the genesis of these grains. This type of provenance analysis has a high potential to support field research with the development of further technologies. However, in our opinion, it cannot not replace it.

Our thanks go to the numerous authors who have made their very valuable research results available.

Wilfried Winkler and Albrecht von Quadt
Editors

minerals

MDPI

Editorial

Editorial for Special Issue "Detrital Mineral U/Pb Age Dating and Geochemistry of magmatic Products in Basin Sequences: State of the Art and Progress"

Wilfried Winkler * and Albrecht von Quadt

Department of Earth Sciences, ETH Zurich, 8092 Zurich, Switzerland; albrecht.vonquadt@erdw.ethz.ch
* Correspondence: wilfried.winkler@erdw.ethz.ch

Citation: Winkler, W.; von Quadt, A. Editorial for Special Issue "Detrital Mineral U/Pb Age Dating and Geochemistry of magmatic Products in Basin Sequences: State of the Art and Progress". *Minerals* **2022**, *12*, 580. https://doi.org/10.3390/min12050580

Received: 29 March 2022
Accepted: 29 April 2022
Published: 4 May 2022

Publisher's Note: MDPI stays neutral with regard to jurisdictional claims in published maps and institutional affiliations.

In general, provenance analysis has developed over the past 70 years into an enormously important tool in sediment investigations, both enabling solving earth science questions in basic research and practically applying it to mineral exploration. The mineral content of mainly siliciclastic sandstones reflects the geological situation of the basin-fringing supply areas. Likewise, associated conglomerates and pelites are used as supporting evidence. In modern times, thanks to methodological and technical advances, we have a variety of methods available to infer paleogeographic, orogenic and plate tectonic processes from the composition of sandstones and selected minerals. It is therefore both timely and interesting to briefly highlight these historical developments.

The first effort was to develop a useful classification scheme for sandstones according to their mineral content and rock fragments. Pioneers in this field were, e.g., Krynine [1] and Folk [2]. These sandstone classifications proved successful for that time but had the disadvantage of the measured mineral and grain content being influenced by the granulometry of the sandstone. A proliferation of classification schemes was the result [3]. However, with the restriction of the statistically point-counted grains to the sand size and the considered rock fragments with exclusively aphanitic texture (coarsely textured rock fragments are included in the classification with their individual mineral grains), the grain-size bias could be limited. This method, developed by Dickinson [4], also allowed an interpretation of the plate tectonic position of the supply areas with the restriction that the age of the volcanic/abyssal rock fragments remained unclear, especially in fossil sandstones. Research tools involving heavy minerals to describe delivery areas go back even further; however, they were definitely established as a recognized analytical method from the 1950s onwards [5]. The high input of the combined use of different analytical methods is emphasized, e.g., by Weltje and von Eynatten [6]. This is impressively manifested in the paper collection by Mange and Wright [7]. In the same volume, the concept of lag-time (in exhumation evaluation) was possibly applied for the first time in the northern Andes of Ecuador using fission-track analysis on detrital zircons [8].

Radiometric dating of detrital minerals represented something of a quantum jump in provenance analysis. Because of its abundance and robustness, zircon was the first choice. In an early study, detrital zircons were dated using the revolutionary SHRIMP (sensitive high resolution ion microprobe) method [9]. Soon thereafter, other researchers, e.g., Jackson [10], were able to prove the suitability of laser ablation ICP-MS (inductively coupled plasma mass spectrometry) techniques, which yielded comparably accurate U-Pb ages and carried the advantage of being far less time-consuming. In addition, geochemical signatures of detrital zircons, and especially the Lu-Hf isotope ratios, allow interpretation of the formation of magma zircons from melts in the Earth's crust, depleted mantle and mixtures thereof [11]. Thus, detrital zircons also provide arguments for the maximum age of the sandstone series, the chronostratigraphic and petrographic variability in the sourcing hinterlands, as well as the formation of magmas in the crust, the depleted mantle and the mixtures thereof, providing arguments for plate tectonic environment [11]. In the following, we present the articles in order of geological age, starting with the youngest.

Interesting evaluations of methods are provided by analyzing modern clastic systems because the relationships between source rocks and sediments are preserved. Three of the papers deal with these topics. Sun et al. [12] compare published muscovite Ar/Ar ages and detrital zircon age populations in modern sediments of the Yangtze River. Pooled age distributions show discrepancies in occurrence in the upper and lower reaches of the river with expected hinterland petrographic/age characteristics. They may be explained by combined effects of, e.g., lower reaches dilution and durability between muscovite and zircon. A very rich set of detrital zircons U-Pb age data from East China Sea, Yellow Sea and Bohai Sea surface samples is presented by Huang et al. [13]. Distinct detrital zircon age populations and mixing modeling support the supply from the framing orogenic belts and strong modification by tidal and ocean currents, which should also be considered in the analysis of fossil sediment series. Kanouo et al. [14] present a detailed study of Archean to Neoproterozoic zircons found in gold-bearing placers from the southern Meiganga area, Cameroon. Analysis of zircon cores by the laser ablation split stream technique (LASS) reveals their trace element composition and U-Pb ages within a single shot. Trace element data suggest that the majority of the zircons are derived from different magmatic sources, but very few mantle sourced or metamorphism-sourced examples are recognized. Supply from the Congo Craton, local Pan-African intrusions and the Cameroon Mobile Belt is interpreted. The latter could be used for tracking gold.

Based on their own original data and a reevaluation of published detrital zircon U-Pb ages and Hf-isotope ratios in early Miocene fine- to medium-grained sediments, Wang et al. [15] successfully reconstruct the sources of detritus in the northwestern South China Sea. The dominant contributor was the Red River system, with minor supply from central Vietnam and the Hainan uplift. This study again underscores the power of combined methods, U/Pb zircon ages and zircon Hf-isotopes, applied to provenance analysis which should be used more often in future. The Silante Formation of the western Cordillera of the North Andes has experienced very different interpretations over the course of more than 40 years of investigation, so that, e.g., its interpreted stratigraphic age was modified from Paleocene to Oligocene. The latest study by Vallejo et al. [16] combines stratigraphic bedding relationships, facies analysis, sandstone petrography, geochemistry of detrital clinopyroxenes, heavy minerals and detrital zircon U-Pb and muscovite Ar/Ar ages. A derivation of clastics from a continental margin arc, presumably the San Juan de Lachas unit, is concluded. The paper of Yin and Wu [17] contributes to understanding the complex sediment supply of the Paleogene Quaidam Basin in the Tibetan Plateau. Heavy minerals, detrital zircon U-Pb chronostratigraphy combined with detailed sediment distribution patterns recognize the Qilian Mountains as the main source area that formed a coeval large catchment area under the control of intensive tectonic activity.

Going backward in time, Di Giulio et al. [18] present a comparative study of the mid-Cretaceous transition from extensional to compressional tectonic settings in two very different plate tectonic settings: the Southern Andes (back-arc system) and the Western Alps (passive European continental margin). A rich data set of detrital zircons U-Pb geochronology, sandstone modal composition and apatite double-dating (in the Andes) supports the paleotectonic discrepancies and the related sediments. The first study of zircon from the Early Cretaceous Verkhneurmiysky granitoids in the Amur Badhzal tin ore district is reported by Machevariani et al. [19]. Raman spectroscopy, morphotype and internal impurities/alterations analysis of zircons allow the reconstruction of distinct stages during their evolution. According to the authors, zircons in Zinnwaldite granites show close affinities with such in Russia, Australia, Germany and the Czech Republic. Lee et al. [20] provide a combined study of the non-marine Late Cretaceous Neungju Basin (southwest Korea) formed in an active continental margin environment. The geochronological story of the basin fill is carried out by detrital zircon U-Pb dating from ash layers. However, the geochemistry and weathering of basin fill mudstones show a quite intriguing picture.

The authors warn that continental basin fill may show strong spatial contrasts in sediment supply because of limited mixing between the sources. The tectonic setting of the Anisian Gejiun alkaline basalts, part of the Eneishan Large Igneous Province, is controversially interpreted. Shang and Chen [21] present new laser ablation zircon U-Pb ages, whole-rock major, trace element and Sr-Nd-Pb isotopic data from outcrops and drill holes. The results imply that the basalts erupted in an extensional environment during the Gejiu-Napo rifting event in the southwestern margin of the South China Block.

Li et al. [22] investigated the Permian series of the west Bogda Shan range between the Junggar and Tarim blocks, the latter representing the southwestern margin of the Central Asian Orogenic Belt. Detrital zircon geochemistry and modal analysis of sandstone allow individualizing a middle Permian basin inversion from Carboniferous early Permian rift and post-rift to a continental arc environment supply. Hadimi et al. [23] analyze the Tiddas Souk Es-Sebt des Ait Ikko Basin, a continental trough in the Central Moroccan Meseta. Based on whole-rock geochemistry, pyroxene and biotite major and trace element data from the volcanic and sub-volcanic rocks calc-alkaline-series characteristics are derived from parental mafic magmas. Moreover, a four-stage basin evolution is recognized, showing, at first, an extension, followed by transpression and compression, respectively, and a final extensional event. The SHRIMP U-Pb dated zircons from a rhyolite dome (ca. 287 Ma) presumably formed during the third stage. Andesites and dacites show similarities with calc-alkaline series rocks, which are unrelated to active subduction. Miao et al. [24] present new research from the Central Asian Orogenic Belt (CAOB) in Mongolia. By means of detrital zircon U-Pb geochronology on Neoproterozoic early Paleozoic sandstones in the Ereendavaa terrane (the later southern margin of the Mongol-Okhotzk ocean), the authors define the earlier Kherlen ocean suture, which is an important new element for understanding the evolution of the CAOB. The existence of the new element is also reflected by Winkler et al. ([25], this volume). The authors investigate the plate tectonic history of the CAOB from Cambrian to Early Jurassic in central and southeastern Mongolia (Gobi). Detrital zircon U-Pb geochronology, Hf-isotope systematics and detrital mode of sandstones are applied. Including earlier results from the Mongoli-Okhotzk belt [26], a paleogeographical varying tectonic system of rifting, drifting and basin inversions (subduction) during the long timespan is reconstructed. For the plate tectonic reconstruction, integration of Hf-isotope systematics proved very efficient. The paper of Kim and Choi [27] deals with Pennsylvanian strata in the Sino-Korean Block. The work is mainly based on detrital zircon U-Pb age data that appear to be distinct from the coeval sediment series in China. A contemporaneous active continental margin setting to the east of the Sino Korean Block is inferred that relates to the westward subduction of the Paleo-Pacific plate.

Our present Special Issue contains a collection of articles that are exemplary for their application and combination of diverse diagnostic techniques. It is also pleasing to note that the case studies come from different parts of our globe, namely South America, Central Asia, Southeast Asia, Africa and Europe. We hope that our Special Issue will serve as a resource for knowledge and as a stimulus for further research and development in this field.

Funding: This research received no external funding.

Conflicts of Interest: The authors declare no conflict of interest.

References

1. Krynine, P.D. The megascopic study and field classification of sedimentary rocks. *J. Geol.* **1948**, *56*, 130–165. [CrossRef]
2. Folk, R.L. *Petrography of Sedimentary Rocks*; Hemphill's Bookstore: Austin, TX, USA, 1964; p. 154.
3. Okada, H. Classification of sandstone: Analysis and proposal. *J. Geol.* **1971**, *79*, 509–525. [CrossRef]
4. Dickinson, W.R. Interpreting detrital modes of arkoses and greywackes. *J. Sediment. Petrol.* **1970**, *40*, 695–707.
5. Mange, M.A.; Maurer, H.F.W. *Heavy Minerals in Colour*; Chapman and Hall: London, UK, 1992; p. 147.
6. Weltje, G.J.; von Eynatten, H. Quantitative provenance analysis of sediments: Review and outlook. *Sediment. Geol.* **2004**, *171*, 1–11. [CrossRef]
7. Mange, M.A.; Wright, D.T. (Eds.) Developments in Sedimentology. In *Heavy Minerals in Use*; Elsevier: Amsterdam, The Netherlands, 2007; p. 1283.

8. Ruiz, G.M.H.; Seward, D.; Winkler, W. Evolution of the Amazon Basin in Ecuador with special reference to hinterland tectonics: Data from zircon fission-track and heavy mineral analysis. In *Heavy Minerals in Use*; Mange, M.A., Wright, D.T., Eds.; Elsevier: Amsterdam, The Netherlands, 2007; Volume 58, pp. 907–934.
9. Morton, A.C.; Claoue´-Long, J.C.; Berge, C. SHRIMP constraints on sediment provenance and transport history in the Mesozoic Statfjord formation, North Sea. *J. Geol. Soc.* **1996**, *153*, 915–929. [CrossRef]
10. Jackson, S.E. The application of Nd:YAG lasers in LA-ICPMS. In *Laser Ablation-ICP-Mass Spectrometry in the Earth Sciences: Principles and Applications, Mineralog. Assoc*; Sylvester, P.J., Ed.; Canada (MAC) Short Course Series; Mineralogical Association of Canada: Québec, QC, Canada, 2001; Volume 29, pp. 29–45.
11. Amelin, Y.; Lee, D.C.; Halliday, A.N.; Pidgeon, R.T. Nature of the Earth's earliest crust from hafnium isotopes in single detrital zircons. *Nature* **1999**, *399*, 252–255. [CrossRef]
12. Sun, X.; Kuiper, K.; Tian, Y.; Li, C.; Zhang, Z.; Wijbrans, J. Comparison of Detrital Zircon U-Pb and Muscovite 40Ar/39Ar Ages in the Yangtze Sediment: Implications for Provenance Studies. *Minerals* **2020**, *10*, 643. [CrossRef]
13. Huang, X.; Song, J.; Yue, W.; Wang, Z.; Mei, X.; Li, Y.; Li, F.; Lian, E.; Yang, S. Detrital Zircon U-Pb Ages in the East China Seas: Implications for Provenance Analysis and Sediment Budgeting. *Minerals* **2020**, *10*, 398. [CrossRef]
14. Kanouo, N.; Kouske, A.; Ngueutchoua, G.; Venkatesh, A.; Sahoo, P.; Basua, E. Eoarchean to Neoproterozoic Detrital Zircons from the South of Meiganga Gold-Bearing Sediments (Adamawa, Cameroon): Their Closeness with Rocks of the Pan-African Cameroon Mobile Belt and Congo Craton. *Minerals* **2021**, *11*, 77. [CrossRef]
15. Wang, C.; Zeng, L.; Lei, Y.; Su, M.; Liang, X. Tracking the Detrital Zircon Provenance of Early Miocene Sediments in the Continental Shelf of the Northwestern South China Sea. *Minerals* **2020**, *10*, 752. [CrossRef]
16. Vallejo, C.; Almagor, S.; Romero, C.; Herrera, J.; Escobar, V.; Spikings, R.; Winkler, W.; Vermeesch, P. Sedimentology, Provenance and Radiometric Dating of the Silante Formation: Implications for the Cenozoic Evolution of the Western Andes of Ecuador. *Minerals* **2020**, *10*, 929. [CrossRef]
17. Yin, J.; Zhang, S.; Wu, Z. Provenance Analysis of the Paleogene Strata in the Northern Qaidam Basin, China: Evidences from Sediment Distribution, Heavy Mineral Assemblages and Detrital Zircon U—Pb Geochronology. *Minerals* **2020**, *10*, 854. [CrossRef]
18. Di Giulio, A.; Amadori, C.; Mueller, P.; Langone, A. Role of the Down-Bending Plate as a Detrital Source in Convergent Systems Revealed by U–Pb Dating of Zircon Grains: Insights from the Southern Andes and Western Italian Alps. *Minerals* **2020**, *10*, 632. [CrossRef]
19. Machevariani, M.; Alekseenko, A.; Bech, J. Complex Characteristic of Zircon from Granitoids of the Verkhneurmiysky Massif (Amur Region). *Minerals* **2021**, *11*, 86. [CrossRef]
20. Lee, H.; Kwon, M.; Shin, S.; Cho, H.; Kim, J.; Roh, Y.; Huh, M.; Choi, T. Relationships between Alluvial Facies/Depositional Environments, Detrital Zircon U-Pb Geochronology, and Bulk-Rock Geochemistry in the Cretaceous Neungju Basin (Southwest Korea). *Minerals* **2020**, *10*, 1023. [CrossRef]
21. Shang, Z.; Chen, Y. Zircon U–Pb Geochronology, Geochemistry and Geological Significance of the Anisian Alkaline Basalts in Gejiu District, Yunnan Province. *Minerals* **2020**, *10*, 1030. [CrossRef]
22. Li, Y.; Yue, W.; Yu, X.; Huang, X.; Yao, Z.; Song, J.; Shan, X.; Yu, X.; Yang, S. Tectonic Evolution of the West Bogeda: Evidences from Zircon U-Pb Geochronology and Geochemistry Proxies, NW China. *Minerals* **2020**, *10*, 341. [CrossRef]
23. Hadimi, I.; Youbi, N.; Ait Lahna, A.; Bensalah, M.; Moutbir, O.; Mata, J.; Doblas, M.; Tassinari, C.; Gaggero, L.; Basei, M.; et al. U–Pb Zircon Geochronological and Petrologic Constraints on the Post-Collisional Variscan Volcanism of the Tiddas-Souk Es-Sebt des Aït Ikko Basin (Western Meseta, Morocco). *Minerals* **2021**, *11*, 1099. [CrossRef]
24. Miao, L.; Zhu, M.; Liu, C.; Baatar, M.; Anaad, C.; Yang, S.; Li, X. Detrital-Zircon Age Spectra of Neoproterozoic-Paleozoic Sedimentary Rocks from the Ereendavaa Terrane in NE Mongolia: Implications for the Early-Stage Evolution of the Ereendavaa Terrane and the Mongol-Okhotsk Ocean. *Minerals* **2020**, *10*, 742. [CrossRef]
25. Winkler, W.; Bussien, D.; Baatar, M.; Anaad, C.; von Quadt, A. Detrital Zircon Provenance Analysis in the Central Asian Orogenic Belt of Central and Southeastern Mongolia—A Palaeotectonic Model for the Mongolian Collage. *Minerals* **2020**, *10*, 880. [CrossRef]
26. Bussien, D.; Gombojav, N.; Winkler, W.; von Quadt, A. The Mongol-Okhotsk Belt in Mongolia—An Appraisal of the Geodynamic Development by the Study of Sandstone Provenance and Detrital Zircons. *Tectonophysics* **2011**, *510*, 132–150. [CrossRef]
27. Kim, M.; Lee, Y.; Choi, T. Tectonic Setting of the Eastern Margin of the Sino-Korean Block in the Pennsylvanian: Constraints from Detrital Zircon Ages. *Minerals* **2020**, *10*, 527. [CrossRef]

minerals MDPI

Article

Comparison of Detrital Zircon U-Pb and Muscovite $^{40}Ar/^{39}Ar$ Ages in the Yangtze Sediment: Implications for Provenance Studies

Xilin Sun [1,2,*], Klaudia F. Kuiper [2], Yuntao Tian [1], Chang'an Li [3], Zengjie Zhang [1] and Jan R. Wijbrans [2]

1 School of Earth Sciences and Engineering, Sun Yat-sen University, Guangzhou 510275, China; tianyuntao@mail.sysu.edu.cn (Y.T.); zhangzj55@mail.sysu.edu.cn (Z.Z.)
2 Department of Earth Sciences, Cluster Geology and Geochemistry, Vrije Universiteit Amsterdam, De Boelelaan 1085, 1081 HV Amsterdam, the Netherlands; k.f.kuiper@vu.nl (K.F.K.); j.r.wijbrans@vu.nl (J.R.W.)
3 School of Geography and Information Engineering, China University of Geosciences, Wuhan 430074, China; chanli@cug.edu.cn
* Correspondence: sunxi-lin@163.com

Received: 30 June 2020; Accepted: 17 July 2020; Published: 20 July 2020

Abstract: Detrital zircon U-Pb and muscovite $^{40}Ar/^{39}Ar$ dating are useful tools for investigating sediment provenance and regional tectonic histories. However, the two types of data from same sample do not necessarily give consistent results. Here, we compare published detrital muscovite $^{40}Ar/^{39}Ar$ and zircon U-Pb ages of modern sands from the Yangtze River to reveal potential factors controlling differences in their provenance age signals. Detrital muscovite $^{40}Ar/^{39}Ar$ ages of the Major tributaries and Main trunk suggest that the Dadu River is a dominant sediment contributor to the lower Yangtze. However, detrital zircon data suggest that the Yalong, Dadu, and Min rivers are the most important sediment suppliers. This difference could be caused by combined effects of lower reaches dilution, laser spot location on zircons and difference in closure temperature and durability between muscovite and zircon. The bias caused by sediment laser spot targeting a core or rim of zircon and zircon reworking should be considered in provenance studies.

Keywords: zircon U-Pb ages; muscovite $^{40}Ar/^{39}Ar$ ages; sediment provenance; Yangtze River

1. Introduction

Detrital zircon U-Pb dating is a much used sediment provenance tool, which is commonly used to constrain Maximum sediment depositional ages [1–4], erosion patterns of a river system [5–7], and orogen development of the catchment area [7–9]. Zircon is formed in Magmatic (plutonic or volcanic) or high-temperature metamorphic rocks. The zircon U-Pb ages are generally crystallization or high-grade metamorphism ages, but age signals can be complicated by multiple age zoning [7,10,11]. Muscovite $^{40}Ar/^{39}Ar$ dating is also widely used for isotope provenance studies and is attractive because muscovite provenance data more-readily can be linked to the more recent processes in the hinterland due to its lower closure temperature and lower resistance to weathering. Muscovite is typically derived from low–medium-grade metamorphic rocks or S-type granites and commonly is present in muddy and sandy sediments. Muscovite $^{40}Ar/^{39}Ar$ ages are cooling ages that record cooling of muscovite grains below the closure temperature (350–425 °C [12]). Because zircon's high closure temperature (>900 °C [13]) and resistance to physical and chemical weathering, it can survive multiple cycles of metamorphic overprinting, weathering, erosion and deposition. In contrast, muscovite is less likely to survive multiple orogenic events because of isotopic resetting or destruction. Detrital zircon ages record old Magmatic or high-grade metamorphic events of source area and often fail to record the most recent metamorphic or Magmatic event [10,14,15]. Detrital muscovite $^{40}Ar/^{39}Ar$ age distributions are commonly simpler and contain the records of the more-recent parts of the tectonic history of a source area. Detrital zircon and muscovite ages of the same sample always give significantly different age distributions and therefore complicate interpretations of sediment provenance [10,16]. The possible factors (difference in closure temperature and durability) have been assessed in previous studies [10,16], but do not fully explain observed differences in sediment provenance based on either of these methods.

The Yangtze River, as the longest river in Asia (Figure 1), is a suitable candidate for assessing factors that control the differences in sediment provenance age signals between zircon and muscovite. Its sediment provenances have been studied intensively by numerous methods, including zircon U-Pb, monazite U-Th-Pb, Pb isotopes of K-feldspar, Nd isotopes, clay mineral composition, and heavy minerals composition [17–24]. Further, previous studies have also reported a large amount of detrital zircon U-Pb and muscovite $^{40}Ar/^{39}Ar$ ages [16,21,24–26] from its trunk and Major tributaries.

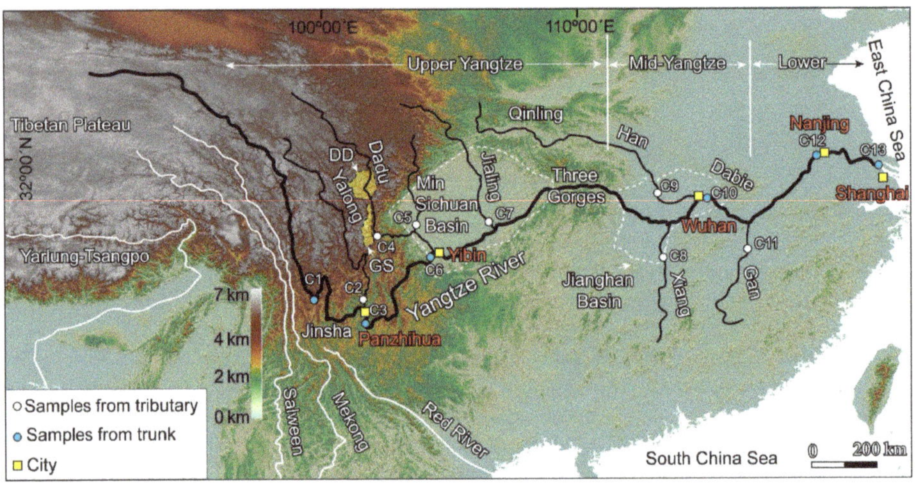

Figure 1. A schematic Map showing the drainage basin, sampling locations and Main distributaries of the Yangtze River. Sample locations of Major tributaries are shown as open circles. Samples from the Main trunk are shown as filled blue circles. The yellow shadow areas represent the Danba dome (DD) and Gongga shan (GS).

In this study, we combine previously reported detrital zircon U-Pb and muscovite $^{40}Ar/^{39}Ar$ data derived from equivalent sampling points along the Yangtze River with the objective to take interpretation beyond the outcome of the individual samples. The comparison of detrital ages from the Yangtze River between muscovite and zircon shows different aspects of the provenance information. We focus on assessing the factors that control differences between data types. These factors should be considered in future sediment provenance studies using the two provenance tools.

2. Geological Setting

The Yangtze River is primarily situated in the Yangtze Block, surrounded by the Yidun Arc to the southwest, the Songpan-Garze Block to the west, the Qinling-Dabie orogen to the northwest and north and the Cathaysia Block to the east (Figure 2). The catchment drains a variety of rocks, including Archean and Proterozoic metamorphic and igneous rocks, Paleozoic carbonate, Mesozoic-Cenozoic igneous and siliciclastic rocks, and Quaternary sediments [27].

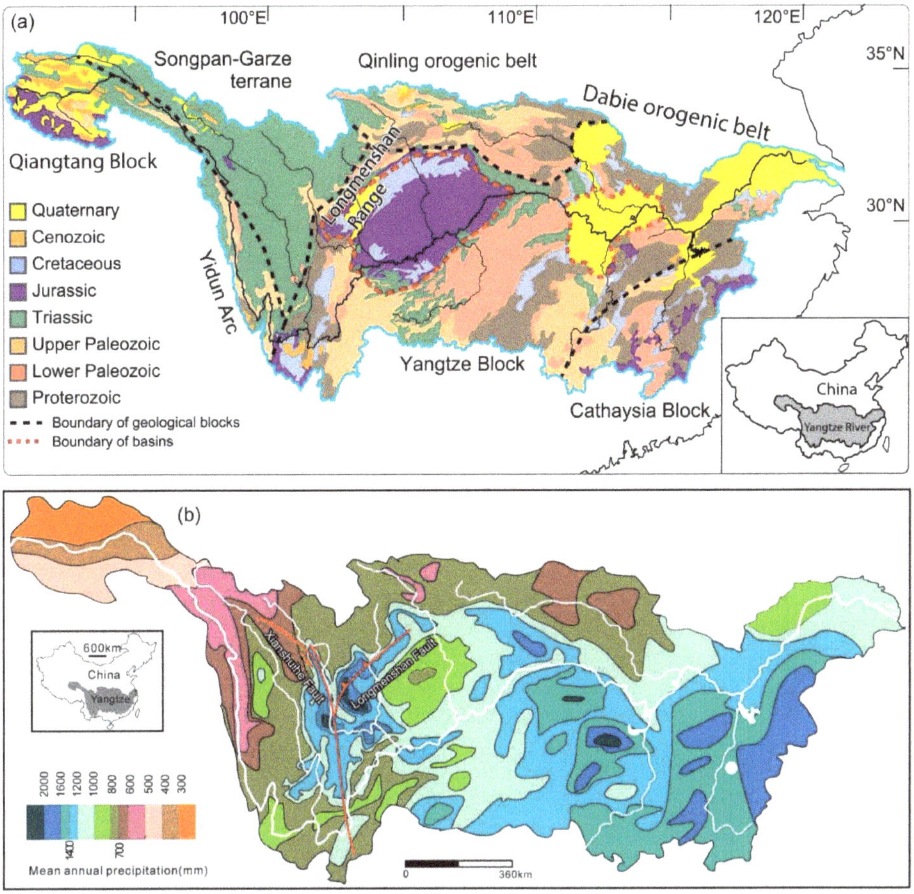

Figure 2. Generalized geological Map (**a**) (Modified from Saito et al. [28]) and the mean annual precipitation distribution of the Yangtze River (**b**). The mean annual precipitation distribution Map modified from Sun et al. [24].

The Major tributaries of the Yangtze River are characterized by different tectonic settings and bedrocks in their respective hinterlands (Figure 2a). From upstream to downstream the Main tributaries and their catchment areas include: (1) The Jinsha River basin, comprising Triassic low-grade metamorphic rocks, Paleozoic carbonate, and clastic and volcanic rocks [29,30]. (2) The Yalong, Min, and Dadu rivers draining the Songpan-Garze Block, which is composed of deformed and locally metamorphosed Triassic turbiditic sedimentary rocks [31,32]. (3) The upper Jialing River drainage area belongs to the South Qinling orogenic belt (Figure 2a), consisting of weakly metamorphosed Meso–Neoproterozoic basement. The central to southeastern Sichuan Basin is drained by the mid-lower Jialing River, where Jurassic-Cretaceous sandstone and mudstone is exposed. (4) The Han River basin in the Qinling orogenic belt is characterized by Neoproterozoic basement and Neoproterozoic—Devonian sediments, and early Paleozoic metamorphic rocks [33]. (5) The Xiang River belong to the Yangtze Block, containing mostly Proterozoic medium-low grade metamorphic and Cambrian and Ordovician carbonate rocks, and Jurassic—Quaternary terrestrial sediments [34]. (6) The Gan River is located in the Cathaysia Block, which is composed of Neoproterozoic conglomerate and sandstone, Jurassic granite, and Quaternary clastic sediments (Figure 2a).

3. Data Sets

We compiled and published data sets of zircon U-Pb ages from He et al. [21] and Yang et al. [26] and of muscovite $^{40}Ar/^{39}Ar$ ages from Sun et al. [16,24,35] and Hoang et al. [25]. Available data cover the Main tributaries of the Yangtze River, including Yalong, Dadu, Min, Jialing, Xiang, Han, and Gan rivers (Figure 1). Six samples are from the trunk of the Yangtze River near Nanjing, Wuhan, Yibin, Panzhihua, and Shanghai cities. More than 91 zircon ages are available for each sample, ensuring with 95% certainty that no fraction greater than 6% was missed from the underlying detrital population [36]. For muscovite, 30–62 analyses per sample ensure a 95% certainty that no fraction greater than 15% was missed [36]. In total, 2557 zircon and 581 muscovite ages were compiled for discussion. All samples were collected from channel deposits. Samples are riverbed sand at least from two locations at each sample site to avoid bias on age distributions caused by hydraulic sorting [16,21,24]. Details of muscovite $^{40}Ar/^{39}Ar$ age determinations are given in Sun et al. [16,24,35] and Hoang et al. [25]. Details of zircon U-Pb age determinations are described in He et al. [21] and Yang et al. [26].

4. Results

4.1. Detrital Zircon

The detrital zircon U-Pb ages of 13 samples are presented in Figures 3b and 4a. Detrital zircon ages define a wide spectrum ranging from ~30 to 3200 Ma. Samples (C1, C3, and C6) from the Jinsha River have five Major age populations, i.e., 2.2–2.6 Ga, 1.5–2.0 Ga, 600–1000 Ma, 300–450 Ma, and 160–300 Ma(Figure 4a-1,-3,-4). For the Major tributaries (Yalong, Dadu and Min) in the upper Yangtze River, two age peaks (100–300 Maand 600–1000 Ma) are displayed (Figure 4a-2,-5,-6). Four Major age peaks are present in the Jialing River (100-400 Ma, 700-1200 Ma, 1.5–2.0 Ga and 2.2–2.6 Ga) (Figure 4a-7). For the midstream segment, five age peaks are displayed in the Mainstream (2.2–2.6 Ga, 1.5–2.0 Ga, 700–1000 Ma, 300–600 Ma, and 200–300 Ma), which is similar to ranges of ages seen in the upper Yangtze.

4.2. Detrital Muscovite

In the upper Yangtze, three samples from the Jinsha River show a dominant muscovite age peak at ~250 Ma(Figure 4b-1,-3,-4). Forty muscovite grains from the Yalong River show a dominant age peak at ~810 Maand a minor peak at ~230 Ma(Figure 4b-2). All dated muscovite grains of the Dadu River are younger than 160 Ma, with a Major muscovite age peak younger than 60 Ma(Figure 4b-5). Muscovite grains from the Min River are dominated by muscovite ages between 80–200 Ma, with a Major peak around 180 Ma(Figure 4b-6). The muscovite grains from the Jialing River show a range from 190 to 300 Ma, with a prominent peak at ~208 Ma(Figure 4b-7). In the mid-lower Yangtze tributaries (Han, Xiang and Gan), most of muscovite ages fall into an age range of 100–300 Ma(Figure 4b-8,-9,-11). None of these grains are younger than 60 Ma. The three modern Yangtze trunk sediments (C10, C12, and C13) yield similar muscovite age distributions (Figure 4b-10,-12,-13). Most of the muscovite grains in samples C10, C12, and C13 are younger than 100 Ma, accounting for ~71%, ~67%, and ~53% of the total dated grains of each sample. These young muscovite ages overlap with the Dadu River provenance pattern (Figure 4b-5,-10,-12,-13).

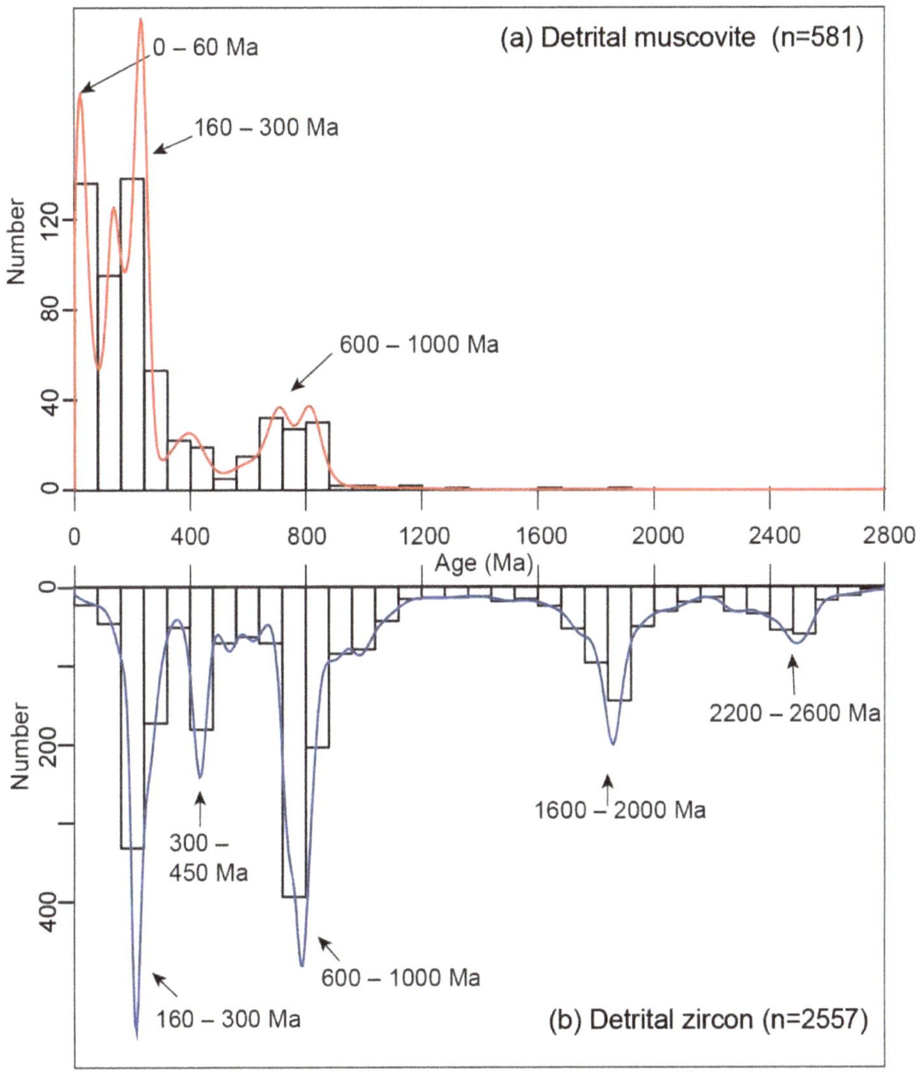

Figure 3. Probability density diagrams for pooled detrital muscovite $^{40}Ar/^{39}Ar$ (**a**) and zircon U-Pb (**b**) ages of the Yangtze River basin. Detrital zircon data from He et al. [21] and Yang et al. [26]. Detrital muscovite data from Sun et al. [16,24,35] and Hoang et al. [25].

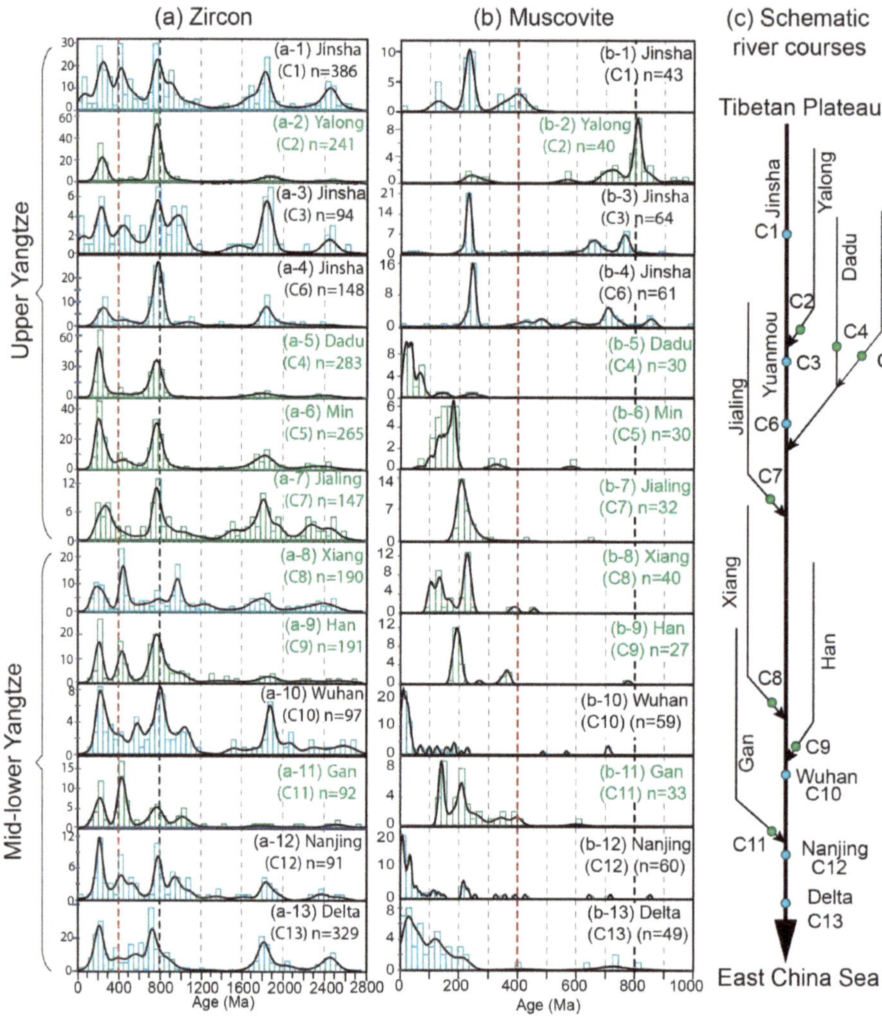

Figure 4. Comparison of age distributions between detrital zircons (**a**) and muscovites (**b**) of modern sands of the Yangtze. (**c**) Schematic river course of the Yangtze. Note the difference in age scale between muscovite (0–1000 Ma) and zircon (0–2800 Ma). The dashed vertical red and black lines Mark 400 and 800 Ma, respectively. The histogram diagrams in blue and green are samples from the Main stream and Major tributaries, respectively. Detrital zircon data from He et al. [21] and Yang et al. [26]. Detrital muscovite data from Sun et al. [16,24] and Hoang et al. [25].

5. Discussion

5.1. Comparison of Detrital Muscovite and Zircon Ages

5.1.1. Comparison of Pooled Age Distributions

In order to assess the general differences in age distribution between detrital muscovites and zircons of the Yangtze River, the muscovite and zircon ages of modern Yangtze sands are pooled and presented in Figure 3. The Major five zircon age peaks (2.2–2.6 Ga, 1.6–2.0 Ga, 600–1000 Ma,

300–450 Ma, and 160–300 Ma) correspond to Major granitoid Magmatic events within the Yangtze River basin [21]. The age distributions of muscovite and zircon are different in the following two aspects: (1) Muscovite grains are lacking in age populations at 1600–2000 Maand 2200–2600 Marecorded by detrital zircons. (2) The height of muscovite age peaks at 0–60 Ma, 160–300 Ma, and 600–1000 Maare different from zircon age peaks in relative probability and number.

5.1.2. Comparison of Major Tributaries with Main Trunk of the Yangtze

The variation in muscovite and zircon age distributions from upstream to downstream in the Yangtze River systems allow us to better understand the Major sediment contributors to the lower Yangtze. The comparison of age distributions between detrital muscovite and zircon grains (Figure 4) shows the following two differences: (1) The Major zircon age peaks (2.2–2.6 Ga, 1.6–2.0 Ga, 600–1000 Ma, 300–450 Ma, and 160–300 Ma, Figures 3b and 4a) can be observed in all trunk samples, but the three Major muscovite age peaks (600–1000 Ma, 160–250 Ma, and 0–60 Ma, Figures 3a and 4b) are not always identified in the trunk samples. (2) The concentration of Cenozoic muscovite grains (>40%) is pronouncedly increased for trunk samples in the mid-lower reaches (C10, C12, and C13), but the detrital zircon age distributions of trunk samples do not significantly change toward lower reaches.

Cenozoic muscovite grains in the mid-lower reaches (JC10, CJ12, and C13) probably were derived from the Dadu River in the upper reaches because Cenozoic muscovites are only present in the Dadu River among all sampled tributaries. This suggests that the Dadu River in the upper reaches is an important muscovite contributor to the sediment of the lower reaches. The reported Cenozoic muscovite $^{40}Ar/^{39}Ar$ ages in the Dadu River basin are from the Gongga shan (3.5–12.1 Ma, [37]) and Danba dome (35.1–104.3 Ma, [38]) (see Figure 1 for location). Therefore, most of the Cenozoic muscovite grains in the lower Yangtze were derived from only these two regions. In contrast, the zircon age distributions show a different picture. Using multiple detrital zircon U-Pb age distribution comparison techniques and a distribution-mixing model, Wissink et al. [6] suggest that the Yalong, Dadu, and Min rivers are the Major sediment contributors to the lower Yangtze.

Main trunk samples (C10, CJ12, and C13) from the mid-lower reaches contain more than 40% Cenozoic muscovite grains which were derived from the Dadu River. It could be argued that the Dadu River cannot supply more than 40% muscovite grains to mid-lower Yangtze because this river is not a dominant water supplier to mid-lower Yangtze. However, the Dadu River drains the Xianshui He Fault area that has already been tectonically active since the Miocene (Figure 2b) [37]. This region experiences strong precipitation and is expected to facilitate more landslides and intense erosion (Figure 2b) and thus water and sediment budgets May be decoupled. Moreover, the exhumation rates of the Xianshui He Fault area inferred from the thermochronometry data are higher than other areas of the upper Yangtze River [39–41]. In addition, Cook et al. [42] report millennial erosion rates based on detrital ^{10}Be data in the Gongga shan of >5 mm/year. Erosion rates increase from west to east more than one hundred-fold from 0.013–0.04 mm/year in the upper Jinsha River to >5 mm/year in the Dadu River basin [42,43].

5.2. *Factors Related to the Difference in Zircon and Muscovite Signals*

To correctly interpret provenance data, we need to understand, which geochronometer or provenance tools best reflects sediment provenance and under what circumstances (e.g., in complex orogens). We argue that the differences in provenance signal between muscovite and zircon could be caused by the following six factors: (1) lithological effect; (2) mineral abundance; (3) dilution of the mid-lower reaches; (4) differences in closure temperature; (5) differences in resistance to weathering; and (6) laser spot location on zircons.

(1) In Figure 4, in some rivers we note remarkable differences in age distributions: For example in the Dadu River we note high abundance of young muscovite ages without an equivalent population of young zircon ages, whereas, in the Jinsha River we see the reverse, young zircon ages that

do not have a complement of young muscovite $^{40}Ar/^{39}Ar$ ages. Such discrepancies point to a lithological effect: Young volcanics and young I-type granites will add a zircon signal to the sediment [44], but because these rock types are generally free of white mica, an equivalent signal in the muscovite population is lacking. Conversely, when we see a young muscovite age signal in the sediment without an equivalent zircon age population, the source rock of the sediment is likely to be more consistent with low to medium grade metamorphic basement where muscovite is abundant, but as the metamorphic grade is too low for zircon crystallization an equivalent age signal in the zircon data will be lacking. This implies that the choice of used provenance tools should depend on the geological composition of the hinterland.

(2) To assess if mineral abundance in the Major tributaries of the Yangtze River causes the dominance of the Dadu River muscovites in the lower Yangtze, we compiled mineral fertility data (mica and zircon) from Vezzoli et al. [45]. Figure 5 shows that the concentration of mica in modern sediment of the Dadu River is much lower than that of the Han River and higher than the Jialing and Jinsha rivers. However, comparison of muscovite age distribution between the Han River and delta (Figure 4b-9,-13) suggests that the Han River is not Major muscovite supplier to the lower Yangtze. The zircon concentration of the Gan and Xiang rivers in the middle reaches are higher than other tributaries. These two rivers are not the Major zircon contributors to the lower reaches. Therefore, we can rule out large concentration variation in mica in the sediment as being a Major control on the dominance of the Dadu River muscovites in the lower reaches.

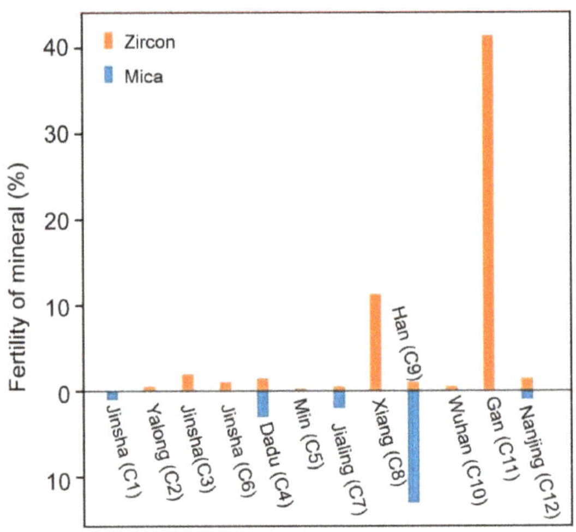

Figure 5. Zircon and mica fertility of Major tributaries. Concentration of zircon grains in the heavy mineral fraction. Mica and zircon data from Vezzoli et al. [45].

(3) The sediment from the upper Yangtze is potentially diluted by sediment from the tributaries in the mid-lower reaches. The effect of this dilution on differences between muscovite and zircon age distributions should be considered. The concentration of Cenozoic muscovite grains (>40%) in trunk samples pronouncedly increased in the mid-lower reaches (C10, C12, and C13), but the detrital zircon age distributions of trunk samples do not significantly change toward lower reaches. The detrital muscovite ages of the Major tributaries in the mid-lower reaches fall into two age peaks at 100–160 Maand 180–260 Ma. The two Major age populations are minor in samples near Wuhan (C10), Nanjing (C12), and delta (C13). This suggests that the Cenozoic muscovite signals of the Dadu River, that is relatively far upstream (Figure 4c), is not significantly affected by the dilution of the mid-lower reaches (Han, Xiang, and Gan rivers). Moreover,

the 1500–2000 Mazircon age peak of the Main trunk (samples C10, C12, and C13) in the mid-lower reaches is much higher than the tributaries (samples C8, C9, and C11) (Figure 4a-10,-12,-13), implying that the effect of dilution of zircons from the mid-lower reaches is also limited. We therefore suggest that the dilution of mid-lower reaches is limited in terms of the muscovite and zircon age signals.

(4) The closure temperature is the temperature at which muscovite $^{40}Ar/^{39}Ar$ or zircon U-Pb systems have cooled so that there is no longer any significant diffusion of the parent and daughter isotopes out of system. The closure temperature of muscovite (350–425 °C, [12,46]) is much lower than zircon (>900 °C, [13]). The zircon U-Pb age gives the crystallization or high-grade metamorphism and the muscovite $^{40}Ar/^{39}Ar$ age is age imparted as terrain cooled through closure temperature. Figure 3a shows that the detrital muscovite grains from the Yangtze are absent of older (>1000 Ma) age peaks, which could be caused by muscovite's lower closure temperature. Potentially older (>1000 Ma) muscovite grains in the Yangtze River could have been reset by younger tectonic events.

(5) The absence of older (>1000 Ma) muscovite grains in the Yangtze River also could be caused by abrasion and chemical dissolution during weathering and transport in multiple phases of sediment recycling. Muscovites in the Triassic Songpan-Garze flysch deposits in the upper Yangtze River have experienced multiple recycling. Previous studies suggest that the detritus in these areas is derived from Paleozoic and pre-Cambrian crystalline and sedimentary rocks of the Qinling-Dabie orogen and South China Block [47–50]. Detrital zircon U-Pb and Hf isotopic data of the Mesozoic sediment in basins of the South China Block (Pingle, Jianghan, and Sichuan basins) also suggest derivation of sediments from the Cathaysia Block [51]. Therefore, current Yangtze sands have experienced complex multiple erosion, transport and deposition processes that May have led to a reduction in size and abundance of the older grains. Detrital muscovite grains from the Yangtze River only record age populations of 600–1000 Ma, 160–300 Ma, and 0–60 Maand lack older populations (2000–2500 Maand 1600–2000 Ma) (Figure 3a). The absence of the two older age components probably results from muscovite's lower resistance to physical and chemical weathering when compared with zircon. The weathering has either completely destroyed the grain or reduced the grains size down to less than 200 μm, which are not selected for age analyses in this study.

(6) The differences in age distribution between zircon and muscovite also could be caused by laser spot location targeted for U-Pb analysis. Owing to the low solubility of zircon in hydrous granitoid melt, zircons can survive anatectic event [11]. Many zircon crystals are zoned with cores reflecting the ages of much older protoliths, which are universally referred to as the inherited component in the zircon population. However, $^{40}Ar/^{39}Ar$ in muscovite in general record only the later, post-peak cooling history. The apparent age of muscovite $^{40}Ar/^{39}Ar$ age system is generally younger than zircon U-Pb system for metamorphic or Magmatic rocks. Muscovite detrital data of trunk samples from the mid-lower reaches suggest that the Gongga shan and Danba dome in the Dadu River basin are two important sediment suppliers to the mid-lower reaches. Unsurprisingly, zircons in the Gongga shan and Danba dome show complex core rim texture (Figure 6c,d). Figure 6c shows that almost all Cenozoic ages are recorded by rims of dated zircons from the Gongga shan. This suggests that the narrow rims grew around the older core during Cenozoic intrusion ((1), (2), and (3) in Figure 6a). The Cenozoic ages recorded by rims of zircons also recorded by the bedrock muscovite $^{40}Ar/^{39}Ar$ ages [37] (Figure 6a). The Danba dome has undergone Barrovian type metamorphism at 200–180 Ma. Zircon growth at low metamorphic grades is minimal ((4) in Figure 6a). In the kyanite and sillimanite zones, narrow zircon growth has been observed in the Danba dome [52]. Therefore, rims of zircons from the Danba dome and Gongga shan are unlikely to be detected in the detrital record, due to laser spot location targeting on core and also the high likelihood that abrasion during transport will remove thin rims. Figure 4a-5 shows that the Cenozoic ages are not detected by the detrital zircons from the Dadu River near the Gongga shan

(sample C4). Figure 6b presents that the cores of detrital zircons from the modern Yangtze sands are always chosen as laser spot position instead of the narrow rims. Combination with dilution of sediment from the tributaries in the mid-lower reaches, the Cenozoic ages of the Gongga shan and Danba dome are not detected by detrital zircon in the mid-lower Yangtze River like muscovites.

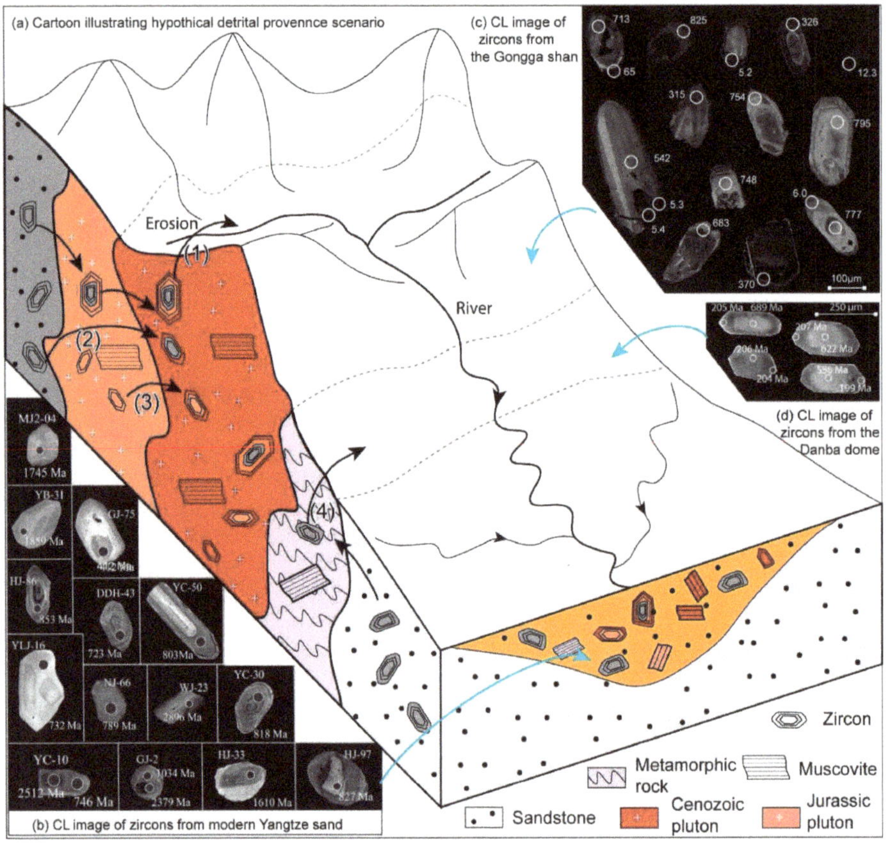

Figure 6. A cartoon showing the possible reasons for lack Cenozoic zircon in the lower reaches of the Yangtze (**a**). (1) Indicates detrital zircons experienced two phases of overgrowth. (2) Indicates overgrowth of detrital zircons in the Cenozoic. Magmatic zircon overgrowth in the Cenozoic (3). (4) Indicates no wide overgrowth of detrital zircon in the low-degrade metamorphism. Note that muscovite Ar system cannot survive Magmatic or metamorphic events. (**b**) Cathodoluminescence images of detrital zircons from the Yangtze River. Images are from He et al. [21]. (**c**,**d**) Cathodoluminescence images of dated zircons from the Gongga shan and Danba dome, respectively. Zircon images of the Gongga shan are from Robert et al. [53]. Zircon images of the Danba dome are from Jolivet et al. [52].

5.3. Implications

The lithological effect on provenance study using detrital mineral ages (e.g., muscovite or zircon) could be reduced by the following two approaches: (1) Carrying out high-resolution petrographic and heavy-mineral analyses. Petrographic and mineralogical data of sediment could provide information for selecting provenance tool (e.g., fertility of target mineral). (2) Applying multi-proxy provenance approach. Multiple chemical and isotopic indicators of single mineral grains can extract more robust information about sediment provenance. Sediment signals missed in one mineral could be detected by other minerals. As we have shown in this study, the Cenozoic signal detected by muscovite grains is missing in zircon ages in the Dadu River. Similarly, the Cenozoic zircon ages in the Jinsha River is not detected by muscovite $^{40}Ar/^{39}Ar$ ages. By combining multiple-proxies approaches, an increased understanding can be obtained of sediment provenance.

The comparison of muscovite and zircon ages suggests that detrital zircons in the Yangtze sediment experienced multiple cycles of deposition and erosion due to its high closure temperature and resistance to weathering. Possibility of zircon reworking should be considered for provenance interpretation. The following four approaches have been used to identify zircon reworking: (1) Direct field observation (e.g., appearance of sandstone gravels) [35]. (2) Time-transgressive similarities in detrital zircon age distribution, especially in old age populations [54]. (3) Supplementary information from other detrital minerals (e.g., apatite fission track ages [55]). (4) He-Pb double dating of detrital zircons [56].

In the case of Magmatic zircons in the Gongga shan, rim overgrowths are typically autocrysts linked to last-stage growth corresponding to crystallization from the final pulses of Magma [53] (Figure 6a). U-Pb rim ages of detrital zircons record the last zircon-forming tectonothermal event in the original source rock, while core ages record inheritance from the local country rocks or co-genetic Magmatic ages derived from earlier melt pulses in the Magma plumbing system [57]. In the Yangtze River, detrital zircon U-Pb dating is widely used to reconstruct evolution of this river in the previous studies [17,18,58–67]. However, no consensus exists on when and how the present drainage pattern formed based on detrital zircon data from various basins. The laser spot positions on rims or cores of zircons are often randomly selected for U-Pb dating. The most important evidence for reconstruction the development of the Yangtze River is the presence of sediment signal from present upper Yangtze River in the ancient Yangtze sediment. Just like the case of modern Yangtze sands in this study, the characteristic Cenozoic signal of the upper Yangtze is difficult to be detected in the ancient sediment in the mid-lower Yangtze using the randomly selecting strategy. By increasing the number of analyses, the chance of detecting Cenozoic signal of the upper Yangtze could be increased. Alternatively, by increasing the number of laser spot on rims of detrital zircons, the characteristic Cenozoic signal of the upper Yangtze could be enhanced. Additionally, multiple isotopic indicators (e.g., detrital mineral $^{40}Ar/^{39}Ar$ dating and apatite and rutile U-Pb) of single mineral grains could provide complementary information.

6. Conclusions

In order to reveal factors controlling the difference in age distributions between muscovite and zircon of same sample, we compared published datasets of detrital muscovite $^{40}Ar/^{39}Ar$ and zircon U-Pb ages of modern sands from the Yangtze River. The comparison of pooled age distributions between muscovite and zircon presents that muscovite grains are absent of age populations of 2000–2500 Maand 1600–2000 Ma, suggesting that zircons in the Yangtze River experienced multiple cycles of erosion and deposition. The detrital muscovite ages of each sample were also compared with corresponding detrital zircon ages. Detrital muscovite ages imply that the Dadu River in the upper Yangtze is an important sediment supplier to the lower Yangtze. However, detrital zircon data suggest that the Yalong, Dadu, and Min rivers are the Major sediment contributors to the lower reaches. These differences could be caused by discrepancies in closure temperature and durability and potentially differences in source rock contributing to the sediment in each of these river branches. In order to reconstruct the development of the Yangtze using detrital U-Pb analysis, increasing the number of analyzed zircon rims is potentially a way to enhance the characteristic sediment signal of the upper Yangtze.

Author Contributions: For research articles with several authors, a short paragraph specifying their individual contributions must be provided. The following statements should be used "Conceptualization, X.S. and J.R.W.; methodology, K.F.K.; formal analysis, X.S.; writing—original draft preparation, X.S., K.F.K.; writing—review and editing, Y.T. and Z.Z.; funding acquisition, J.R.W., K.F.K., Y.T. and C.L.", please turn to the CRediT taxonomy for the term explanation. Authorship must be limited to those who have contributed substantially to the work reported. All authors have read and agreed to the published version of the manuscript.

Funding: This study is financially supported by the National Natural Science Foundation of China (41671011, 41672355, 41888101, and 41801003) and the Fundamental Research Funds for the Central Universities (19lgpy72). K.F.K. is supported by NWO-ALW grant 864.12.005.

Acknowledgments: This work was supported by the argon geochronology laboratory of the Vrije Universiteit Amsterdam.

Conflicts of Interest: The authors declare no conflict of interest.

References

1. Copeland, P. On the use of geochronology of detrital grains in determining the time of deposition of clastic sedimentary strata. *Basin Res.* **2020**. [CrossRef]
2. Dickinson, W.R.; Gehrels, G.E. Use of U–Pb ages of detrital zircons to infer Maximum depositional ages of strata: A test against a Colorado Plateau Mesozoic database. *Earth Planet. Sci. Lett.* **2009**, *288*, 115–125. [CrossRef]
3. Coutts, D.S.; Matthews, W.A.; Hubbard, S.M. Assessment of widely used methods to derive depositional ages from detrital zircon populations. *Geosci. Front.* **2019**, *10*, 1421–1435. [CrossRef]
4. Fedo, C.M.; Sircombe, K.N.; Rainbird, R.H. Detrital zircon analysis of the sedimentary record. *Rev. Mineral. Geochem.* **2003**, *53*, 277–303. [CrossRef]
5. He, M.; Zheng, H.; Bookhagen, B.; Clift, P.D. Controls on erosion intensity in the Yangtze River basin tracked by U–Pb detrital zircon dating. *Earth Sci. Rev.* **2014**, *136*, 121–140. [CrossRef]
6. Wissink, G.K.; Hoke, G.D. Eastern Margin of Tibet supplies most sediment to the Yangtze River. *Lithosphere* **2016**. [CrossRef]
7. Gehrels, G. Detrital zircon U-Pb geochronology applied to tectonics. *Annu. Rev. Earth Planet. Sci.* **2014**, *42*, 127–149. [CrossRef]
8. Najman, Y. The detrital record of orogenesis: A review of approaches and techniques used in the Himalayan sedimentary basins. *Earth Sci. Rev.* **2006**, *74*, 1–72. [CrossRef]
9. Zhu, G.; Wang, Y.S.; Wang, W.; Zhang, S.; Liu, C.; Gu, C.C.; Li, Y.J. An accreted micro-continent in the north of the Dabie Orogen, East China: Evidence from detrital zircon dating. *Tectonophysics* **2017**, *698*, 47–64. [CrossRef]
10. Haines, P.W.; Turner, S.P.; Kelley, S.P.; Wartho, J.A.; Sherlock, S.C. $^{40}Ar/^{39}Ar$ dating of detrital muscovite in provenance investigations: A case study from the Adelaide Rift Complex, South Australia. *Earth Planet. Sci. Lett.* **2004**, *227*, 297–311. [CrossRef]
11. Miller, J.S.; Matzel, J.E.P.; Miller, C.F.; Burgess, S.D.; Miller, R.B. Zircon growth and recycling during the assembly of large, composite arc plutons. *J. Volcanol. Geotherm. Res.* **2007**, *167*, 282–299. [CrossRef]
12. Harrison, T.M.; Célérier, J.; Aikman, A.B.; Hermann, J.; Heizler, M.T. Diffusion of ^{40}Ar in muscovite. *Geochim. Cosmochim. Acta* **2009**, *73*, 1039–1051. [CrossRef]
13. Lee, J.K.W.; Williams, I.S.; Ellis, D.J. Pb, U and Th diffusion in natural zircon. *Nature* **1997**, *390*, 159–162. [CrossRef]
14. Garzanti, E.; Vermeesch, P.; Andò, S.; Vezzoli, G.; Valagussa, M.; Allen, K.; Kadi, K.A.; Al-Juboury, A.I.A. Provenance and recycling of Arabian desert sand. *Earth Sci. Rev.* **2013**, *120*, 1–19. [CrossRef]
15. Thomas, W.A. Detrital-zircon geochronology and sedimentary provenance. *Lithosphere* **2011**, *3*, 304–308. [CrossRef]
16. Sun, X.L.; Li, C.A.; Kuiper, K.F.; Wang, J.T.; Tian, Y.T.; Vermeesch, P.; Zhang, Z.J.; Zhao, J.X.; Wijbrans, J.R. Geochronology of detrital muscovite and zircon constrains the sediment provenance changes in the Yangtze River during the late Cenozoic. *Basin Res.* **2018**, *30*, 636–649. [CrossRef]
17. Jia, J.T.; Zheng, H.B.; Huang, X.T.; Wu, F.Y.; Yang, S.Y.; Wang, K.; He, M.Y. Detrital zircon U-Pb ages of late cenozoic sediments from the Yangtze delta: Implication for the evolution of the Yangtze River. *Chin. Sci. Bull.* **2010**, *55*, 1520–1528. [CrossRef]
18. Wang, J.T.; Li, C.A.; Yang, Y.; Shao, L. Detrital zircon geochronology and provenance of core sediments in zhoulao town, Jianghan Plain, China. *J. Earth Sci.* **2010**, *21*, 257–271. [CrossRef]
19. Yang, S.Y.; Wang, Z.B.; Guo, Y.; Li, C.X.; Cai, J.G. Heavy mineral compositions of the Changjiang (Yangtze River) sediments and their provenance-tracing implication. *J. Asian Earth Sci.* **2009**, *35*, 56–65. [CrossRef]

20. Zhang, Z.J.; Tyrrell, S.; Li, C.A.; Daly, J.S.; Sun, X.L.; Blowick, A.; Lin, X. Provenance of detrital K-feldspar in Jianghan Basin sheds new light on the Pliocene–Pleistocene evolution of the Yangtze River. *Geol. Soc. Am. Bull.* **2016**, *128*, B31445.1. [CrossRef]
21. He, M.Y.; Zheng, H.B.; Clift, P.D. Zircon U–Pb geochronology and Hf isotope data from the Yangtze River sands: Implications for Major Magmatic events and crustal evolution in central China. *Chem. Geol.* **2013**, *360–361*, 186–203. [CrossRef]
22. Shao, L.; Li, C.A.; Yuan, S.Y.; Kang, C.G.; Wang, J.T.; Li, T. Neodymium isotopic variations of the late Cenozoic sediments in the Jianghan Basin: Implications for sediment source and evolution of the Yangtze River. *J. Asian Earth Sci.* **2012**, *45*, 57–64. [CrossRef]
23. He, M.Y.; Zheng, H.B.; Huang, X.T.; Jia, J.T.; Li, L. Yangtze River sediments from source to sink traced with clay mineralogy. *J. Asian Earth Sci.* **2013**, *69*, 60–69. [CrossRef]
24. Sun, X.L.; Li, C.A.; Kuiper, K.F.; Zhang, Z.J.; Gao, J.H.; Wijbrans, J.R. Human impact on erosion patterns and sediment transport in the Yangtze River. *Glob. Planet. Chang.* **2016**, *143*, 88–99. [CrossRef]
25. Hoang, L.V.; Clift, P.D.; Mark, D.; Zheng, H.B.; Mai Thanh, T. Ar-Ar muscovite dating as a constraint on sediment provenance and erosion processes in the Red and Yangtze river systems, SE Asia. *Earth Planet. Sci. Lett.* **2010**, *295*, 379–389. [CrossRef]
26. Yang, S.Y.; Zhang, F.; Wang, Z.B. Grain size distribution and age population of detrital zircons from the Changjiang (Yangtze) River system, China. *Chem. Geol.* **2012**, *296–297*, 26–38. [CrossRef]
27. *Geological Map of China (1:2,500,000)*; China Geological Map Press: Beijing, China, 2004.
28. Saito, K.; Tada, R.; Zheng, H.B.; Irino, T.; Luo, C.; He, M.Y.; Wang, K.; Suzuki, Y. ESR signal intensity of quartz in the fine-silt fraction of riverbed sediments from the Yangtze River: A provenance tracer for suspended particulate Matter. *Prog. Earth Planet. Sci.* **2017**, *4*. [CrossRef]
29. Reid, A.J.; Wilson, C.J.L.; Phillips, D.; Liu, S. Mesozoic cooling across the Yidun Arc, central-eastern Tibetan plateau: A reconnaissance $^{40}Ar/^{39}Ar$ study. *Tectonophysics* **2005**, *398*, 45–66. [CrossRef]
30. Wu, W.H.; Xu, S.J.; Yang, J.D.; Yin, H.W.; Tao, X.C. Sr fluxes and isotopic compositions in the headwaters of the Yangtze River, Tongtian River and Jinsha River originating from the Qinghai–Tibet Plateau. *Chem. Geol.* **2009**, *260*, 63–72. [CrossRef]
31. Roger, F.; Jolivet, M.; Malavieille, J. The tectonic evolution of the Songpan-Garze (north Tibet) and adjacent areas from proterozoic to present: A synthesis. *J. Asian Earth Sci.* **2010**, *39*, 254–269. [CrossRef]
32. Xu, Z.Q.; Hou, L.W.; Wang, Z.X.; Fu, X.F.; Huang, M.H. *Orogenic Processes of the Songpan-Garze Orogenic Belt of China*; Geological Publishing House: Beijing, China, 1992; pp. 1–235.
33. Dong, Y.P.; Zhang, G.W.; Neubauer, F.; Liu, X.M.; Genser, J.; Hauzenberger, C. Tectonic evolution of the Qinling orogen, China: Review and synthesis. *J. Asian Earth Sci.* **2011**, *41*, 213–237. [CrossRef]
34. Shu, L.S.; Faure, M.; Yu, J.H.; Jahn, B.M. Geochronological and geochemical features of the Cathaysia Block (South China): New evidence for the Neoproterozoic breakup of Rodinia. *Precambrian Res.* **2011**, *187*, 263–276. [CrossRef]
35. Sun, X.L.; Kuiper, K.F.; Tian, Y.T.; Li, C.A.; Zhang, Z.J.; Gemignani, L.; Guo, R.J.; de Breij, V.H.L.; Wijbrans, J.R. $^{40}Ar/^{39}Ar$ mica dating of late Cenozoic sediments in SE Tibet: Implications for sediment recycling and drainage evolution. *J. Geol. Soc.* **2020**, *177*, 843–854. [CrossRef]
36. Vermeesch, P. How Many grains are needed for a provenance study? *Earth Planet. Sci. Lett.* **2004**, *224*, 441–451. [CrossRef]
37. Zhang, Y.Q.; Chen, W.; Yang, N. $^{40}Ar/^{39}Ar$ dating of shear deformation of the Xianshuihe fault zone in west Sichuan and its tectonic significance. *Sci. China Ser. D-Earth Sci.* **2004**, *47*, 794–803.
38. Itaya, T.; Hyodo, H.; Tsujimori, T.; Wallis, S.; Aoya, M.; Kawakami, T.; Gouzu, C. Regional-scale excess Ar wave in a Barrovian type metamorphic belt, eastern Tibetan Plateau. *Island Arc* **2009**, *18*, 293–305. [CrossRef]
39. Kirby, E. *Geomorphic Insights into the Growth of Eastern Tibet and Implications for the Recurrence of Great Earthquakes*; AGU Fall Meeting Abstracts: Washington, DC, USA, 2008; p. 3.
40. Kirby, E.; Reiners, P.W.; Krol, M.A.; Whipple, K.X.; Hodges, K.V.; Farley, K.A.; Tang, W.Q.; Chen, Z.L. Late cenozoic evolution of the eastern Margin of the Tibetan Plateau: Inferences from $^{40}Ar/^{39}Ar$ and (U-Th)/He thermochronology. *Tectonics* **2002**, *21*. [CrossRef]
41. Godard, V.; Pik, R.; Lave, J.; Cattin, R.; Tibari, B.; de Sigoyer, J.; Pubellier, M.; Zhu, J. Late cenozoic evolution of the central Longmen shan, eastern Tibet: Insight from (U-Th)/He thermochronometry. *Tectonics* **2009**, *28*. [CrossRef]

42. Cook, K.L.; Hovius, N.; Wittmann, H.; Heimsath, A.M.; Lee, Y.-H. Causes of rapid uplift and exceptional topography of Gongga shan on the eastern Margin of the Tibetan Plateau. *Earth Planet. Sci. Lett.* **2018**, *481*, 328–337. [CrossRef]
43. Henck, A.C.; Huntington, K.W.; Stone, J.O.; Montgomery, D.R.; Hallet, B. Spatial controls on erosion in the Three Rivers region, southeastern Tibet and southwestern China. *Earth Planet. Sci. Lett.* **2011**, *303*, 71–83. [CrossRef]
44. Lu, Y.J.; Kerrich, R.; Cawood, P.A.; McCuaig, T.C.; Hart, C.J.R.; Li, Z.X.; Hou, Z.Q.; Bagas, L. Zircon shrimp U–Pb geochronology of potassic felsic intrusions in western Yunnan, SW China: Constraints on the relationship of Magmatism to the Jinsha suture. *Gondwana Res.* **2012**, *22*, 737–747. [CrossRef]
45. Vezzoli, G.; Garzanti, E.; Limonta, M.; Andò, S.; Yang, S.Y. Erosion patterns in the Changjiang (Yangtze River) catchment revealed by bulk-sample versus single-mineral provenance budgets. *Geomorphology* **2016**, *261*, 177–192. [CrossRef]
46. McDougall, I.; Harrison, T.M. *Geochronology and Thermochronology by the $^{40}Ar/^{39}Ar$ Method*; Oxford University Press on Demand: Oxford, UK, 1999.
47. Enkelmann, E.; Weislogel, A.; Ratschbacher, L.; Eide, E.; Renno, A.; Wooden, J. How was the triassic Songpan-Ganzi basin filled? A provenance study. *Tectonics* **2007**, *26*. [CrossRef]
48. Weislogel, A.L.; Graham, S.A.; Chang, E.Z.; Wooden, J.L.; Gehrels, G.E. Detrital zircon provenance from three turbidite depocenters of the middle-upper triassic Songpan-Ganzi complex, central China: Record of collisional tectonics, erosional exhumation, and sediment production. *Geol. Soc. Am. Bull.* **2010**, *122*, 2041–2062. [CrossRef]
49. Weislogel, A.L.; Graham, S.A.; Chang, E.Z.; Wooden, J.L.; Gehrels, G.E.; Yang, H.S. Detrital zircon provenance of the late triassic Songpan-Ganzi complex: Sedimentary record of collision of the north and south China blocks. *Geology* **2006**, *34*, 97–100. [CrossRef]
50. Yan, Z.K.; Tian, Y.T.; Li, R.; Vermeesch, P.; Sun, X.L.; Li, Y.; Rittner, M.; Carter, A.; Shao, C.J.; Huang, H.; et al. Late Triassic tectonic inversion in the upper Yangtze Block: Insights from detrital zircon U–Pb geochronology from south-western Sichuan basin. *Basin Res.* **2019**, *31*, 92–113. [CrossRef]
51. She, Z.B.; Ma, C.Q.; Wan, Y.S.; Zhang, J.Y.; Li, M.; Chen, L.; Xu, W.J.; Li, Y.Q.; Ye, L.F.; Gao, J. An early Mesozoic transcontinental palaeoriver in south China: Evidence from detrital zircon U–Pb geochronology and Hf isotopes. *J. Geol. Soc.* **2012**, *169*, 353–362. [CrossRef]
52. Jolivet, M.; Roger, F.; Xu, Z.Q.; Paquette, J.L.; Cao, H. Mesozoic–Cenozoic evolution of the Danba dome (Songpan Garzê, east Tibet) as inferred from LA-ICPMS U–Pb and fission-track data. *J. Asian Earth Sci.* **2015**, *102*, 180–204. [CrossRef]
53. Roberts, N.M.W.; Searle, M.P. Zircon U–Pb–Hf constraints from Gongga Shan granites on young crustal melting in eastern Tibet. *Geosci. Front.* **2019**, *10*, 885–894. [CrossRef]
54. Schwartz, T.M.; Schwartz, R.K.; Weislogel, A.L. Orogenic recycling of detrital zircons characterizes age distributions of north Aamerican cordilleran strata. *Tectonics* **2019**, *38*, 4320–4334. [CrossRef]
55. Wang, W.T.; Zheng, W.J.; Zhang, P.Z.; Li, Q.; Kirby, E.; Yuan, D.Y.; Zheng, D.W.; Liu, C.C.; Wang, Z.C.; Zhang, H.P.; et al. Expansion of the Tibetan Plateau during the Neogene. *Nat. Commun.* **2017**, *8*, 15887. [CrossRef] [PubMed]
56. Campbell, I.H.; Reiners, P.W.; Allen, C.M.; Nicolescu, S.; Upadhyay, R. He–Pb double dating of detrital zircons from the Ganges and Indus Rivers: Implication for quantifying sediment recycling and provenance studies. *Earth Planet. Sci. Lett.* **2005**, *237*, 402–432. [CrossRef]
57. Zimmermann, S.; Mark, C.; Chew, D.; Voice, P.J. Maximising data and precision from detrital zircon U-Pb analysis by LA-ICPMS: The use of core-rim ages and the single-analysis concordia age. *Sediment. Geol.* **2018**, *375*, 5–13. [CrossRef]
58. Hoang, L.V.; Wu, F.-Y.; Clift, P.D.; Wysocka, A.; Swierczewska, A. Evaluating the evolution of the Red River system based on in situ U-Pb dating and Hf isotope analysis of zircons. *Geochem. Geophys. Geosyst.* **2009**, *10*. [CrossRef]
59. Kong, P.; Zheng, Y.; Fu, B.H. Cosmogenic nuclide burial ages and provenance of late cenozoic deposits in the Sichuan Basin: Implications for early Quaternary glaciations in east Tibet. *Quat. Geochronol.* **2011**, *6*, 304–312. [CrossRef]
60. Zheng, H.B.; Clift, P.D.; Wang, P.; Tada, R.; Jia, J.T.; He, M.Y.; Jourdan, F. Pre-Miocene birth of the Yangtze River. *Proc. Natl. Acad. Sci. USA* **2013**, *110*, 7556–7561. [CrossRef]

61. Wang, P.; Zheng, H.; Liu, S.; Hoke, G. Late Cretaceous drainage reorganization of the middle Yangtze River. *Lithosphere* **2018**, *10*, 392–405. [CrossRef]
62. Wang, P.; Zheng, H.B.; Chen, L.; Chen, J.; Xu, Y.; Wei, X.; Yao, X. Exhumation of the Huangling anticline in the Three Gorges region: Cenozoic sedimentary record from the western Jianghan basin, China. *Basin Res.* **2014**, *26*, 505–522. [CrossRef]
63. Wissink, G.K.; Hoke, G.D.; Garzione, C.N.; Liu-Zeng, J. Temporal and spatial patterns of sediment routing across the southeast Margin of the Tibetan Plateau: Insights from detrital zircon. *Tectonics* **2016**, *35*. [CrossRef]
64. Chen, Y.; Yan, M.D.; Fang, X.M.; Song, C.H.; Zhang, W.L.; Zan, J.B.; Zhang, Z.G.; Li, B.S.; Yang, Y.P.; Zhang, D.W. Detrital zircon U–Pb geochronological and sedimentological study of the Simao Basin, Yunnan: Implications for the early Cenozoic evolution of the Red River. *Earth Planet. Sci. Lett.* **2017**, *476*, 22–33. [CrossRef]
65. Kong, P.; Zheng, Y.; Caffee, M.W. Provenance and time constraints on the formation of the first bend of the Yangtze River. *Geochem. Geophys. Geosyst.* **2012**, *13*. [CrossRef]
66. Deng, B.; Chew, D.; Jiang, L.; Mark, C.; Cogné, N.; Wang, Z.J.; Liu, S.G. Heavy mineral analysis and detrital U-Pb ages of the intracontinental Paleo-Yangzte basin: Implications for a transcontinental source-to-sink system during late Cretaceous time. *GSA Bull.* **2018**, *130*, 2087–2109. [CrossRef]
67. Yang, C.Q.; Shen, C.B.; Zattin, M.; Yu, W.; Shi, S.X.; Mei, L.F. Provenances of Cenozoic sediments in the Jianghan Basin and implications for the formation of the Three Gorges. *Int. Geol. Rev.* **2019**, *61*, 1980–1999. [CrossRef]

minerals

Article

Detrital Zircon U-Pb Ages in the East China Seas: Implications for Provenance Analysis and Sediment Budgeting

Xiangtong Huang [1,*], Jiaze Song [1], Wei Yue [2], Zhongbo Wang [3], Xi Mei [3], Yalong Li [1], Fangliang Li [1], Ergang Lian [1] and Shouye Yang [1,*]

1 State Key Laboratory of Marine Geology, Tongji University, Shanghai 200092, China; jz_song@tongji.edu.cn (J.S.); 1710583@tongji.edu.cn (Y.L.); lifl17@tongji.edu.cn (F.L.); eglian@tongji.edu.cn (E.L.)

2 School of Geography, Geomatics and Planning, Jiangsu Normal University, Xuzhou 221116, China; yuexiyuan@126.com

3 Qingdao Institute of Marine Geology, China Geological Survey, Ministry of Natural Resources, Qingdao 266071, China; wangzhongbo@mail.cgs.gov.cn (Z.W.); meixi@mail.cgs.gov.cn (X.M.)

* Correspondence: xiangtong@tongji.edu.cn (X.H.); syyang@tongji.edu.cn (S.Y.)

Received: 5 April 2020; Accepted: 29 April 2020; Published: 29 April 2020

Abstract: Linking marine sinks to potential terrestrial sources is one of most intriguing but challenging aspects of sediment source-to-sink studies. In this study, we analyzed 23 zircon samples (3271 filtered best ages) from surface sediments of the east China seas (ECSs) that cover a large portion of the Bohai Sea, Yellow Sea, East China Sea to part of the northeastern South China Sea. The results of U-Pb age distributions exhibit variable signatures in different seas. The Bohai Sea is characterized by 4 age populations at 203–286 Ma, 383–481 Ma, 1830–1940 Ma and 2480–2548 Ma, whereas the southern Yellow Sea and the East China Sea are featured by 5 age populations at 176–223 Ma, 383–481 Ma, 732–830 Ma, 1830–1940 Ma and 2480–2548 Ma. We propose that the presence or absence of the population of 732–830 Ma in the Yangtze Craton (YC) and the North China Craton (NCC) is a possible geochronological signature to distinguish zircon grains derived from the two source regions. Furthermore, on the basis of multidimensional scaling (MDS), U-Pb ages in the sediments of the Bohai Sea, East China Sea and the Taiwan Strait could be correspondently linked to those of the Yellow River, the Yangtze River and Taiwan rivers. The good linkages support the view that U-Pb age distributions of detrital zircons in the margin seas are mainly controlled by fluvial discharges, and ultimately, by the tectonic history of the corresponding source regions. Using a sediment forward mixing model, we obtained the relative sediment contributions and spatial variations of five most important river discharges in the region. The mixing results suggest that the major rivers in the region, i.e., the Yangtze and the Yellow Rivers, are the dominant sediment contributors to the continental margin, and their mixing coefficients could be used to infer relative sediment budgeting. In addition, spatial variations in mixing coefficient in the East China Sea indicate that sediment mixing and partitioning processes in the marine depositional environment have played a part role in propagating the provenance signals as a result of interaction of oceanic currents and tides. The combined method between provenance analysis and mixing modeling provides a feasible way to appreciate sediment budgeting in the geological past.

Keywords: U-Pb geochronology; detrital zircon; source-to-sink; provenance analysis; mixing model; sediment budgeting; east China seas

1. Introduction

Continental margin sea is the most important sink for terrestrial sediments due to large accommodation space and rapid sediment accumulation rates [1]. Linking marine sinks to terrestrial sources using sedimentological or geochemical methods is one of the most intriguing but challenging aspects of sediment routing system studies [2]. From a perspective of source-to-sink systems, it is very important to know where sediments have been derived from and how associated environmental signals, such as provenance and sediment fluxes related to tectonics and climate change, are propagated and modified within the systems [3]. To estimate sediment budgeting influenced by natural forcing or by human activities is therefore fundamental to understand source-to-sink processes in sedimentary systems [4,5]. With the rapid development of in situ micro-analytical techniques in Earth science since the last two decades, the detrital zircon U-Pb geochronology has been widely applied to examine various sedimentary questions, such as sediment provenance, ages and rock characteristics of source terranes, stratigraphic correlation and determination of maximum depositional ages [6–9]. It has been regarded as a real progress in provenance analysis by providing a time dimension to assist the interpretation of detrital modes [10]. However, many previous studies were focused on long-term provenance analysis and based on an untested premise that detrital zircon ages are mainly controlled by tectonic factors and are less susceptible to hydrological sorting and other factors. Some recent studies argued that detrital zircon ages could be applied to investigate sedimentary mixing processes on short-term scales [11,12], and the age distributions may be influenced by factors other than tectonics, such as fertility [13], grain-size sorting [14] and the number of analyses [15]. Due to the on-going debate on the factors that control detrital zircon U-Pb ages in sediments, it is necessary to examine it from continental margin seas, which are the important sinks for terrestrial sediment.

The east China seas (hereafter referred to as ECSs) are favored in this contribution for several reasons. First, the ECSs contain the broadest continental shelf in the world. Annually, there are large amounts of sediments discharged into them by large rivers like the Yangtze and Yellow River that are originated from the Tibetan Plateau and numerous moderate and small mountainous rivers draining in tectonically stable areas like SE China and active one like Taiwan [16–19]. Due to the remarkable contrasts in tectonic setting and lithology, fluvial sediments with distinguished terrestrial U-Pb age signatures could be preserved in marine sediments. Second, a substantial of detrital zircon U-Pb ages have been published from many modern fluvial sediments in East Asia [20–25], making it possible to link terrestrial sources to marine sinks by U-Pb geochronological methods. Lastly, despite that the sandy deposits in the ECSs cover a larger extent in the region (Figure 1), less attention has been paid to them when compared with the muddy deposits [26–28]. As a result, the sources of marine sediments and to which extent ocean hydrodynamics, such as currents, tides or waves, can play a role in the dispersion of terrestrial sediments and propagation of provenance signals in the seas are still not clear. Accordingly, there are three primary objectives of this study: (1) link marine sediments in the ECSs to the present-day fluvial sediments using detrital zircon U-Pb geochronology; (2) examine how marine hydrological processes may modify provenance signatures of sediments in different seas; (3) estimate the sediment budgeting in the seas quantitatively or semi-quantitatively using sediment mixing models developed in the community [8,29].

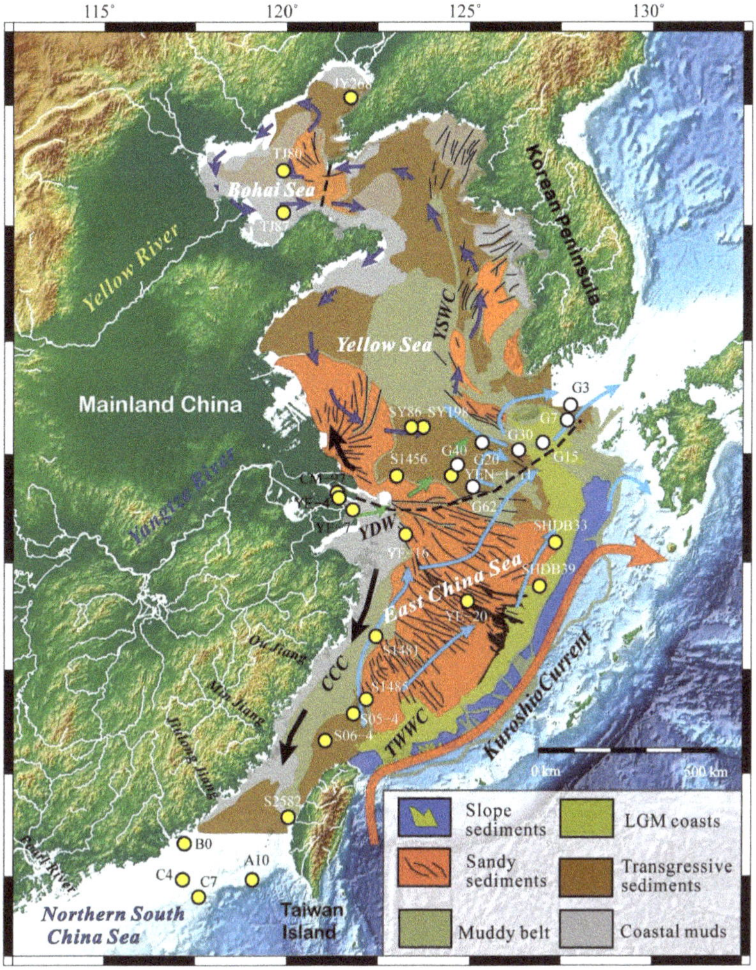

Figure 1. Map showing surface sediment types, ocean currents and sampling sites in the east China seas. The surface sediment distribution and ocean currents are from the literature [30,31], respectively. Ocean currents are denoted by solid lines with different colors (red: Kuroshio Current; light blue: Taiwan Warm Current (TWWC); black: China Coast Current (CCC); green: Yangtze Diluted Water (YDW); dark blue: Yellow Sea Warm Current (YSWC)). Circles filled by yellow are sites analyzed in this study; the ones filled by white are from the literature [32].

2. Sedimentary and Tectonic Setting

2.1. Sediment Fluxes and Oceanic Hydrodynamics

The ECSs in this study are defined as an area consisting of the Bohai Sea, Yellow Sea, East China Sea and a part of the northern South China Sea (Figure 1). This extensive area is surrounded by mainland China to the west and the north, by the Korean Peninsula and Kyushu and Ryukyu Islands to the east and by Taiwan Island to the south. There are large but variable discharges of sediment into the seas.

The highest sediment discharge was from the Yellow River with annual suspended sediments reaching to 1086 Mt in the history [33], but it has reduced to less than 100 Mt since the last decade possibly due to human activities. The second highest sediment discharge is from the Yangtze River, which had a pre-dam annual discharge of about 500 Mt but has dropped to less than 130 Mt nowadays [34]. Although mountainous rivers in Taiwan Island are smaller in terms of catchment area and river length, the annual sediment discharge could be as high as 300 Mt [35]. A number of Zhe-Min (Zhejiang and Fujian provinces) rivers draining in SE China (Figure 1) have a total annual sediment discharge of 23 Mt [16], which is relatively less important compared with Taiwan. The annual sediment discharge of rivers in the Korean Peninsula is around 20 Mt [36], which is equivalent to that of the Zhe-Min rivers.

The sediment distribution in the ECSs is mainly controlled by the dynamic climate and ocean settings in the region [30,37]. According to the result of oceanographic observations, the surface and bottom current systems in the ECSs are strongly influenced by a large-scale anti-clockwise circulation in the western Pacific [38] (Figure 1). As a result, there are several persistent currents in the region, including the Kuroshio Current, Taiwan Warm Current (TWWC), Tsushima Warm Current, Yellow Sea Warm Current (YSWC) and the China Coastal Current (CCC). In contrast, the Yangtze Diluted Water Current (DWC) is stronger in summer than in winter. In addition to the ocean currents, there are very strong tidal currents and waves with the great axis of the tidal ellipse-oriented NW to SE in the East China Sea [39]. It is believed that sand ridges are formed in the areas with strong rectilinear tidal currents, sand sheets in areas are dominated by strong rotatory tidal currents, and clay sediments are deposited mainly in the areas of weak tidal currents [40].

2.2. Tectonic Setting

Geologically, the rivers flowing into the ECSs drain different first-order tectonic units, containing the North China Craton (NCC) in the north, the Yangtze Craton (YC) and the Cathaysia Fold Belt (CF) in the south and the Songpan–Ganzi Fold Belt (SPGF) in the west (Figure 2). Each of the tectonic units has a different geological history in terms of the basement, sedimentary cover and igneous rocks (Figure 2).

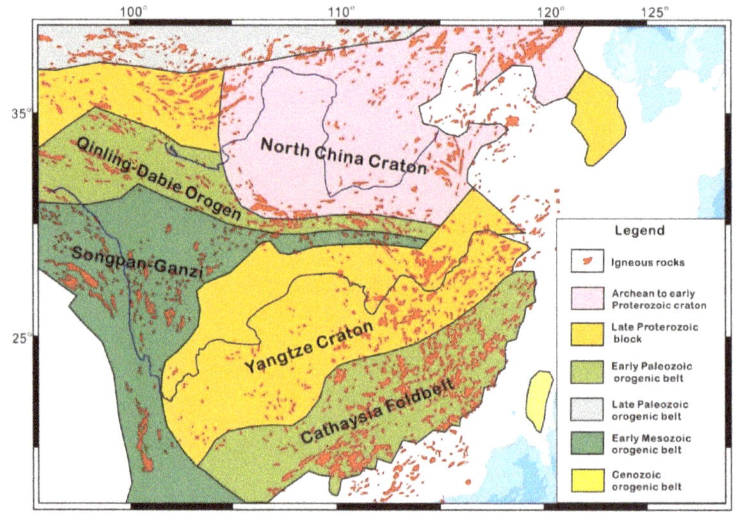

Figure 2. Schematic map illustrating ancient orogenic belts, first-order tectonic units and the distribution of igneous rocks in East Asia (modified after [41,42]).

The NCC has a widespread Neoarchean basement formed around 2.7–2.5 Ga, which covers 85% of the total exposed basement [43,44]. The cratonization of the NCC started during the Paleoproterozoic (1.9–1.8 Ga), and there were no significant tectonic activities until the Paleozoic Era. According to a compilation in [45], there were six stages of magmatism in the NCC from the Paleozoic to the Cenozoic, with age populations around the Carboniferous to Early Permian (324–270 Ma), Late Permian to Middle Triassic (262–236 Ma), Late Triassic (231–199 Ma), Jurassic (189–146 Ma), Early Cretaceous (141–108 Ma) and the Cenozoic (65 Ma–present).

The YC is separated from the NCC by the Qinling–Dabie Orogen, which was formed during the period from the late Paleozoic to middle Mesozoic (Figure 2). In contrast to the NCC, rocks of the Archean and Paleoproterozoic are sporadically outcropped in the YC, and most of the exposed basements were formed during the Neoproterozoic. This is supported by abundant 790–900 Ma granitoids observed in the Jiangnan Orogenic Belt in the southern margin of the YC and 740–830 Ma felsic plutonic and volcanic rocks at the western margin [46]. The tectonic activity in the YC became reactivated during the Paleozoic and the Mesozoic, which is characterized by the presence of late Paleozoic Emeishan basalts (260 Ma) in the upper YC [47] and Early Cretaceous igneous rocks (134–125 Ma) in the lower YC [48].

The CF is located southeast to the YC (Figure 2). The two blocks were assembled together during the Neoproterozoic (1.00–0.85 Ga). Unlike to the YC, which has a widespread exposure of the Neoproterozoic rocks, the CF is characterized by voluminous outcrops of the Paleozoic to Mesozoic igneous rocks [44,49]. The more distinct age populations are related to the Paleozoic Kwangsian granites (420–442 Ma) and Yanshanian granitoids and volcanic rocks (80–190 Ma).

The SPGF is located at eastern edge of the Qinghai–Tibetan Plateau (Figure 2), which is the headwater region of the present Yellow and Yangtze rivers. The block is mainly covered by a thick Triassic flysch sequence and was a part of Permo–Triassic orogenic belts situated at the northeast Tibetan Plateau and southwest China [50]. The detrital zircons in this region are explained to have been derived from two different sources [51]. The first one is the YC, which is characterized by a Neoproterozoic U-Pb age population at 720–900 Ma, while the second is the Qinling Mountains (Figure 2), which contribute older U-Pb ages at 2.2–1.4 Ga.

Although Taiwan Island is relatively small in area, it has a very complicated geological history. It is marked by a Mesozoic basement and a mélange assembly in the Tananao Complex, slate formations of the Eocene-early Oligocene in the Hsuehsan Range and the late Oligocene to Pleistocene at the West Foothills [52]. Similarly, the Korean fluvial sediments may be derived mainly from the Precambrian igneous and metamorphic rocks [53] and Jurassic–Cretaceous granites and schists [54].

3. Samples and Methods

3.1. Samples

We carried out a detrital zircon U-Pb analysis on 23 samples in this study, including 22 surface samples taken from the ECSs and one early Holocene sample from a borehole at the Yangtze Delta (Figure 1). Most samples are mainly composed of silty sands except for those obtained from the Bohai Sea and the Yellow Sea, which are silty muds. A detailed information on water depths and the numbers of U-Pb best age are given in Table 1.

Using standard mineral separation techniques, which included heavy liquids, magnetic separation and handy picking concentrates, zircon grains were first separated from the sediments. Afterwards, about 300 grains for each sample were randomly chosen and mounted in the epoxy within 1-inch mounting cups. After drying in an oven, the mounts were sanded down and polished to expose interiors of most zircon grains. Images of transmitted and reflected light of the zircon grains were taken using an optical microscope. A field emission scanning electron microscope (TESCAN, MIRA 3LMH, Nanjing Hongchuang Geological Exploration Technology Service Co., Ltd., Nanjing, China) was used to obtain the cathodoluminescence (CL) images. With help of these images, spots for the laser ablation were chosen by avoiding inclusions, fractures and inherited cores in zircons.

Table 1. Detrital U-Pb sample locations, water depths and the number of best U-Pb ages in the east China seas (ECSs).

Sample ID	Latitude	Longitude	Water Depth (m)	Area	Best Age Number	Sources
JY268	40°9′47″	121°44′57″	8	Bohai Sea	106	This study
TJ80	38°39′29″	119°54′6″	26	Bohai Sea	115	This study
TJ87	37°46′39″	119°54′6″	26	Bohai Sea	80	This study
SY198	33°5′40″	123°44′2″	46	Southern Yellow Sea	112	This study
SY86	33°5′40″	123°24′17″	39	Southern Yellow Sea	110	This study
S1456	32°0′3″	124°29′60″	47	Southern Yellow Sea	241	This study
YEN-1-11	31°59′52″	122°59′60″	32	East China Sea/Yellow Sea	109	This study
YE-16	30°39′17″	123°15′1″	53	East China Sea	109	This study
YE-20	29°6′47″	124°56′23″	87	East China Sea	112	This study
YE-4	31°30′20″	121°25′50″	10	Yangtze Estuary	194	This study
YE-7	31°13′36″	121°48′14″	9	Yangtze Estuary	108	This study
CM-97	31°37′0″	121°22′60″	−60	Yangtze Estuary	112	This study
S05-4	26°27′0″	121°49′12″	89	East China Sea	106	This study
S06-4	25°48′12″	121°3′36″	82	East China Sea	105	This study
S1481	28°17′20″	122°26′2″	70	East China Sea	111	This study
S1485	26°47′29″	122°10′0″	90	East China Sea	231	This study
S2582	23°56′27″	120°2′58″	33	Taiwan Straight	267	This study
SHDB33	30°28′29″	127°20′41″	117	East China Sea	227	This study
SHDB39	29°27′43″	126°54′40″	120	East China Sea	113	This study
A10	22°24′37″	119°3′54″	124	Northern South China Sea	119	This study
B0	23°17′33″	117°12′21″	30	Northern South China Sea	117	This study
C4	22°24′34″	117°9′43″	41	Northern South China Sea	249	This study
C7	21°58′55″	117°35′59″	127	Northern South China Sea	118	This study
G3	33°35′5″	127°45′5″	109	Southeastern Yellow Sea	61	[32]
G7	33°15′7″	127°39′51″	128.2	Southeastern Yellow Sea	61	[32]
G15	32°45′3″	127°0′1″	115.3	Southeastern Yellow Sea	67	[32]
G20	32°45′3″	125°20′6″	115.3	Southeastern Yellow Sea	60	[32]
G30	32°34′57″	126°19′51″	105.4	Southeastern Yellow Sea	67	[32]
G40	32°14′59″	124°39′56″	45	Southeastern Yellow Sea	74	[32]
G62	31°44′55″	125°5′5″	45.7	Southeastern Yellow Sea	62	[32]

3.2. LA-ICP-MS

Mounted zircon grains were ablated at Tongji University using a Resonetics RESOlution M50 193 nm excimer laser system connected to a quadrupole inductively coupled plasma mass spectrometry (LA-ICP-MS, Agilent 7900, Santa Clara, CA, USA). Before data acquisition, a standard reference material glass (SRM 612) produced by the National Institute of Standards and Technology (NIST) was used to tune the ICP-MS. The glass was ablated in a routine condition with a beam spot size of 40 μm and a laser repletion of 6 Hz at fluence of 4 J·cm^{-2}. After tuning, the sensitivity of mass ^{238}U reached to 40,000 counts or 10,000 cps/ppm with a fractionation between ^{232}Th and ^{238}U less than 2% and oxide production rate (ThO/Th) less than 0.3%. Data were obtained for masses 204, 206, 207, 208, 232, 235 and 238 using the ion counting modes of the detector and the integration time for each mass was set to 20 ms. The data acquisition sequence for each unknown or reference material consisted of 15 s blank, 40 s ablation and 15 s washout. Reference zircon materials 91500 [55] and Plešovice [56] were measured periodically in a sample sequence to perform external U-Pb age calibration and monitor the quality of measurements. Except for very fine grain-size samples, which were ablated at a spot size of 18 μm, most of the samples were ablated at a spot size of 26 μm. The spot size differences between samples and the tuning were due to the size limitation of zircon sizes in samples. Despite of the differences, this did not cause significant age deviations in the measurements as supported by the consistent age results from the reference zircon materials.

Data reduction was conducted using the method of mean of isotopic ratios [57], which consisted of the blank subtraction, isotopic ratio calculation, normalization by primary zircon isotopic ratios and the instrumental drift correction in a sequence. The data reduction, age calculation and uncertainty propagation were performed on an in-house Shiny server website software developed by the first author (Huang X.T.), which is accessible at http://60.205.227.89:3838/LaUPb. The weighted mean concordia ages of zircon 91500 (n = 754) and Plešovice (n = 374) are 1061.5 ± 0.25 and 334.9 ± 1.6 Ma (2 se, standard error), respectively, which are consistent with the published U-Pb ages (1062 Ma [55], 337 Ma [56]) within the uncertainty.

In total we obtained 3664 zircon U-Pb ages and filtered out 3271 so-called best ages used in the following discussion. The filtering was mainly based on a 10% cutoff of the discordance of a given $^{206}Pb/^{238}U$ age [9]. The discordance of $^{206}Pb/^{238}U$ age less than 1.4 Ga is defined as 100 × (1 − $^{206}Pb/^{238}U/^{207}Pb/^{235}U$) and the discordance of $^{206}Pb/^{238}U$ age greater than 1.4 Ga is defined as 100 × (1 − $^{206}Pb/^{238}U/^{207}Pb/^{206}Pb$). Additionally, when a $^{206}Pb/^{238}U$ age is less than 300 Ma, the $^{206}Pb/U^{238}$ age is set as the best age due to the low precision of ^{207}Pb [14]. The dataset of this study is provided in the Supplementary Table S1.

3.3. Multidimensional Scaling (MDS)

In order to trace the sediment sources in the East China Sea, a robust statistical method known as multidimensional scaling (MDS) was applied to evaluate the similarity between different U-Pb age distributions [58]. Compared to the other visual or statistical methods, such as cross correlation coefficient, Kolmogorov-Smirnov test (K-S) and Kuiper test [59], the advantage of the MDS lies in its effectiveness to remove redundant features of age distributions while preserving and amplifying the significant differences between them [60]. This is particularly efficient for the pairwise comparison between a large number of samples. Using the DZmds, a MATLAB graphical user interface developed by [59], we chose the kernel density estimates (KDEs) at a bandwidth of 25 M yrs to perform a metric cross-correlation comparison mainly due to the differences of the best ages among the samples and potential sources. We produced a MDS plot for a group of marine and fluvial samples produced in this study and cited from the literature (the Yellow River [23,61], the Yangtze River [20], Korean rivers [24], the Choshui River in western Taiwan [25], Ou Jiang [21] (Jiang means river in Chinese), Min Jiang [22], Jiulong Jiang [21] and Pearl River [62]). The goodness of fit is suggested by a Shepard stress of 0.16, which is a fair result according to rules of thumb [60]. When the wide range of our samples and the potential sources is considered, the results are believed to be good enough to distinguish the dissimilarity among these distributions and more details will be discussed in Section 5.1.

3.4. Mixing Model of Detrital Zircon U-Pb Age Distribution

To quantitatively estimate the relative contribution of sediments or zircon grains from different fluvial systems, we applied a sediment linear mixing model for detrital zircon U-Pb age distributions [29]:

$$D_m = \sum_{i=1}^{n} M_i P_i \tag{1}$$

where P_i is the ith parent or fluvial source age distribution and M_i is the corresponding mixing coefficient or relative contribution and D_m is the modeled mixed daughter distribution, which is the best fit for a measured sample distribution.

To the best of our knowledge, there are generally two ways to solve mixing models [4,8,11,29]. The first are forward models in which the number and distributions of sources are known or specified through geological interpretation and assumption. The second are reverse models in which both the number and distributions are unknown, but they can be solved through optimization methods. In this study, we favor the former one because: (1) the age distributions of fluvial sources can be readily achieved in the literature and (2) they are intimately associated with the geological settings and (3) the source signals are relatively homogeneous spatially and temporally. For example, the U-Pb age distributions become homogenized in the lower reaches of modern Yangtze [20,63] and Yellow rivers [23]. The uniform distributions have been also observed on glacial-interglacial and tectonic time scales. For instance, zircon U-Pb age distributions of the Mangshan loess-palaeosol profile [64], which have been deposited in the flood plain of the Yellow River since the late Pleistocene, are similar to those of the modern Yellow River [23]. In addition, the present Yangtze-like age distributions have been reported from the Pleistocene to Miocene sedimentary records in the Yangtze delta [65] and catchment [63]. As a result, we treated the age distributions in lower reaches of modern rivers as source signals for marine sediments (Figure 3) and applied the forward mixing model (AnalySize) developed by [66] to obtain the mixing coefficients from different sources. We did no use the default non-parametric approach in the mixing model [66] to obtain the parent distributions because it is difficult for the model to get geologically meaningful results due to the uncertainty on the determination of the number of the endmembers and a very large variability in the KDE distributions. Instead, we performed the mixing modeling by using the KDEs shown in Figure 3 as the defined endmembers and the KDEs of our samples as the mixed daughter distributions (Equation (1)). The mixing results are given in Table 2, in which we combined the mixing coefficients of the Zhe-Min Rivers together. The coefficients of determination (R^2) of this forward mixing modeling are generally higher than 0.75 (Table 2), indicating a goodness-of-fit of the modeling. More confidently, most of the mixing results are consistent with the regional sedimentary and oceanic setting. Please see more details in Sections 4.2 and 5.4.

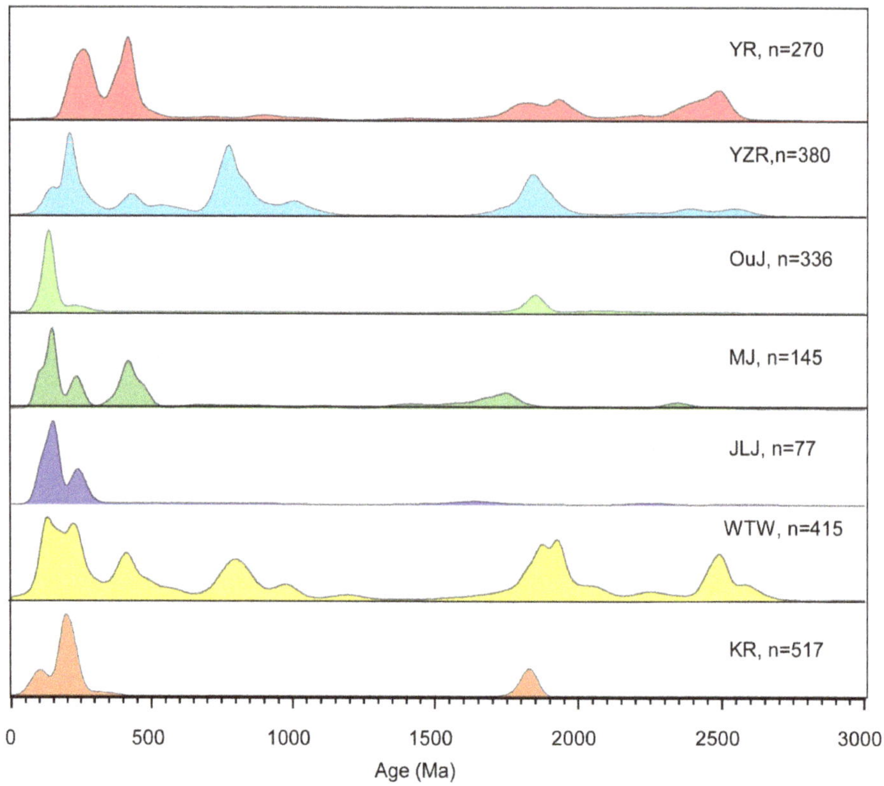

Figure 3. Normalized kernel density estimates (KDEs) of U-Pb age distribution in sediments of seven rivers flowing into the ECSs. The age numbers (n) are shown along the KDE distributions. Abbreviations: YR (Yellow River), YZR (Yangtze River), OuJ (Ou Jiang), MJ (Min Jiang), WTW (rivers in western Taiwan) and KR (Rivers in South Korea).

Table 2. Mixing coefficients of five important river sources in the east China seas.

Sample ID	R^2 *	YR	YZR	ZMR *	WTW	KR
JY268	0.65	87.7	0.0	12.3	0.0	0.0
TJ80	0.58	90.9	0.0	1.8	7.4	0.0
TJ87	0.47	82.8	3.7	13.5	0.0	0.0
SY198	0.63	7.2	58.7	23.6	0.0	10.5
SY86	0.76	38.2	45.0	3.9	0.0	13.0
S1456	0.85	8.3	67.8	0.0	12.1	11.8
YEN-1-11	0.69	0.0	72.8	5.9	0.0	21.3
CM97	0.70	11.5	79.0	0.0	2.0	7.5
YE-4	0.75	26.2	60.7	0.0	13.2	0.0
YE-7	0.79	0.0	97.6	2.4	0.0	0.0
YE-16	0.74	18.9	81.1	0.0	0.0	0.0
SHDB33-M	0.91	9.2	66.8	0.1	0.0	23.9
SHDB39-M	0.95	0.0	43.2	0.0	0.0	56.8
YE-20	0.75	0.0	77.0	0.0	0.0	23.0
S1481	0.85	0.0	88.9	11.1	0.0	0.0
S1485	0.76	27.8	60.6	0.0	11.6	0.0
S05-4	0.75	6.6	65.2	28.1	0.0	0.0
S06-4	0.88	27.0	39.7	21.2	0.0	12.2
S2582	0.79	0.0	8.0	0.0	81.6	10.4
B0	0.85	0.0	0.0	100.0	0.0	0.0
A10	0.71	25.0	63.6	6.7	4.7	0.0
C4	0.94	0.0	0.0	89.9	0.0	10.1
C7	0.77	0.0	0.0	77.2	14.8	8.0
G20	0.72	0.0	86.0	2.9	0.0	11.1
G3	0.81	0.0	0.0	4.0	11.5	84.5
G30	0.63	0.0	67.1	4.6	0.0	28.3
G40	0.53	0.0	89.9	2.6	0.0	7.5
G62	0.63	4.8	73.1	22.1	0.0	0.0
G7	0.96	0.0	0.0	4.7	0.0	95.3
G15	0.83	0.0	16.8	0.0	15.4	67.7

* R^2 is the linear determination coefficient to indicate the goodness-of-fit of the mixing modeling; ZMR represents rivers of Ou Jiang, Min Jiang and Jiulong Jiang.

4. Results

4.1. U-Pb Age Distribution

Because the numbers of best age are variable among different samples (Table 1) and the standard technique kernel density estimation (KDE) is a more robust alternative to conventional age probability distribution [67], we used a KDE plot (Figure 4) to visualize the U-Pb age distributions in this contribution. There are four primary age populations for samples in the Bohai Sea, which are at 203–286 Ma with a peak around 252 Ma, 383–481 Ma with peaks around 420 Ma and 460 Ma, 1830–1940 Ma with a peak around 1860 Ma and 2480–2548 Ma with a peak around 2500 Ma (Figure 4a–c). The most abundant age population is the Mesozoic to Paleozoic one at 203–286 Ma, accounting for a detrital zircon population of 25–40%. Despite that the basement of the NCC formed mainly during the Neoarchean and Paleoproterozoic [45,68,69], there is a minor grain contribution (<10%) from the populations at 1830–1940 and 2480–2548 Ma. The signature of the dominant Mesozoic to Paleozoic age populations in the Bohai Sea samples is similar to that observed from Chinese loess sediments [64], fluvial and desert sediments in the Yellow River catchment [23]. This provides strong evidence from detrital zircon to argue that sediments in the Bohai Sea were derived probably from the Yellow River catchment, and this is true—at least in recent geological history, like the late Pleistocene.

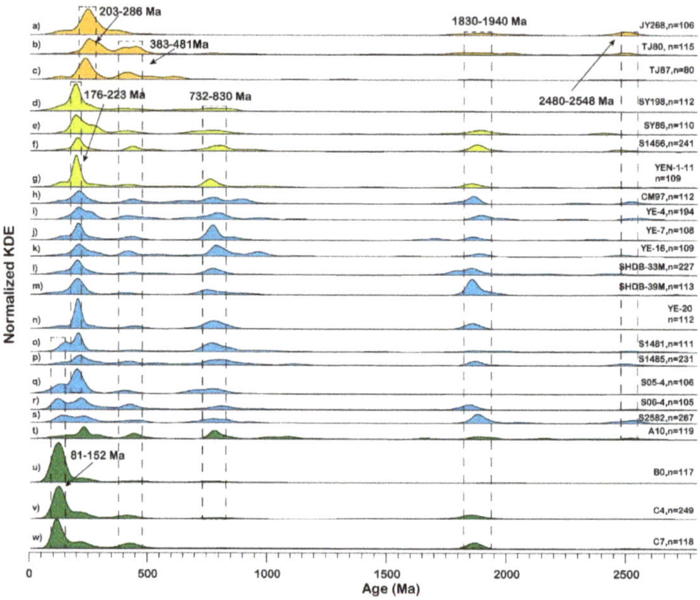

Figure 4. KDEs of detrital zircon U-Pb ages of in the ECSs. Different colors are used to represent the samples taken from different seas, with brown for the Bohai Sea (**a–c**), green-yellow for the Yellow Sea (**d–g**), blue for the East China Sea (**h–s**) and green for the northern South China Sea (**t–w**). The visually distinctive age populations are highlighted by the vertical dashed bars.

In contrast to samples in the Bohai Sea, the age distributions in the Yellow Sea are obviously different (Figure 4d–g). First, the most abundant age population is at 176–223 Ma with a peak around 200 Ma, which is obviously younger than the age population at 203–286 Ma in the Bohai Sea. Second, there are no Neoproterozoic ages in samples of the Bohai Sea but a very remarkable population at 732–830 Ma in samples of the Yellow Sea occurs. Because that the Yellow River flowed into the Yellow Sea during some periods in the history, the differences of zircon signatures between the Bohai Sea and the Yellow Sea may imply that their sources were different and there existed alternative sediment sources for the sediments in the Yellow Sea other than the Yellow River.

Similar to those of the Yellow Sea sediments, the primary age populations in the East China Sea sediments contain the one at 176–223 Ma with peaks around 200 Ma, 383–481 Ma with peaks around 420 Ma and 440 Ma, 732–830 Ma with a peak around 780 Ma, 1830–1940 Ma with peaks around 1850 Ma and 1900 Ma and 2480–2548 Ma with a primary peak about 2500 Ma (Figure 4h–s). In addition to these five most pronounced populations, for samples in the southern East China Sea there is noticeable one at 81–152 Ma peaking around 130 Ma (Figure 4o–s).

With the exception of sample A10, samples in the northern South China Sea have relatively simple age distributions compared with the other seas. Up to 50% of zircon grains are in the range of 81–152 Ma that peaks around 130 Ma (Figure 4u–w).

4.2. Spatial Variability of Mixing Coefficient

The results of mixing coefficients are shown in Figure 5, which indicate relative variations in sediment contribution from different fluvial sources. Samples in the Bohai Sea are dominated by the contribution of the Yellow River, with mixing coefficients in the range of 80–90%. Except for sample TJ87, there seems to be no contribution from the Yangtze River. However, it seems that there are 7–14% contributions of small rivers in Taiwan and Zhe-Min provinces, SE China (Figure 4). We think that this is an artifact either due to that we input irrelevant endmembers in the mixing model for samples in the Bohai Sea, or due to the uncertainty of the mixing model to distinguish age overlapping between different samples.

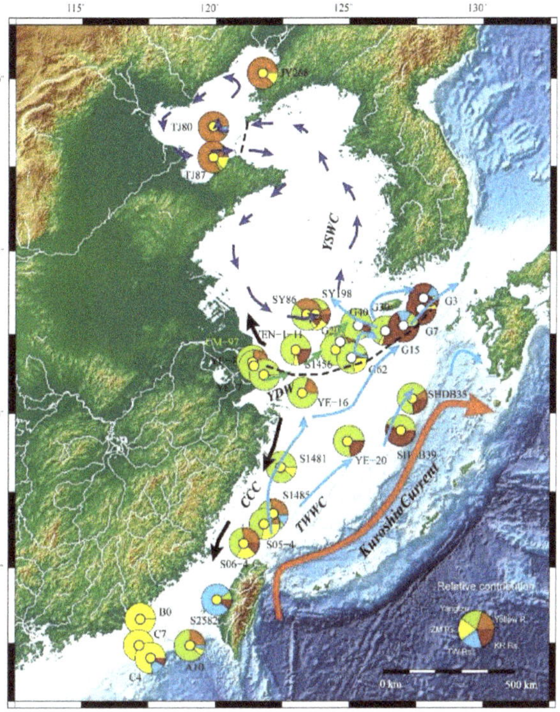

Figure 5. Spatial distribution of the mixing coefficients illustrating the relative contributions of zircons in the ECSs from five major river sources. The relative contributions from different sources at sampling sites are shown in the pie plots.

In contrast, samples in the Yellow Sea present increased contributions from the Yangtze River and Korean rivers (Figure 5). The mixing coefficients of the Yangtze river are in the range of 40–70%, while a minor 10–20% for Korean rivers. In contrast, the contributions of the Yellow River in the Yellow Sea reduce to 10–40%. Samples in the East China Sea are dominated by the Yangtze source signal with mixing coefficients in the range of 40–98% except for Sample S2582, which has a mixing coefficient of 8% for the Yangtze River. This exception can be explained by enhanced contributions from Taiwan rivers, such as the Choshui River [25]. The contributions from the Yellow River to the East China Sea are generally lower than 30%, indicating a dominant Yangtze contribution in the sea. In addition, the contributions of Zhe-Min rivers can only be observed in the southern part of the East China Sea in the range of 10–30%. In contrast, their contributions in the northern South China Sea increase greatly up to 50–90%.

5. Discussion

5.1. Provenance Linkage between Fluvial and Marine Surface Sediments

The analyses of detrital zircons U-Pb ages in the ECSs provide us a way to link various fluvial age signals to the those in the continental margin seas. The MDS plot in Figure 6 produced a 'map' of points on which similar samples cluster closely together, and dissimilar samples stay far apart, which provides a straightforward way to distinguish provenance similarity and linkage. As shown in the plot of MDS (Figure 6), samples in the Bohai Sea are very close to those in the lower Yellow River reaches [23,61]. This agrees with the fact that modern sediments in the Bohai Sea are mainly derived from the Yellow River catchment.

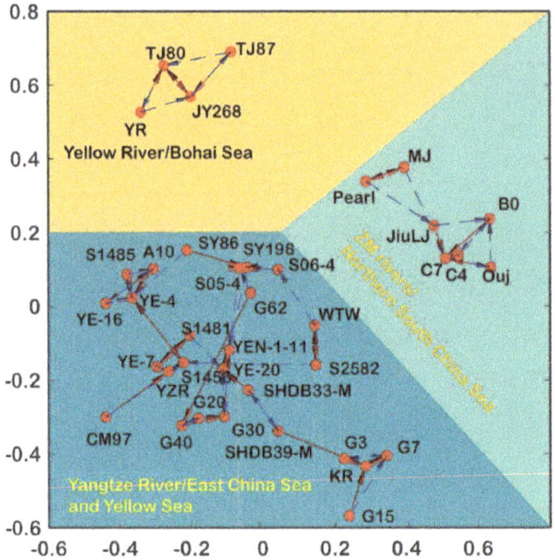

Figure 6. Multidimensional scaling (MDS) configuration of samples in the ECSs. There are three distinct groups in the MDS configuration, which could be linked to the potential sources. The solid (red) and dash (blue) lines denote the first and second closest neighbors, which indicate more similar age distributions and closer provenance linkages.

Similarly, samples in the northern South China Sea are clustered around the fluvial endmembers, including Ou Jiang [21], Min Jiang [22], Jiulong Jiang [21] and the Pearl River [62] (Figure 6). This implies that the rivers draining the CF (Figure 1) have contributed most of sediments deposited in the northern South China Sea. This is at least correct for sediments in the shallow sea. On the other hand, the contribution from Taiwan Island would be significant when samples taken from slope and deep sea are examined (e.g., sample A10).

Samples S2582 and S06-4 exhibit a short distance to the Choshui River in western Taiwan, suggesting a source-to-sink connection (Figure 5). In contrast, samples A10 and S05-4 have closest neighbors of samples in the Yangtze delta (YE-4) and the Yellow Sea (SY198). This is difficult to explain on current geographic context and we propose two contrasting explanations. The first is that it could be caused by the recycling processes of zircon grains in which some zircon grains in Taiwan Strait may have been derived from the YC and NCC in geological past [25]. The recycling processes have made the signal less distinguished. If this is true, the discrimination between Taiwan and Yangtze sources by the method of zircon U-Pb geochronology alone would be difficult. Another explanation is that it could have been caused by sediment mixing between those derived from Taiwan Island and the Cathaysia. This is more likely because the U-Pb age distributions of the Yangtze River and Choshui River can be slightly distinguished by the presence or absence of a ~130 Ma population [20,25], which is the most distinct age population in fluvial sediments from the CF [21,22]. As shown in Figure 4, our results are in agreement with the latter explanation.

In contrast to samples taken from the other seas, which show clear source-to-sink connections, there exists a more complicated pattern for sediments in the East China Sea and the Yellow Sea (Figure 5). Despite of that, most samples exhibit good correlations with the Yangtze fluvial sediments as indicated by the first and second closest neighbor lines in Figure 5, which evidently suggest that sediments derived from the Yangtze Catchment have played a very important role in determining detrital zircon signatures over this very broad continental shelf. On the other hand, the very scattered points on the MDS plot for samples in the East China Sea also imply that the fluvial source signals have been modified in the shallow sea environment relative to the Yangtze fluvial source [20]. We consider that this pattern is related to the hydrodynamical system in the region, which will be discussed in detail in Section 5.3.

5.2. Provenance Discrimination from A Tectonic Perspective

In order to discuss our results from a geological perspective, we classified the age distributions in Figure 6 based on the most pronounced tectonic events in the geological history of China and East Asia [71]. The samples from the Bohai Sea are characterized by high proportions of zircon grains of the Paleozoic and Paleoproterozoic-Neoarchean eras. Originally, the Paleozoic grains were associated with four tectonic events, such as Indianian (200–260 Ma), Tianshanian (260–397 Ma), Qilianian (397–513 Ma) and Sinian (513–680 Ma). The Paleoproterozoic-Neoarchean grains may be related to the tectonic events of Lvliang (1800–2500 Ma) and Wutai (2500–2800 Ma) which are related to the formation of the NCC. By taking the U-Pb ages in the Yellow River catchment [23,61] into account, we argue that there are two potential sources of the zircon grains for the modern sediments of interest. The first is the Neoarchean basement of the NCC, which is characterized by the presences of 1.8 Ga and 2.5 Ga zircons. The second is the northeastern Tibetan Plateau, which is marked by a dominant Paleozoic zircon U-Pb at populations of 200–350 Ma and 350–550 Ma (Figure 7).

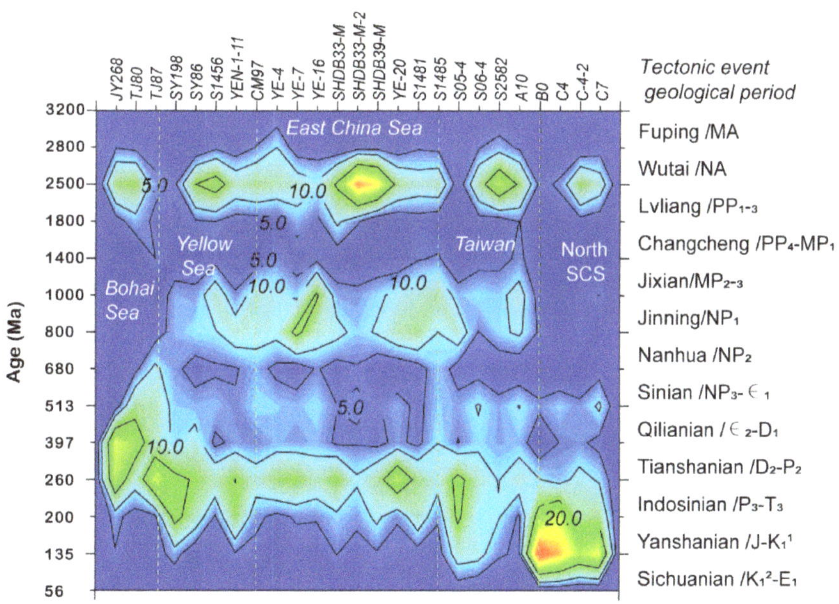

Figure 7. Spatial variation in the relative abundances of different zircon age populations in the east China seas. Zircon age populations are classified according to tectonic events that happened in the geological history of East Asia [70]. The yellow dashed lines are used to mark samples taken from different seas.

Different to the Bohai zircon ages, the presence of Neoproterozoic zircons is very typical in the Yellow and East China Sea. We think the Neoproterozoic zircons can be related to the formation of the basement of Yangtze Craton in the Jinning Orogeny during 700–900 Ma [70]. Although the high proportions of Paleozoic zircons in the Yellow and East China Seas appear to be similar to those in the Bohai Sea (Figure 7), the sporadic distribution of Mesozoic igneous rocks in the lower Yangtze due to the subduction of Paleo–Pacific plate to the Eurasian plate may have provided extra grains [48].

The provenance of zircon in Taiwan Island is a matter of controversy. Potential sources include the YC [71], CF [22], NCC [25] and a subducted unknown microcontinent [52]. As shown in Figure 6, the zircon U-Pb age proportions of samples close to Taiwan Island in our study looks similar to those from the Yangtze source. In contrast, they have relatively low content of zircons of the Indosinian and Jinning periods but relatively high of the Wutai period (2500–2800 Ma). Moreover, in samples of S05-4 and S06-4, there are distinctive contributions from the Yanshanian period (135–200 Ma). This suggests that the zircon grains of these sites have multiple sources, which is possibly due to sediment mixing between Taiwan Island and the CF. According to the present-day configuration of ocean currents (Figure 1), it is less likely for zircons from the Yangtze River to be transported to the Taiwan Strait or further southward to the South China Sea.

The U-Pb age compositions in the northern South China Sea are featured by a dominance of Mesozoic Yanshanian zircon grains and a paucity of Neoproterozoic grains. This allows us to speculate that their source area may consist of relatively young rocks and are very proximal to the place of deposition. The widely exposed Mesozoic Yanshanian igneous rocks in the CF are therefore the most likely sources [21,22].

5.3. *Hydrodynamic Influence on Zircon U-Pb Age Distribution*

Because zircon is generally believed to be resistant to chemical and physical weathering during sediment transport, it has been regarded one of the most reliable approach for provenance analysis [9]. However, this view has been challenged because an age distribution may be biased by factors like fertility [13], grain size [14] or sampling and analytical procedures [72]. To discuss these factors is far beyond the scope of this study, however, we would expect less grain-size sorting on our samples as we just focused on the very fine fraction of the zircons (60–125 μm) and there is no statistically significant correlation between zircon sizes and U-Pb ages on this fraction [14]. As the result, we would like to just focus on the question of spatial variability of the Yangtze mixing coefficients in the East China Sea.

As shown in Figure 8, there is a gradual decreasing trend of the Yangtze mixing coefficients from the estuary to the continental shelf along directions of 60° NE and 200° SW. The decreasing trend implies the fluvial provenance signal has changed in the marine environment. It seems contradictory to our argument above that the provenance or fluvial source is the dominant factor controlling U-Pb age distribution of the sediments in the seas. This can be attributed to the influences of oceanic hydrodynamics on zircon age populations. Spatial variations in mixing coefficients of the Yangtze River are generally consistent with seasonal current change in the East China Sea (Figure 8). The surface and deep currents of the East China Sea are mainly driven by the Asian summer/winter monsoons. During summer as the result of the northeastward movements of surface and bottom currents (Figures 1 and 8), the surface and bottom sediments are transported in the primary directions. The coastal currents—driven by the northeast winter monsoon—become strengthened in winter, which may cause part of sediments transporting southward. Moreover, a recent study using the regional ocean circulation model (ROMS) also suggested that the sediments derived from the Yangtze River are not only deposited in the East China Sea but also in the Yellow Sea [73]. In addition, a tide–current model suggested that a very pronounced trend of erosion/deposition of sediments along the direction from the Yangtze Estuary to the Cheju Island due to the influence of tides [40]. Therefore, we argue that the gradual decreasing trends of mixing coefficients of the Yangtze River in the East China Sea are resulted from the combined effects of currents and tides (Figures 1 and 8).

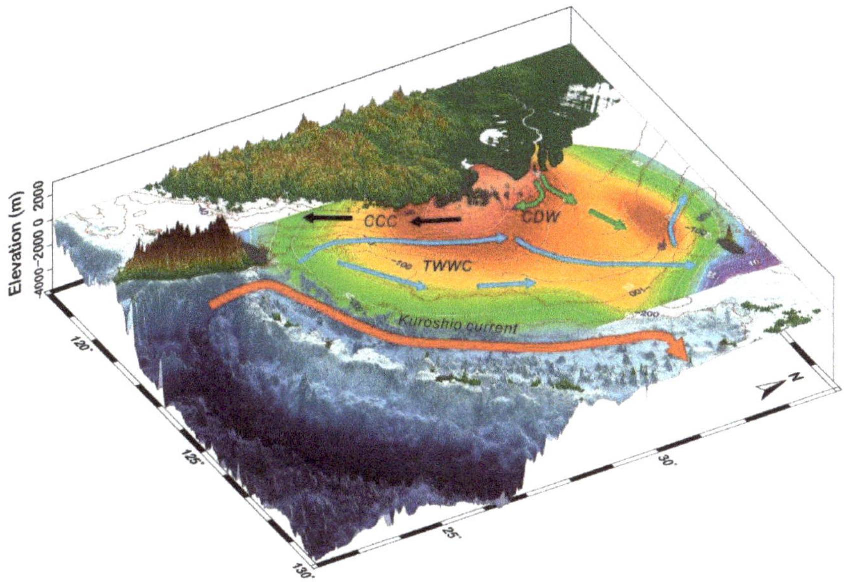

Figure 8. Spatial distribution of mixing coefficients of the Yangtze River in the East China Sea and the relationship with regional ocean currents. The color-filled area are contours of mixing coefficient of the Yangtze River while the red dashed denote the bathymetric contours. The arrows are ocean currents denoted in Figure 1.

5.4. Sediment Budgeting

Mixing coefficients modeled from U-Pb age distributions have been used in some studies as proxies for relative erosion rates in various fluvial catchments [4,8,11,74]. From a perspective of mass balance, when a long-term equilibrium between sediment erosion and deposition is reached, the mixing coefficients in sink area can be used to infer relative depositional fluxes/rates and sediment budgeting [8]. To test whether this is true in the marine environment, we compared the mixing coefficients of our study with sediment budgets estimated from nuclides methods such as ^{210}Pb and ^{137}Cs from short-cores in the ECSs [75,76].

According to the sediment budgeting model in [75], the historical annual sediment flux of the Yellow River can reach to 1086 Mt per year and ~70% of the sediments are deposited in the Bohai Sea while ~30% are transported into the Yellow Sea. This means that most sediments derived from the Yellow River are deposited in the Bohai Sea and only 30% of them, equivalent to 376 Mt sediments per year, are transported into the Yellow Sea. This accounts for 57% of total input sediments of the Yellow Sea [75]. Despite it is still not clear which are the sources of the rest of input (~40%), there are 414 Mt sediments deposited in the Yellow Sea per year [75]. This means that the relative sediment contribution from the Yellow River in the Yellow Sea could reduce to around 22%, which coincides broadly with the mixing coefficients of samples SY198 and SY86 that are in the range of 7–38% (Table 2). The decreasing trend of the mixing coefficients of the Yellow River from Bohai Sea to the Yellow Sea helps us to catch a glimpse of provenance signal modification along different sedimentary systems as the result of sediment mixing.

A similar decreasing trend of mixing coefficients of the Yangtze River is also observed in the East China Sea. The mean sediment contribution from the Yangtze River in the East China Sea is estimated to be about 72% [75], which is strikingly consistent with the averaged mixing coefficient (72.8%) in the East China Sea (Table 2). According to the study [76], the annual sediment contribution from the Zhe-Min rivers in the East China Sea is about 17–20 Mt, the contribution from the Taiwan rivers is in the range 70–150 Mt and the rest 55 Mt may be from the Yangtze River or the northern East China Sea. If this is correct, this means that 33–67% of sediments are derived from the Taiwan rivers. However, it is different with our results that give a very high mixing coefficient value (81.6%) in the sample of S2582 but very low coefficients (0–11%) in the other samples close to the Taiwan Island in the East China Sea (Figure 5). This could be partly attributed to the scale differences between the two studies.

According to the above discussion, we argue that the mixing coefficients of a specific source obtained from zircon U-Pb age mixing models could be applied to infer relative sediment budgeting when the present source signal is well defined. From the mixing coefficients of the Yangtze River distributed in the East China Sea, we can observe that most of surface sandy sediments in the East China Sea are primarily derived from the Yangtze River and the contributions from the Yellow River are limited to the region of the Bohai and Yellow seas. This argument seems different if compared with a recent provenance study in the Okinawa Trench [28], in which the authors suggested that the main source of terrestrial sediments in the Okinawa Trench was the Yellow River during the period of the last deglaciation. It is difficult to compare results of the two studies directly due to the differences in methods, sediment grain-sizes and time scales. It needs to be studied further in the future, especially when the technique of small-volume U-Pb zircon geochronology by laser ablation multi-collector ICP-MS [77] and the large-n analysis [15,72] become routinely feasible, which will minimize the uncertainties in terms of source and sink signals of the mixing model and disentangle provenance signature from signals of hydrodynamics [78].

6. Conclusions

In this study, we obtained over three thousand detrital zircon U-Pb ages from 23 samples in the ECSs and a number of conclusions can be drawn as follows:

(1) Sediments in Bohai Sea, East China Sea and Taiwan Strait can be readily linked to fluvial inputs from the Yellow River, Yangtze River and Taiwan rivers, respectively. The presence or absence of the population of 732–830 Ma in the YC and NCC is the most reliable age signature to distinguish zircon grains derived from the two source regions. Detrital zircon U-Pb geochronology is very robust approach to perform provenance analysis for surface sediments in the ECSs. The age distributions of detrital zircon in the margin seas are mainly controlled by tectonic settings of the continent.

(2) Due to the combined effects of currents and tides, the sediments derived from the Yangtze River could be distributed over a very wide area in the East China Sea, resulting in two dominant transport trends as viewed from the mixing coefficient pattern. The ocean hydrodynamics play an important role in partitioning sediments and modifying associated provenance signals in the marine source-to-sink system. The spatial variations of mixing coefficient of a special source endmember could provide a quantitative or semi-quantitative way to understand how the provenance signal change in the seas due to the influence of hydrodynamics.

(3) Lastly, the mixing modeling results of the Yangtze River are generally comparative to the sedimentation rate pattern of the sea. This enables us to argue that the sediment discharge from large rivers (e.g., the Yangtze and Yellow River) have contributed more sediments in the continental margin. As a result, we conclude that the combination of the U-Pb zircon geochronology with mixing models could provide us a feasible way to infer relative changes of sediment budgets in the geological past when the U-Pb distributions of present rivers are well defined and their distinctive signatures have been preserved in sediment records.

Supplementary Materials: The following are available online at http://www.mdpi.com/2075-163X/10/5/398/s1, Table S1: U-Pb ratios and ages of surface sediments in the east China seas.

Author Contributions: Conceptualization, X.H. and S.Y.; methodology, J.S., W.Y., Y.L. and F.L.; resources, Z.W. and X.M.; writing—original draft preparation, X.H.; writing—review and editing, E.L. and S.Y.; funding acquisition, X.H., S.Y. and Z.W. All authors have read and agreed to the published version of the manuscript.

Funding: This research was funded by the National Natural Science Foundation of China (grant No. 41730531, 41991324, 41876059), Shanghai Committee of Science and Technology, China (grant No. 18ZR1440400) and China Geology Survey (DD20190208).

Conflicts of Interest: The authors declare no conflict of interest.

References

1. Nittrouer, C.A.; Austin, J.A.; Field, M.E.; Kravitz, J.H.; Syvitski, J.P.M.; Wiberg, P.L. Writing a Rosetta Stone: Insights into Continental-Margin Sedimentary Processes and Strata. In *Continental Margin Sedimentation*; Jarvis, I., Nittrouer, C.A., Austin, J.A., Field, M.E., Kravitz, J.H., Syvitski, J.P.M., Wiberg, P.L., Eds.; Wiley: Malden, MA, USA, 2007; pp. 1–48.
2. Allen, P.A. From landscapes into geological history. *Nature* **2008**, *451*, 274–276. [CrossRef]
3. Romans, B.W.; Castelltort, S.; Covault, J.A.; Fildani, A.; Walsh, J.P. Environmental signal propagation in sedimentary systems across timescales. *Earth-Sci. Rev.* **2016**, *153*, 7–29. [CrossRef]
4. Mason, C.C.; Fildani, A.; Gerber, T.; Blum, M.D.; Clark, J.D.; Dykstra, M. Climatic and anthropogenic influences on sediment mixing in the Mississippi source-to-sink system using detrital zircons: Late Pleistocene to recent. *Earth Planet. Sci. Lett.* **2017**, *466*, 70–79. [CrossRef]
5. Hinderer, M. From gullies to mountain belts: A review of sediment budgets at various scales. *Sediment. Geol.* **2012**, *280*, 21–59. [CrossRef]
6. Cawood, P.A.; Hawkesworth, C.J.; Dhuime, B. The continental record and the generation of continental crust. *Geol. Soc. Am. Bull.* **2013**, *125*, 14–32. [CrossRef]
7. Cawood, P.A.; Nemchin, A.A.; Freeman, M.; Sircombe, K. Linking source and sedimentary basin: Detrital zircon record of sediment flux along a modern river system and implications for provenance studies. *Earth Planet. Sci. Lett.* **2003**, *210*, 259–268. [CrossRef]
8. Amidon, W.H.; Burbank, D.W.; Gehrels, G.E. Construction of detrital mineral populations: Insights from mixing of U-Pb zircon ages in Himalayan Rivers. *Basin Res.* **2005**, *17*, 463–485. [CrossRef]
9. Gehrels, G. Detrital zircon U-Pb geochronology applied to tectonics. *Annu. Rev. Earth Planet. Sci.* **2014**, *42*, 127–149. [CrossRef]
10. Garzanti, E.; Vermeesch, P.; Rittner, M.; Simmons, M. The zircon story of the Nile: Time-structure maps of source rocks and discontinuous propagation of detrital signals. *Basin Res.* **2018**, *30*, 1098–1117. [CrossRef]
11. Saylor, J.E.; Knowles, J.N.; Horton, B.K.; Nie, J.; Mora, A. Mixing of Source Populations Recorded in Detrital Zircon U-Pb Age Spectra of Modern River Sands. *J. Geol.* **2013**, *121*, 17–33. [CrossRef]
12. Sickmann, Z.T.; Paull, C.K.; Graham, S.A. Detrital-Zircon Mixing and Partitioning in Fluvial to Deep Marine Systems, Central California, USA. *J. Sediment. Res.* **2016**, *86*, 1298–1307. [CrossRef]
13. Moecher, D.P.; Samson, S.D. Differential zircon fertility of source terranes and natural bias in the detrital zircon record: Implications for sedimentary provenance analysis. *Earth Planet. Sci. Lett.* **2006**, *247*, 252–266. [CrossRef]
14. Lawrence, R.L.; Cox, R.; Mapes, R.W.; Coleman, D.S. Hydrodynamic fractionation of zircon age populations. *Geol. Soc. Am. Bull.* **2011**, *123*, 295–305. [CrossRef]
15. Pullen, A.; Ibanez-Mejia, M.; Gehrels, G.E.; Ibanez-Mejia, J.C.; Pecha, M. What happens when n = 1000? Creating large-n geochronological datasets with LA-ICP-MS for geologic investigations. *J. Anal. Atom. Spectrom.* **2014**, *29*, 971–980. [CrossRef]
16. Zhang, J.; Liu, C.L. Riverine composition and estuarine geochemistry of particulate metals in China—Weathering features, anthropogenic impact and chemical fluxes. *Estuar. Coast. Shelf Sci.* **2002**, *54*, 1051–1070. [CrossRef]
17. Dadson, S.J.; Hovius, N.; Chen, H.; Dade, W.B.; Lin, J.C.; Hsu, M.L.; Lin, C.W.; Horng, M.J.; Chen, T.C.; Milliman, J.; et al. Earthquake-triggered increase in sediment delivery from an active mountain belt. *Geology* **2004**, *32*, 733–736. [CrossRef]
18. Yang, S.L.; Belkin, I.M.; Belkina, A.I.; Zhao, Q.Y.; Zhu, J.; Ding, P. Delta response to decline in sediment supply from the Yangtze River: Evidence of the recent four decades and expectations for the next half-century. *Estuar. Coast. Shelf Sci.* **2003**, *57*, 689–699. [CrossRef]
19. Wang, H.J.; Saito, Y.; Zhang, Y.; Bi, N.S.; Sun, X.X.; Yang, Z.S. Recent changes of sediment flux to the western Pacific Ocean from major rivers in East and Southeast Asia. *Earth-Sci. Rev.* **2011**, *108*, 80–100. [CrossRef]

20. He, M.Y.; Zheng, H.B.; Clift, P.D. Zircon U-Pb geochronology and Hf isotope data from the Yangtze River sands: Implications for major magmatic events and crustal evolution in Central China. *Chem. Geol.* **2013**, *360*, 186–203. [CrossRef]
21. Xu, X.S.; O'Reilly, S.Y.; Griffin, W.L.; Wang, X.L.; Pearson, N.J.; He, Z.Y. The crust of Cathaysia: Age, assembly and reworking of two terranes. *Precambrian Res.* **2007**, *158*, 51–78. [CrossRef]
22. Xu, Y.H.; Sun, Q.Q.; Yi, L.; Yin, X.J.; Wang, A.; Li, Y.H.; Chen, J. Detrital Zircons U-Pb Age and Hf Isotope from the Western Side of the Taiwan Strait: Implications for Sediment Provenance and Crustal Evolution of the Northeast Cathaysia Block. *Terr. Atmos. Ocean. Sci.* **2014**, *25*, 505–535. [CrossRef]
23. Nie, J.S.; Stevens, T.; Rittner, M.; Stockli, D.; Garzanti, E.; Limonta, M.; Bird, A.; Ando, S.; Vermeesch, P.; Saylor, J.; et al. Loess Plateau storage of Northeastern Tibetan Plateau-derived Yellow River sediment. *Nat. Commun.* **2015**, *6*, 1–10. [CrossRef] [PubMed]
24. Choi, T.; Lee, Y.I.; Orihashi, Y. Crustal growth history of the Korean Peninsula: Constraints from detrital zircon ages in modern river sediments. *Geosci. Front.* **2016**, *7*, 707–714. [CrossRef]
25. Deng, K.; Yang, S.Y.; Li, C.; Su, N.; Bi, L.; Chang, Y.P.; Chang, S.C. Detrital zircon geochronology of river sands from Taiwan: Implications for sedimentary provenance of Taiwan and its source link with the east China mainland. *Earth-Sci. Rev.* **2017**, *164*, 31–47. [CrossRef]
26. Xu, K.; Li, A.; Liu, J.P.; Milliman, J.D.; Yang, Z.; Liu, C.-S.; Kao, S.-J.; Wan, S.; Xu, F. Provenance, structure, and formation of the mud wedge along inner continental shelf of the East China Sea: A synthesis of the Yangtze dispersal system. *Mar. Geol.* **2012**, *291*, 176–191. [CrossRef]
27. Xu, K.; Milliman, J.D.; Li, A.; Liu, J.P.; Kao, S.-J.; Wan, S. Yangtze- and Taiwan-derived sediments on the inner shelf of East China Sea. *Cont. Shelf Res.* **2009**, *29*, 2240–2256. [CrossRef]
28. Li, Q.; Zhang, Q.; Li, G.; Liu, Q.; Chen, M.-T.; Xu, J.; Li, J. A new perspective for the sediment provenance evolution of the middle Okinawa Trough since the last deglaciation based on integrated methods. *Earth Planet. Sci. Lett.* **2019**, *528*, 115839. [CrossRef]
29. Sharman, G.R.; Johnstone, S.A. Sediment unmixing using detrital geochronology. *Earth Planet. Sci. Lett.* **2017**, *477*, 183–194. [CrossRef]
30. Li, G.X.; Li, P.; Liu, Y.; Qiao, L.L.; Ma, Y.Y.; Xu, J.S.; Yang, Z.G. Sedimentary system response to the global sea level change in the East China Seas since the last glacial maximum. *Earth-Sci. Rev.* **2014**, *139*, 390–405. [CrossRef]
31. Hickox, R.; Belkin, I.; Cornillon, P.; Shan, Z. Climatology and seasonal variability of ocean fronts in the east China, Yellow and Bohai Seas from satellite SST data. *Geophys. Res. Lett.* **2000**, *27*, 2945–2948. [CrossRef]
32. Choi, T.; Lee, Y.I.; Orihashi, Y.; Yi, H.I. The provenance of the southeastern Yellow Sea sediments constrained by detrital zircon U-Pb age. *Mar. Geol.* **2013**, *337*, 182–194. [CrossRef]
33. Milliman, J.D.; Syvitski, J.P.M. Geomorphic Tectonic Control of Sediment Discharge to the Ocean—The Importance of Small Mountainous Rivers. *J. Geol.* **1992**, *100*, 525–544. [CrossRef]
34. Yang, S.L.; Zhang, J.; Zhu, J.; Smith, J.P.; Dai, S.B.; Gao, A.; Li, P. Impact of dams on Yangtze River sediment supply to the sea and delta intertidal wetland response. *J. Geophys. Res.-Earth Surf.* **2005**, *110*, 1–12. [CrossRef]
35. Dadson, S.J.; Hovius, N.; Chen, H.G.; Dade, W.B.; Hsieh, M.L.; Willett, S.D.; Hu, J.C.; Horng, M.J.; Chen, M.C.; Stark, C.P.; et al. Links between erosion, runoff variability and seismicity in the Taiwan orogen. *Nature* **2003**, *426*, 648–651. [CrossRef]
36. Yang, S.Y.; Jung, H.S.; Lim, D.I.; Li, C.X. A review on the provenance discrimination of sediments in the Yellow Sea. *Earth-Sci. Rev.* **2003**, *63*, 93–120. [CrossRef]
37. Liu, J.P.; Xu, K.H.; Li, A.C.; Milliman, J.D.; Velozzi, D.M.; Xiao, S.B.; Yang, Z.S. Flux and fate of Yangtze river sediment delivered to the East China Sea. *Geomorphology* **2007**, *85*, 208–224. [CrossRef]
38. Isobe, A. Recent advances in ocean-circulation research on the Yellow Sea and East China Sea shelves. *J. Oceanogr.* **2008**, *64*, 569–584. [CrossRef]
39. Berne, S.; Vagner, P.; Guichard, F.; Lericolais, G.; Liu, Z.X.; Trentesaux, A.; Yin, P.; Yi, H.I. Pleistocene forced regressions and tidal sand ridges in the East China Sea. *Mar. Geol.* **2002**, *188*, 293–315. [CrossRef]
40. Zhu, Y.R.; Chang, R.F. On the relationships between the radial tidal current field and the radial sand ridges in the southern Yellow Sea: A numerical simulation. *Geo-Mar. Lett.* **2001**, *21*, 59–65. [CrossRef]
41. Zhao, G.C.; Wilde, S.A.; Cawood, P.A.; Sun, M. Archean blocks and their boundaries in the North China Craton: Lithological, geochemical, structural and P-T path constraints and tectonic evolution. *Precambrian Res.* **2001**, *107*, 45–73. [CrossRef]

42. Xiong, S.Q.; Yang, H.; Ding, Y.Y.; Li, Z.K.; Li, W. Distribution of igneous rocks in China revealed by aeromagnetic data. *J. Asian Earth Sci.* **2016**, *129*, 231–242. [CrossRef]
43. Zhai, M.G.; Santosh, M. The early Precambrian odyssey of the North China Craton: A synoptic overview. *Gondwana Res.* **2011**, *20*, 6–25. [CrossRef]
44. Zheng, Y.F.; Xiao, W.J.; Zhao, G.C. Introduction to tectonics of China. *Gondwana Res.* **2013**, *23*, 1189–1206. [CrossRef]
45. Zhu, R.X.; Yang, J.H.; Wu, F.Y. Timing of destruction of the North China Craton. *Lithos* **2012**, *149*, 51–60. [CrossRef]
46. Zhou, X.M.; Sun, T.; Shen, W.Z.; Shu, L.S.; Niu, Y.L. Petrogenesis of Mesozoic granitoids and volcanic rocks in South China: A response to tectonic evolution. *Episodes* **2006**, *29*, 26–33. [CrossRef]
47. Xu, Y.-G.; Luo, Z.-Y.; Huang, X.-L.; He, B.; Xiao, L.; Xie, L.-W.; Shi, Y.-R. Zircon U-Pb and Hf isotope constraints on crustal melting associated with the Emeishan mantle plume. *Geochim. Cosmochim. Acta* **2008**, *72*, 3084–3104. [CrossRef]
48. Chen, L.; Zheng, Y.F.; Zhao, Z.F. A common crustal component in the sources of bimodal magmatism: Geochemical evidence from Mesozoic volcanics in the Middle-Lower Yangtze Valley, South China. *Geol. Soc. Am. Bull.* **2018**, *130*, 1959–1980. [CrossRef]
49. Wang, Y.; Fan, W.; Zhang, G.; Zhang, Y. Phanerozoic tectonics of the South China Block: Key observations and controversies. *Gondwana Res.* **2013**, *23*, 1273–1305. [CrossRef]
50. She, Z.; Ma, C.; Mason, R.; Li, J.; Wang, G.; Lei, Y. Provenance of the Triassic Songpan-Ganzi flysch, west China. *Chem. Geol.* **2006**, *231*, 159–175. [CrossRef]
51. Chen, Y.L.; Li, D.P.; Zhou, J.; Liu, F.; Zhang, H.F.; Nie, L.S.; Jiang, L.T.; Song, B.; Liu, X.M.; Wang, Z. U-Pb dating, geochemistry, and tectonic implications of the Songpan-Ganzi block and the Longmen Shan, China. *Geochem. J.* **2009**, *43*, 77–99. [CrossRef]
52. Chen, C.H.; Lee, C.Y.; Lin, J.W.; Chu, M.F. Provenance of sediments in western Foothills and Hsuehshan Range (Taiwan): A new view based on the EMP monazite versus LA-ICPMS zircon geochronology of detrital grains. *Earth-Sci. Rev.* **2019**, *190*, 224–246. [CrossRef]
53. Sagong, H.; Cheong, C.S.; Kwon, S.T. Paleoproterozoic orogeny in South Korea: Evidence from Sm-Nd and Pb step-leaching garnet ages of Precambrian basement rocks. *Precambrian Res.* **2003**, *122*, 275–295. [CrossRef]
54. Kee, W.S.; Kim, S.W.; Jeong, Y.J.; Kwon, S. Characteristics of Jurassic Continental Arc Magmatism in South Korea: Tectonic Implications. *J. Geol.* **2010**, *118*, 305–323. [CrossRef]
55. Wiedenbeck, M.; Alle, P.; Corfu, F.; Griffin, W.L.; Meier, M.; Oberli, F.; Vonquadt, A.; Roddick, J.C.; Speigel, W. Three Natural Zircon Standards for U-Th-Pb, Lu-Hf, Trace-Element and Ree Analyses. *Geostand. Newsl.* **1995**, *19*, 1–23. [CrossRef]
56. Slama, J.; Kosler, J.; Condon, D.J.; Crowley, J.L.; Gerdes, A.; Hanchar, J.M.; Horstwood, M.S.A.; Morris, G.A.; Nasdala, L.; Norberg, N.; et al. Plesovice zircon—A new natural reference material for U-Pb and Hf isotopic microanalysis. *Chem. Geol.* **2008**, *249*, 1–35. [CrossRef]
57. Fisher, C.M.; Longerich, H.P.; Jackson, S.E.; Hanchar, J.M. Data acquisition and calculation of U-Pb isotopic analyses using laser ablation (single collector) inductively coupled plasma mass spectrometry. *J. Anal. Atom. Spectrom.* **2010**, *25*, 1905–1920. [CrossRef]
58. Spencer, C.J.; Kirkland, C.L.; Taylor, R.J.M. Strategies towards statistically robust interpretations of in situ U-Pb zircon geochronology. *Geosci. Front.* **2016**, *7*, 581–589. [CrossRef]
59. Saylor, J.E.; Sundell, K.E. Quantifying comparison of large detrital geochronology data sets. *Geosphere* **2016**, *12*, 203–220. [CrossRef]
60. Vermeesch, P. Multi-sample comparison of detrital age distributions. *Chem. Geol.* **2013**, *341*, 140–146. [CrossRef]
61. Yang, J.; Gao, S.; Chen, C.; Tang, Y.Y.; Yuan, H.L.; Gong, H.J.; Xie, S.W.; Wang, J.Q. Episodic crustal growth of North China as revealed by U-Pb age and Hf isotopes of detrital zircons from modern rivers. *Geochim. Cosmochim. Acta* **2009**, *73*, 2660–2673. [CrossRef]
62. Shao, L.; Meng, A.; Li, Q.; Qiao, P.; Cui, Y.; Cao, L.; Chen, S. Detrital zircon ages and elemental characteristics of the Eocene sequence in IODP Hole U1435A: Implications for rifting and environmental changes before the opening of the South China Sea. *Mar. Geol.* **2017**, *394*, 39–51. [CrossRef]
63. Zheng, H.B.; Clift, P.D.; Wang, P.; Tada, R.J.; Jia, J.T.; He, M.Y.; Jourdan, F. Pre-Miocene birth of the Yangtze River. *Proc. Natl. Acad. Sci. USA* **2013**, *110*, 7556–7561. [CrossRef] [PubMed]

64. Shang, Y.; Prins, M.A.; Beets, C.J.; Kaakinen, A.; Lahaye, Y.; Dijkstra, N.; Rits, D.S.; Wang, B.; Zheng, H.B.; van Balen, R.T. Aeolian dust supply from the Yellow River floodplain to the Pleistocene loess deposits of the Mangshan Plateau, central China: Evidence from zircon U-Pb age spectra. *Quat. Sci. Rev.* **2018**, *182*, 131–143. [CrossRef]
65. Jia, J.T.; Zheng, H.B.; Huang, X.T.; Wu, F.Y.; Yang, S.Y.; Wang, K.; He, M.Y. Detrital zircon U-Pb ages of Late Cenozoic sediments from the Yangtze delta: Implication for the evolution of the Yangtze River. *Chin. Sci. Bull.* **2010**, *55*, 1520–1528. [CrossRef]
66. Paterson, G.A.; Heslop, D. New methods for unmixing sediment grain size data. *Geochem. Geophys. Geosyst.* **2015**, *16*, 4494–4506. [CrossRef]
67. Vermeesch, P. On the visualisation of detrital age distributions. *Chem. Geol.* **2012**, *312*, 190–194. [CrossRef]
68. Zhai, M.; Peng, P. Paleoproterozoic events in the north China craton. *Acta Petrol. Sin.* **2007**, *23*, 2665–2682.
69. Kusky, T.M.; Windley, B.F.; Zhai, M.-G. Tectonic evolution of the North China Block: From orogen to craton to orogen. *Geol. Soc. Lond. Spec. Publ.* **2007**, *280*, 1–34. [CrossRef]
70. Wan, T. Introduction. In *The Tectonics of China: Data, Maps and Evolution*; Wan, T., Ed.; Springer: Berlin/Heidelberg, Germany, 2012; pp. 1–26.
71. Zhang, R.Q.; Sun, W.D.; Li, C.Y.; Lehmann, B.; Seltmann, R. Constraints on tin mineralization events by cassiterite LA-ICP-MS U-Pb dating. *Miner. Resour. Discov.* **2017**, *3*, 1009–1012.
72. Ibanez-Mejia, M.; Pullen, A.; Pepper, M.; Urbani, F.; Ghoshal, G.; Ibanez-Mejia, J.C. Use and abuse of detrital zircon U-Pb geochronology-A case from the Rio Orinoco delta, eastern Venezuela. *Geology* **2018**, *46*, 1019–1022. [CrossRef]
73. Bian, C.; Jiang, W.; Greatbatch, R.J. An exploratory model study of sediment transport sources and deposits in the Bohai Sea, Yellow Sea, and East China Sea. *J. Geophys. Res.-Ocean.* **2013**, *118*, 5908–5923. [CrossRef]
74. Sharman, G.R.; Covault, J.A.; Stockli, D.F.; Wroblewski, A.F.J.; Bush, M.A. Early Cenozoic drainage reorganization of the United States Western Interior-Gulf of Mexico sediment routing system. *Geology* **2017**, *45*, 187–190. [CrossRef]
75. Qiao, S.Q.; Shi, X.F.; Wang, G.Q.; Zhou, L.; Hu, B.Q.; Hu, L.M.; Yang, G.; Liu, Y.G.; Yao, Z.Q.; Liu, S.F. Sediment accumulation and budget in the Bohai Sea, Yellow Sea and East China Sea. *Mar. Geol.* **2017**, *390*, 270–281. [CrossRef]
76. Jia, J.J.; Gao, J.H.; Cai, T.L.; Li, Y.; Yang, Y.; Wang, Y.P.; Xia, X.M.; Li, J.; Wang, A.J.; Gao, S. Sediment accumulation and retention of the Changjiang (Yangtze River) subaqueous delta and its distal muds over the last century. *Mar. Geol.* **2018**, *401*, 2–16. [CrossRef]
77. Johnston, S.; Gehrels, G.; Valencia, V.; Ruiz, J. Small-volume U-Pb zircon geochronology by laser ablation-multicollector-ICP-MS. *Chem. Geol.* **2009**, *259*, 218–229. [CrossRef]
78. Blum, M.; Rogers, K.; Gleason, J.; Najman, Y.; Cruz, J.; Fox, L. Allogenic and Autogenic Signals in the Stratigraphic Record of the Deep-Sea Bengal Fan. *Sci. Rep.* **2018**, *8*, 7973. [CrossRef]

minerals

Article

Eoarchean to Neoproterozoic Detrital Zircons from the South of Meiganga Gold-Bearing Sediments (Adamawa, Cameroon): Their Closeness with Rocks of the Pan-African Cameroon Mobile Belt and Congo Craton

Nguo Sylvestre Kanouo [1,*], Arnaud Patrice Kouske [2], Gabriel Ngueutchoua [3], Akella Satya Venkatesh [4], Prabodha Ranjan Sahoo [4] and Emmanuel Archelaus Afanga Basua [5]

1 Department of Mining Engineering and Mineral Processing, Faculty of Mines and Petroleum Industries, University of Maroua, Maroua 46, Cameroon
2 Department of Civil Engineering, University Institute of Technology (UIT), University of Douala, Douala 1623, Cameroon; arnaudpatricek@gmail.com
3 Department of Earth Sciences, University of Yaoundé I, Yaoundé 812, Cameroon; ngueutchoua2@yahoo.fr
4 Department of Applied Geology, Indian Institute of Technology (Indian School of Mines), Dhanbad 826004, India; asvenkatesh@iitism.ac.in (A.S.V.); prabodha@iitism.ac.in (P.R.S.)
5 Faculty of Earth Sciences, School of Earth Sciences, China University of Geosciences, Wuhan 430074, China; basfanga@yahoo.com
* Correspondence: sylvestrekanouo@yahoo.fr; Tel.: +237-678966240

Abstract: The core of detrital zircons from the southern Meiganga gold-bearing placers were analyzed by Laser Ablation Split Stream analytical techniques to determine their trace element abundances and U-Pb ages. The obtained data were used to characterize each grain, determine its formation condition, and try to trace the provenance. The Hf (5980 to 12,010 ppm), Y (27–1650 ppm), U (25–954 ppm), Th (8–674 ppm), Ti (2–256 ppm), Ta, Nb, and Sr (mainly <5 ppm), Th/U (0.06–2.35), Ti zircon temperature (617–1180 °C), ∑REE (total rare earth element) (98–1030 ppm), and Eu/Eu* (0.03 to <1.35) are predominant values for igneous crustal-derived zircons, with very few from mantle sources and of metamorphic origin. Crustal igneous zircons are mainly inherited grains crystallized in granitic magmas (with some charnockitic and tonalitic affinities) and a few from syenitic melts. Mantle zircons were crystallized in trace element depleted mantle source magmatic intrusion during crustal opening. Metamorphic zircons grown in sub-solidus solution in equilibrium with garnet "syn-metamorphic zircons" and in equilibrium with anatectic melts "anatectic zircons" during crustal tectono-metamorphic events. The U-Pb (3671 ± 23–612 ± 11 Ma) ages distinguish: Eoarchean to Neoproterozoic igneous zircons; Neoarchean to Mid Paleoproterozoic anatectic zircons; and Late Neoproterozoic syn-metamorphic grains. The Mesoarchean to Middle Paleoproterozoic igneous zircons are probably inherited from pyroxene-amphibole-bearing gneiss (TTGs composition) and amphibole-biotite gneiss, whose features are similar to those of the granites, granodiorites, TTG, and charnockites found in the Congo Craton, south Cameroon. The youngest igneous zircons could be grains eroded from Pan-African intrusion(s) found locally. Anatectic and syn-metamorphic zircons could have originated from amphibole-biotite gneiss underlying the zircon-gold bearing placers and from locally found migmatized rocks that are from the Cameroon mobile belt, which could be used as proxies for tracking gold.

Keywords: Cameroon; Meiganga; gold placer; detrital zircon; trace element; geochronology; Archean-Proterozoic origins

Citation: Kanouo, N.S.; Kouske, A.P.; Ngueutchoua, G.; Venkatesh, A.S.; Sahoo, P.R.; Basua, E.A.A. Eoarchean to Neoproterozoic Detrital Zircons from the South of Meiganga Gold-Bearing Sediments (Adamawa, Cameroon): Their Closeness with Rocks of the Pan-African Cameroon Mobile Belt and Congo Craton. *Minerals* **2021**, *11*, 77. https://dx.doi.org/10.3390/min11010077

Received: 30 August 2020
Accepted: 11 November 2020
Published: 15 January 2021

1. Introduction

Meiganga is one of the key areas for small scale gold mining activities in Cameroon. As in many areas in this country, gold is extracted from supergene assemblages and terrigenous sediments (alluvium, eluvium, colluvium, and terrace) of mostly unknown primary sources (Figure 1). The clastic gold particles (very fine to coarse grained) are generally associated with some heavy minerals (e.g., zircon, magnetite, kyanite, ilmenite, and tourmaline) [1,2]. These weathering-resistant minerals are very useful in provenance studies, as they can register important information on their source rock petrogenesis, paleoenvironment, and tectonic reconstitution [3–9]. Some of these heavy minerals often fingerprint information on the chemistry of their environment of crystallization, the nature of their source rocks, and on the pre-existing tectonic settings [4,5,8–11]. Zircon in particular is an important mineral in fingerprinting source parameters [5,12–15]. One of the key tools to determine zircon source parameters is the combination of zircon geochemistry and U-Pb dating [4,8–10]. Each zircon has a characteristic age reflecting its genesis, and the population of detrital zircons in a sediment is a function of the age signature of source rocks in the proto-source terranes [12].

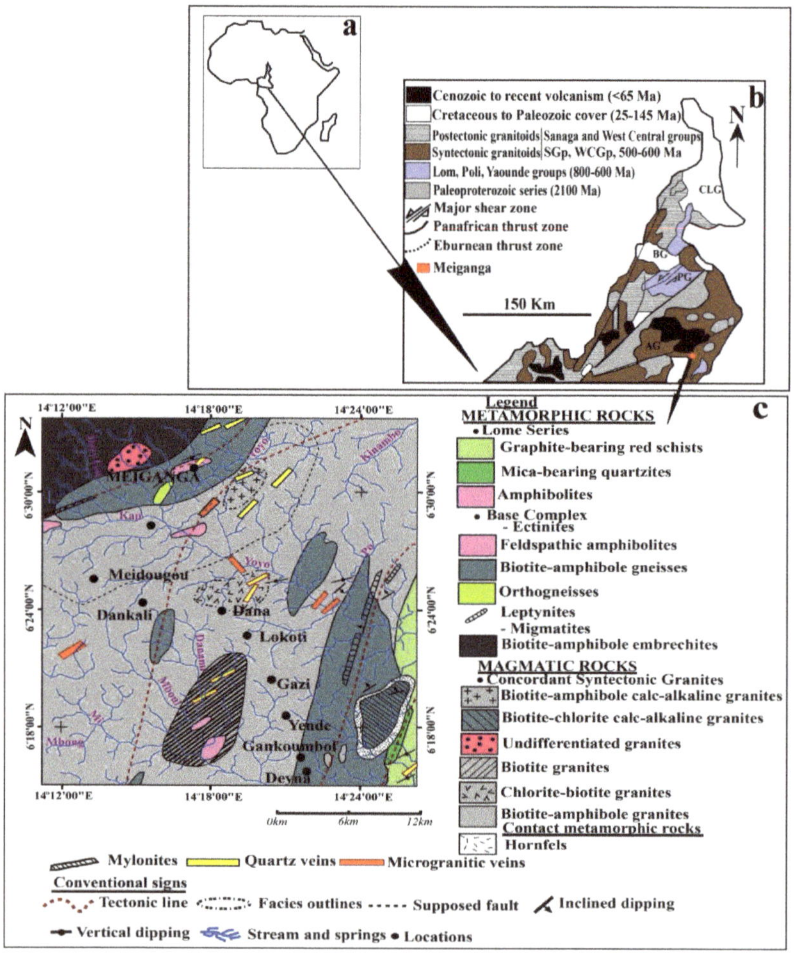

Figure 1. Geological sketch map of regional and local settings; (**a**) Cameroon in Africa, (**b**) geologic map of Adamawa, North and Far North Cameroon with the location of the south of Meiganga and (**c**) geological map of the south of Meiganga with the sample location.

Meiganga is in the Central part of the Cameroon mobile belt [16], a mega-tectonic structure that formed during the Neoproterozoic, from a collision between the Saharan meta-craton and Congo Craton [17,18]. Recent research works carried out on rocks found in the west of Meiganga have revealed the existence of Archean and Paleoproterozoic inheritance [19–21]. Trace element geochemistry and U-Pb ages of detrital zircon from gold-bearing placers in the west of Meiganga show that they were mainly crystallized and sourced from Archean to Precambrian granitoids [9].

Gold-bearing sediments were found in some streams in the southern part of Meiganga. The gold particles are associated with zircon, tourmaline, magnetite, kyanite, and ilmenite [1]. The source rocks and crystallization processes of most of these heavy minerals are poorly constrained. Djou [1] suggested a gneissic origin for a part of the deposited clasts, based on the presence of kyanite within the heavy mineral suites. Detailed analyses have not yet been carried on those minerals to help understand their source history and constrain their provenance. In this paper, we present trace element abundance and U-Pb core age for zircons from this gold placer. These data are used to characterize each grain, understand its formation history, and try to locate its proto-source and source rock within the local and regional settings.

2. Overview on the Regional and Local Geologic Settings

2.1. *Brief Review of the Regional Geology*

Basement formations in Cameroon (Figure 2) comprise Archean, Paleoproterozoic (Eburnean), and Pan-African rocks (Table 1). Archean units (>2500 Ma) constitute the Congo Craton, while the Paleoproterozoic ones (2400–1800 Ma) include the West Central African Belt and Pan-African/Cambrian units that constitute the Oubanguide Belt in the Mobile Zone [22,23].

2.1.1. The Ntem Complex

The Archean Craton or Ntem Complex (Figure 2) located in the northwestern end of the Congo Craton [24] is mainly composed of Archean rocks (Table 1) with some reworked Paleoproterozoic material that formed in early Proterozoic times [25]. It is structurally made up of two main units: the Ntem (at the center and south) and the Nyong (in the northwestern) [24]. The Ntem unit is essentially made up of tonalite, trondhjemite, and granodioritic suites (TTGs) and charnockites with TTGs cutting across charnockites and greenstone belts [26]. TTGs and charnockites enclose xenolithic remnants (3.1 Ga) of greenstone belts (banded iron formations and sillimanite-bearing paragneisses) [27,28]. Bounded Iron Formations found in the Ntem unit are locally intercalated with meta-siltstones and meta-sandstones [29]. The TTGs and charnockites were intruded by K-rich granitoids (monzogranite and syenogranite) during the Archean [30,31] and cross-cut by metadoleritic dykes during the Eburnean [32,33] or Late Archean time [34]. The Nyong unit, ranging in age from Archean to Paleoproterozoic [32,35–37], and part of the West Central African Fold Belt [23], is composed of migmatitic orthogneisses (TTGs), meta-gabbro, amphibolite, garnetite, eclogite, felsic gneiss of volcanic to volcano-sedimentary origin, quartzite, charnockite, meta-syenite, and BIF [22,23,37]. Migmatites, charnockites, and meta-sedimentary rocks are Archean in age [32,35].

Table 1. Summarized ages of plutonic and metamorphic rocks found in Congo Craton, Cameroon mobile belt, and Meiganga.

Regional Scale						Local Scale	
Congo Craton (Ntem Complex)				Cameroon Mobile Belt			
Ntem Unit		Nyong Unit					
Rock Type	Age and Author(s)	Rock Type	Age and Author(s)	Rock Type	Age and Author (s)	Rock Type	Age and Author (s)
Charnockites (north of the Ntem unit)	2900 ± 44 and 2818 ± 48 Ma (Rb/Sr [39])	Magnetite-bearing quartzites (BIF) in Eseka	2776 ± 34 Ma and 2126 ± 136 Ma (SHRIMP U-Pb zircon: [37])	Panafrican granitoids	Poli (520 ± 20 Ma) and Lom (498 ± 5 (Ma) (Rb-Sr age on whole –rock, [35]) Nkambe (530 ± 10 Ma and 510 ± 25 Ma; 569 ± 12 to 558 ± 24 Ma, and 533 ± 12 to 524 ± 28 Ma) [40,41] Ngondo (600 Ma) [42] Tonga (618 Ma) [43]	Pyroxene-amphibole-bearing gneiss (TTG composition) (SW of Meiganga)	1711.7 ± 4 to 2602.2 ± 13.6 Ma ($^{207}Pb/^{206}Pb$ zircon evaporation: [44]) 1738 ± 14 to 2987 ± 28 Ma ($^{206}Pb/^{238}U$) 1999 ± 2 to 2884 ± 4 Ma ($^{207}Pb/^{206}Pb$) [20]
Charnockites (north of the Ntem unit)	2882 ± 70 Ma (Rb/Sr: [32]) 2912 ± 25 Ma ($^{207}Pb/^{206}Pb$ zircon evaporation: [26])	Garnet-bearing gneiss in Eseka	2761–2790 Ma (SHRIMP U-Pb zircon: [37])	Granulite in Yaoundé	620 ± 10 Ma [45]	Amphibole-biotite gneiss	1887 to 2339.4 Ma and 675 to 889 Ma ($^{207}Pb/^{206}Pb$ zircon evaporation: [19])
Charnockites (Ebolowa)	2896 ± 7 Ma (U/Pb zircon: [32])	Kribi metquartzites	2948 ± 47 Ma and 2049 ± 36 Ma (U-Pb zircon: [32])	Monzodiorite in Bafia	600 Ma [46]	Metadiorite (NE of Meiganga)	614.1 ± 3.9 Ma and 619.8 ± 9.8 Ma ($^{207}Pb/^{206}Pb$ zircon evaporation: [47]) 562 ± 6 to 637 ± 5 Ma (U-Pb zircon: [48])
Charnockites (Sangmelima)	3010 ± 10 to 2756 ± 14 Ma ($^{207}Pb/^{206}Pb$ zircon evaporation: [49])	Amphibolite in Eseka	2000–2010 Ma (U-Pb zircon: [32])	Metasediments in Bafia	1617 ± 16 [36]	Two micas granite in Doua (West of Meiganga)	647 ± 46 Ma (U-Pb zircon: [21])

Table 1. *Cont.*

Regional Scale						Local Scale	
Congo Craton (Ntem Complex)				Cameroon Mobile Belt			
Ntem Unit		Nyong Unit					
Rock Type	**Age and Author(s)**	**Rock Type**	**Age and Author(s)**	**Rock Type**	**Age and Author (s)**	**Rock Type**	**Age and Author (s)**
Tonalites (Sangmelima)	2825 ± 11 to 2678 ± 17 Ma ($^{207}Pb/^{206}Pb$ zircon evaporation: [49])	Amphibolite in Kopongo	2037± 10 Ma and 626 ± 26 Ma Ma (U-Pb zircon: [32])	-	-	Amphibole-biotite granite in Doua (West of Meiganga)	607 ± 3.9 (U-Pb zircon: [21])
Granodiorites (Sangmelima)	2999 ± 10 to 2671 ± 25 ($^{207}Pb/^{206}Pb$ zircon evaporation: [49])	Orthopyroxene-garnet geneiss (charnockitic) Eseka	3174 ± 4 Ma, 3129 ± 10 Ma, 3064 ± 4 Ma 2086 ± 8 Ma, 2300 ± 17 Ma (SHRIMP U-Pb zircon: [37])	-	-	-	–
High-K granites (Sangmelima)	2717 ± 9 to 2724 ± 3 Ma ($^{207}Pb/^{206}Pb$ zircon evaporation: [31])	Bienkok charnockite	2051 ± 10 Ma and 2043 ± 22 Ma (SHRIMP U-Pb zircon: [37])	-	-	-	-
Granodiorites	2880 ± 70 Ma (Rb/Sr isochrones: [35])	Bonguen metagranodiorites	2066 ± 4 Ma (SHRIMP U-Pb zircon: [37])	-	-	-	-
Granodiorites (north of the Ntem unit)	2.97 Ga ($^{207}Pb/^{206}Pb$ zircon evaporation: [26])	Rocher du Loup Panafrican metasyenites	591 ± 19Ma (SHRIMP U-Pb zircon: [37])	-	-	-	–
Tonalites (north of the Ntem unit)	3.10–2.97 Ga ($^{207}Pb/^{206}Pb$ zircon evaporation: [26])	Lolodorph metasyenites	2055 ± 5 Ma (SHRIMP U-Pb zircon: [37])	-	-	-	-
Pyroxene-bearing gneisses	2980 ± 45 Ma (Rb/Sr isochrones: [35])	Nkonlong and Akom syenites	525 and 807 Ma (K/Ar dating on hornblende: [50])	-	-	-	-
Xenoliths from greenstone belts	3.1 Ga ($^{207}Pb/^{206}Pb$ zircon evaporation: [28])	Lolodorf-Doum clinopyroxene syenites	2837 ± 1–2349 ± 1 ($^{207}Pb/^{206}Pb$ zircon evaporation: [51])	-	-	-	-

Table 1. *Cont.*

Regional Scale						Local Scale	
Congo Craton (Ntem Complex)				Cameroon Mobile Belt			
Ntem Unit		Nyong Unit					
Rock Type	**Age and Author(s)**	**Rock Type**	**Age and Author(s)**	**Rock Type**	**Age and Author (s)**	**Rock Type**	**Age and Author (s)**
K-rich granitoids (Ebolowa)	2.7–2.5 Ga ($^{207}Pb/^{206}Pb$ zircon evaporation: [25])	-	-	-	-	-	-
Metadoleritic dykes	2.1 Ga (U-Pb zircon: [32])	-	-	-	-	-	-
Mengueme two pyroxenes syenites	2321 ± 1 ($^{207}Pb/^{206}Pb$ Pb–Pb zircon evaporation: [51])	-	-	-	-	-	-
Njweng metasandstone (Mbalam iron ore)	3000–1000 Ma [29]	-	-	-	-	-	-

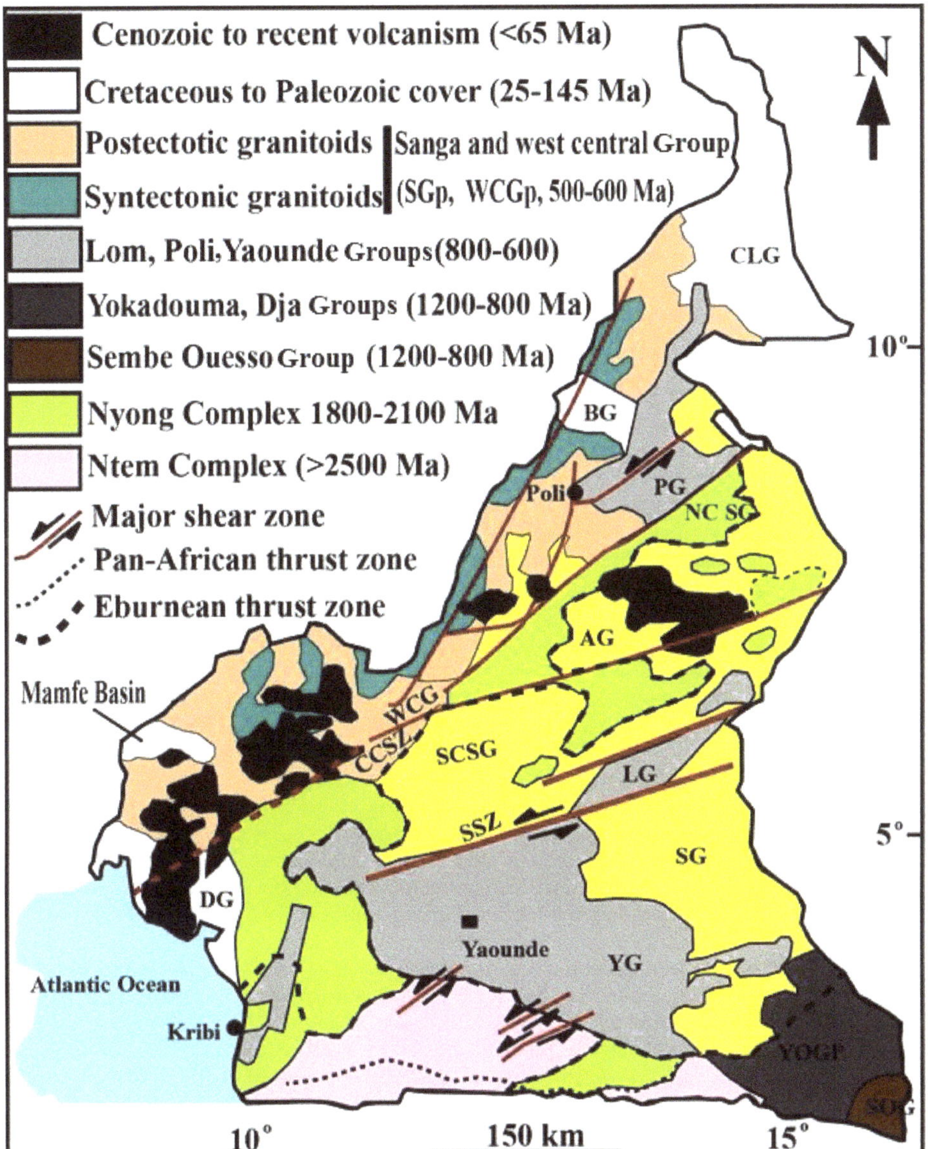

Figure 2. Sketch geological map of Cameroon (adapted from [38]). Oubanguide Complex: NCSG: Northern Cameroon SGp (PG: Poli Group, AG: Adamawa Group, WCG: West Cameroon Group); SCSG: Southern Cameroon SGp (YG: Yaoundé Group, LG: Lom Group, SG: Sanaga Group); SECSGp: Southeastern Cameroon SGp (DG: Dja, YoGroup: Yokadouma, S.O.Group: Sembe Ouesso Group); CCSZ: Centre Cameroon shear zone; SSZ: Sanaga shear zone; Sedimentary cover: (CLG: Chad Lake Group; BG: Benue Group; MG: Manfe Group; DG: Douala Group); B: Cameroon main litho-structural units.

2.1.2. The Cameroon Mobile Belt

The Cameroon mobile zone or Central African Fold belt is a mega-tectonic structure underlying Cameroon, Chad, and the Central African Republic between the Congo Craton to the south and the Nigerian shield to the north [52]. It was formed during the Neoproterozoic, from the collision between the Saharan meta-craton and the Congo craton [17,18]. In the Cameroonian territory, the Central African Fold belt is made up of three main structural units: the Poli Group in its northern part, the Adamawa at the center, and the Yaoundé Group in the south [18]. Within the central African fold belt, several domains are recognized on the basis of field, petrographic, structural, and isotopic studies. These include the Paleoproterozoic gneissic basement, Mesoproterozic to Neoproterozoic schists, and gneisses of Poli, Yaoundé, and Lom, and Pan-African granitoids whose ages range from the early stage of the deformation (orthogneisses) to the late uplit stages of the belts [36]. Examples of geochronological data of some of these rocks are summarized in Table 1.

2.2. Local Geology

The Meiganga part of the Adamawa-Yadé domain (AYD) (Figure 2) is situated between the Congo Craton and the Sahara Metacratron [20]. Basement rocks in Meiganga are composed of paragneisses, orthogneisses, amphibolites, granulites, migmatites, quartzites, metadiorites, schists, and granites [1,2,16,21]. Some gneisses and amphibolites underwent retrograde metamorphism that led to the formation of greenschist facies overprints [16]. Partial melting of gneiss led to the crystallization of leucogranites found in the northern part of Meiganga [16]. Magmatism, cataclastic deformation, rock fracturing, and partial melting of some basement rocks led to the formation of mafic dykes and dykelets, syenitic, micro-granitic, quartzo-feldspathic and quartz-rich veins, brecciated shear zones, and mylonites [1,2,16]. The basement rocks are locally overlain by basaltic flows, lithified clastic sediments (sandstones and conglomerates), unconsolidated detritus (e.g., colluvium, eluvium, alluvium), or red soil [1,2,53]. Skarnoids (hornfels) are visible at the contact between some intrusions and overlying sedimentary rocks.

The western part of Meiganga is made up of pyroxene-amphibole orthogneiss, amphibole gneiss, biotite gneiss, amphibole-biotite gneiss, amphibolite, calc-alkaline and two mica-granites, and amphibole-biotite granite [16,19–21,44,47,53]. Pyroxene-amphibole orthogneiss locally enclose mafic xenoliths [44]. The geochemical features of the orthogneiss and U-Pb zircon ages are similar to those of many TTGs and charnockites outcropping within the Archean Ntem complex in the south of Cameroon [20,44]. The ages of some rocks in the west of Meiganga presented in Table 1 range from Archean to Neoproterozoic. Zircons occurring in a gold-bearing placer in the west of Meiganga are inherited grains crystallized from Archean to Precambrian magmatic crustal evens with part of their source rocks being granitoids, TTG, and charnockites [9].

The southern part of Meiganga (Figure 1c) from where the studied zircon were sampled are composed of mainly undated graphite schists, amphibolites, mica-rich quartzites, amphibole-biotite gneisses, orthogneisses, migmatites, calc-alkaline granitic rocks, biotite-amphibole, and biotite-chlorite granites whose formation periods are assumed to be Precambrian as they also belong to the central part of the Cameroon mobile belt [53]. Hornfels are found at the contact between calc-alkaline biotite-chlorite granite and biotite-chlorite granite at the south eastern part of the locality (Figure 1c). Rocks found in valleys are locally covered by alluvial flats and terraces with part of the alluvium hosting gold.

3. Materials and Methods

In total, 111 zircons from gold-bearing alluviums in two areas (Gankoumbol and Yende: Figure 1) were analyzed to determine the trace elements composition and U-Pb core age at the University of California, Santa Barbara, CA, USA. The results from each zircon core were acquired by Laser Ablation Split Stream analytical techniques. The analyzed zircons were sampled upstream and in small size streams to be close to the source area.

They were separated from pre-concentrated heavy and light minerals mixtures obtained from 50 L of mainly very coarse-grained alluvium, at the bottom of the gold-bearing pits. Heavy mineral fractions were separated from light minerals using bromoform (Density: 2.7 g/cm^{-3}) at the Department of Earth Sciences of the University of Yaoundé I, Cameroon. The separation procedure is similar to the one described in [54,55].

The gold-bearing heavy mineral fractions were sent for zircon separation, trace element analysis, and U-Pb dating at the Department of Earth Sciences of the University of California. The analytical procedures used to obtain the zircon trace element and U-Pb age data are the same as those presented in [9]. Each mounted grain is polished and analyzed following standard procedures using a laser ablation "split stream" setup consisting of a Photon Machines Excimer 193 nm laser ablation unit coupled to a Nu Instruments, "Nu Plasma" multi-collector inductively coupled a plasma-mass spectrometer and an Agilent 7700S quadrupole inductively coupled plasma-mass spectrometer (for detailed methodology see [56–58]. Samples were abraded for 20 s using a fluence of 1.5 J/cm^2, a frequency of 4 Hz, and a spot size of 20 μm diameter, resulting in crater depths of ~9 μm. Utilizing a standard-sample bracketing technique, analyses of reference materials with known isotopic compositions were measured before and after each set of the seven unknown analyses. Data reduction, including corrections for baseline, instrumental drift, mass bias, down-hole fractionation, and age and trace element concentration calculations were carried out using Iolite v. 2.1.2 [59]. "91500" zircon (1065.4 ± 0.3 Ma ^{207}Pb/^{206}Pb ID-TIMS age and 1062.4 ± 0.4 Ma ^{206}Pb/^{238}U ID-TIMS age: [60]) served as the primary reference material to monitor and correct for mass bias, as well as Pb/U down-hole fractionation and to calibrate concentration data, while "GJ-1" zircon (608.5 ± 0.4 Ma ^{207}Pb/^{206}Pb and 601.7 ± 1.3 Ma ^{206}Pb/^{238}U ID-TIMS ages: [61]) was treated as an unknown in order to assess accuracy and precision. Twenty-three analyses of GJ-1 zircon throughout the analytical session yield a weighted mean ^{207}Pb/^{206}Pb date of 593 ± 5 Ma, MSWD = 0.8 and a weighted mean ^{206}Pb/^{238}U date of 603 ± 2 Ma, MSWD = 1.0. Concordia and Kernal Density Estimate (KDE) plots were calculated in Isoplot version 2.4 [62] and Density Plotter [63], respectively, using the ^{238}U and ^{235}U decay constants of [64]. All uncertainties are quoted at 95% confidence levels or 2 s level and include contributions from the external reproducibility of the primary reference material for the ^{207}Pb/^{206}Pb and ^{206}Pb/^{238}U ratios. For plotting and age interpretation purposes, the ^{207}Pb/^{206}Pb dates are used for analyses older than 1000 Ma, whereas the ^{206}Pb/^{238}U dates are used for analyses younger than 1000 Ma.

4. Results

4.1. Zircon Geochemistry

4.1.1. Minor Elements

The relatively high Y and Hf contents in part of the studied zircons can reflect a crystallization in Hf-Y-rich melts with favorable conditions for Hf and Y to substitute Zr. The Hf/Y ratios (5.0−293.0) are mainly less than 30, with the highest values exclusively being those of zircons with very low Y contents.

4.1.2. Trace Elements

The trace element (U, Th, Ti, Ta, Nb, Pb, and Sr) abundances (<1000 ppm) and Th/U ratios are heterogeneous with similar values found in some grains (Tables 2 and 3). Within U and Th elemental suites, U contents (25−954 ppm) mainly exceed 219 ppm, and Th contents (8–674 ppm) are mostly greater than 100 ppm. The Th/U ratios (0.06 to >2.0) are larger than 0.4. Four groups can be distinguished (Figure 3): (1) zircons with Th/U ratios (<0.2) (lowest proportion); (2) zircons with Th/U ratios (≥0.2 to ≤0.5) (highest proportion); (3) zircons with Th/U ratios (>0.5 to ≤1.0); and (4) zircons with Th/U ratios >1.0. The plotted data in Th versus U binary diagram (Figure 4a) show a pronounced positive correlation (the increase in U content when Th content increases) for group 2–4 zircons; however, no correlation is found for group 1 zircons, as their plots are scattered. This correlation is less pronounced (as part of the plots are scattered) in the Th/U versus U(Figure 4b) and Th/U versus Th (Figure 4c) plot diagrams. The Th/U ratios for group 2 to group 4 are within the range of igneous zircons as presented in [4,5,11,65,66]. Those of group 1 (e.g., MSDZ015, MSDZ040, MSDZ045, MSDZ067, MSDZ082, and MSDZ093) characterize metamorphic zircons if based on the criteria of [5], [67], and [68].

Table 2. Minor and trace element concentrations (ppm) in the southern Meiganga detrital zircons.

Sample Spot-Name	Hf	Th	Ti	Ta	Nb	Sr	Hf/Y	Nb/Ta	Ti-in-Zircon T (°C) $\pm$ 2σ	
MSDZ001	10,750	154	8.7	1.16	1	0.11	38.256	0.842	733	50
MSDZ002	7820	79	30	1.56	2.2	0.27	5.6259	1.421	861	88
MSDZ003	7210	48	7.3	0.46	0.6	0.18	7.2755	1.202	717	79
MSDZ004	9860	341	10	3.46	3.3	0.18	10.956	0.966	746	40
MSDZ005	10,260	177	20.9	1.75	2.1	7.22	18.127	1.222	820	76
MSDZ006	8930	165	8.9	1.92	1.8	0.22	13.696	0.961	735	66
MSDZ007	6750	16	6	0.35	0.6	0.15	12.546	1.848	700	60
MSDZ008	11,440	268	47	1.78	2.3	0.31	21.749	1.283	915	135
MSDZ009	7090	70	9.8	0.16	0.5	0.11	10.381	3.315	744	87
MSDZ010	8910	230	8.1	1.06	1.4	0.15	35.217	1.303	726	71
MSDZ011	8310	217	14.8	1.23	1.5	0.23	13.317	1.18	784	102
MSDZ012	7600	42	2.5	0.59	0.9	0.14	15.05	1.58	630	84
MSDZ013	11,390	299	16.3	1.54	1.1	0.03	54.76	0.7	794	44
MSDZ014	11,520	116	11	0.35	0.4	1.86	44.24	1.213	755	61
MSDZ015	10,380	123	10.9	2.19	1.8	0.31	18.703	0.831	754	39
MSDZ016	5960	12	10.7	0.37	0.3	0.08	21.06	0.885	752	60
MSDZ017	9630	217	7.1	3.53	3.8	0.16	14.038	1.085	715	54
MSDZ018	7030	44	5	0.75	1.1	0.16	7.8547	1.53	685	98
MSDZ019	7500	22	6.1	0.65	0.9	0.17	13.915	1.316	701	63
MSDZ020	7040	38	10.1	0.27	0.4	0.09	22.564	1.603	747	51
MSDZ021	9110	87	5.4	1.33	1.5	0.16	14.717	1.155	691	65
MSDZ022	9110	264	6.5	3.91	4.6	0.2	11.016	1.169	707	74
MSDZ023	9190	253	11.6	1.58	2.2	0.07	22.36	1.397	760	50
MSDZ024	9630	161	7	3.44	3.6	0.16	14.658	1.041	713	68
MSDZ025	10,340	269	35.7	0.76	0.9	1.36	26.649	1.24	881	48
MSDZ026	9570	263	8.1	3.17	4.4	0.19	9.071	1.394	726	76
MSDZ027	10,520	155	8	2.25	2.8	0.14	23.223	1.259	725	79
MSDZ028	10,470	155	10.1	3.11	3.9	0.11	21.904	1.25	747	69
MSDZ029	7480	22	3.6	0.32	0.4	0.07	19.947	1.187	658	122
MSDZ030	7790	70	7.9	1.44	0.8	0.03	292.857	0.58	724	70
MSDZ031	7760	9	5.4	0.14	0.3	0.08	26.667	2.385	691	75
MSDZ032	7470	78	10	1.12	1	0.13	14.792	0.868	746	61
MSDZ033	9030	71	6.7	2.33	2.2	0.21	16.329	0.958	710	40
MSDZ034	9020	19	5.2	0.63	0.9	0.11	21.683	1.355	688	56
MSDZ035	9080	466	25.8	2.25	1.6	0.14	10.425	0.703	843	42
MSDZ036	8360	160	7.2	0.75	0.5	0.16	23.158	0.726	716	77
MSDZ037	10,630	298	8.6	1.6	2.1	0.14	15.842	1.311	732	46
MSDZ038	8700	16	5.2	0.46	0.8	0.09	30.742	1.659	688	50
MSDZ039	8090	266	11.8	1.14	1.5	0.07	16.116	1.338	762	72
MSDZ040	8910	13	3.5	0.16	0.3	0.09	26.053	2.056	656	60
MSDZ041	10,120	334	19	1.32	2.2	0.14	10.87	1.681	810	40
MSDZ042	9560	334	8.2	2.09	3.5	0.15	14.914	1.669	728	49
MSDZ043	8790	97	15.8	0.51	1	0.18	10.68	1.883	791	61
MSDZ044	9150	91	35.9	0.26	0.4	0.05	48.93	1.53	882	66
MSDZ045	7100	45	2.5	0.77	0.8	0.12	9.9162	1.089	630	84
MSDZ046	9790	26	5.3	1.76	1.1	0.04	63.161	0.637	690	59
MSDZ047	9680	180	9.9	1.24	1.6	0.08	40.844	1.295	745	54
MSDZ048	9630	203	6.2	2.66	3.5	0.19	10.201	1.327	703	62
MSDZ049	9960	145	20.9	1.8	1.7	0.1	39.059	0.965	820	97
MSDZ050	7980	36	2.7	0.46	0.8	0.16	12.587	1.702	636	73
MSDZ051	11,750	100	5.6	1.67	1.8	0.18	11.75	1.064	694	73

Table 2. *Cont.*

Sample Spot-Name	Hf	Th	Ti	Ta	Nb	Sr	Hf/Y	Nb/Ta	Ti-in-Zircon T (°C) ± 2σ	
MSDZ052	9930	347	7.1	4.43	4.4	0.23	14.56	0.982	715	78
MSDZ053	10,450	122	11.5	0.68	1	0.07	51.225	1.412	759	56
MSDZ054	8030	91	6.7	0.46	1.5	0.28	6.323	3.213	710	67
MSDZ055	9810	153	256	1.23	2.9	15.99	15.046	2.393	1180	47
MSDZ056	10,190	363	4.4	1.93	2	0.2	8.291	1.032	674	48
MSDZ057	8230	20	6.4	0.38	0.5	0.07	32.789	1.38	706	49
MSDZ058	9990	208	6	2.66	4.1	0.2	8.339	1.534	700	60
MSDZ059	10,110	138	7.5	0.71	1.3	0.02	273.984	1.842	720	57
MSDZ060	8710	19	3	0.44	0.4	0.12	16.75	0.913	644	60
MSDZ061	10,540	17	2.1	0.6	0.4	0.09	57.127	0.747	618	81
MSDZ062	8080	17	4.3	0.35	0.3	0.09	33.115	0.777	672	52
MSDZ063	9890	67	28.7	0.67	1.1	0.14	32.426	1.695	856	70
MSDZ064	12,010	186	10.9	1.8	1.8	0.06	29.728	0.974	754	60
MSDZ065	9780	245	13.3	1.29	1.6	0.16	18.178	1.198	773	68
MSDZ066	10,220	304	36	1.26	1.6	0.15	14.898	1.245	882	234
MSDZ067	10,570	103	15.2	2.37	1.5	0.37	33.987	0.647	787	44
MSDZ068	7490	29	5.1	0.88	1.2	0.08	21.039	1.405	686	61
MSDZ069	10,130	399	4.2	1	1.6	0.16	8.393	1.59	670	61
MSDZ070	7850	31	2.4	0.84	1.4	0.17	12.189	1.672	627	73
MSDZ071	10,570	122	5.3	0.88	1.3	0.06	34.542	1.439	690	69
MSDZ072	8680	36	4	1.18	2.3	0.22	9.333	1.963	667	43
MSDZ073	8860	231	12.3	2.42	3	0.18	10.56	1.241	766	65
MSDZ074	7760	63	35.2	5.95	3.6	0.06	11.811	0.602	880	39
MSDZ075	10,260	209	4.7	2.48	4.4	0.21	8.472	1.755	680	72
MSDZ076	6510	51	5	0.5	0.9	0.2	5.54	1.711	685	58
MSDZ077	7590	25	3.5	0.29	0.4	0.14	16.013	1.495	656	65
MSDZ078	7600	21	2.1	0.63	0.7	0.07	22.823	1.116	617	76
MSDZ079	9740	238	5.3	2.67	4.5	0.14	11.513	1.672	690	56
MSDZ080	10,000	99	8.8	0.67	0.7	0.08	36.364	1.02	734	41
MSDZ081	10,430	213	8.1	2.21	2.6	0.14	15.072	1.181	726	68
MSDZ082	9290	38	7.6	1.44	1.3	0.15	5.63	0.869	721	40
MSDZ083	11,030	318	9.2	1.82	2	0.16	19.25	1.084	738	58
MSDZ084	9300	154	2.8	2.8	4.1	0.13	10.852	1.452	639	77
MSDZ085	9140	674	33.2	25.1	19.6	0.11	6.201	0.781	873	42
MSDZ086	9330	261	6	2.12	3.5	0.14	13.64	1.643	700	76
MSDZ087	9130	150	4.3	2.92	2.9	0.09	14.221	0.993	672	56
MSDZ088	8620	34	5.1	1.35	1.2	0.07	46.596	0.88	686	41
MSDZ089	10,930	110	4.8	1.6	1.4	0.11	51.801	0.899	681	40
MSDZ090	9070	100	7	0.52	0.7	0.09	35.43	1.377	713	76
MSDZ091	8460	485	7	2.78	2.8	0.16	7.05	0.99	713	49
MSDZ092	8300	182	5.5	2.44	3.9	0.16	7.431	1.608	693	54
MSDZ093	10,400	78	7.3	0.94	0.8	0.22	57.143	0.861	717	76
MSDZ094	9750	457	4.5	1.78	2.9	0.15	11.861	1.636	676	62
MSDZ095	5980	46	3.4	0.7	0.6	0.14	14.071	0.829	653	50
MSDZ096	7060	49	3.8	0.58	1.2	0.16	9.605	2.086	662	80
MSDZ097	10,890	275	10	2.21	1.5	0.31	18.272	0.656	746	90
MSDZ098	10,040	163	12.7	0.64	1	0.05	21.004	1.608	769	48
MSDZ099	8830	80	6.8	0.75	0.8	0.09	16.598	1.007	711	89
MSDZ100	9080	137	6.7	1.37	1.4	0.11	24.809	1.047	710	43
MSDZ101	9230	30	12	1.08	1.2	0.17	18.76	1.085	763	244
MSDZ102	11,620	132	8.4	0.54	0.8	0.07	45.214	1.549	730	44
MSDZ103	9550	149	10.9	1.06	1.9	0.12	17.884	1.785	754	60
MSDZ104	7330	26	4.2	0.39	0.9	0.15	9.338	2.234	670	78
MSDZ105	8990	230	10.6	0.93	1.5	0.07	19.168	1.627	751	61
MSDZ106	7980	8	3.5	0.33	0.3	0.07	29.446	1.01	656	90
MSDZ107	10,040	273	6.5	1.03	1.7	0.18	12.395	1.627	707	60
MSDZ108	9140	52	29.3	1.19	1.9	0.1	17.116	1.614	858	53
MSDZ109	7320	21	17	0.54	0.9	0.08	18.164	1.698	798	115
MSDZ110	9170	68	1.7	0.24	0.8	0.05	80.439	3.187	824	54
MSDZ111	9460	31	4.6	1.11	1.1	0.07	29.379	0.999	678	61

Table 3. U, Th, and Pb abundance (in ppm), isotopic geochemical data, and U-Pb core ages (in Ma) for the southern Meiganga detrital zircons.

Sample Spot-Number	U	Th	Pb	Th/U	^{207}Pb/^{206}Pb	2s%	^{207}Pb/^{235}U	2s%	^{206}Pb/^{238}U	2s%	6/38–7/35 Rho	^{207}Pb/^{206}Pb (Ma) Age	2s. Abs.	^{207}Pb/^{235}U Age (Ma)	2s. Abs.	^{206}Pb/^{238}U Age (Ma)	2s. Abs.	% Discordance (6/38–7/35)	% Discordance (6/38–7/6)
MSDZ001	522	154	122	0.296	0.12425	1.0287	4.889	2.1541	0.2869	1.8925	0.88	2018	18	1800	39	1626	31	9.7	19.4
MSDZ002	145	79	119	0.537	0.26727	1.0095	21.49	2.0761	0.5848	1.8141	0.87	3290	16	3160	66	2968	54	6.1	9.8
MSDZ003	69	48	74	0.704	0.22325	1.0507	17.77	2.0178	0.5797	1.7226	0.85	3004	17	2977	60	2948	51	1	1.9
MSDZ004	259	341	407	1.315	0.17496	1.0996	10.468	2.1677	0.4356	1.8681	0.86	2605	18	2477	54	2330	44	5.9	10.6
MSDZ005	124	177	498	1.431	0.315	3.6324	24.7	4.1461	0.5676	1.9988	0.48	3531	56	3283	136	2902	58	11.6	17.8
MSDZ006	598	165	114	0.275	0.15845	1.0314	5.201	2.5306	0.2399	2.3109	0.91	2439	17	1852	47	1386	32	25.2	43.2
MSDZ007	54	16	23	0.287	0.21296	1.065	15.34	2.0274	0.5269	1.7252	0.85	2928	17	2836	58	2728	47	3.8	6.8
MSDZ008	488	268	259	0.55	0.13435	1.1417	5.984	2.0884	0.3259	1.7487	0.84	2155	20	1973	41	1818	32	7.9	15.6
MSDZ009	66	70	108	1.046	0.221	1.0243	17.08	1.9976	0.5669	1.715	0.86	2988	16	2939	59	2895	50	1.5	3.1
MSDZ010	495	230	207	0.462	0.13476	1.0858	5.779	2.1565	0.3142	1.8632	0.86	2161	19	1943	42	1761	33	9.4	18.5
MSDZ011	338	217	174	0.632	0.1226	1.3435	4.43	2.2949	0.2633	1.8605	0.81	1993	24	1719	39	1507	28	12.3	24.4
MSDZ012	105	42	55	0.403	0.20116	1.0216	13.09	2.0777	0.4736	1.8092	0.87	2836	17	2685	56	2502	45	6.8	11.8
MSDZ013	338	299	323	0.88	0.13567	1.0905	6.657	2.0879	0.3585	1.7805	0.85	2174	19	2067	43	1975	35	4.4	9.1
MSDZ014	509	116	109	0.228	0.13695	1.1887	6.03	2.2048	0.3183	1.857	0.84	2189	21	1980	44	1781	33	10.1	18.6
MSDZ015	954	123	46	0.128	0.08105	1.2155	1.2248	2.1535	0.10972	1.7777	0.83	1224	24	812	17	671	12	17.3	45.2
MSDZ016	27	12	19	0.451	0.22252	1.0873	17.52	2.1095	0.5705	1.8077	0.86	2999	17	2963	63	2910	53	1.8	3
MSDZ017	125	217	238	1.715	0.16046	1.0369	8.15	2.2115	0.3659	1.9533	0.88	2461	18	2247	50	2010	39	10.5	18.3
MSDZ018	77	44	67	0.568	0.22195	1.0271	16.69	2.0589	0.5443	1.7844	0.87	2995	17	2920	60	2801	50	4.1	6.5
MSDZ019	79	22	35	0.279	0.21649	1.0693	16.41	2.0471	0.5445	1.7456	0.85	2955	17	2901	59	2802	49	3.4	5.2
MSDZ020	47	38	62	0.799	0.21889	1.0345	17.29	2.014	0.567	1.728	0.86	2973	17	2951	59	2895	50	1.9	2.6
MSDZ021	189	87	123	0.46	0.19979	1.063	12.675	2.0103	0.4554	1.7063	0.85	2824	17	2656	53	2419	41	8.9	14.4
MSDZ022	219	264	365	1.203	0.17656	1.1383	10.966	2.1316	0.4461	1.8022	0.85	2620	19	2520	54	2380	43	5.6	9.2
MSDZ023	192	253	270	1.319	0.13305	1.0486	6.229	2.0321	0.3355	1.7407	0.86	2139	18	2008	41	1865	32	7.1	12.8
MSDZ024	313	161	168	0.51	0.1345	1.3402	5.942	2.1893	0.3177	1.7312	0.79	2156	23	1967	43	1778	31	9.6	17.5
MSDZ025	734	269	112	0.365	0.08579	1.1816	1.445	2.1	0,.1217	1.7361	0.83	1333	23	908	19	740	13	18.5	44.5
MSDZ026	229	263	374	1.136	0.17876	1.1196	11.09	2.158	0.4464	1.8449	0.85	2641	19	2530	55	2379	44	6	9.9
MSDZ027	238	155	166	0.653	0.13304	1.1592	6.056	2.1298	0.327	1.7867	0.84	2138	20	1984	42	1824	33	8	14.7
MSDZ028	166	155	208	0.923	0.17506	1.0553	10.65	2.0759	0.4389	1.7877	0.86	2607	18	2493	52	2345	42	5.9	10
MSDZ029	38	22	38	0.585	0.21003	1.0526	15.66	2.186	0.5409	1.9159	0.88	2906	17	2855	62	2791	53	2.2	3.9
MSDZ030	125	70	84	0.553	0.13283	1.0466	6.889	2.1274	0.3751	1.8522	0.87	2136	18	2097	45	2053	38	2.1	3.9
MSDZ031	32	9	14	0.277	0.1965	1.122	13.7	2.2204	0.5028	1.916	0.86	2797	18	2729	61	2625	50	3.8	6.1
MSDZ032	240	78	112	0.321	0.2201	1.139	14.35	2.4437	0.4745	2.1621	0.88	2981	18	2772	68	2502	54	9.7	16.1
MSDZ033	95	71	74	0.748	0.12894	1.067	5.737	2.4103	0.3235	2.1612	0.9	2083	19	1938	47	1806	39	6.8	13.3
MSDZ034	48	19	32	0.383	0.22028	1.0578	17.57	2.4692	0.5788	2.2311	0.9	2983	17	2965	73	2943	66	0.7	1.3
MSDZ035	330	466	164	1.393	0.06227	1.0503	0.9296	2.0932	0.1086	1.8106	0.86	683	22	667	14	664	12	0.4	2.8
MSDZ036	449	160	147	0.35	0.12971	1.122	5.145	2.1521	0.2884	1.8365	0.85	2094	20	1843	40	1635	30	11.3	21.9
MSDZ037	150	298	334	1.976	0.13175	1.0836	6.657	2.1093	0.3666	1.8097	0.86	2123	19	2067	44	2013	36	2.6	5.2
MSDZ038	32	16	24	0.5	0.19623	1.0583	13.83	2.0647	0.5109	1.7728	0.86	2795	17	2738	57	2660	47	2.9	4.8

Table 3. *Cont.*

Sample Spot-Number	U	Th	Pb	Th/U	^{207}Pb/^{206}Pb	2s%	^{207}Pb/^{235}U	2s%	^{206}Pb/^{238}U	2s%	6/38–7/35 Rho	^{207}Pb/^{206}Pb (Ma) Age	2s. Abs.	^{207}Pb/^{235}U Age (Ma)	2s. Abs.	^{206}Pb/^{238}U Age (Ma)	2s. Abs.	% Discordance (6/38–7/35)	% Discordance (6/38–7/6)
MSDZ039	237	266	339	1.122	0.17017	1.039	9.84	2.2355	0.4205	1.9794	0.89	2559	17	2420	54	2262	45	6.5	11.6
MSDZ040	76	13	20	0.172	0.2141	1.1699	15.57	2.3122	0.5291	1.9944	0.86	2937	19	2855	66	2737	55	4.1	6.8
MSDZ041	306	334	346	1.088	0.13399	1.0674	6.614	2.144	0.3589	1.8594	0.87	2151	19	2062	44	1977	37	4.1	8.1
MSDZ042	137	334	373	2.381	0.13342	1.044	6.981	2.1077	0.3819	1.831	0.87	2143	18	2109	44	2085	38	1.1	2.7
MSDZ043	130	97	34	0.741	0.06126	1.1855	0.8834	2.1325	0.1049	1.7726	0.83	648	25	643	14	643	11	0	0.8
MSDZ044	433	91	87	0.207	0.13238	1.1884	5.563	2.2081	0.3051	1.861	0.84	2129	21	1910	42	1716	32	10.2	19.4
MSDZ045	69	45	63	0.647	0.19738	1.056	13.14	2.1316	0.4829	1.8516	0.87	2805	17	2690	57	2543	47	5.4	9.3
MSDZ046	396	26	27	0.065	0.13906	1.1228	6.707	2.3098	0.3494	2.0186	0.87	2215	19	2073	48	1931	39	6.9	12.8
MSDZ047	204	180	188	0.873	0.13205	1.0892	6.139	2.1421	0.3378	1.8445	0.86	2125	19	1997	43	1876	35	6	11.7
MSDZ048	146	203	270	1.37	0.17176	1.1007	10.69	2.1434	0.4521	1.8391	0.86	2575	18	2497	54	2404	44	3.7	6.6
MSDZ049	160	145	140	0.897	0.12653	1.1206	5.585	2.0822	0,.3191	1.7549	0.84	2050	20	1913	40	1785	31	6.7	12.9
MSDZ050	112	36	59	0.321	0.22315	1.0536	17.16	2.1308	0.5574	1.8521	0.87	3004	17	2944	63	2859	53	2.9	4.8
MSDZ051	297	100	109	0.333	0.13008	1.1179	6.321	2.2149	0.3534	1.9121	0.86	2099	20	2021	45	1951	37	3.5	7
MSDZ052	360	347	374	0.956	0.13702	1.14	6.766	2.1137	0.359	1.78	0.84	2190	20	2082	44	1977	35	5	97
MSDZ053	294	122	117	0.405	0.1303	1.0854	5.516	2.3289	0.3081	2.0605	0.88	2102	19	1902	44	1731	36	9	17.6
MSDZ054	236	91	131	0.381	0.2234	1.1147	15.49	2.0905	0.5043	1.7686	0.85	3005	18	2846	59	2632	47	7.5	12.4
MSDZ055	93	153	465	1.621	0.3421	1.4946	28.27	2.2929	0.6021	1.7389	0.76	3671	23	3428	79	3038	53	11.4	17.2
MSDZ056	191	363	484	1.887	0.17582	1.1157	11.461	2.1782	0.4745	1.8708	0.86	2614	19	2561	56	2503	47	2.3	4.2
MSDZ057	44	20	29	0.464	0.19934	1.0387	13.85	2.5014	0.5068	2.2755	0.91	2821	17	2738	68	2642	60	3.5	6.3
MSDZ058	169	208	265	1.222	0.16976	1.0316	10.47	2.2061	0.4485	1.9501	0.88	2555	17	2476	55	2388	47	3.6	6.5
MSDZ059	151	138	143	0.907	0.13006	1.0462	6.428	2.1954	0.3613	1.9301	0.88	2099	18	2037	45	1988	38	2.4	5.3
MSDZ060	44	19	31	0.438	0.2195	1.0496	17.87	2.1974	0.5925	1.9305	0.88	2978	17	2982	66	2999	58	−0.6	−0.7
MSDZ061	55	17	21	0.313	0.18958	1.0624	11.65	2.2321	0.4484	1.9631	0.88	2738	17	2578	58	2391	47	7.3	12.7
MSDZ062	38	17	27	0.448	0.21604	1.0484	17.2	2.1184	0.5823	1.8408	0.87	2951	17	2946	62	2958	54	−0.4	−0.2
MSDZ063	224	67	88	0.301	0.19233	1.066	12.28	2.2693	0.4645	2.0034	0.88	2763	17	2625	60	2459	49	6.3	11
MSDZ064	339	186	164	0.545	0.12804	1.0388	5.534	2.2585	0.3143	2.0054	0.89	2071	18	1906	43	1762	35	7.5	14.9
MSDZ065	229	245	240	1.058	0.13116	1.1061	6.202	2.2089	0.3439	1.912	0.87	2113	19	2004	44	1905	36	5	9.9
MSDZ066	277	304	248	1.103	0.12381	1.0218	5.26	2.1466	0.3094	1.8878	0.88	2012	18	1862	40	1738	33	6.7	13.6
MSDZ067	576	103	63	0.179	0.1355	1.4918	3.806	2.3417	0.2035	1.805	0.77	2169	26	1593	37	1194	22	25	45
MSDZ068	62	29	41	0.463	0.18511	1.0242	12.67	2.0947	0.4986	1.8272	0.87	2699	17	2656	56	2607	48	1.9	3.4
MSDZ069	140	399	527	2.817	0.16994	1.0529	11.05	2.1488	0.4723	1.8731	0.87	2557	18	2528	54	2493	47	1.4	2.5
MSDZ070	71	31	43	0.43	0.19918	1.0429	13.31	2.3448	0.4852	2.1001	0.9	2820	17	2703	63	2549	54	5.7	9.6
MSDZ071	240	122	121	0.505	0.13048	1.0975	6.104	2.184	0.3391	1.8882	0.86	2104	19	1990	43	1882	36	5.4	10.6
MSDZ072	134	36	53	0.256	0.21533	1.0968	15.99	2.1918	0.5391	1.8976	0.87	2947	18	2878	63	2779	53	3.4	5.7
MSDZ073	97	231	299	2.331	0.1689	1.0517	10.58	2.1776	0.4531	1.9069	0.88	2547	18	2487	54	2409	46	3.1	5.4
MSDZ074	182	63	21	0.341	0.065	1.1662	0.9	2.1599	0.1008	1.818	0.84	774	25	652	14	619	11	5	20

Table 3. *Cont.*

Sample Spot-Number	U	Th	Pb	Th/U	^{207}Pb/^{206}Pb	2s%	^{207}Pb/^{235}U	2s%	^{206}Pb/^{238}U	2s%	6/38–7/35 Rho	^{207}Pb/^{206}Pb (Ma) Age	2s. Abs.	^{207}Pb/^{235}U Age (Ma)	2s. Abs.	^{206}Pb/^{238}U Age (Ma)	2s. Abs.	% Discordance (6/38–7/35)	% Discordance (6/38–7/6)
MSDZ075	161	209	282	1.274	0.17388	1.0868	11.62	2.2744	0.4847	1.998	0.88	2595	18	2574	59	2547	51	1	1.9
MSDZ076	93	51	79	0.535	0.21857	1.037	17.28	2.1931	0.5721	1.9325	0.88	2970	17	2950	65	2916	56	1.2	1.8
MSDZ077	96	25	31	0.257	0.20631	1.0225	13.89	2.5455	0.4896	2.3311	0.92	2877	17	2742	70	2569	60	6.3	10.7
MSDZ078	43	21	23	0.5	0.14038	1.1174	6.704	2.071	0.3463	1.7436	0.84	2232	19	2073	43	1917	33	7.5	14.1
MSDZ079	198	238	297	1.189	0.17122	1.0635	10.3	2.3104	0.4364	2.051	0.89	2569	18	2461	57	2334	48	5.2	9.2
MSDZ080	183	99	103	0.525	0.1309	1.0705	6.319	2.2667	0.351	1.998	0.88	2110	19	2021	46	1939	39	4	8.1
MSDZ081	338	213	212	0.618	0.13361	1.1431	6.257	2.3372	0.3404	2.0386	0.87	2146	20	2014	47	1888	38	6.2	12
MSDZ082	265	38	48	0.142	0.1964	1.2583	11.22	2.3151	0.4171	1.9433	0.84	2796	21	2543	59	2247	44	11.6	19.6
MSDZ083	464	318	316	0.676	0.1366	1.2139	6.22	2.4767	0.3312	2.1588	0.87	2184	21	2007	50	1844	40	8.1	15.6
MSDZ084	137	154	211	1.099	0.17027	1.0564	10.781	2.1134	0.4612	1.8304	0.87	2560	18	2504	53	2445	45	2.4	4.5
MSDZ085	461	674	219	1.441	0.06096	1.1145	0.8321	2.1662	0.0996	1.8575	0.86	637	24	615	13	612	11	0.4	3.9
MSDZ086	100	261	369	2.564	0.17089	1.0598	11.4	2.1486	0.4865	1.869	0.87	2566	18	2556	55	2555	48	0	0.4
MSDZ087	94	150	216	1.55	0.17263	1.0727	11.48	2.1035	0.4852	1.8094	0.86	2583	18	2563	54	2549	46	0.5	1.3
MSDZ088	161	34	47	0.205	0.1992	1.2223	13.16	2.4488	0.4807	2.122	0.87	2819	20	2693	66	2530	54	6.1	10.3
MSDZ089	299	110	126	0.36	0.163	1.2791	8.47	2.5914	0.378	2.2538	0.87	2486	22	2282	59	2066	47	9.5	16.9
MSDZ090	311	100	109	0.319	0.1353	1.2435	6.19	2.6886	0.3339	2.3838	0.89	2168	22	2001	54	1856	44	7.2	14.4
MSDZ091	549	485	570	0.88	0.1463	1.5855	6.59	3.149	0.3272	2.7207	0.86	2302	27	2057	65	1824	50	11.3	20.8
MSDZ092	141	182	267	1.264	0.17642	1.1076	11.81	2.3478	0.4872	2.0702	0.88	2619	18	2589	61	2558	53	1.2	2.3
MSDZ093	431	78	86	0.175	0.1648	1.3121	7.959	2.5126	0.3508	2.1428	0.85	2504	22	2226	56	1942	42	12.8	22.4
MSDZ094	240	457	509	1.876	0.13638	1.2104	6.991	2.4474	0.371	2.1271	0.87	2181	21	2110	52	2034	43	3.6	6.7
MSDZ095	101	46	77	0.442	0.2282	1.1967	17.74	2.5354	0.5638	2.2352	0.88	3041	19	2974	75	2881	64	3.1	5.3
MSDZ096	162	49	70	0.292	0.1981	1.1961	12.82	2.3474	0.4705	2.0198	0.86	2810	20	2668	63	2485	50	6.9	11.6
MSDZ097	516	275	248	0.526	0.1349	1.3888	5.735	2.498	0.3098	2.0764	0.83	2164	24	1936	48	1739	36	10.2	19.6
MSDZ098	183	163	155	0.874	0.12897	1.0566	6.096	2.3265	0.3415	2.0728	0.89	2084	19	1989	46	1893	39	4.8	9.2
MSDZ099	70	80	93	1.119	0.14679	1.0909	6.476	2.1416	0.3171	1.8429	0.86	2309	19	2042	44	1776	33	13	23.1
MSDZ100	119	137	134	1.133	0.12738	1.0554	5.962	2.1	0.338	1.8155	0.86	2062	19	1971	41	1877	34	4.8	9
MSDZ101	103	30	32	0.289	0.15633	1.0461	7.902	2.0903	0.3666	1.8097	0.87	2416	18	2220	46	2013	36	9.3	16.7
MSDZ102	379	132	131	0.345	0.13019	1.1109	5.72	2.2848	0.3182	1.9965	0.87	2100	20	1936	44	1780	36	8	15.2
MSDZ103	184	149	157	0.804	0.12858	1.0218	6.667	2.249	0.3732	2.0035	0.89	2079	18	2068	47	2044	41	1.2	1.7
MSDZ104	84	26	34	0.314	0.17747	1.0574	10.85	2.2772	0.4398	2.0168	0.89	2629	18	2511	57	2353	47	6.3	10.5
MSDZ105	149	230	316	1.538	0.18392	1.0681	12.86	2.1207	0.5042	1.8321	0.86	2688	18	2669	57	2632	48	1.4	2.1
MSDZ106	25	8	8	0.316	0.1316	1.3033	6.12	2.7694	0.3357	2.4436	0.88	2121	23	1994	55	1865	46	6.5	12.1
MSDZ107	244	273	275	1.05	0.12861	1.0786	6.128	2.1541	0.3447	1.8645	0.87	2079	19	1995	43	1909	36	4.3	8.2
MSDZ108	81	52	72	0.63	0.17535	1.0746	12.35	2.8957	0.509	2.6889	0.93	2609	18	2636	76	2649	71	−0.5	−1.5
MSDZ109	33	21	25	0.624	0.1671	1.3438	9.35	2.3794	0.4041	1.9636	0.83	2528	23	2375	57	2188	43	7.9	13.4
MSDZ110	100	68	70	0.67	0.12757	1.037	6.272	2.1746	0.3538	1.9115	0.88	2065	18	2014	44	1952	37	3.1	5.5
MSDZ111	92	31	40	0.332	0.16737	1.0728	10.878	2.0171	0.4681	1.7082	0.85	2531	18	2513	51	2475	42	1.5	2.2

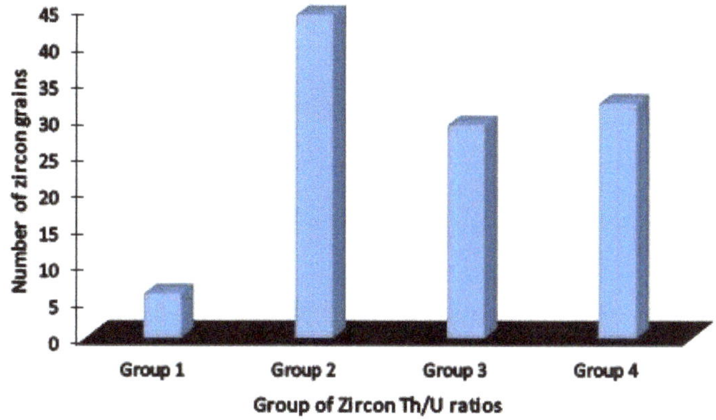

Figure 3. Number of zircon grains in various classified Th/U ratio's groups (group 1: Th/U < 0.2, group 2: Th/U [0.2–0.5], group 3: Th/U [0.5–1.0], and group 4: Th/U > 1.0).

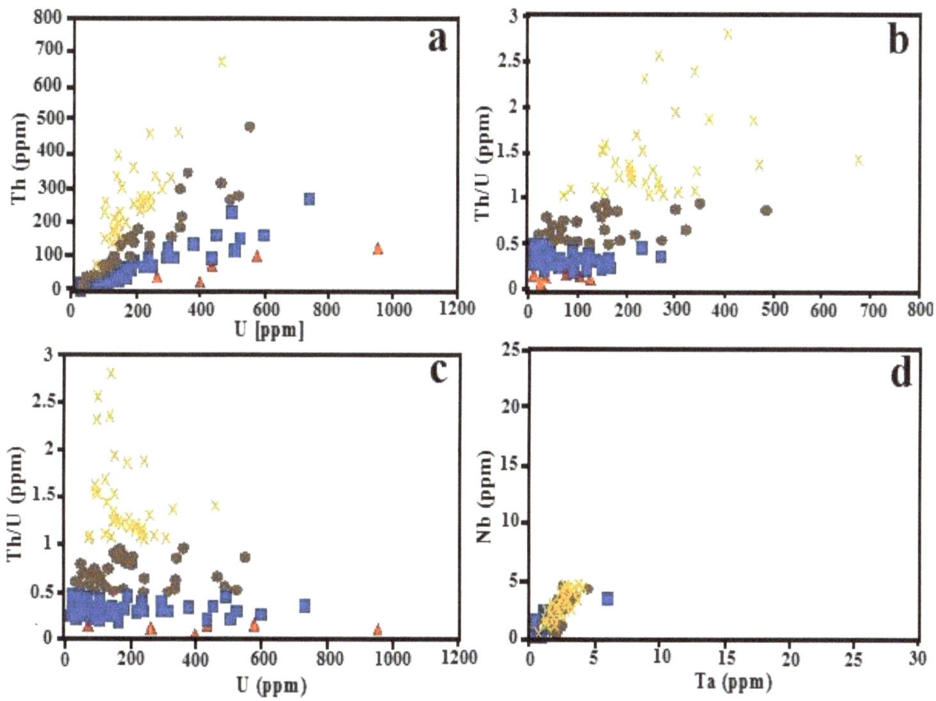

Figure 4. Geochemical correction of various elements/ratios within the southern Meiganga detrital zircons: (**a**) Th versus U; (**b**) Th/U versus Th; (**c**) Th/U versus U; (**d**) Nb versus Ta (red triangles represent zircon with Th/U < 0.2; blue squares represent zircon with Th/U ratios [0.2 to 0.5]; brown circles represent zircons with Th/U ratios [>0.5 to 0.956]; and yellow stars represent zircons with Th/U > 1.0).

The Ti, Ta, Nb, and Sr contents (<48 ppm) are generally low. This indicates low degrees of substitution of these elements within the crystal structure of zircon. Within these element suites, Ti contents (2 to 256 ppm) are globally less than 12 ppm; with the highest value (256 ppm) being that of MSDZ055. Low Ti-zircons generally have low Th and U, which clearly differentiate them from others. The calculated Ti-zircon temperatures

(617 to 1180 °C) (Table 2) are mainly more than 700 °C, with the predominance of zircons whose temperatures ranging from 700–717 °C, 680–694 °C, and 720–728 °C. The highest temperatures are that of MSDZ008 and MSDZ055. The Ta and Nb abundances are generally very close (Figure 4d), and vary from 0.2 to 25.1 ppm and 0.3 to 19.6 ppm, respectively. The highest Ta (25.1 ppm) and Nb (19.6 ppm) values were found in MSDZ085, which also has the highest Th (674 ppm) and relatively high Y (1474 ppm), U (461 ppm), and Ti (33.2 ppm) contents. The Nb/Ta ratios vary from 0.6 to 3.4, with the highest values generally being for zircons with Hf contents (<9000 ppm) and Y contents (>361 to 931 ppm).

4.1.3. Rare Earth Elements (REE)

The REE abundances (Table 4) are variable with the values of total light rare earth elements (LREE: La-Pr) being generally less than those of middle rare earth elements (MREE: Nd-Gd) and heavy rare earth elements (HREE: Tb-Lu). The total rare elements (∑REE) range from 43 to 1030 ppm, with most values being less 400 ppm. Lowest ∑REE contents are that of MSDZ030 (43 ppm), MSDZ046 (98 ppm), and MSDZ059 (79 ppm), of which the normalized patterns (Figure 5a–g) are different from others.

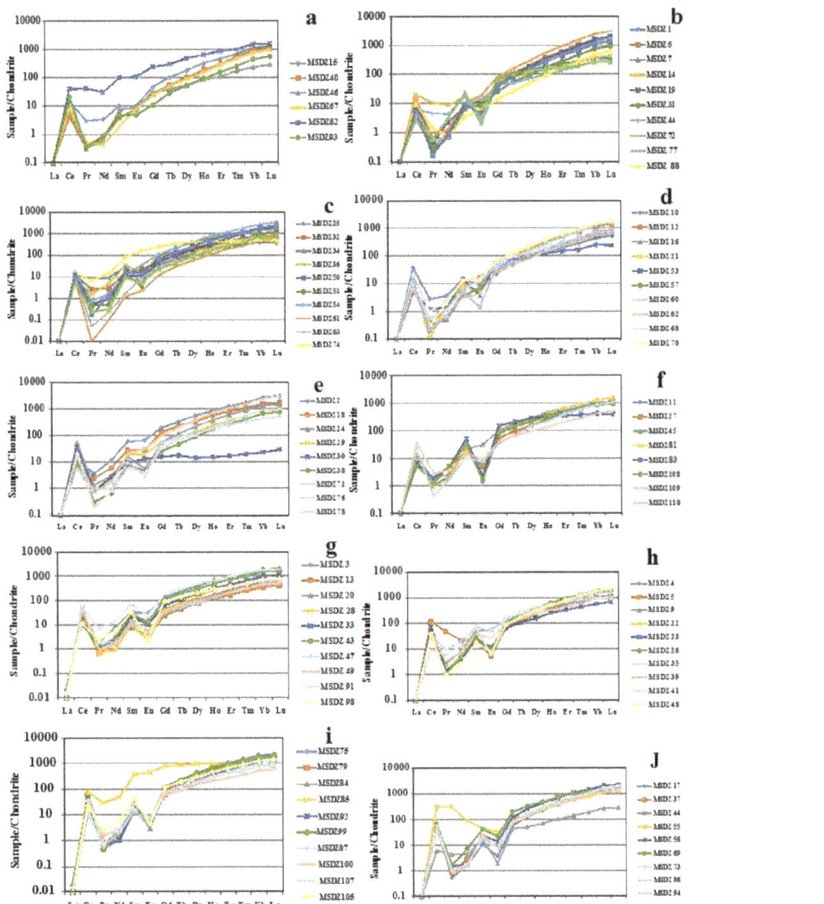

Figure 5. REE patterns for southern Meiganga zircons normalized to [69], chondrite values versus element (La–Lu) diagrams ((**a**) zircon with Th/U < 0.2; (**b–d**) zircon with Th/U [0.2 to 0.5]; (**e–g**) zircon with Th/U [>0.5 to 1.0]; (**h–j**) zircon with Th/U > 1.0).

Table 4. Rare earth element abundance (in ppm) in the southern Meiganga detrital zircons. REE (rare earth elements).

Sample Spot-Name	La	Ce	Pr	Nd	Sm	Eu	Gd	Tb	Dy	Ho	Er	Tm	Yb	Lu	∑REE	Gd/Yb	Lu/Hf	Sm/La_N	Ce/Ce*	Eu/Eu*
MSDZ001	bdl	2	0	1.5	3.2	0.13	14	3.6	32	10	36	6	57	9	173	0.246	0.001	-	-	0.06
MSDZ002	bdl	9	0.3	5.7	9.1	3.87	40	12.1	139	47	210	43	432	78	1029	0.092	0.01	-	-	0.62
MSDZ003	0.01	6	0.2	3.6	4.9	1.73	27	8.5	102	37	158	30	310	51	741	0.087	0.007	1569.299	50.443	0.46
MSDZ004	bdl	30	0.1	2.1	3.3	0.49	20	7.1	82	29	132	28	264	47	645	0.075	0.005	-	-	0.19
MSDZ005	bdl	76	4.4	10.3	3.1	0.3	14	4.3	53	20	90	19	174	29	497	0.082	0.003	-	-	0.14
MSDZ006	bdl	9	0	0.5	1	0.71	8	3.8	48	20	101	26	257	51	526	0.032	0.006	-	-	0.77
MSDZ007	bdl	3	0	0.4	1.1	0.67	8	3.3	44	18	87	21	226	45	457	0.036	0.007	-	-	0.68
MSDZ008	bdl	4	0.1	1.5	3.4	0.14	16	5.2	52	17	68	13	116	17	313	0.138	0.002	-	-	0.06
MSDZ009	bdl	21	0.2	3.9	7.4	3.26	27	7.1	77	25	100	19	171	30	492	0.158	0.004	-	-	0.71
MSDZ010	bdl	23	0.3	1.7	2.2	0.52	8	2.3	24	8	34	7	67	13	191	0.121	0.001	-	-	0.37
MSDZ011	bdl	11	0.1	1	3.4	0.26	17	5.8	62	22	87	17	172	28	426	0.101	0.003	-	-	0.1
MSDZ012	bdl	7	0	0.6	1.5	0.95	9	3.5	41	19	85	19	183	36	404	0.048	0.005	-	-	0.8
MSDZ013	bdl	13	0.1	0.8	1.2	0.19	5	1.6	20	7	27	6	57	11	149	0.083	0.001	-	-	0.24
MSDZ014	bdl	13	1	4.2	2.6	0.37	12	3.2	27	8	32	7	64	11	185	0.181	0.001	-	-	0.21
MSDZ015	bdl	9	0.3	1.6	1.5	0.57	10	3.7	45	17	74	17	183	32	394	0.052	0.003	-	-	0.46
MSDZ016	bdl	3	0.1	0.6	1	0.68	6	2.2	26	10	43	10	101	17	220	0.059	0.003	-	-	0.85
MSDZ017	bdl	29	0.1	0.8	1.7	0.22	12	4.7	56	22	99	23	245	40	533	0.047	0.004	-	-	0.15
MSDZ018	bdl	5	0.2	2.8	4.5	1.59	28	8.6	86	31	130	28	259	43	627	0.106	0.006	-	-	0.44
MSDZ019	bdl	6	0	1.1	1.4	0.74	8	3.2	43	18	94	23	261	52	512	0.032	0.007	-	-	0.67
MSDZ020	bdl	19	0.1	1.2	2.1	0.84	9	2.7	28	10	45	10	96	16	239	0.092	0.002	-	-	0.6
MSDZ021	bdl	9	0	0.8	1.3	0.28	10	4	49	21	97	22	232	45	491	0.043	0.005	-	-	0.24
MSDZ022	bdl	47	0.1	2.2	3.9	0.55	19	6.8	71	27	115	26	250	44	613	0.076	0.005	-	-	0.2
MSDZ023	bdl	43	0.1	1.8	3.7	0.64	13	3.9	39	14	54	11	94	17	296	0.132	0.002	-	-	0.29
MSDZ024	bdl	35	0.1	1.3	2.4	0.32	11	4.2	53	21	96	22	218	36	499	0.051	0.004	-	-	0.19
MSDZ025	bdl	7	0.8	4.2	4.7	0.29	22	5.4	50	13	42	8	66	10	233	0.329	0.001	-	-	0.09
MSDZ026	bdl	26	0.1	2.3	4.1	0.54	24	8.3	98	35	157	32	321	53	761	0.075	0.006	-	-	0.17
MSDZ027	bdl	5	0.2	2.2	3	0.31	11	3.5	38	14	68	14	149	28	337	0.072	0.003	-	-	0.17
MSDZ028	bdl	34	0.1	0.4	1.5	0.13	8	2.8	37	16	69	16	164	29	379	0.051	0.003	-	-	0.12
MSDZ029	bdl	6	0.1	0.8	2.7	1.14	11	3	32	13	56	13	138	23	300	0.081	0.003	-	-	0.63
MSDZ030	bdl	24	0.1	1.2	1.5	0.77	3	0.6	4	1	3	0	4	1	43	0.836	9.628	-	-	1.09
MSDZ031	bdl	3	0	0.3	1	0.5	5	2	23	10	45	11	123	24	248	0.042	0.003\	-	-	0.65
MSDZ032	bdl	8	0.3	1.4	2.6	1.32	11	3.4	43	16	71	15	157	28	358	0.073	0.004	-	-	0.75
MSDZ033	bdl	29	0.1	1.1	3.3	0.58	13	4.2	48	18	75	17	158	28	394	0.082	0.003	-	-	0.27

Table 4. *Cont.*

Sample Spot-Name	La	Ce	Pr	Nd	Sm	Eu	Gd	Tb	Dy	Ho	Er	Tm	Yb	Lu	∑REE	Gd/Yb	Lu/Hf	Sm/La$_N$	Ce/Ce*	Eu/Eu*
MSDZ034	bdl	8	0	0.7	1.6	0.6	8	2.4	31	14	66	14	169	27	343	0.045	0.003	-	-	0.53
MSDZ035	bdl	22	0.9	9.4	11.2	3.02	29	9	81	27	123	26	254	41	636	0.113	0.004	-	-	0.51
MSDZ036	bdl	5	0.2	2.3	5.4	0.9	19	4.5	38	11	47	10	95	17	255	0.195	0.002	-	-	0.27
MSDZ037	bdl	29	0.1	1.3	2.6	0.48	13	5.1	61	23	110	23	234	42	545	0.057	0.004	-	-	0.25
MSDZ038	bdl	7	0	0.3	1.2	0.26	5	1.7	23	9	43	10	110	19	230	0.05	0.002	-	-	0.32
MSDZ039	bdl	27	0.3	3.5	5.7	1.49	20	5.3	57	18	71	15	149	21	394	0.136	0.003	-	-	0.42
MSDZ040	bdl	3	0	0.3	0.6	0.43	5	2	25	11	48	14	154	29	291	0.035	0.003	-	-	0.73
MSDZ041	bdl	6	0.6	6.1	10	0.34	35	9.9	101	31	114	21	162	27	523	0.217	0.003	-	-	0.06
MSDZ042	bdl	30	0.1	1.1	2.9	0.57	15	4.8	56	20	88	19	184	32	454	0.08	0.003	-	-	0.27
MSDZ043	bdl	8	0.1	1.4	4.3	0.75	24	7.5	76	29	121	26	270	44	612	0.088	0.005	-	-	0.23
MSDZ044	bdl	4	0.4	2.1	2.7	0.11	9	1.9	18	6	23	5	45	7	123	0.2	0.001	-	-	0.07
MSDZ045	bdl	7	0.2	1.2	3.3	1.73	21	6.4	69	25	103	22	207	38	504	0.1	0.005	-	-	0.64
MSDZ046	bdl	5	0	0.3	1.1	0.51	6	1.4	14	5	17	4	37	7	98	0.163	0.001	-	-	0.59
MSDZ047	bdl	26	0.2	0.9	1.4	0.35	6	1.8	19	7	35	8	83	15	205	0.074	0.002	-	-	0.36
MSDZ048	bdl	24	0.1	1.2	2.8	0.42	17	6.3	76	30	138	31	315	58	699	0.054	0.006	-	-	0.19
MSDZ049	bdl	32	0.1	0.7	2	0.27	7	2.3	25	8	35	8	70	13	203	0.097	0.001	-	-	0.22
MSDZ050	bdl	7	0.1	0.4	1.7	0.65	10	3.7	50	18	95	25	250	50	511	0.038	0.006	-	-	0.49
MSDZ051	bdl	5	0	0.2	2.1	0.21	21	8.2	99	35	146	29	253	43	640	0.081	0.004	-	-	0.1
MSDZ052	bdl	32	0.1	0.6	1.8	0.61	13	4.5	56	22	95	20	208	38	492	0.064	0.004	-	-	0.38
MSDZ053	bdl	3	0.1	0.6	2.1	0.22	11	2.7	25	7	24	4	40	6	126	0.274	0.001	-	-	0.14
MSDZ054	bdl	11	0.1	0.8	2	2.11	24	8.3	103	43	199	46	461	88	989	0.052	0.011	-	-	0.94
MSDZ055	bdl	187	28.8	41.3	6.7	1.88	21	5.1	56	20	94	19	184	32	698	0.115	0.003	-	-	0.48
MSDZ056	bdl	30	0.1	0.9	3.5	0.8	25	9.4	107	37	172	37	339	54	815	0.073	0.005	-	-	0.26
MSDZ057	bdl	5	0	0.2	0.5	0.4	5	1.8	23	8	37	9	79	15	184	0.058	0.002	-	-	0.78
MSDZ058	bdl	30	0.1	1.1	2.5	0.49	24	10.2	113	41	174	38	348	60	841	0.068	0.006	-	-	0.19
MSDZ059	bdl	53	0.2	1.6	2	1.13	6	1	6	1	3	1	3	0	79	2.37	4.055	-	-	0.97
MSDZ060	bdl	6	0.1	0.5	1	0.69	9	3.5	39	16	78	19	210	41	422	0.04	0.005	-	-	0.73
MSDZ061	bdl	5	bdl	0	0.2	0.13	3	1.1	15	6	32	7	76	16	161	0.033	0.002	-	-	0.58
MSDZ062	bdl	4	0	0.3	0.5	0.25	5	1.8	19	8	37	8	84	16	185	0.062	0.002	-	-	0.49
MSDZ063	bdl	6	0	0.1	0.3	0.53	5	2.1	24	9	43	11	109	20	230	0.05	0.002	-	-	1.33
MSDZ064	bdl	3	0.1	1.5	3.2	0.06	22	5.1	46	12	40	7	47	7	194	0.463	0.001	-	-	0.02
MSDZ065	bdl	15	0.1	1.8	3.1	0.33	18	5.7	56	19	74	15	136	23	367	0.134	0.002	-	-	0.13
MSDZ066	bdl	6	0.2	1.8	4	1.71	21	6.8	69	21	88	17	150	27	413	0.14	0.003	-	-	0.57

Table 4. *Cont.*

Sample Spot-Name	La	Ce	Pr	Nd	Sm	Eu	Gd	Tb	Dy	Ho	Er	Tm	Yb	Lu	∑REE	Gd/Yb	Lu/Hf	Sm/La_N	Ce/Ce*	Eu/Eu*
MSDZ067	bdl	6	0.1	0.2	0.3	0.47	5	1.9	23	9	46	12	126	25	254	0.039	0.002	-	-	1.21
MSDZ068	bdl	13	0.1	1.1	2.1	0.89	11	3.5	34	13	46	11	106	19	259	0.105	0.003	-	-	0.57
MSDZ069	bdl	41	0.2	3.4	6.1	1.23	39	12	118	39	162	31	274	46	773	0.142	0.005	-	-	0.24
MSDZ070	bdl	7	0	0.9	1.9	0.79	13	4.5	55	21	101	23	246	45	519	0.053	0.006	-	-	0.48
MSDZ071	bdl	3	0.1	0.5	1.5	0.17	11	3.2	29	10	41	8	68	12	188	0.155	0.001	-	-	0.13
MSDZ072	bdl	10	0.1	0.7	1.6	1.09	16	5.8	75	32	161	40	412	77	832	0.038	0.009	-	-	0.65
MSDZ073	bdl	23	0.1	0.8	3.4	0.45	17	5.8	73	27	126	25	255	43	599	0.066	0.005	-	-	0.18
MSDZ074	bdl	9	0.7	6.4	12.7	10.15	51	12.9	101	24	73	11	83	11	406	0.609	0.001	-	-	1.22
MSDZ075	bdl	32	0.1	1.2	4	0.3	25	8.4	106	38	173	36	339	57	820	0.074	0.006	-	-	0.09
MSDZ076	bdl	10	0.1	0.8	3.1	1.79	20	7.4	94	39	178	39	401	77	870	0.049	0.012	-	-	0.7
MSDZ077	bdl	4	0.1	0.3	1.4	0.77	8	3	38	15	77	17	181	41	387	0.043	0.005	-	-	0.71
MSDZ078	bdl	13	0	0.4	0.9	0.23	7	2.7	25	11	53	13	142	24	292	0.047	0.003	-	-	0.29
MSDZ079	bdl	37	0.1	0.7	3.1	0.34	19	6.4	80	30	124	26	244	45	615	0.08	0.005	-	-	0.14
MSDZ080	bdl	3	0.1	0.8	2.4	0.09	12	3.2	30	8	34	7	65	11	177	0.188	0.001	-	-	0.05
MSDZ081	bdl	6	0.1	0.9	3.2	0.14	17	5.5	66	23	105	22	202	39	488	0.083	0.004	-	-	0.06
MSDZ082	bdl	25	4	14.3	15	6.26	49	11	120	36	142	25	247	40	735	0.198	0.004	-	-	0.7
MSDZ083	bdl	4	0.2	2.4	7.6	0.11	30	7.5	71	19	56	10	69	9	286	0.443	0.001	-	-	0.02
MSDZ084	0,01	34	0	0.5	2.5	0.18	16	5.6	82	27	125	27	261	43	624	0.061	0.005	813.47	523.962	0.08
MSDZ085	bdl	50	2.9	23.2	58.7	25.3	161	32.5	248	54	144	21	153	17	990	1.05	0.002	-	-	0.79
MSDZ086	bdl	37	0.1	0.7	3.9	0.39	16	5.5	65	22	94	23	189	35	492	0.086	0.004	-	-	0.15
MSDZ087	bdl	20	0.1	0.5	2.4	0.32	13	4.4	58	20	93	21	200	37	468	0.062	0.004	-	-	0.18
MSDZ088	bdl	6	0.1	0.4	0.5	0.31	3	1	12	5	27	7	77	16	155	0.036	0.002	-	-	0.78
MSDZ089	0,01	17	0	0.2	0.8	0.34	3	1.4	15	6	31	8	86	19	188	0.037	0.002	246.603	436.072	0.65
MSDZ090	bdl	3	0.1	0.6	2.3	0.15	9	2.7	28	9	33	6	59	10	162	0.15	0.001	-	-	0.1
MSDZ091	bdl	37	0.7	6.5	10.5	1.31	37	10.9	122	39	158	34	298	47	802	0.124	0.006	-	-	0.2
MSDZ092	bdl	29	0	1.1	3.6	0.3	25	7.7	106	37	152	34	294	51	739	0.083	0.006	-	-	0.1
MSDZ093	bdl	13	0	0.4	0.7	0.28	2	1.1	13	5	26	6	71	14	153	0.031	0.001	-	-	0.66
MSDZ094	bdl	37	0.1	0.9	2.9	0.41	15	5	69	25	119	27	282	52	635	0.052	0.005	-	-	0.19
MSDZ095	bdl	5	0	0.6	1.3	0.67	7	2.4	34	12	66	16	177	36	359	0.041	0.006	-	-	0.67
MSDZ096	bdl	9	0	0.5	1.6	0.96	14	4.3	63	25	113	28	336	55	651	0.042	0.008	-	-	0.62
MSDZ097	bdl	8	0.5	5.6	14.3	0.88	44	9.5	83	19	60	10	77	10	341	0.573	0.001	-	-	0.11
MSDZ098	bdl	7	0.2	2.7	4.9	0.16	19	5	52	17	63	13	117	20	320	0.159	0.002	-	-	0.05
MSDZ099	bdl	7	0.1	1	2.4	0.27	13	3.8	51	18	77	17	154	28	373	0.084	0.003	-	-	0.15

Table 4. *Cont.*

Sample Spot-Name	La	Ce	Pr	Nd	Sm	Eu	Gd	Tb	Dy	Ho	Er	Tm	Yb	Lu	∑REE	Gd/Yb	Lu/Hf	Sm/La$_N$	Ce/Ce*	Eu/Eu*
MSDZ100	bdl	23	0.1	1.5	4.3	0.43	12	3.3	39	12	44	9	86	15	250	0.14	0.002	-	-	0.18
MSDZ101	bdl	4	0	0.4	1.4	0.25	7	2.9	38	15	75	17	186	35	382	0.039	0.004	-	-	0.24
MSDZ102	bdl	2	0.1	1.5	3.5	0.16	15	3.8	34	8	27	5	44	7	149	0.333	0.001	-	-	0.07
MSDZ103	bdl	8	0.1	2.8	5.9	0.3	21	5.7	56	17	72	15	133	22	360	0.159	0.002	-	-	0.08
MSDZ104	bdl	7	0	0.6	2.4	0.92	13	5.1	66	26	128	30	308	61	648	0.044	0.008	-	-	0.49
MSDZ105	bdl	22	0.1	1.5	4.1	0.41	16	4.6	50	16	63	13	125	21	336	0.128	0.002	-	-	0.15
MSDZ106	bdl	3	0	0.2	1.4	0.32	5	1.6	23	9	41	10	116	20	230	0.041	0.003	-	-	0.38
MSDZ107	bdl	15	0.2	2.9	7.1	0.27	24	7.3	77	26	107	22	203	36	528	0.12	0.004	-	-	0.06
MSDZ108	bdl	3	0.1	2.4	4.7	0.09	16	5.2	53	18	70	15	136	23	347	0.117	0.003	-	-	0.03
MSDZ109	bdl	14	0	0.8	1.7	0.42	7	2.5	37	12	64	14	145	27	325	0.048	0.004	-	-	0.38
MSDZ110	bdl	45	0.6	10.8	18.2	4.34	26	4.3	22	4	10	1	11	1	159	2.476	0.0002	-	-	0.61
MSDZ111	bdl	8	0	0.3	0.7	0.08	7	2.1	26	9	46	11	116	21	247	0.059	0.002	-	-	0.11

Within the LREE suites, the Ce content (2–187 ppm) is dominant. Significantly high Ce content (187 ppm) was obtained in MSDZ055, which has the highest Pr (28.8 ppm), Nd (41.3 ppm), Ti (256 ppm), and Sr (15.99 ppm) contents. The calculated Ce/Ce* anomalies for a few zircons MSDZ003, MSDZ084, and MSDZ089, are 50, 524, and 436, respectively (Table 4). MREE suites show the predominance of Gd contents (3–161 ppm) over those of Nd (≤41.3 ppm), Sm (≤58.7 ppm), and Eu (≤23.50 ppm). The calculated Eu/Eu* (0.03 to <1.33) (Table 4) and normalized plots (Figure 5), mainly show negative anomalies with just a few slightly pronounced positive anomalies (Figure 5a,d,g). The calculated Sm/LaN ratios range from 246 to 1569. Within the HREE suites, Yb contents (2–461 ppm) are generally higher than the contents of Er (3–210 ppm), Dy (4–248 ppm), Lu (1–78 ppm), Ho (1.0–54 ppm), Tm (≤46 ppm), and Tb (≤32.5 ppm). The HREE normalized patterns (Figure 5) generally show an increase from Tb to Lu, except for a few grains (e.g., MSDZ074, MSDZ090, and MSDZ102) whose plots are almost flat. The calculated Gd/Yb and Lu/Hf ratios are <3.0 and <4.1, respectively, with the highest values in MSDZ059 also having the highest Hf/Y ratio (≈274).

4.2. U-Pb Dating

The U-Pb zircon core ages (Table 3, Figures 6 and 7) show Eoarchean to Late Neoproterozoic $^{207}Pb/^{206}Pb$ (3671 ± 23 to 637 ± 24 Ma), $^{207}Pb/^{235}U$ (3428 ± 79 to 615 ± 13 Ma), and $^{206}Pb/^{238}U$ (3038 ± 53 to 612 ± 11 Ma) ages. They are highly heterogeneous, with some different grains having the same age. For plotting and age interpretation purposes, the $^{207}Pb/^{206}Pb$ dates are used for analyses older than 1000 Ma, whereas the $^{206}Pb/^{238}U$ dates are used for analyses younger than 1000 Ma. The $^{207}Pb/^{206}Pb$ data plots for ages greater than 1000 Ma (Figure 7) show the predominance of Middle Paleoproterozoic ages (2050–1993 Ma and 2232–2062 Ma) with the peak at 2130 Ma. Neoarchean age zircons (2797–2531 Ma with the peak at 2700 Ma), in addition to Mesoarchean zircons (3041–2805 Ma) are also abundant. Three grains have ages >3100 Ma (one of Paleo-archean, 3290 Ma, and two of Eo-archean, >3500 Ma). Two grains are of Middle Mesoproterozoic age (>1200 Ma). The $^{206}Pb/^{238}U$ ages (less than 1000 Ma) for younger zircons show the predominance of the Middle Neoproterozoic ages (Cryogenian) (740–643 Ma), with a few Late Neoproterozoic ages (Ediacarian) (612 and 613 Ma).

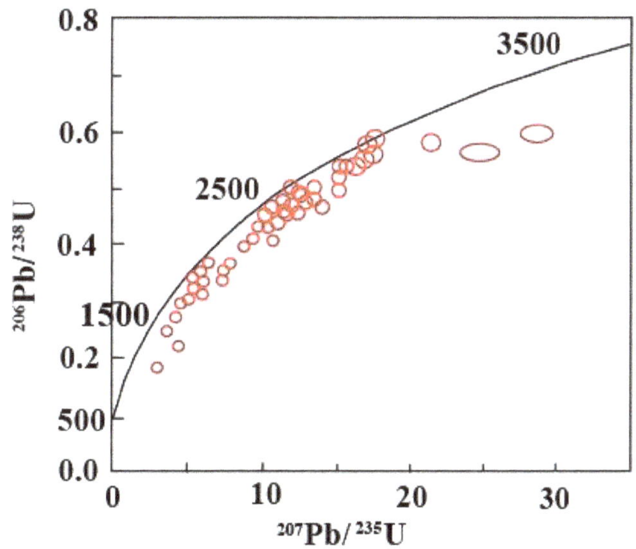

Figure 6. U-Pb discordia diagram for the southern Meiganga detrital zircons.

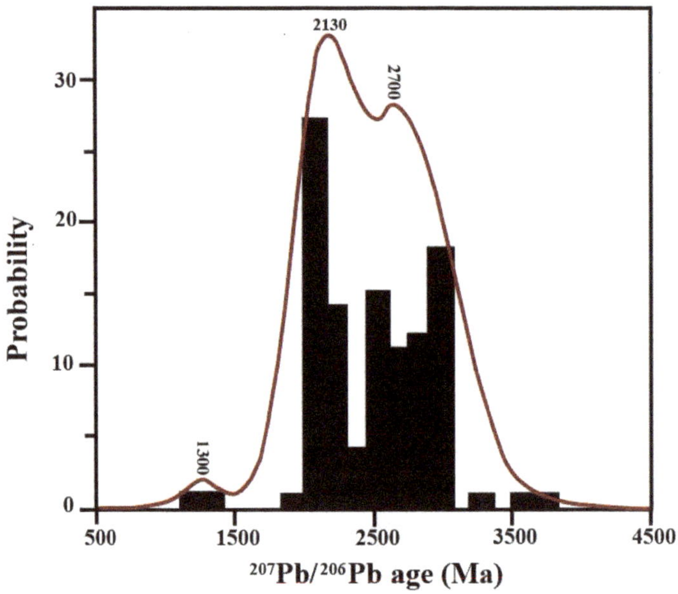

Figure 7. Plot showing the spatial distribution of $^{207}Pb/^{206}Pb$ ages (>1000 Ma) for the southern Meiganga detrital zircons.

5. Discussion

5.1. Zircon Geochemistry, Characterization, Classification, and Environment of Crystallization

The Hf, Y, U, Th, Ti, Nb, Ta, Sr, and REE contents, Th/U ratios, and the Ti-in-zircon temperature are variable, and mainly show a crystallization in different environments. The Th/U ratios distinguish four groups: (1) Th/U < 0.2; (2) Th/U [0.2–0.5]; (3) Th/U [0.5–1.0]; and (4) Th/U > 1.0. They were re-organized into two main groups: (1) igneous affiliated zircons (Th/U ratios $\geq$ 0.2) and (2) metamorphic affiliated zircons (Th/U ratios < 0.2).

5.1.1. Igneous Affiliated Zircons

The trace and rare earth element abundances in the studied igneous affiliated zircons are generally less than those in some zircons found in Cameroon (e.g., [8,9,70]). For example, zircon inclusions in Mayo Kila gem corundum found in the NW region of Cameroon are composed of Hf ($\leq$26,238 ppm), U ($\leq$17,175 ppm), and Th ($\leq$45,584 ppm) [70]. The $\sum$REE contents obtained for detrital zircons occurring with gem corundum in the Mamfe Basin, SW region of Cameroon are up to 1470 ppm [8]. These values are largely greater than those of the studied igneous zircons (Tables 2–4). They can, therefore, be classified as Hf-U-Th-REE-low zircons. Their Hf values are mainly close to those of magmatic zircons found in the western Meiganga gold-bearing placers (cf. [9]), and might show closeness in their crystallization history. The Hf contents in part of the studied zircons are compatible with the values (<11,000 ppm) in zircons crystallized in alkaline magmas [4,71], suggesting a crystallization in alkaline melts. Their plotted data in Figures 4 and 8 show some correlations, as some zircons are plotted together, suggesting a cogenesis and crystallization in the same/similar magma or in different magmas with similar features. This similarity is supported by the closeness of the values of other trace elements, Th/U ratios, and Ti zircon temperatures, and Eu/Eu*.

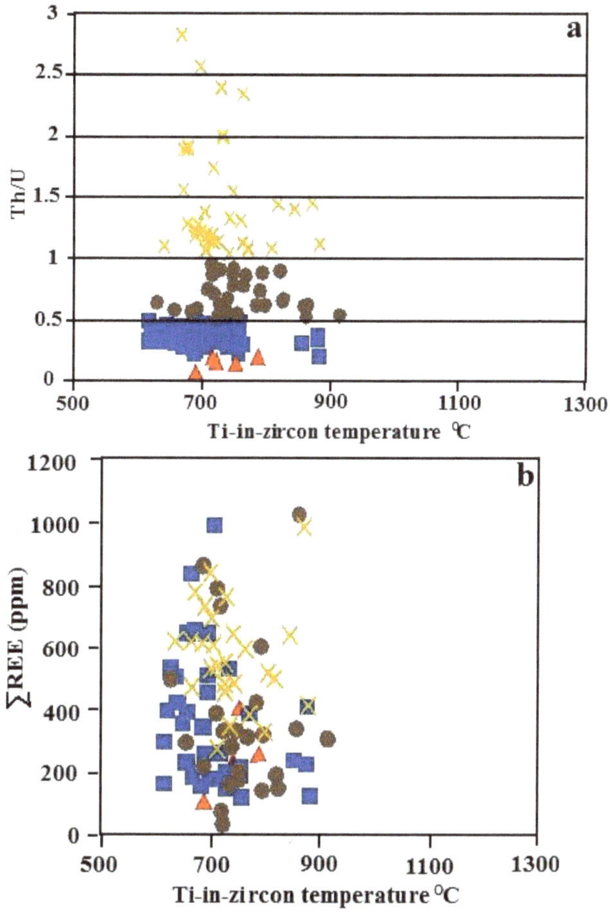

Figure 8. (**a**) Th/U ratio versus Ti-in-zircon temperature (°C) with some grouping of plots, and (**b**) ∑REE (ppm) versus Ti zircon temperature (°C) showing correlations within some zircons from the southern Meiganga gold-bearing placers. Some plots are close while others scattered. Red triangles represent zircon with Th/U < 0.2; blue squares represent zircon with Th/U ratios [0.2 to 0.5]; brown circles represent zircons with Th/U ratios [>0.5 to 1.0]; and stars represent zircons with Th/U > 1.0.

The elemental abundances, Th/U ratios, and Ti-zircon temperatures (617–1180 °C) in the igneous zircons distinguished those with relatively high and relatively low values. Relatively high elemental abundance zircons were probably crystallized in trace elements and REE-enriched melts, with favorable conditions for these elements to substitute Zr in each forming crystal. They are probably crustal-derived zircons, as zircon from crustal rocks generally have elevated contents of some trace elements (notably U, Th, and Y) and REE [4,5]. The relatively high Th (674 ppm), Y (1474 ppm), U (461 ppm), Ti (33 ppm), Nb (25 ppm), and Ta (20 ppm) in MSDZ085, for example, can relate its crystallization in a Y-Th-U-Ti-Nb-Ta-rich magma. The Zr substitution by Nb and Ta during this zircon crystallization was probably governed by Nb, Ta, and REE coupled mechanism (cf. [5]), as this grain also has significant total REE (990 ppm). The relatively high Ti (256 ppm in MSDZ055) may be due to crystallization in Ti-enriched environment with sufficient temperature for Ti to substitute Zr; alternatively, it can be due to Ti-rich mineral inclusion. Relating the high Ti content to a mineral inclusion is difficult, as no inclusion was visualized. The relatively high-elemental zircons are generally from granitoids, as their plots fall essentially in granitoid fields in Figures 9–11. The granitoid origin of those zircons is confirmed by the Y, U, Th, and Yb abundances, largely within the range limit in granitic zircons [4,11,71].

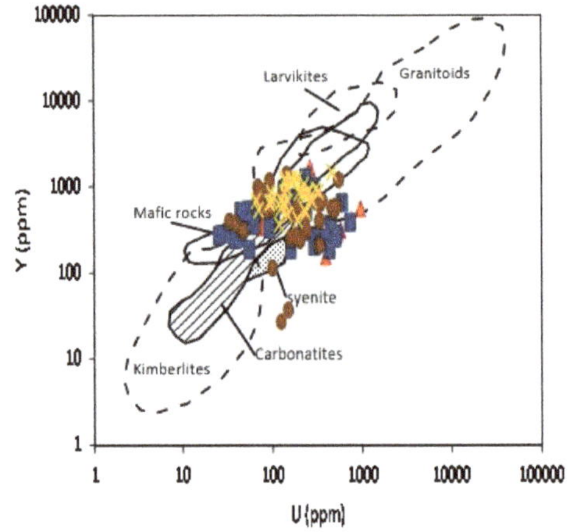

Figure 9. Y versus U plot for southern Meiganga detrital zircons (the lithounit fields are from [4,71]). Red triangles represent zircon with Th/U < 0.2; blue squares represent zircon with Th/U ratios [0.2 to 0.5]; brown circles represent zircons with Th/U ratios [>0.5 to 1.0]; and stars represent zircons with Th/U > 1.0.

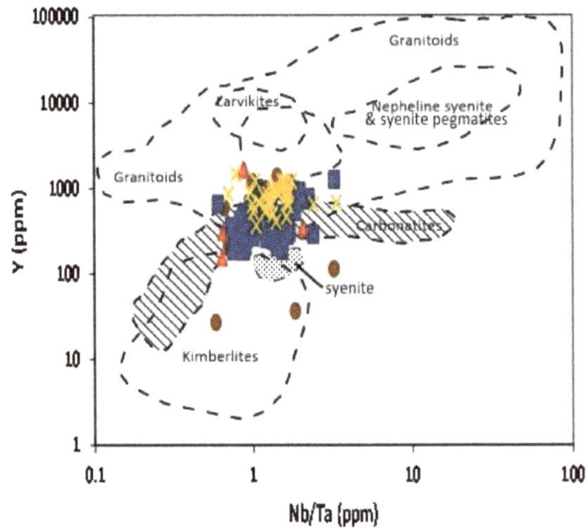

Figure 10. Y versus Nb/Ta plot for southern Meiganga detrital zircons (the lithounit fields are from [4,71]). Red triangles represent zircon with Th/U < 0.2; blue squares represent zircon with Th/U ratios [0.2 to 0.5]; brown circles represent zircons with Th/U ratios [>0.5 to 1.0]; and stars represent zircons with Th/U > 1.0).

Interpretations for very low to low elemental contents in part of the studied zircons can be approached in three ways: (1) elements in those zircon's forming melts are present, but good conditions to ensure that these elements go into their structure are lacking; (2) the depletion or absence of some elements in those zircons' environment of crystallization; or (3) the presence of other accessory minerals (e.g., apatite, xenotime, monazites, allanite, and titanite) [66,72,73] crystallizing in the same melt and competing for REE and other trace elements. The lowest Hf contents in relatively low elemental zircons are within the range limit (4576–6500 ppm) in magmatic zircons found in the western Mamfe corundum gem placers [6–8] and zircon mega-crysts found in alluvial gem corundum deposits associated with alkali basalts (e.g., [74]). These values are also within the range limit (Hf < 9000 ppm) in [10] magmatic zircons crystallized during tectonic rifting. Rifting cannot yet be suggested, as Hf isotopic data are lacking for a detailed interpretation. Zircons from basic and ultrabasic igneous rocks (mantle zircons) are generally depleted in U, Th, Y, and REE [10,75,76]; it is possible that part of the southern Meiganga zircons (e.g., MSDZ016, MSDZ031, MSDZ038, and MSDZ106) were crystallized in mantle source magma(s) as their features, namely U < 30 ppm, Th < 10 ppm, and some plots falling in the mafic rocks field (Figures 9–11), are within the range limit in mantle zircons. The Ti-zircon-temperature (<850 °C) for part of the very low U and Th zircons is less than the temperature (>1300 °C: [77]) for the primary mantle source magma. This temperature difference can complicate the affiliation of part of the very low U and Th zircons to mantle sources. They could be crystals that crystallized at the last stage of cooling mantle source magmas or crystals formed in cooling magmas that originated from the partial fusion of pre-existing mafic rocks. A mafic granulitic origin can be suggested, as part the temperatures are within the range limit (816 ± 12 °C to 798 ± 13 °C) proposed by [78]. Based on the plots of very low to low elemental contents zircons in Figures 9–11, three protosources are distinguished: granitoids, syenites, and mafic rocks.

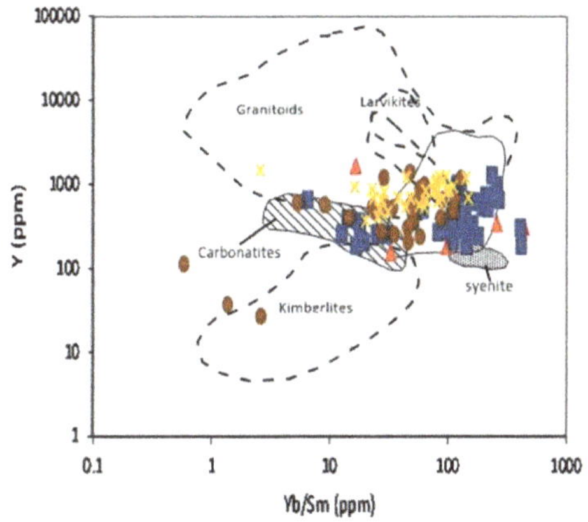

Figure 11. Y versus Yb/Sm plot for southern Meiganga detrital zircons (the lithounit fields are from [4,71]). Red triangles represent zircon with Th/U < 0.2; blue squares represent zircon with Th/U ratios [0.2 to 0.5]; brown circles represent zircons with Th/U ratios [>0.5 to 1.0]; and stars represent zircons with Th/U > 1.0).

5.1.2. Metamorphic Affiliated Zircons

The geochemical features in part of the southern Meiganga detrital zircons are compatible with those of metamorphic zircons grown in equilibrium with garnet (Th/U < 0.07, depletion in REE, Eu/Eu*: 0.24–0.63) (cf. [68]) and crystals grown in equilibrium with an anatectic melt (Th/U < 0.2; relatively trace element-enriched, depleted in MREE, steep REE patterns, positive Ce, and negative Eu anomalies) (cf. [5,9,11,67]). Only one zircon (MSDZ046) with Hf: 9790 ppm, Y: 155 ppm, U: 396 ppm, Th: 26 ppm, Th/U: 0.065, and ∑REE: 98.15 ppm, and Ti temperature: 690 °C, has features close to that of [68] metamorphic zircon grown in subsolidus solution in equilibrium with garnet. This zircon may have crystallized during syn-metamorphic crustal even in low-Th-REE melt. The other zircons (MSDZ015, MSDZ040, MSDZ067, and MSDZ093) have features of zircon grown in equilibrium with anatectic melts, as presented above. The Ti-in-zircon temperatures for these zircons range from 656 ± 60 °C to 778 ± 44 °C, with some values being close to experimental values for granulitic facies metamorphism presented in [78]. Relating their sources to granulitic facies metamorphism is difficult, as some analyses are still needed. MSDZ082, with its positive normalized Pr and Gd plots, is different from the other zircons, as it also has the highest Y (1650 ppm). This grain was plotted in granitoid fields in Figures 9–11, and its other features are close to those of granitoid zircons. It was probably crystallized in an anatectic melt of a granitic composition, as a geochemical feature of a metamorphic zircon grown in equilibrium with anatectic melt does not differ from that of igneous zircons (cf. [5]).

5.2. Detrital Zircon Geochronology and Fingerprinted Magmatic-Metamorphic Events

The recorded U-Pb ages (Table 3, Figures 6 and 7) are mainly heterogeneous with some similarities. The heterogeneity of most of the ages show that they were crystallized at different periods and probably sourced from different protosources and/or source rocks. The crystallization periods of igneous crustal derived zircons, ranging from Eoarchean to Late Neoproterozoic (Figure 7), is composed of three main periods with the following peaks: (1) 1300 Ma of Late Mesoproterozoic; (2) 2130 Ma of Early to Middle Paleoproterozoic; and (3) 2700 Ma of mainly Mesoarchean to Neoarchean. The other magmatic zircon crystallization period is Middle to Late Neoproterozoic (740–612 Ma). This could date four main magmatic episodes linked to crustal fusion; the Mesoarchean to Neoarchean, Early to Middle Paleoproterozoic, Late Mesoproterozoic, and Middle to Late Neoproterozoic events. The obtained group of ages for crustal igneous zircons from the southern Meiganga gold placers show that different magmatic protosources and source rocks provided detritus forming these gold placers. The closeness in ages of some zircons (e.g., MSDZ018: 2801 Ma and MSDZ019: 2802 Ma; MSZD032: 2502 Ma; and MSDZ056: 2503 Ma) may show cogenesis and crystallization at the same time and in the same magma, as their plots overlap (in Figure 12) and fall in the same rock type field (in Figures 9–11).

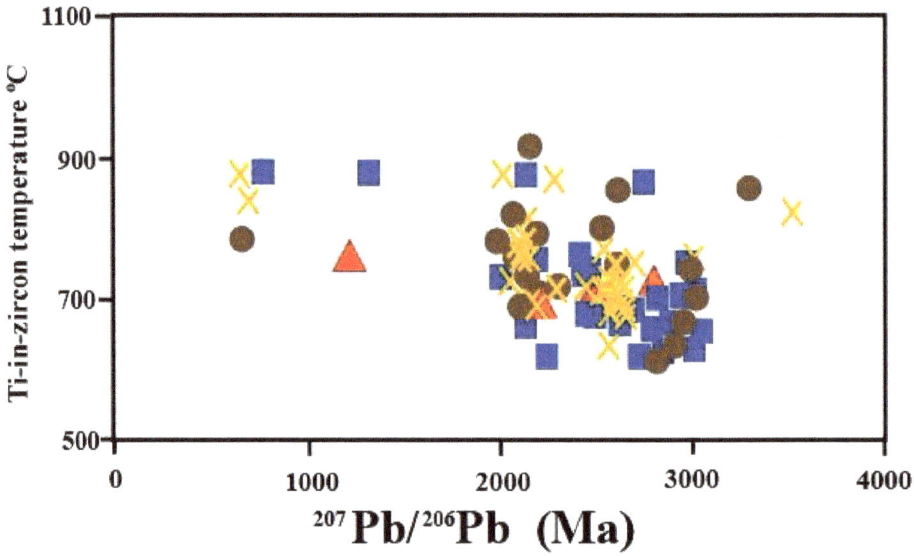

Figure 12. Ti-in-zircon temperature (°C) versus $^{207}Pb/^{206}Pb$ ages (Ma) showing correlations within some zircons from the southern Meiganga gold-bearing placers.

Mafic rock's zircons (MSDZ016: 2999 Ma, MSDZ031: 2797 Ma, MSDZ038: 2795 Ma, and MSDZ106: 2121 Ma), are Early Neoarchean, Late Mesoarchean, and Middle Paleoproterozoic (Rhyacian) mantle source crystals formed probably in magmatic intrusions during crustal opening. The closeness in age between MSDZ031 and MSDZ038 can show that they crystallized at the same time, and probably in the same magma, as their data are plotted together in Figures 2, 4, 7 and 12. Their trace element abundances, and their calculated values are also very close (see Tables 2–4). The Hf contents in this group of zircons (mantle zircons) are all bellow 9000 ppm, and therefore, within the limit proposed by [10] for zircons crystallized during rifting. Rifting and mantle magmatic intrusion cannot be demonstrated easily, as Hf isotopic data are lacking.

The metamorphic zircons $^{207}Pb/^{206}Pb$ ages (2796 Ma, 2559 Ma, 2504 Ma, 2215 Ma, and 2169 Ma) and $^{206}Pb/^{238}U$ age (671 Ma) date three main events: the Neoarchean, Middle Paleoproterozoic, and Middle Neoproterozoic. The Neoarchean and Middle Paleoproterozoic zircons with anatectic melt zircon characteristics, could be grains whose proto-sources underwent metamorphism and partial melting (migmatization). They could be syngenetic zircon crystallized in migrating melts during the Neoarchean and Middle Paleoproterozoic periods. The 671 $\pm$ 12 Ma age of MSDZ015 and its geochemical features are similar to those of zircons grown in equilibrium with garnet, which shows that this syngenetic zircon was crystallized in a garnet-rich rock during Middle Neoproterozoic event, probably the Pan-African orogeny, which affected the Cameroon Mobile Belt. This age is close to those of some Pan-African rocks within the Cameroon Mobile Belt presented in Table 1.

5.3. *Age Correlation, Potential Sources Rocks, and Deposition*

The southern part of Meiganga from where the studied zircons were sampled is mostly made up of undated biotite-amphibole granites; biotite-amphibole gneisses; biotite granites; biotite-chlorite granitic rocks; and few amphibolites and hornfels (see [53]). With a lack of available data dating those rocks, it is difficult to do a local correlation to locate nearby proto-source(s) and source rocks for the southern detrital zircon Meiganga. However, at local and regional scales, the obtained ages are partly similar to those of zircons occurring in the western Meiganga gold-bearing placer presented in [9] and to the ages of some rocks outcropping in the southwest, northeast, and west of Meiganga, and Congo Craton (see Table 1).

Crustal-derived igneous zircons with ages ranging from 3671 to 612 Ma have some age similarities with those of zircons from some igneous and meta-magmatic rocks found in other parts of Meiganga and in the Congo Craton in South Region of Cameroon (Table 5). For these examples, $^{207}Pb/^{206}Pb$ ages (2605 $\pm$ 14 Ma: MSDZ004, 2988 $\pm$ 16 Ma: MSDZ009, and 2877 $\pm$ 17 Ma: MSDZ077) are close to zircon inherited ages (2602.2 Ma, 2987 Ma, and 2884 Ma) for pyroxene-amphibole-bearing gneiss (TTGs composition: [20,32]) found in the SW of Meiganga. The $^{206}Pb/^{238}U$ ages (612 $\pm$ 11 Ma: MSDZ085 and 619 $\pm$ 11 Ma: MSDZ074) are similar to zircon ages (614.1 $\pm$ 3.9 and 619.8 $\pm$ 9.8 Ma: [37]) for meta-diorite outcropping in the NE of Meiganga. The age of MSDZ043 (643 $\pm$ 11 Ma) is close to that of two micas granite (647 $\pm$ 46 Ma: [21]) outcropping in Doua, west of Meiganga. It is not easy in the current geologic setting to consider these rocks to be source rocks of the southern Meiganga crustal-derived magmatic zircons, as those rocks are often found very far from the sampling points of the studied zircons and their host gold bearing placer. They could be detritus from a nearby undated proto-source and source rock or could be polycyclic detritus from the above rocks.

Mesoarchean, as well as Neoarchean ages of crustal derived igneous zircons are often similar to those from rocks (e.g., charnockite, tonalite, granodiorite, syenite, and granite) found in the Ntem complex (Northern Congo Craton), with just a few links with those from rocks (e.g., garnet-bearing gneiss, meta-quartzite, clinopyroxene syenite, and orthopyroxene-garnet gneiss) of the Nyong Unit (Tables 1 and 5). Early and Middle Paleoproterozoic aged zircons are mainly similar to those from rocks (e.g., amphibolite, charnockite, meta-granodiorite, meta-syenite, and orthopyroxene-garnet gneiss) found within the Nyong Unit (Tables 1 and 5). Their presence in the studied area (within the Cameroon Mobile Belt) shows Archean to Paleoproterozoic inheritance, and post-Archean reworking. The Archean to Paleoproterozic igneous zircons inheritance in some metamorphic rocks found at the west of Meiganga was proven by [20] (see Table 1). Plotted in granitoid field

(Figures 9–11), they could be inherited grains from granite and granodiorite proto-sources with features similar to those of granitoids in the Congo Craton. Those zircons whose plots fall out the various discriminating fields could be inherited grains crystallized in charnockitic and tonalitic magmas, as their ages are close to that of charnockite and tonalite found in the Congo Craton. Those old Archean and Paleoproterozoic rocks were probably reworked with the conservation of some inherited zircons, during the two main tectono-magmatic and metamorphic events (the Eburnean and Pan-African) registered within the Cameroon Mobile Belt.

Table 5. Age correlations between the southern Meiganga detrital zircons and other lithounits in the Congo Craton and in Meiganga.

Age of the Southern Meiganga Detrital Zircons	Age Inheritance and Possible Proto-Source(s) within Rocks in the Congo Craton	Age and Possible Source Rocks in Meiganga
Crustal derived igneous zircons	-	-
Mesoarchean (3041–2805 Ma)	3010–2880 Ma (charnockites, tonalities, clinopyroxene syenites, granodiorites, pyroxene bearing gneiss, and orthopyroxene-garnet gneiss)	2984–2884 Ma: Pyroxene-amphibole-bearing gneiss (TTG composition)
Neoarchean (2796–2531 Ma)	2790–2671 Ma (charnockites, tonalities, granodiorites, High-K granites, garnet-bearing gneiss, and magnetite-bearing quartzites)	≤2602 Ma: pyroxene-amphibole-bearing gneiss (TTG composition)
Early Paleoproterozoic (2486–2302 Ma)	2349 and 2300 Ma (orthopyroxene-garnet gneiss and clinopyroxene syenite)	1999–2339 Ma: pyroxene-amphibole-bearing gneiss (TTG composition) and amphibole-biotite gneiss
Middle Paleoproterozoic (2050–1993 Ma and 2232–2062 Ma)	2126–2000 Ma (meta-syenite, orthopyroxene-garnet gneiss, charnockite, amphibolite, metagranodiorites, and magnetite-bearing quartzite)	
Middle to Late Neoproterozoic (740–612 Ma)	-	614–647 Ma: meta-diorite and two micas granite
Mantle derived zircons 1-Middle Mesoarchean (2999) 2-Early Neoarchean (2797 and 2795 Ma) 3-Middle Paleoproterozoic (2121 Ma)	-	1999–2984 Ma: pyroxene-amphibole-bearing gneiss (TTG composition) and amphibole-biotite gneiss
Metamorphic zircons	-	-
1-Zircon grown in equilibrium with garnet (671 Ma)	-	675 Ma: amphibole-biotite gneiss
2-Zircon grown in equilibrium with an anatectic melt *Neoarchean (2796–2504 Ma) *Middle Paleoproterozoic (2215 and 2169 Ma)	-	≤2884 Ma: Pyroxene-amphibole-bearing gneiss (TTG composition) 1999–2339 Ma: Pyroxene-amphibole-bearing gneiss (TTG composition) and amphibole-biotite gneiss

The age (671 ± 12 Ma) of a metamorphic zircon grown in equilibrium with garnet (MSDZ015) is close to the youngest zircon age (675 Ma: [19]) for amphibole-biotite gneiss found in the west of Meiganga. This rock also hosts Early Paleoproterozic age zircons with some similarities to those of the studied zircons (Table 5). Undated amphibole-biotite gneiss cropping in the south of Meiganga (Figure 1) is the bed-rock of the studied zircon-gold bearing placers. If age extrapolation is possible, it can be suggested that amphibole-biotite gneiss found in the south of Meiganga may be the source rock of part of the detritus forming zircon-gold bearing placers. Indirect sources of the placers can also be pyroxene-amphibole-bearing gneiss, meta-diorite, and two micas granites found within the local settings.

The primary sources of host gold grains are difficult to be directly constrained as gold crystals were not found in placer's rock fragments or in the underlying and surrounding rocks. The depositional periods of the studied zircons are also not easy to constrain. The unconsolidated nature of their host-sediments and their location in streams may suggest post-Neoproterozoic to recent deposition.

6. Conclusions

The southern Meiganga detrital zircon-gold bearing placers are composed of igneous (crustal derived and mantle origin) and metamorphic zircons (grown in equilibrium with garnet and those grown in equilibrium with an anatectic melt) with different histories of crystallization and from mainly different sources.

Crustal derived igneous zircons were crystallized in granitic magmas with some charnockitic and tonalitic affinities during Eoarchean to Late Neoproterozoic periods. Mantle igneous zircons were crystallized from mantle source magmas during Early Neoarchean to Middle Paleoproterozoic times.

The inherited igneous zircons of Mesoarchean to Middle Paleoproterozoic were probably sorted from pyroxene-amphibole-bearing and amphibole-biotite gneiss, with their features similar to those of rocks in the Congo Craton. Late Neoproterozoic zircons, with ages close to those of meta-diorite and two mica granite found in the NE and west of Meiganga, were probably eroded from unidentified nearby rocks formed in the same periods.

Metamorphic zircons grown in equilibrium with garnet were crystallized in low-Th-REE subsolidus solution during the Pan-African syn-metamorphic crustal event. Metamorphic zircons grown in equilibrium with an anatectic melt were probably crystallized during the Neoarchean and Middle Paleoproterozoic in migrated melts from partial fusion of metamorphic protoliths. These inherited zircons were probably sourced from amphibole-biotite gneiss underlying the zircon-gold-bearing placers.

Author Contributions: Conceptualization, N.S.K.; Data curation, A.P.K.; Funding acquisition, N.S.K.; Methodology, P.R.S.; Project administration, N.S.K.; Software, E.A.A.B.; Validation, G.N.; Writing—review and editing, N.S.K. and A.S.V. All authors have read and agreed to the published version of the manuscript.

Funding: The research was partly funded by Professor John Cottle of the Department of Earth Sciences, and the Earth Research Institute, University of California, Santa Barbara, CA, USA, who financed the laboratory analyses.

Acknowledgments: The authors thank John Cottle of the Department of Earth Sciences, and the Earth Research Institute, University of California, Santa Barbara, CA, USA, for financial support and laboratory facilities. We also thank the personnel at the Department Earth Sciences at this same University, as they analyzed the zircons in this study. Our gratitude to Elena Belousova, Tom Andersen, and David Richard Lentz, whose reviews and comments have helped to improve the original manuscript. The authors also extend their gratitude to the two anonymous reviewers whose detailed and constructive comments helped to improve the manuscript.

Conflicts of Interest: The authors declare no conflict of interest.

References

1. Djou, E.S. Prospection des Indices de Minéralisation d'or Dans la Zone de Méiganga-Sud (Nord-Est Cameroun). Master's Thesis, University Yaoundé I, Yaoundé, Cameroon, 2015.
2. Kenna, H.S. Prospection des Indices de Minéralisations d'or Dans la Zone de Méiganga Ouest. Master's Thesis, University Yaoundé I, Yaoundé, Cameroon, 2015.
3. Morton, C.A.; Claoue-Long, C.J.; Hallsworth, C.R. Zircon age and heavy mineral constraints of North Sea Carboniferous sandstones. *Mar. Pet. Geol.* **2001**, *18*, 319–337. [CrossRef]
4. Belousova, A.E.; Griffin, L.W.; O'Reilly, Y.S.; Fisher, I.N. Igneous zircon: Trace element composition as an indicator of source rock type. *J. Miner. Petrol.* **2002**, *143*, 602–622. [CrossRef]
5. Hoskin, P.W.O.; Schaltegger, U. The composition of zircon and igneous and metamorphic petrogenesis. In *Reviews in Mineralogy and Geochemistry*; Mineralogical Society of America: Washington, DC, USA, 2003; Volume 53, pp. 27–55.

6. Kanouo, S.N.; Zaw, K.; Yongue, F.R.; Sutherland, L.F.; Meffre, S.; Njonfang, E.; Ma, C.; Tchouatcha, S.T. U-Pb zircon age constraining the source and provenance of gem-bearing Late Cenozoic detrital deposit, Mamfe Basin, SW Cameroon. *Resour. Geol.* **2012**, *62*, 316–324. [CrossRef]
7. Kanouo, S.N.; Yongue, F.R.; Ekomane, E.; Njonfang, E.; Ma, C.; Lentz, D.R.; She, Z.; Zaw, K.; Venkatesh, A.S. U-Pb ages for zircon grains from Nsanaragati Alluvial Gem Placers: Its correlation to the source rocks. *Resour. Geol.* **2015**, *65*, 103–121. [CrossRef]
8. Kanouo, S.N.; Ekomane, E.; Yongue, F.R.; Njonfang, E.; Zaw, K.; Ma, C.; Ghogomu, R.T.; Lentz, D.R.; Venkatesh, A.S. Trace elements in corundum, chrysoberyl, and zircon: Application to mineral exploration and provenance study of the western Mamfe gem clastic deposits (SW Cameroon, Central Africa). *J. Afr. Earth Sci.* **2016**, *113*, 35–50. [CrossRef]
9. Kanouo, S.N.; Ngueutchoua, G.; Kouske, P.A.; Yongue, F.R.; Venkatesh, A.S. Trace element and U-Pb core age for zircons from western Meiganga gold placer, Cameroon: Their genesis and Archean-Proterozoic sources. *Minerals* **2018**, *8*, 194. [CrossRef]
10. Heaman, L.M.; Bowins, R.; Crocket, J. The chemical composition of igneous zircon suites: Implications for geochemical tracer studies. *Geochem. Cosmochem. Acta.* **1990**, *54*, 1597–1607. [CrossRef]
11. Kanouo, S.N.; Njonfang, E.; Kouské, P.A.; Yongue, F.R.; Ngueutchoua, G. U-Pb zircon age: Preliminary data evaluating the Earth history recorded by two basement rocks (granitic pegmatite and mica-schist) in Mamfe Basin (SW Cameroon, Central Africa). *J. Geol. Geophys.* **2017**, *6*, 1–9. [CrossRef]
12. Cawood, A.P.; Nemchin, A.A. Provenance record of a rift basin: U/Pb ages of detrital zircons from the Perth Basin, Western Australia. *J. Sed. Geol.* **2000**, *134*, 209–234. [CrossRef]
13. Zeh, A.; Gerdes, A.; Klemd, R.; Barton-Jr, M.J. U-Pb and Lu-Hf isotope record of detrital zircon grains from the Limpopo Belt-Evidence for crustal recycling at the Hadean to early-Archean transition. *Geochem. Cosmochem. Acta* **2008**, *72*, 5304–5329. [CrossRef]
14. Gehrels, G. Detrital zircon U-Pb geochronology applied to tectonics. *Annu. Rev. Earth Planet. Sci.* **2014**, *42*, 127–149. [CrossRef]
15. Andersen, T.; Kristoffersen, M.; Elburg, A.M. How far can we trust provenance and crustal evolution information from detrital zircons? A South African case study. *Gondwana Res.* **2016**, *34*, 129–148. [CrossRef]
16. Ganwa, A.A. Les Granitoïdes de Méiganga: Étude Pétrographique, Géochimique, Structurale et Géochronologique: Leur Place dans la Chaîne Panafricaine. Ph.D. Thesis, University Yaoundé I, Yaoundé, Cameroon, 2005.
17. Abbelsalam, G.M.; Liégeois, P.J.; Stern, J.R. The Saharan Metacraton. *J. Afr. Earth Sci.* **2002**, *34*, 119–136. [CrossRef]
18. Ngako, V.; Njonfang, E.; Affaton, P. Pan-African tectonics in northwestern Cameroon: Implication for the history of the western Gondwana. *Gondwana Res.* **2008**, *14*, 509–522. [CrossRef]
19. Ganwa, A.A.; Siebel, W.; Shang, K.C.; Naimou, S.; Ekodeck, G.E. New constraints from Pb-evaporation zircon ages of the Méiganga amphibole-biotite gneiss, Central Cameroon, on Proterozoic crustal evolution. *Int. J. Geosci.* **2011**, *2*, 138–147.
20. Ganwa, A.A.; Klotzli, S.U.; Hauzenberger, C. Evidence for Archean inheritance in the pre-Pan-African crust of Central Cameroon: Insight from zircon internal structure and LA-MC-ICP-MS U-Pb ages. *J. Afr. Earth Sci.* **2016**, *120*, 12–22. [CrossRef]
21. Kepnamou, D.A.; Ganwa, A.A.; Klötzli, S.U.; Hauzenberger, C.; Ngounouno, I.; Naïmou, S. The Pan-African biotite-muscovite granite and amphibole-biotite granite of Doua (Central Cameroon): Zircon features, LA-MC-ICP-MS U-Pb dating and implication on their tectonic setting. *J. Geosci. Geom.* **2017**, *5*, 119–129.
22. Nédélec, A.; Minyem, D.; Barbey, P. High-P-high-T anatexis of Archean tonalitic grey gneisses: The Eseka migmatites, Cameroon. *Precambrian Res.* **1993**, *62*, 191–205. [CrossRef]
23. Penaye, J.; Toteu, S.F.; Tchameni, R.; Van Schmus, W.R.; Tchakounte, J.; Ganwa, A.; Minyem, D.; Nsifa, E.N. The 2.1 Ga West Central African Belt in Cameroon: Extension and evolution. *J. Afr. Earth Sci.* **2004**, *39*, 159–164. [CrossRef]
24. Maurizot, P.; Abessolo, A.; Feybesse, A.; Johan, J.L.; Lecompte, P. *Etude et prospection minière au Sud Ouest Cameroun. Synthèse des travaux de 1978–1985*; Technical Report for BRGM; BRGM: Orléans, France, 1986; CMR 066; p. 274.
25. Tchameni, R.; Mezger, K.; Nsifa, E.; Pouclet, A. Neoarchaean evolution in the Congo Craton: Evidence from K-rich granitoids of the Ntem Complex, Southern Cameroon. *J. Afr. Earth Sci.* **2000**, *30*, 133–147. [CrossRef]
26. Pouclet, A.; Tchameni, R.; Mezger, K.; Vidal, M.; Nsifa, E.; Shang, C.; Penaye, J. Archean crustal accretion at the northern border of the Congo Craton (South Cameroon). The charnockite-TTG link. *Bull. Soc.* **2007**, *5*, 331–342.
27. Nsifa, E.N.; Tchameni, R.; Belinga, S.M.E. *De l'existence de FORMATION Catarchéennes Dans le Complexe Cratonique du Ntem (Sud-Cameroun). Project No. 273, Archean Cratonic Rocks of Africa*; 1993; Volume 23.
28. Tchameni, R.; Pouclet, A.; Mezger, K.; Nsifa, N.E.; Vicat, J.P. Monozircon and Sm-Nd whole rock ages from the Ebolowa greenstone belts: Evidence for the terranes older than 2.9 Ga in the Ntem Complex (Congo craton, South Cameroon). *J. Cameroon Acad. Sci.* **2004**, *4*, 213–224.
29. Chombong, N.N.; Suh, C.E.; Ilouga, C.D.C. New detrital zircon U-Pb ages from BIF-related metasediments in the Ntem Complex (Congo craton) of southern Cameroon, West Africa. *Nat. Sci.* **2013**, *5*, 835–847. [CrossRef]
30. Shang, C.K.; Satir, M.; Siebel, W.; Taubald, H.; Nsifa, E.N.; Westphal, M.; Reitter, E. Genesis of K-rich granitoids in the Sangmelima region, Ntem Complex (Congo craton), Cameroon. *Terra Nostra* **2001**, *5*, 60–63.
31. Shang, K.C.; Liégeois, P.J.; Satir, M.; Frisch, W.; Nsifa, E.N. Late Archaean high-K granite geochronology of the northern metacratonic margin of the Archaean Congo craton, Southern Cameroon: Evidence for Pb-loss due to non-metamorphic causes. *Gondwana Res.* **2010**, *18*, 337–355. [CrossRef]
32. Toteu, S.F.; Van Schmus, W.R.; Penaye, J.; Nyobé, J.B. U-Pb and Sm-Nd evidence for Eburnean and Pan-African high-grade metamorphism in cratonic rocks of southern Cameroon. *Precambrian Res.* **1994**, *108*, 45–73. [CrossRef]

33. Vicat, J.P.; Leger, J.M.; Nsifa, N.E.; Piguet, P.; Nzenti, J.P.; Tchameni, R.; Pouclet, A. Distinction au sein du craton congolais du Sud-Ouest du Cameroun, de deux épisodes doléritiques initiant les cycless orogéniques éburnéen (Paléoprotérozoïque) et panafricain (Néoprotérozoïque). *C. R. Acad. Sci. Paris* **1996**, *323*, 575–582.
34. Shang, K.C.; Satir, M.; Nsifa, N.E.; Liégeois, P.J.; Siebel, W.; Taubald, H. Archaean high-K granitoids produced by remelting of earlier Tonalite–Trondhjemite–Granodiorite (TTG) in the Sangmelima region of the Ntem complex of the Congo craton, southern Cameroon. *Int. J. Earth Sci.* **2007**, *96*, 817–841. [CrossRef]
35. Lasserre, M.; Soba, D. Age Libérien des granodiorites et des gneiss à pyroxènes du Cameroun Méridional. Bull. *BRGM* **1976**, *2*, 17–32.
36. Toteu, S.F.; Penaye, J.; Michard, A. New U-Pb and Sm-Nd data from North-Central Cameroon and its bearing on the pre-Pan-African history of central Africa. *Precambrian Res.* **2001**, *108*, 45–73. [CrossRef]
37. Lerouge, C.; Cocherie, A.; Toteu, F.S.; Penaye, J.; Mile, P.J.; Tchameni, R.; Nsifa, N.E.; Fanning, M.C.; Deloule, E. Shrimp U-Pb zircon age evidence for Paleoproterozoic sedimentation and 2.05 Ga syntectonic plutonism in the Nyong Group, South-Western Cameroon: Consequences for the Eburnean-Transamazonian belt of NE Brazil and Central Africa. *J. Afr. Earth Sci.* **2006**, *44*, 412–427. [CrossRef]
38. Owona, S. Archean-Eburnean and Pan-African Features and Relationships in Their Junction Zone in the South of Yaoundé (Cameroon). Ph.D. Thesis, University Douala, Douala, Cameroon, 2008.
39. Delhal, J.; Ledent, D. *Données Géochronologiques sur le Complexe Calco-Magnésien du Sud-Cameroun*; Technical Report for Department of Geology and Mineralogy of the Royal Museum for Central Africa: Tervuren, Belgium, 1975; pp. 71–76.
40. Tetsopgang, S.; Suzuki, K.; Adachi, M. Preliminary CHIME dating of granites from Nkambe area, northwestern Cameroon. *J. Earth Planet. Sci. Nagoya Univ.* **1999**, *46*, 57–70.
41. Tetsopgang, S.; Suzuki, K.; Njonfang, E. Petrology and CHIME geochronology of Pan-African high K and Sr/Y granitoids in the Nkambe area, Cameroon. *Gondwana Res.* **2008**, *14*, 686–699. [CrossRef]
42. Tagne-Kamga, G. Petrogenesis of the Neoproterozoic Ngondo igneous plutonic complex (Cameroon, west central Africa): A case of late collisional ferro-potassic magmatism. *J. Afr. Earth Sci.* **2003**, *36*, 149–171. [CrossRef]
43. Njiosseu, T.L.E.; Nzenti, P.J.; Njanko, T.; Kapajika, B.; Nédélec, A. New U-Pb zircon ages from Tonga (Cameroon): Coexisting Eburnian-Transmazonian (2.1 Ga) and Pan-African (0.6 Ga) imprints. *C. R. Geosci.* **2005**, *337*, 551–562. [CrossRef]
44. Ganwa, A.A.; Frisch, W.; Siebel, W.; Ekodeck, E.G.; Shang, K.C.; Ngako, V. Archean inheritances in the pyroxene-amphibole-bearing gneiss of the Méiganga area (Central North Cameroon): Geochemical and $^{207}Pb/^{206}Pb$ age imprints. *C. R. Geosci.* **2008**, *340*, 211–222. [CrossRef]
45. Nzenti, J.P.; Barbey, P.; Macaudiere, J.; Soba, D. Origin and evolution of the Late Precambrian high-grade Yaoundé gneisses (Cameroon). *Precambrian Res.* **1988**, *38*, 91–109. [CrossRef]
46. Tchakounté, N.J.; Toteu, F.S.; Van Schmus, R.W.; Penaye, J.; Deloule, E.; Ondoua, M.J.; Houketchang, B.M.; Ganwa, A.A.; White, M.W. Evidence of ca 1.6 Ga detrital zircon in Bafia group (Cameroon): Implication for chronostratigraphy of the Pan-African Belt north of the Congo craton. *C. R. Geosci.* **2007**, *339*, 132–142. [CrossRef]
47. Ganwa, A.A.; Frisch, W.; Siebel, W.; Shang, C.K. Geochemistry of magmatic rocks and time constraints on deformational phases and shear zone slip in the Méiganga area, Central Cameroon. *Int. Geol. Rev.* **2011**, *53*, 759–784. [CrossRef]
48. Ganwa, A.A.; Klötzli, S.U.; Kepnamou, D.A.; Hauzenberger, C. Multiple Ediacaran tectono-metamorphic events in the Adamawa-Yadé Domain of the Central Africa Fold Belt: Insight from the zircon U-Pb LAM-ICP-MS geochronology of the metadiorite of Meiganga (Central Cameroon). *Geol. J.* **2018**, 1–14. [CrossRef]
49. Shang, C.K.; Siebel, W.; Satir, M.; Chen, F.; Mvondo, J.O. Zircon Pb-Pb and U-Pb systematics of TTG rocks in the Congo craton: Constraints of crustal formation, crystallization and Pan-African lead loss. *Bull. Geosci.* **2004**, *79*, 205–219.
50. Kornprobst, J.; Cantagrel, J.M.; Fabries, J.; Lasserre, M.; Rollet, M.; Soba, D. Existence in Cameroon of a Pan-African or older alkaline magmatism, the nepheline syenite with mboziite of Nkonglong; comparison with the known alkaline rocks of the same region. *Bull. Soci. Geol. France* **1976**, *18*, 1295–1305. [CrossRef]
51. Tchameni, R.; Mezger, K.; Nsifa, N.E.; Pouclet, A. Crustal origin of Early Proterozoic syenites in the Congo craton (Ntem complex), South Cameroon. *Lithos* **2001**, *57*, 23–42. [CrossRef]
52. Toteu, S.F.; Penaye, J.; Poudjom Djomani, Y.H. Geodynamic evolution of the Pan-African belt in Central Africa with special reference to Cameroon. *Can. J. Earth Sci.* **2004**, *41*, 73–85. [CrossRef]
53. Lassere, M. Carte Géologique de Reconnaissance à l'échelle du 1/500 000 Territoire du Cameroun, Ngaoundéré Est. Dir. Mines Géol. Cameroun, Yaoundé, 1 carte et notice. 1961.
54. Parfenoff, A.; Pomerol, C.; Tourenq, J. *Les Minéraux en Grains. Méthodes D'étude et Détermination*; Masson et Cie Edition: Paris, France, 1970.
55. Kanouo, S.N. Geology of the Western Mamfe Corundum Deposits, SW Region Cameroon: Petrography, Geochemistry, Geochronology, Genesis, and Origin. Ph.D. Thesis, University Yaoundé I, Yaoundé, Cameroon, 2014.
56. Cottle, J.M.; Waters, D.J.; Riley, D. Metamorphic history of the South Tibetan Detachment System, Mt. Everest region, revealed by RSCM thermometry and phase equilibria modelling. *J. Metamorph. Geol.* **2011**, *29*, 561–582. [CrossRef]
57. Cottle, J.M.; Burrows, A.J.; Kylander-Clark, A.; Freedman, P.A.; Cohen, R. Enhanced sensitivity in laser ablation multi-collector inductively coupled plasma mass spectrometry. *J. Anal. At. Spectrom.* **2013**, *28*, 1700–1706. [CrossRef]

58. Kylander-Clark, A.R.C.; Hacker, B.R.; Cottle, J.M. Laser-ablation split-stream ICP petrochronology. *Chem. Geol.* **2013**, *345*, 99–112. [CrossRef]
59. Paton, C.; Woodhead, J.; Hellstrom, J.; Hergt, J.; Greig, A.; Maas, R. Improved laser ablation U-Pb zircon geochronology through robust down hole fractionation correction. *Geochem. Geophys. Geosyst.* **2010**, *11*, 1–36. [CrossRef]
60. Wiedenbeck, M.; Alle, P.; Corfu, F.; Grin, W.; Meier, M.; Oberli, F.; Von Quadt, A.; Roddick, J.; Spiegel, W. Three natural zircon standards for U-Th-Pb, Lu-Hf, trace element and REE analyses. *Geostand. Newslett.* **1995**, *19*, 1–23. [CrossRef]
61. Jackson, S.; Pearson, N.; Griffin, W.; Belousova, E. The application of laser ablation-inductively coupled plasma-mass spectrometry to in situ U/Pb zircon geochronology. *Chem. Geol.* **2004**, *657*, 47–69. [CrossRef]
62. Ludwig, K. *User's Manual for Isoplot 2.4: A Geochronological Toolkit for Microsoft Excel*; Berkeley Geochronology Center: Berkeley, CA, USA, 2012.
63. Vermeesch, P. On the visualisation of detrital age distributions. *Chem. Geol.* **2012**, *312–313*, 190–194. [CrossRef]
64. Steiger, R.; Jager, E. Subcommission on geochronology: Convention on the use of decay constants in geo and cosmochronology. *Earth Planet. Sci. Lett.* **1977**, *36*, 359–362. [CrossRef]
65. Konzett, J.; Armstrong, R.A.; Sweeny, R.J.; Compston, W. The timing of Marid suite metasomatism in the Kaapvaal mantle: An ion probe study of zircons from Marid xenoliths. *Earth Planet. Sci. Lett.* **1998**, *160*, 133–145. [CrossRef]
66. Kirkland, C.L.; Smithies, R.H.; Taylor, R.J.M.; Evans, N.; McDonald, B. Zircon Th/U ratios in magmatic environs. *Lithos* **2015**, *212–215*, 397–414. [CrossRef]
67. Hiess, J.; Nutman, A.P.; Bennett, V.C.; Holden, P. Ti-in-zircon thermometry applied to contrasting Archean metamorphic and igneous systems. *Chem. Geol.* **2008**, *247*, 323–338. [CrossRef]
68. Rubatto, D. Zircon trace element geochemistry: Partitioning with garnet and link between U-Pb ages and metamorphism. *Chem. Geol.* **2002**, *184*, 123–138. [CrossRef]
69. McDonough, F.W.; Sun, S.S. The composition of the Earth. *Chem. Geol.* **1995**, *120*, 223–253. [CrossRef]
70. Mbih, P.K.; Sebastien Meffre, S.; Yongue, F.R.; Kanouo, S.N.; Jay, T. Chemistry and origin of the Mayo Kila sapphires, NW region Cameroon (Central Africa): Their possible relationship with the Cameroon volcanic line. *J. Afr. Earth Sci.* **2016**, *118*, 263–273. [CrossRef]
71. Veevers, J.J.; Belousova, A.E.; Saeed, A.; Sircombe, K.; Cooper, F.A.; Read, E.S. Pan-Gondwanaland detrital zircons from Australia analyzed for Hf-isotopes and trace elements reflect an ice-covered Antarctic provenance of 700–500 Ma age, TDM of 2.0–1.0 Ga, and alkaline affinity. *Earth Sci. Rev.* **2006**, *76*, 135–174. [CrossRef]
72. Gromet, P.L.; Silver, T.L. Rare earth element distributions among minerals in a granodiorite and their petrogenetic implications. *Geochem. Cosmochem. Acta.* **1983**, *47*, 925–939. [CrossRef]
73. Pettke, T.; Audétat, A.; Schaltegger, U.; Heinrich, C.A. Magmatic-to-hydrothermal crystallization in the W-Sn mineralized Mole Granite (NSW, Australia): Part II: Evolving zircon and thorite trace element chemistry. *Chem. Geol.* **2005**, *220*, 191–213. [CrossRef]
74. Abduriyim, A.; Sutherland, L.F.; Belousova, A.E. U-Pb age and origin of gem zircon from the New England sapphire fields, New South Wales, Australia. *Aust. J. Earth Sci.* **2012**, *59*, 1067–1081. [CrossRef]
75. Belousova, A.E.; Griffin, L.W.; Pearson, J.N. Trace element composition and cathodoluminesence properties of southern African kimberlitic zircons. *Mineral. Mag.* **1998**, *62*, 355–366. [CrossRef]
76. Hoskin, P.W.O.; Black, L.P. Metamorphic zircon formation by solid-state recrystallization of protolith ignous zircon. *J. Met. Petrol.* **2000**, *18*, 423–439. [CrossRef]
77. Green, H.D.; Falloon, J.T.; Eggins, M.S.; Yaxley, M.G. Primary magmas and mantle temperatures. *Eur. J. Mineral.* **2001**, *13*, 437–451. [CrossRef]
78. Pattison, M.R.D.; Chacko, T.; Farquhar, J.; Mcfarlane, M.R.C. Temperatures of granulite-facies metamorphism: Constraints from experimental phase equilibria and thermobarometry corrected for retrograde exchange. *J. Petrol.* **2003**, *44*, 867–900. [CrossRef]

minerals

MDPI

Article

Tracking the Detrital Zircon Provenance of Early Miocene Sediments in the Continental Shelf of the Northwestern South China Sea

Ce Wang [1,2,3], Letian Zeng [1,2,3], Yaping Lei [1,2,3], Ming Su [1,2,3,*] and Xinquan Liang [4]

1 School of Marine Sciences, Sun Yat-sen University, Zhuhai 519082, China; wangce@mail.sysu.edu.cn (C.W.); zenglt5@mail2.sysu.edu.cn (L.Z.); eeslyp@mail.sysu.edu.cn (Y.L.)
2 Southern Marine Science and Engineering Guangdong Laboratory (Zhuhai), Zhuhai 519080, China
3 Guangdong Provincial Key Laboratory of Marine Resources and Coastal Engineering, Guangzhou 510006, China
4 State Key Laboratory of Isotope Geochemistry, Guangzhou Institute of Geochemistry, Chinese Academy of Sciences, Guangzhou 510640, China; liangxq@gig.ac.cn
* Correspondence: suming3@mail.sysu.edu.cn

Received: 7 July 2020; Accepted: 22 August 2020; Published: 25 August 2020

Abstract: Sediment provenance studies have become a major theme for source-to-sink systems and provide an important tool for assessing paleogeographic reconstruction, characterizing the depositional system, and predicting reservoir quality. The lower Miocene is an important stratigraphic unit for deciphering sediment evolution in the continental shelf of the northwestern South China Sea, but the provenance characteristics of this strata remain unclear. In this study, detrital zircon U-Pb geochronology and Lu-Hf isotopes from the lower Miocene Sanya Formation in the Yinggehai-Song Hong Basin were examined to study the provenance and its variation in the early Miocene. U-Pb dating of detrital zircons yielded ages ranging from Archean to Cenozoic (3313 to 39 Ma) and displayed age distributions with multiple peaks and a wide range of $\varepsilon_{Hf}(t)$ values (from −27.2 to +8.5). Multi-proxy sediment provenance analysis indicates that the Red River system was the major source for the sediments in the northern basin, with additional contribution from central Vietnam, and the Hainan played the most important role in contributing detritus to the eastern margin of the basin in the middle Miocene. This paper highlights the provenance of early Miocene sediments and contributes to paleogeographic reconstruction and reservoir evaluation.

Keywords: provenance; detrital zircon; Lu-Hf isotopes; early Miocene; Yinggehai-Song Hong Basin; South China Sea

1. Introduction

Tracing sediment delivery from terrigenous sources to oceanic sinks is crucial for understanding the sedimentary process at ocean margins. Clastic sedimentary rocks contain key information relating to their provenance and the sedimentary process and can therefore be used to reveal source-to-sink systems during deposition [1–3]. The provenance study of sediments provides an important tool for assessing paleogeographic reconstruction, characterizing the depositional system, and predicting reservoir quality [4,5]. Sediment provenance studies have become a major theme for source-to-sink studies and have promoted efforts to link terrestrial and marine segments via the sediments produced, transported, and deposited within systems [6–8]. Over the last decades, geochronology applied to detrital minerals, such as zircon, apatite, monazite, rutile, and titanite, have received increased attention for their ability to provide insight into provenance, erosion, and tectonic processes [9,10].

Zircon has received considerable attention for its stability and has many uses within earth science systems, as zircon can withstand the effects of weathering, erosion, and thermal alteration during transportation, deposition, and diagenesis. U-Pb geochronology of detrital zircon is a robust tool that has been widely used to evaluate sediment provenance, tectonics, climate, and source-to-sink processes in margins [11]. In addition, studies of Lu-Hf isotopic compositions in detrital zircon are another distinctive feature of zircon other than U-Pb age. Their use can make source identification more robust [12].

The Yinggehai-Song Hong (Y-SH) Basin, which is located on the continental shelf of the northwestern South China Sea (Figure 1) and was formed at the southeastern termination of the Red River Fault in the Cenozoic era, is one of the most gas-rich sedimentary basins in this region. The marine sediments in this region provide an important case for understanding the provenance and transport processes from terrigenous source to oceanic sink. Over the last decade, diverse methods have been used to study the provenance of sediments, ranging from detrital minerals, geochemistry, seismic data, and isotopes to geochronology in the basin [13–18]. The provenance studies have been conducted using an integrated approach, which allows us to quantitatively assess the contributions of different potential source terranes of the Red River, central Vietnam, and Hainan Island (Figure 1). Most provenance studies in the northwestern shelf of the South China Sea have been focused on the middle Miocene to Quaternary sediments; by contrast, only a few studies have examined the lower Miocene and underlying strata. The lower Miocene Sanya Formation is an important potential reservoir in the Y-SH Basin [19]. The understanding of sediment provenance of Sanya Formation and its transport pathway is a key point to reconstruct the paleogeography and establish the reservoir presence in clastic petroleum systems of the basin. However, up to now, the provenance of this stratigraphic unit was not well investigated, and a more effective method should be conducted to reveal the provenance of sediments.

The amount of zircon U-Pb geochronological data around the northwestern margin of the South China Sea has substantially increased over the last decades [17,20–22]. For instance, Hoang et al. (2009) [23] used the U-Pb dating and Hf isotopic analysis of detrital zircons from Red River sands in order to constrain the provenance of the modern river and to decipher drainage evolution through time. U-Pb dating of detrital zircon along the main drainage systems in the western Hainan was carried out by Wang et al. (2015) [24] with aims to reveal the provenance and tectonic evolution of Hainan Island. Jonell et al. (2017) [25] used multiple geochemical datasets, including the detrital zircon U-Pb geochronology from the Gianh River in the central Vietnam, to understand the sediment transport and erosion patterns of the eastern Indochina Block. Wang et al. (2018) [20] and Fyhn et al. (2019) [17] investigated the detrital zircon geochronology of each source terranes around the Y-SH Basin and identified the differences in the age composition of detrital zircons. This abundance of comparative data for each source terrane makes the detrital zircon method especially suitable for provenance study in the region. Here, a new set of detrital zircon U-Pb dating and Lu-Hf isotopes from the lower Miocene Sanyan Formation in the Y-SH Basin was used to study the sediment provenance of early Miocene sediments, elucidate the spatial and temporal variation in provenance, and understand the source-to-sink process in the northwestern margin of the South China Sea.

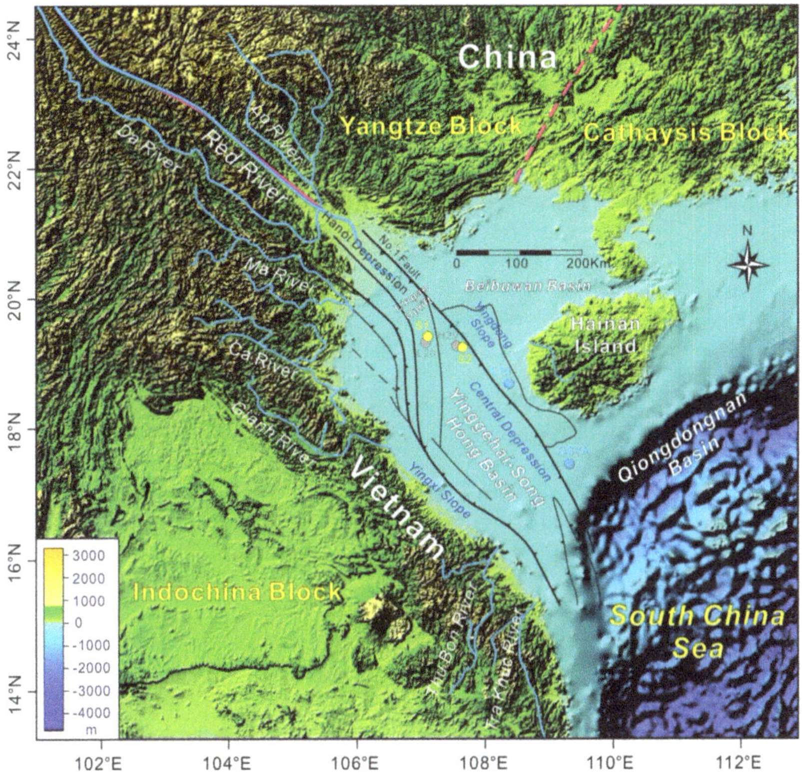

Figure 1. Topographic map and major drainage systems around the Yinggehai-Song Hong Basin. Yellow dots indicate the borehole locations in this study. Borehole locations of samples QSYA and L15 are from Yan et al. (2011) [13] and Wang et al. (2019) [26], respectively. Borehole locations of L20 and H29 are from Wang et al. (2016) [27].

2. Materials and Methods

Two sedimentary samples (S1-1 and S2-1) of the Sanya Formation in this study were collected from two boreholes that were drilled by the China National Offshore Oil Corporation in the Y-SH Basin (Figure 1). The primary strata boundaries can be effectively marked from the analysis and correlation of seismic sequences and nanofossil biostratigraphy. Samples S1-1 and S2-1 were obtained from the boreholes at depths of 3400 m and 3618 m, respectively. One published sample (L15-1) with dated zircon U-Pb ages by Wang et al. (2019) [26] was used only for Lu-Hf isotopic analyses. The selected samples are core chips and are fine- to medium-grained sandstone.

The zircons were separated from these two samples using conventional heavy liquids and magnetic separation techniques. After hand sorting using the binocular stereoscope, the zircon grains were mounted in transparent epoxy and polished down to approximately half sections. The internal structures of zircons were examined using cathodoluminescence (CL) images taken by a Mono CL3 detector attached to an electron microprobe (JXA-8100, JEOL, Tokyo, Japan) attached to an electron microprobe. The analytical spots without inclusions and fractures were selected according to the CL images. Zircon grains were randomly picked from the selected spots for laser ablation analyses.

The zircons were dated using laser ablation inductively coupled plasma mass spectrometry (LA-ICP-MS) at the Guangzhou Institute of Geochemistry, Chinese Academy of Sciences. Sample mounts were placed in the two-volume sample cell flushed with Argon and Helium.

Laser ablation was performed at a constant energy of 80 mJ and at 8 Hz, with a spot diameter of 31 μm. The ablated ions were carried by Helium to the Agilent 7500a ICP-MS (Agilent Technologies, Santa Clara, CA, USA). Isotopic ratios of U-Th-Pb were calculated by software of ICPMSDataCal (Version 11.0) [28]. The details of precision, accuracy, and analytical procedure are described by Zhang et al. (2018) [29]. Common Pb was corrected by ComPbCorr#3_151 using the method of Andersen (2002) [30]. $^{206}Pb/^{238}U$ ages of less than 1000 Ma were used, and $^{207}Pb/^{206}Pb$ ages were selected if the $^{206}Pb/^{238}U$ ages were more than 1000 Ma.

Zircon Lu-Hf isotopic analyses were carried out on a Neptune Plus multi-collector ICP-MS (Thermo Scientific, Bremen, Germany) equipped with a RESOlution M-50 laser-ablation system (Resonetics, Nashua, NH, USA) at the same institute. The Lu-Hf isotopic measurements were made on the same spots previously analyzed for U-Pb isotopes. The laser parameters were as follows: spot size, 45 μm; repetition rate, 8 Hz; energy, 80 mJ. Helium was used as a carrier gas and a small flow of nitrogen was added to the gas line to enhance the sample signal. ^{173}Yb and ^{175}Lu were used to correct the isobaric interference of ^{176}Yb and ^{176}Lu on ^{176}Hf. The $^{176}Hf/^{177}Hf$ was normalized to $^{179}Hf/^{177}Hf = 0.7325$ using an exponential law for mass bias correction. The initial Hf isotope ratios were denoted as $\varepsilon_{Hf}(t)$ values that were calculated with the chondritic uniform reservoir (CHUR) at the time of zircon crystallization, and the present-day $^{176}Hf/^{177}Hf$ and $^{176}Lu/^{177}Hf$ ratios of chondrite and depleted mantle were at 0.28277 and 0.0332, 0.28325, and 0.0384, respectively [31]. Initial $^{176}Hf/^{177}Hf$ values were calculated based on a ^{176}Lu decay constant of 1.865×10^{-11} year^{-1} [32]. Two-stage model Hf ages (T_{DM2}) were computed using a $^{176}Lu/^{177}Hf$ value of 0.015 for the average continental crust [33].

3. Results

A total of 170 zircon analyses were performed using LA-ICP-MS. The U-Pb concordia diagrams and CL images of the representative zircons are shown in Figure 2. Most of the zircon grains were colorless to transparent, with a wide range of prismatic to oval-shaped crystals. The morphologies of the zircon grains were distinctive for early Miocene sediments. Samples S1-1 and S2-1 were mostly concentrated in the range of 70–150 μm, and most of them had sub-rounded or rounded corners, which generally implied prolonged or multi-cycle transport. By contrast, most zircons from the sediments near Hainan offshore (L15-1, Figure 1) were complete crystals without obvious rounded corners [26], suggestive of proximal transport.

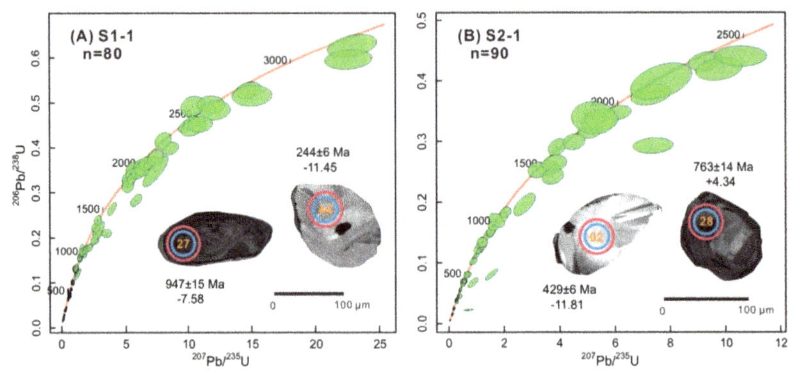

Figure 2. U-Pb concordia diagrams and representative CL images for zircons from the lower Miocene sediments. IsoplotR was used for concordia diagrams [34]. The small and large circles denote the analytical spots for U-Pb ages and Lu-Hf isotopes, respectively. Numbers near the circles are U-Pb ages and the corresponding ε_{Hf} (t) values.

The original U-Pb isotopic data of the zircon grains are shown in Table S1. Discordances smaller than 10% were considered in this study. The samples varied greatly in Th/U values, ranging from 0.05

to 3.77, and only four spots were less than 0.1 (Figure 3). Compared with the age probability model, kernel density estimators (KDEs) were more robust for displaying age populations and peak ages [35]. KDEs were used for displaying age populations and the ages of peaks (Figure 4). The Lu-Hf isotopic results for the detrital zircons are provided in Table S2 and are illustrated in Figure 5.

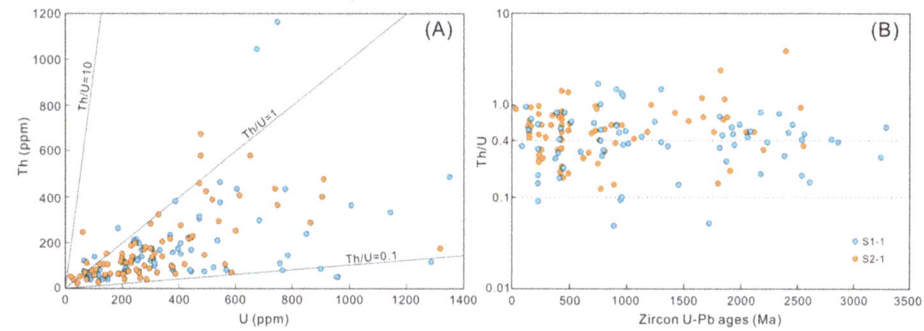

Figure 3. (**A**) Th versus U and (**B**) Th/U versus U-Pb diagrams for the concordant zircons from the S1-1 and S2-1 samples in the Yinggehai-Song Hong Basin.

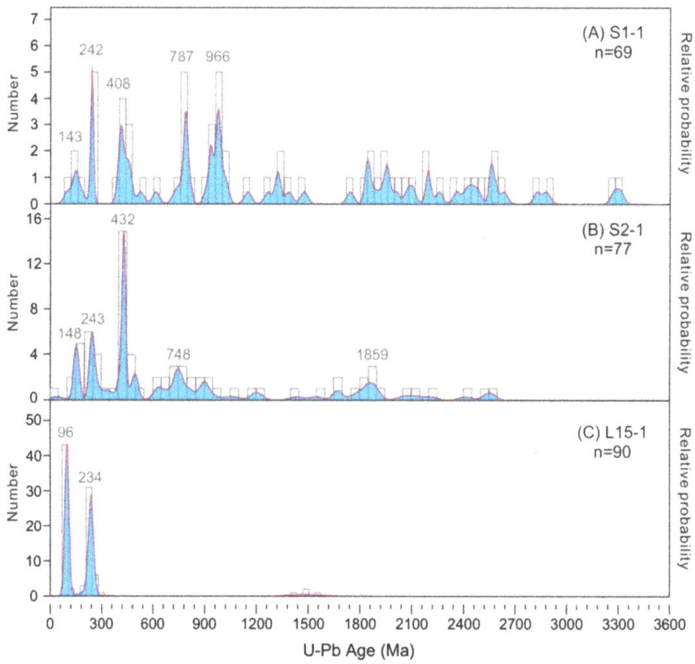

Figure 4. Kernel density estimation plots of detrital zircon U-Pb ages from the lower Miocene sediments in the Yinggehai-Song Hong Basin (**A**–**C**). The data of sample L15-1 was collected from Wang et al. (2019) [26].

3.1. Sample S1

Most of the zircon grains in sample S1 showed obvious oscillatory growth zoning under CL (Figure 2), along with Th/U values greater than 0.4 (Figure 3), suggesting that they have a magmatic origin [36]. In addition, only four zircons had a Th/U ratio smaller than 0.1 (Figure 3), implying a potential metamorphic origin only for these few grains. A total of 80 zircon grains were analyzed, and 69 concordant ages were obtained for the sample. Measured U-Pb ages ranged from 3313.3 Ma to 99.6 Ma (Table S1). The concordant zircons showed five major age peaks at ca. 143 Ma, 242 Ma, 408 Ma, 787 Ma, and 966 Ma (Figure 4A).

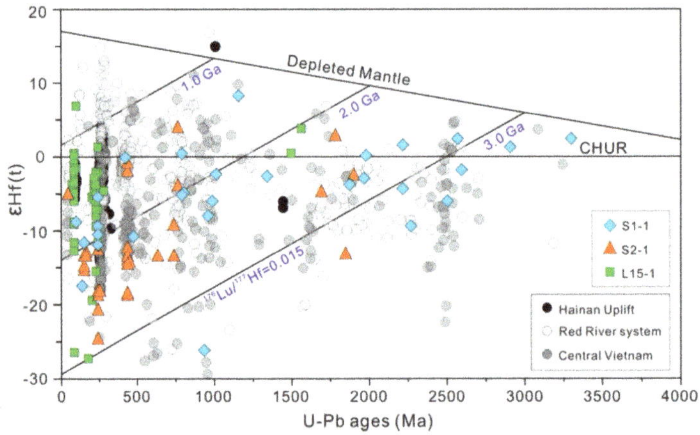

Figure 5. Diagrams showing the relationships between $\varepsilon_{Hf}(t)$ values and zircon U-Pb ages. The data of Red River and rivers in central Vietnam are from Wang et al. (2018) [20] and references therein. The data of bedrock in Hainan Uplift are from Wen (2013) [37], Wen et al. (2013) [38] and Sun et al. (2018) [39]. CHUR: chondritic uniform reservoir.

Twenty-nine dated zircon grains from sample S1 were selected for Lu-Hf isotope analysis. The analysis revealed a broad range of $^{176}Hf/^{177}Hf$ ratios from 0.280773 to 0.282527, with $\varepsilon_{Hf}(t)$ values from −25.7 to +8.5 (Figure 5). The two-stage Hf isotopic model ages (T_{DM2}) for these zircons ranged from 3414 to 1184 Ma (Table S2).

3.2. Sample S2

The crystals were euhedral with partially round corners for most zircon grains (Figure 2). These zircons showed oscillatory growth zoning under CL (Figure 2). A total of 90 zircon grains were analyzed, and 77 concordant ages were obtained for the sample. The Th/U ratios of the zircons ranged from 0.12 to 3.77 and were concentrated in the range of 0.5–0.7 (Table S1), indicating that these zircons were magmatic in origin. Measured U-Pb ages showed a wide age span ranging from 2564.8 to 38.8 Ma (Table S1). The concordant zircons displayed five major age peaks at ca. 148 Ma, 243 Ma, 432 Ma, 748 Ma, and 1859 Ma (Figure 4B).

Thirty-one dated zircon grains from the sample were selected for Lu-Hf isotope analysis. The analysis revealed a range of $^{176}Hf/^{177}Hf$ ratios from 0.281269 to 0.282621, with $\varepsilon_{Hf}(t)$ values from −24.1 to +4.3 (Figure 5). The two-stage Hf isotopic model ages (T_{DM2}) for these zircons ranged from 2988 to 1102 Ma (Table S2).

3.3. Sample L15-1

The detrital zircon U-Pb geochronological data of this sample were described in detail by Wang et al. (2019) [26]. This sample had two major age peaks at ca. 96 and 234 Ma (Figure 4C). Sixty dated zircon grains from the sample were selected for Lu-Hf isotope analysis. The analysis revealed a range of $^{176}Hf/^{177}Hf$ ratios from 0.281872 to 0.282901, with $\varepsilon_{Hf}(t)$ values from −27.2 to +6.8 (Figure 5). The two-stage Hf isotopic model ages (T_{DM2}) for these zircons ranged from 2338 to 596 Ma (Table S2).

4. Discussion

4.1. Potential Source Characteristics of the Sediments

Previous studies have suggested three potential source terranes for the Y-SH Basin fill, including the Red River source (Southern Yangtze Block), central Vietnam (Eastern Indochina Block), and the Hainan Uplift [13,14]. These sources have completely different tectonic-magmatic evolutions and have produced zircons with distinct age spectra, which can be illustrated by the age distributions of the bedrock and modern drainage systems (Figure 6).

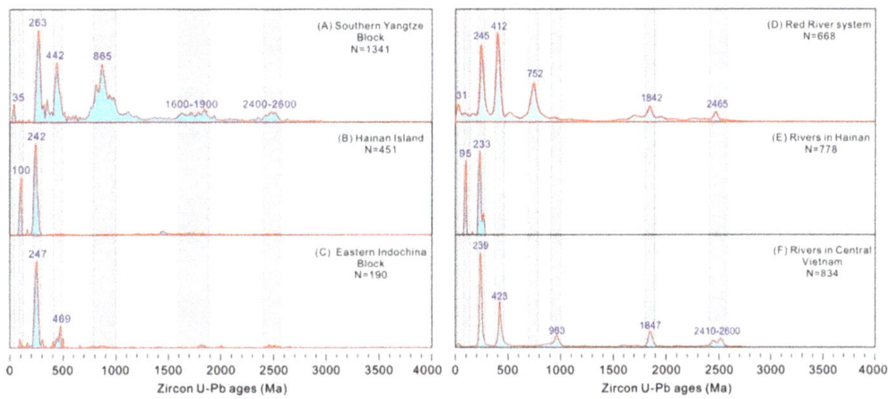

Figure 6. Summary of zircon U-Pb ages of the potential sources (**A–C**) and drainage systems discharge to the Yinggehai-Song Hong Basin (**D–F**). The original data of basement rocks of the potential sources were compiled by Wang et al. (2018) [21] and references therein. The data for the Red River and drainage systems in Central Vietnam are from Jonell et al., (2018) [25] and Wang et al. (2018) [20]. The data for the drainage systems in Hainan Uplift are from Cao et al. (2015) [22] and Wang et al. (2015) [24].

The Yangtze Block is an important component of the South China Block, which is separated from the Indochina Block by the Song Ma suture to the south and bounded by the Cathaysia Block to the east (Figure 1). Compared with the Hainan and eastern Indochina Block, the bedrock of the southern Yangtze Block displays a wide U-Pb age distribution, with multiple peaks from the Archean to Cenozoic (Figure 6A). The Archean age population of 2600–2400 Ma accounts for a small portion of the total, which was confirmed to be the basement of the Yangtze Block [40]. The Paleoproterozoic age range from 1900 to 1600 Ma was consistent with the ages for the coeval granitoids and sedimentary rocks in the Yangtze Block [41], such as the granites in the Kongling terrain, which yielded a U-Pb age of ca. 1850 Ma [42]. Ages around 865 Ma are related to the middle Neoproterozoic magmatic rocks and are widespread in the Yangtze Block [43]. Previous studies have shown that ca. 830–820 Ma granites can be found in the southeastern Yangtze Block [44]. The age peaks of 442 and 263 Ma are associated with the Late Paleozoic and Permian-Triassic tectonic-thermal events, respectively, which produced large amounts of igneous and metamorphic rocks. Cenozoic zircon ages ranging from 60 to 20 Ma have been reported from intrusions along the Red River Fault Zone [27,45]. The age peak of 35 Ma might represent an important signature for igneous activity associated with the fault zone [14]. The modern Red River system displays a similar age feature with the bedrock in the southern Yangtze Block but also shows different age peaks, potentially suggesting that there is a small difference in provenance (Figure 6A,D). The Red River system displays a wide range of $\varepsilon_{Hf}(t)$ values in different age populations (Figure 5). The Phanerozoic zircons with significant positive $\varepsilon_{Hf}(t)$ values suggests that the protolith was formed in part from the juvenile component. In addition, the Precambrian zircons generally had negative $\varepsilon_{Hf}(t)$ values, indicating that they were formed mainly by remelting of truly ancient crust.

The Hainan Uplift is located on the southern margin of the Cathaysia Block. Mesoproterozoic granitic rocks have been discovered isolated in the island's west with zircon U-Pb ages of 1460–1400 [46,47], which can also be found in modern rivers in western Hainan [24]. Hainan is characterized by widespread Permian to Triassic granitoids with a range of dated zircon U-Pb ages from 280 to 220 Ma [38], yielding a dominant age peak at ca. 242 Ma (Figure 6B). Cretaceous granitic and volcanic rocks with age peaks of ca. 100 Ma are primarily distributed in southwestern Hainan [39,48]. Both bedrock and modern rivers in Hainan show relatively simple age structures and are characterized by two age populations of 100–95 and 242–233 Ma (Figure 6B,E, respectively). Compared to the Red River and central Vietnam, the Hainan source shows relatively narrow $\varepsilon_{Hf}(t)$ values for the Triassic to Cretaceous zircons (Figure 5).

Central Vietnam, which is located in the eastern Indochina Block, is characterized by Paleozoic to Mesozoic granitic and metamorphic rocks [49,50]. Recent studies have shown that the Ordovician–Silurian and Permian-Late Triassic magmatic activities in the Truong Son Belt and Kontum Massif have ages of 470–420 and 300–200 Ma, respectively [49,51–53]. The summarized geochronological data of the bedrock showed two major age peaks at ca. 247 and 469 Ma (Figure 6C). Compared with the bedrock, the drainage systems in central Vietnam show a relatively complicated age structure, which was reflected by multiple peaks at ca. 2600–2410 Ma, 1847 Ma, 963 Ma, 423 Ma, and 239 Ma (Figure 6F). Although the Hf isotopic composition of zircons in the central Vietnam is overlapped by counterparts from the source of Red River, the Phanerozoic zircons with more negative values (Figure 5) indicates that the protolith was formed primarily by an ancient crustal component.

4.2. Provenance of the Detrital Zircons

The detrital zircon age distributions of the two cored samples from the lower Miocene Sanya Formation form multiple peaks, which may be characteristic of the age signatures of specific bedrock and drainage systems around the basin.

Only one dated zircon was of Cenozoic age (38.8 Ma) in sample S2-1, with the $\varepsilon_{Hf}(t)$ value of −4.5. Although Cenozoic zircons have been widely reported in the Red River Fault Zone, they are practically absent from our sampling; however, this can be distinctively tied to that distal area. Zircon age was consistent with the Cenozoic granitic and metamorphic rocks in the fault zone [45,54], suggesting that these zircons may be derived from the southwestern margin of the Yangtze Block through the Red River system [14].

Samples S1-1 and S2-1 had Late Jurassic age peaks at ca. 143 and 148 Ma, respectively, ages consistent with the widespread Yanshanian granitoids in the South China Block [55], implying a source from the hinterland of the Cathaysia Block or the Hainan Uplift. A recent study of detrital zircons from the Cretaceous strata on Hainan Island yielded a major age at ca. 155 Ma, an age close to the peak observed in the samples [56]. The Cretaceous zircons in sample L15-1 displayed little variation in $\varepsilon_{Hf}(t)$ values (most in −10 to 0), suggesting that they were formed from the remelting of the ancient crust, which was consistent with the $\varepsilon_{Hf}(t)$ values observed in Hainan Island work (Figure 5).

The Permian to Triassic zircon grains, which had age peaks around 243 Ma (Figure 4), match known exposures of the Indosinian igneous (and metamorphic) activity in the different source terranes. The Permian to Triassic granitic rocks is widely distributed in the Indochina Block and South China Block [49,57], with dated zircon U-Pb ages mainly concentrated between 280 to 220 Ma (Figure 6), making specific matching of these peaks difficult. The age peaks of 408 and 432 Ma in the samples are indicative of sources in the Yangtze and Indochina blocks. The zircons had negative $\varepsilon_{Hf}(t)$ values, suggesting that they originated from the ancient crust. These age ranges are rarely reported in the Hainan Uplift (Figure 6B) and thus the sources may be the Red River region and central Vietnam. Although the zircons with $\varepsilon_{Hf}(t)$ values $\leq$10 fell within the ranges of both the Red River system and drainage systems in the central Vietnamese rivers, other zircons with $\varepsilon_{Hf}(t)$ values $\geq$5 were more consistent with the Red River (Figure 5), implying that these grains might be derived mainly from the source of the Red River, although there is a lack of zircons with $\varepsilon_{Hf}(t)$ positive values (>10) sampled in those terranes.

Neoproterozoic age groups with age peaks at ca. 748 Ma, 787 Ma, and 966 Ma were similar to those observed in the Southern Yangtze Block and the Red River (Figure 6A,D). In addition, the drainage systems in central Vietnam could have also provided a small amount of the Neoproterozoic-aged zircon grains (963 Ma) (Figure 6F). The zircons with age population between 800 and 700 Ma in the S1 and S2 samples displayed a wide range of $\varepsilon_{Hf}(t)$ values (Figure 5), which are consistent with the modern Red River system. The Neoarchean to Paleoproterozoic ages can account for a small part of the total ages, which can be regarded as the basement of the Yangtze and Indochina blocks and eroded from the granitic rocks or recycled from the sedimentary rocks that had previously formed. A recent study indicated that the Neoarchean and Paleoproterozoic K-rich granites in the Phan Si Pan Complex of north Vietnam yielded zircon U-Pb ages of 2.85 to 1.86 Ga [58], which may also provide the detritus to the basin through the Red River system. The Precambrian zircons had variable $\varepsilon_{Hf}(t)$ values, ranging from negative to positive (Figure 5), indicating that they were formed by both a juvenile source and the remelting of ancient crust. These zircons basically fall within the range of the Red River system, representing the important influence of Red River on sediments.

To examine the degrees of similarity among zircon U-Pb ages, the newly developed Multi-Dimensional Scaling (MDS) plots were used to compare the genetic relationships between samples and source terranes. The MDS plot based on the Kolmogorov–Smirnov (K-S) statistical method [59] is a dissimilarity matrix that constructs a map on which similar samples cluster closely together and dissimilar samples plot far apart [60]. In the MDS plots (Figure 7), the early Miocene samples showed a distinct affinity with the source terrane and also the drainage systems. The samples close to Hainan (L15-1 and QSYA) generally showed a greater affinity with the bedrock and rivers in Hainan, suggesting that the sediments were primarily derived from nearby sources (Figure 8). By contrast, samples S1 and S2 in this study plot were relatively far from Hainan (Figure 7A) but close to both rivers in central Vietnam and the Red River, indicating that they had an affinity with these drainage systems. However, compared with the bedrock, the affinity of these two samples with the southern Yangtze Block was greater than that of the Indochina Block, suggesting a major contribution from the southern Yangtze Block and a small contribution from the eastern Indochina Block (Figure 8).

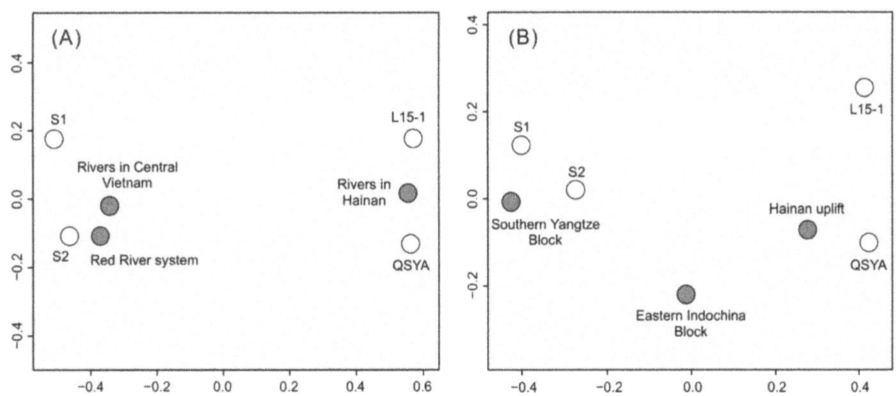

Figure 7. Multi-dimensional scaling plots using Kolmogorov–Smirnov (K-S) statistic for detrital zircon U-Pb data [59,60]. The plots show the relationships between the core samples with regard to (**A**) the major drainage systems as defined from river sands and (**B**) potential source terranes around the Yinggehai-Song Hong Basin sampled from the outcrop. Axes are in dimensionless "K-S units" of distance between samples. Samples QSYA and L15-1 are from Yan et al. (2011) [13] and Wang et al. (2019) [26], respectively.

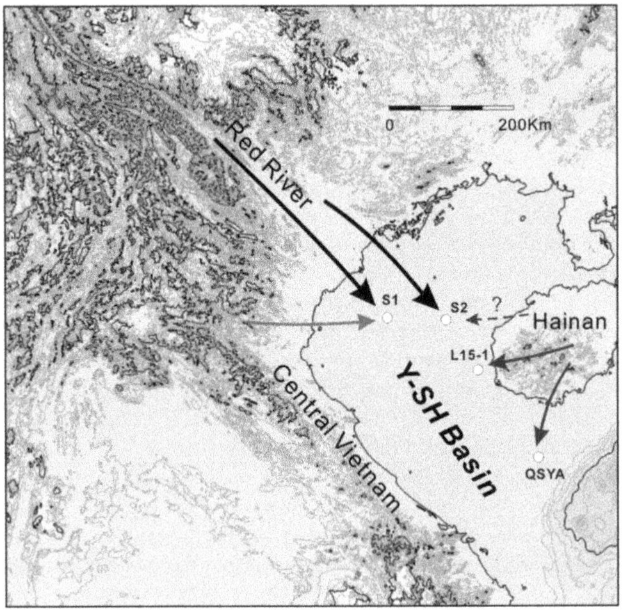

Figure 8. Provenance and possible transport pathways of the early Miocene sediments in the Yinggehai-Song Hong Basin. The arrows indicate the possible direction of the supplied sources.

In summary, detrital zircon U-Pb ages from the lower Miocene Sanya Formation suggest that the sediments in this study were derived from multiple source terranes (Figure 8). The southern Yangtze Block played an important role in contributing detritus to the northern Y-SH Basin, and the eastern Indochina Block may also have a minor influence in this region. By contrast, the Hainan Uplift is a proximal source, which apparently had limited influence on the northern Y-SH Basin; it dominates eastern samples only.

4.3. Provenance Variations in the Miocene

Detrital zircon geochronological data presented here are suggestive of spatial variation in provenance in the early Miocene, which was primarily controlled by the location and distance of the source terranes. In order to characterize the variations in the provenance through time, from the early to middle Miocene sediments, these data were compared with published data of the middle Miocene sediments in the boreholes (L20 and H29, Figure 1) [27] at the northern basin, and were examined to understand temporal changes in provenance (Figure 9). The zircon age distributions for the lower to middle Miocene sediments were diverse, with some slight changes appearing in the proportions of the different age groups (Figure 9).

Samples in the Lingao uplift of the Y-SH Basin showed a range of age populations (Figure 9A–C), which may reflect the changing source terranes that supplied the sample location. The early Miocene sample in this study was more diverse than the middle Miocene samples, especially for ages older than 1800 Ma (Figure 9C). Thus, sediments contained more Paleoproterozoic sources, which might be derived from the coeval magmatic rocks or recycled from the sedimentary rocks in the source region. The older Neoproterozoic zircon grains in samples L20-3 and S1-1, which showed a similar characteristic to the drainage systems in central Vietnam, suggested a possible contribution originating from the eastern Indochina Block. By contrast, the middle Miocene (13.8–10.5 Ma) showed more Silurian and less Early Triassic ages (Figure 9A), which was more similar to the Red River system, and revealed that the sample was primarily derived from the southern Yangtze Block.

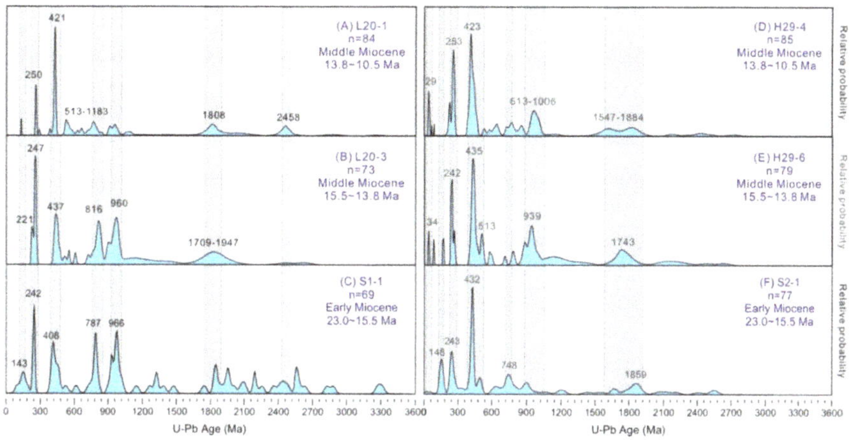

Figure 9. Detrital zircon U-Pb age characteristics of the early-middle Miocene sediments in the Yinggehai-Song Hong Basin (**A–F**). The borehole L20 is located in the Lingao area and H29 is located in the Haikou area of the northern Y-SH Basin (Figure 1). Detailed sampling information of L20-1, L20-3, H29-4, and H29-6 can be found in the reference by Wang et al. (2016) [27].

Samples in the Haikou area of the Y-SH Basin displayed a similar age spectrum from the early to middle Miocene (Figure 9D–F), indicating that a relatively stable provenance was supplied in this sedimentary period. In spite of this, there was still variation in provenance at different geological periods. Compared with the early Miocene (23.0–15.5 Ma) (Figure 9F), samples of the middle Miocene (15.5–10.5 Ma) (Figure 9D,E) showed similar age distributions, which differed from the early Miocene sample by minor age peaks at ca. 29–34 Ma, and may thus represent an increased supply from the Red River Fault Zone after the early Miocene. The Neoproterozoic grains in the middle Miocene (>900 Ma) were older than those in the early Miocene (748 Ma), suggesting a different source for the Neoproterozoic rocks. In addition, middle Miocene samples were absent in the Late Jurassic age population as well as in the early Miocene sample (148 Ma), implying that there was no influence from the Hainan or the hinterland Cathaysia Block in the middle Miocene.

Overall, the study of detrital zircon U-Pb ages revealed that there were no dramatic changes during the early to middle Miocene in the northern Y-SH Basin, suggesting that this was a relatively variational source supply at the time. Slight variation in detrital zircon and provenance was probably related to the amount of detritus materials inputted from central Vietnam, and may also be influenced by the evolution of the Red River system [23,61].

5. Conclusions

The following conclusions have been made based on the new dataset of U-Pb dating and Lu-Hf isotopes for detrital zircons from the lower Miocene Sanya Formation in the Y-SH Basin in the northwestern South China Sea.

(1) Detrital zircons analyzed in this study revealed several age clusters at ca. 148–143 Ma, 243–242 Ma, 432–408 Ma, 787–748 Ma, 966 Ma, and 1859 Ma. These zircon grains showed sub-rounded or rounded corners, implying prolonged or multi-cycle transport.

(2) The zircons had a broad range of $^{176}Hf/^{177}Hf$ ratios from 0.280773 to 0.282901, with $\varepsilon_{Hf}(t)$ values of −27.2 to +8.5, suggesting that there were multiple sources for the sediments. Most of the zircon grains had negative $\varepsilon_{Hf}(t)$ values, implying that the ancient crustal component was involved in the formation of zircon. Precambrian zircons were highly variable, with values ranging from negative to positive, indicating that they were formed by both a juvenile source and the remelting of ancient crust.

(3) Based on comprehensive analyses of the results and previously published data, we confirmed that the Red River system was a dominant sediment contributor to the northern Yinggehai-Song Hong Basin during the early Miocene and that central Vietnam may also be a minor source terrane. The Hainan was the exclusive contributor controlling sediments in the eastern margin of the basin.

(4) This study showed that there was some variational source supply during the early to middle Miocene in the northern basin. The slight provenance variation depended on the amount of detritus materials inputted by central Vietnam, and may also be influenced by the evolution of the Red River system.

Supplementary Materials: The following are available online at http://www.mdpi.com/2075-163X/10/9/752/s1, Table S1: LAICPMS UPb dating data of zircons in this study; Table S2: LA-MC-ICP-MS Hf isotopic compositions of zircon in this study.

Author Contributions: Conceptualization, C.W. and X.L.; data curation, L.Z.; formal analysis, L.Z.; methodology, X.L.; project administration, M.S.; writing—original draft, C.W.; writing—review and editing, Y.L. and M.S. All authors have read and agreed to the published version of the manuscript.

Funding: This research was funded by the National Natural Science Foundation of China, grant number 41702115, the Guangdong Basic and Applied Basic Research Foundation, grant numbers 2017A030313250 and 2019A1515012017, the Fundamental Research Funds for the Central Universities, grant number 20lgpy61.

Acknowledgments: We appreciate L. Zhang and D. Wu for their assistance in sample analyses in GIGCAS. Thanks to China National Offshore Oil Corporation Ltd., Zhanjiang, China, for supplying the borehole samples. We wish to thank the editors Wilfried Winkler and Albrecht von Quadt and the four anonymous reviewers for their helpful comments and constructive suggestions.

Conflicts of Interest: The authors declare no conflict of interest.

References

1. Liu, Z.; Zhao, Y.; Colin, C.; Stattegger, K.; Wiesner, M.G.; Huh, C.A.; Zhang, Y.; Li, X.; Sompongchaiyakul, P.; You, C.F. Source-to-sink transport processes of fluvial sediments in the South China Sea. *Earth-Sci. Rev.* **2016**, *153*, 238–273. [CrossRef]
2. Kuehl, S.A.; Alexander, C.R.; Blair, N.E.; Harris, C.K.; Marsaglia, K.M.; Ogston, A.S.; Orpin, A.R.; Roering, J.J.; Bever, A.J.; Bilderback, E.L. A source-to-sink perspective of the Waipaoa River margin. *Earth-Sci. Rev.* **2016**, *153*, 301–334. [CrossRef]
3. Marsaglia, K.M.; DeVaughn, A.M.; James, D.E.; Marden, M. Provenance of fluvial terrace sediments within the Waipaoa sedimentary system and their importance to New Zealand source-to-sink studies. *Mar. Geol.* **2010**, *270*, 84–93. [CrossRef]
4. Fleming, E.J.; Flowerdew, M.J.; Smyth, H.R.; Scott, R.A.; Morton, A.C.; Omma, J.E.; Frei, D.; Whitehouse, M.J. Provenance of Triassic sandstones on the southwest Barents Shelf and the implication for sediment dispersal patterns in northwest Pangaea. *Mar. Pet. Geol.* **2016**, *78*, 516–535. [CrossRef]
5. Vincent, S.J.; Morton, A.C.; Hyden, F.; Fanning, M. Insights from petrography, mineralogy and U–Pb zircon geochronology into the provenance and reservoir potential of Cenozoic siliciclastic depositional systems supplying the northern margin of the Eastern Black Sea. *Mar. Pet. Geol.* **2013**, *45*, 331–348. [CrossRef]
6. Parra, J.G.; Marsaglia, K.M.; Rivera, K.S.; Dawson, S.T.; Walsh, J.P. Provenance of sand on the Poverty Bay shelf, the link between source and sink sectors of the Waipaoa River sedimentary system. *Sediment. Geol.* **2012**, *280*, 208–233. [CrossRef]
7. Warrick, J.A.; Bountry, J.A.; East, A.E.; Magirl, C.S.; Randle, T.J.; Gelfenbaum, G.; Ritchie, A.C.; Pess, G.R.; Leung, V.; Duda, J.J. Large-scale dam removal on the Elwha River, Washington, USA: Source-to-sink sediment budget and synthesis. *Geomorphol.* **2015**, *246*, 729–750. [CrossRef]
8. Carter, L.; Orpin, A.R.; Kuehl, S.A. From mountain source to ocean sink—The passage of sediment across an active margin, Waipaoa Sedimentary System, New Zealand. *Mar. Geol.* **2010**, *270*, 1–10. [CrossRef]
9. Carrapa, B. Resolving tectonic problems by dating detrital minerals. *Geology* **2010**, *38*, 191–192. [CrossRef]
10. Avigad, D.; Morag, N.; Abbo, A.; Gerdes, A. Detrital rutile U-Pb perspective on the origin of the great Cambro-Ordovician sandstone of North Gondwana and its linkage to orogeny. *Gondwana Res.* **2017**. [CrossRef]

11. Mason, C.C.; Fildani, A.; Gerber, T.; Blum, M.D.; Clark, J.D.; Dykstra, M. Climatic and anthropogenic influences on sediment mixing in the Mississippi source-to-sink system using detrital zircons: Late Pleistocene to recent. *Earth Planet. Sci. Lett.* **2017**, *466*, 70–79. [CrossRef]
12. Sevastjanova, I.; Clements, B.; Hall, R.; Belousova, E.A.; Griffin, W.L.; Pearson, N. Granitic magmatism, basement ages, and provenance indicators in the Malay Peninsula: Insights from detrital zircon U–Pb and Hf-isotope data. *Gondwana Res.* **2011**, *19*, 1024–1039. [CrossRef]
13. Yan, Y.; Carter, A.; Palk, C.; Brichau, S.; Hu, X. Understanding sedimentation in the Song Hong–Yinggehai Basin, South China Sea. *Geochem. Geophys. Geosyst.* **2011**, *12*, Q06014. [CrossRef]
14. Wang, C.; Liang, X.; Xie, Y.; Tong, C.; Pei, J.; Zhou, Y.; Jiang, Y.; Fu, J.; Dong, C.; Liu, P. Provenance of Upper Miocene to Quaternary Sediments in the Yinggehai-Song Hong Basin, South China Sea: Evidence from Detrital Zircon U–Pb ages. *Mar. Geol.* **2014**, *355*, 202–217. [CrossRef]
15. Jiang, T.; Cao, L.; Xie, X.; Wang, Z.; Li, X.; Zhang, Y.; Zhang, D.; Sun, H. Insights from heavy minerals and zircon U–Pb ages into the middle Miocene–Pliocene provenance evolution of the Yinggehai Basin, northwestern South China Sea. *Sediment. Geol.* **2015**, *327*, 32–42. [CrossRef]
16. Zhao, M.; Shao, L.; Liang, J.; Li, Q. No Red River capture since the late Oligocene: Geochemical evidence from the Northwestern South China Sea. *Deep Sea Res. Part II Top. Stud. Oceanogr.* **2015**, *122*, 185–194. [CrossRef]
17. Fyhn, M.B.W.; Thomsen, T.B.; Keulen, N.; Knudsen, C.; Rizzi, M.; Bojesen-Koefoed, J.; Olivarius, M.; Tri, T.V.; Phach, P.V.; Minh, N.Q.; et al. Detrital zircon ages and heavy mineral composition along the Gulf of Tonkin—Implication for sand provenance in the Yinggehai-Song Hong and Qiongdongnan basins. *Mar. Pet. Geol.* **2019**, *101*, 162–179. [CrossRef]
18. Clift, P.D.; Sun, Z. The sedimentary and tectonic evolution of the Yinggehai–Song Hong basin and the southern Hainan margin, South China Sea: Implications for Tibetan uplift and monsoon intensification. *J. Geophys. Res. Solid Earth* **2006**, *111*, B06405. [CrossRef]
19. Huang, B.; Xiao, X.; Li, X. Geochemistry and origins of natural gases in the Yinggehai and Qiongdongnan basins, offshore South China Sea. *Org. Geochem.* **2003**, *34*, 1009–1025. [CrossRef]
20. Wang, C.; Liang, X.; Foster, D.A.; Tong, C.; Liu, P.; Liang, X.; Zhang, L. Linking source and sink: Detrital zircon provenance record of drainage systems in Vietnam and the Yinggehai–Song Hong Basin, South China Sea. *GSA Bull.* **2018**, *131*, 191–204. [CrossRef]
21. Wang, C.; Wen, S.; Liang, X.; Shi, H.; Liang, X. Detrital zircon provenance record of the Oligocene Zhuhai Formation in the Pearl River Mouth Basin, northern South China Sea. *Mar. Pet. Geol.* **2018**, *98*, 448–461. [CrossRef]
22. Cao, L.; Jiang, T.; Wang, Z.; Zhang, Y.; Sun, H. Provenance of Upper Miocene sediments in the Yinggehai and Qiongdongnan basins, northwestern South China Sea: Evidence from REE, heavy minerals and zircon U–Pb ages. *Mar. Geol.* **2015**, *361*, 136–146. [CrossRef]
23. Van Hoang, L.; Wu, F.; Clift, P.D.; Wysocka, A.; Swierczewska, A. Evaluating the evolution of the Red River system based on in situ U-Pb dating and Hf isotope analysis of zircons. *Geochem. Geophys. Geosyst.* **2009**, *10*, Q11008. [CrossRef]
24. Wang, C.; Liang, X.; Zhou, Y.; Fu, J.; Jiang, Y.; Dong, C.; Xie, Y.; Tong, C.; Pei, J.; Liu, P. Construction of age frequencies of provenances on the eastern side of the Yinggehai Basin: Studies of LA-ICP-MS U-Pb ages of detrital zircons from six modern rivers, western Hainan, China. *Earth Sci. Front.* **2015**, *22*, 277–289. [CrossRef]
25. Jonell, T.N.; Clift, P.D.; Hoang, L.V.; Hoang, T.; Carter, A.; Wittmann, H.; Böning, P.; Pahnke, K.; Rittenour, T. Controls on erosion patterns and sediment transport in a monsoonal, tectonically quiescent drainage, Song Gianh, central Vietnam. *Basin Res.* **2017**, *29*, 659–683. [CrossRef]
26. Wang, C.; Liang, X.; Foster, D.A.; Liang, X.; Tong, C.; Liu, P. Detrital zircon ages: A key to unraveling provenance variations in the eastern Yinggehai–Song Hong Basin, South China Sea. *Am. Assoc. Pet. Geol. Bull.* **2019**, *103*, 1525–1552. [CrossRef]
27. Wang, C.; Liang, X.; Foster, D.A.; Xie, Y.; Tong, C.; Pei, J.; Fu, J.; Jiang, Y.; Dong, C.; Zhou, Y.; et al. Zircon U-Pb geochronology and heavy mineral composition constraints on the provenance of the middle Miocene deep-water reservoir sedimentary rocks in the Yinggehai-Song Hong Basin, South China Sea. *Mar. Pet. Geol.* **2016**, *77*, 819–834. [CrossRef]
28. Liu, Y.; Gao, S.; Hu, Z.; Gao, C.; Zong, K.; Wang, D. Continental and oceanic crust recycling-induced melt-peridotite interactions in the Trans-North China Orogen: U-Pb dating, Hf isotopes and trace elements in zircons from mantle xenoliths. *J. Petrol.* **2010**, *51*, 537–571. [CrossRef]

29. Zhang, L.; Ren, Z.-Y.; Xia, X.-P.; Wang, C.; Qian, S.-P. An improved U-Pb age dating method for detrital zircon by LA-MC-ICP-MS. *Geochem. J.* **2018**, *52*, 433–439. [CrossRef]
30. Andersen, T. Detrital zircons as tracers of sedimentary provenance: Limiting conditions from statistics and numerical simulation. *Chem. Geol.* **2005**, *216*, 249–270. [CrossRef]
31. Blichert-Toft, J.; Albarède, F. The Lu-Hf isotope geochemistry of chondrites and the evolution of the mantle-crust system. *Earth Planet. Sci. Lett.* **1997**, *148*, 243–258. [CrossRef]
32. Scherer, E. Calibration of the Lutetium-Hafnium Clock. *Science* **2001**, *293*, 683–687. [CrossRef] [PubMed]
33. Griffin, W.L.; Wang, X.; Jackson, S.E.; Pearson, N.J.; O'Reilly, S.Y.; Xu, X.; Zhou, X. Zircon chemistry and magma mixing, SE China: In-situ analysis of Hf isotopes, Tonglu and Pingtan igneous complexes. *Lithos* **2002**, *61*, 237–269. [CrossRef]
34. Vermeesch, P. IsoplotR: A free and open toolbox for geochronology. *Geosci. Front.* **2018**, *9*, 1479–1493. [CrossRef]
35. Vermeesch, P. On the visualisation of detrital age distributions. *Chem. Geol.* **2012**, *312–313*, 190–194. [CrossRef]
36. Wu, Y.; Zheng, Y. Genesis of zircon and its constraints on interpretation of U-Pb age. *Chinese Sci. Bull.* **2004**, *49*, 1554–1569. [CrossRef]
37. Wen, S. Geochronologic and Geochemical Studies of Permian-Triassic Magmatism in Hainan Island, South China. Ph.D. Thesis, University of Chinses Academy of Sciences, Beijing, China, 2013.
38. Wen, S.; Liang, X.; Fan, W.; Wang, Y.; Chi, G.; Liang, X.; Zhou, Y.; Jiang, Y. Zircon U-Pb ages, Hf isotopic composition of Zhizhong granitic intrusion in Ledong area of Hainan island and their tectonic implications. *Geotecton. Metallog.* **2013**, *37*, 294–307.
39. Sun, S.J.; Zhang, L.P.; Zhang, R.Q.; Ding, X.; Zhu, H.L.; Zhang, Z.F.; Sun, W.D. Mid–Late Cretaceous igneous activity in South China: The Qianjia example, Hainan Island. *Int. Geol. Rev.* **2018**, *60*, 1665–1683. [CrossRef]
40. Zheng, J.; Griffin, W.L.; O'Reilly, S.Y.; Zhang, M.; Pearson, N.; Pan, Y. Widespread Archean basement beneath the Yangtze craton. *Geology* **2006**, *34*, 417–420. [CrossRef]
41. Zhao, X.F.; Zhou, M.F.; Li, J.W.; Sun, M.; Gao, J.F.; Sun, W.H.; Yang, J.H. Late Paleoproterozoic to early Mesoproterozoic Dongchuan Group in Yunnan, SW China: Implications for tectonic evolution of the Yangtze Block. *Precambrian Res.* **2010**, *182*, 57–69. [CrossRef]
42. Peng, M.; Wu, Y.; Gao, S.; Zhang, H.; Wang, J.; Liu, X.; Gong, H.; Zhou, L.; Hu, Z.; Liu, Y.; et al. Geochemistry, zircon U-Pb age and Hf isotope compositions of Paleoproterozoic aluminous A-type granites from the Kongling terrain, Yangtze Block: Constraints on petrogenesis and geologic implications. *Gondwana Res.* **2012**, *22*, 140–151. [CrossRef]
43. Zhou, M.F.; Yan, D.P.; Kennedy, A.K.; Li, Y.; Ding, J. SHRIMP U–Pb zircon geochronological and geochemical evidence for Neoproterozoic arc-magmatism along the western margin of the Yangtze Block, South China. *Earth Planet. Sci. Lett.* **2002**, *196*, 51–67. [CrossRef]
44. Wu, R.X.; Zheng, Y.F.; Wu, Y.B.; Zhao, Z.F.; Zhang, S.B.; Liu, X.; Wu, F.Y. Reworking of juvenile crust: Element and isotope evidence from Neoproterozoic granodiorite in South China. *Precambrian Res.* **2006**, *146*, 179–212. [CrossRef]
45. Zhang, L.S.; Schärer, U. Age and origin of magmatism along the Cenozoic Red River shear belt, China. *Contrib. Mineral. Petrol.* **1999**, *134*, 67–85. [CrossRef]
46. Xu, D.R.; Xia, B.; Li, P.C.; Chen, G.H.; Ma, C.; Zhang, Y.Q. Protolith natures and U-Pb sensitive high mass-resolution ion microprobe (SHRIMP) zircon ages of the metabasites in Hainan Island, South China: Implications for geodynamic evolution since the late Precambrian. *Isl. Arc* **2007**, *16*, 575–597. [CrossRef]
47. Li, Z.X.; Li, X.H.; Li, W.X.; Ding, S. Was Cathaysia part of Proterozoic Laurentia?—New data from Hainan Island, South China. *Terra Nov.* **2008**, *20*, 154–164. [CrossRef]
48. Zhou, Y.; Liang, X.; Kröner, A.; Cai, Y.; Shao, T.; Wen, S.; Jiang, Y.; Fu, J.; Wang, C.; Dong, C. Late Cretaceous lithospheric extension in SE China: Constraints from volcanic rocks in Hainan Island. *Lithos* **2015**, *232*, 100–110. [CrossRef]
49. Faure, M.; Nguyen, V.V.; Hoai, L.T.T.; Lepvrier, C. Early Paleozoic or Early-Middle Triassic collision between the South China and Indochina Blocks: The controversy resolved? Structural insights from the Kon Tum massif (Central Vietnam). *J. Asian Earth Sci.* **2018**, *166*, 162–180. [CrossRef]
50. Van Tran, T.; Faure, M.; Van Nguyen, V.; Bui, H.H.; Fyhn, M.B.W.; Nguyen, T.Q.; Lepvrier, C.; Thomsen, T.B.; Tani, K.; Charusiri, P. Neoproterozoic to Early Triassic tectono-stratigraphic evolution of Indochina and adjacent areas: A review with new data. *J. Asian Earth Sci.* **2020**, *191*, 104231. [CrossRef]

51. Liu, J.; Tran, M.-D.; Tang, Y.; Nguyen, Q.-L.; Tran, T.-H.; Wu, W.; Chen, J.; Zhang, Z.; Zhao, Z. Permo-Triassic granitoids in the northern part of the Truong Son belt, NW Vietnam: Geochronology, geochemistry and tectonic implications. *Gondwana Res.* **2012**, *22*, 628–644. [CrossRef]
52. Shi, M.F.; Lin, F.C.; Fan, W.Y.; Deng, Q.; Cong, F.; Tran, M.D.; Zhu, H.P.; Wang, H. Zircon U–Pb ages and geochemistry of granitoids in the Truong Son terrane, Vietnam: Tectonic and metallogenic implications. *J. Asian Earth Sci.* **2015**, *101*, 101–120. [CrossRef]
53. Hieu, P.T.; Yang, Y.Z.; Binh, D.Q.; Nguyen, T.B.T.; Dung, L.T.; Chen, F. Late Permian to Early Triassic crustal evolution of the Kontum massif, central Vietnam: Zircon U-Pb ages and geochemical and Nd-Hf isotopic composition of the Hai Van granitoid complex. *Int. Geol. Rev.* **2015**. [CrossRef]
54. Gilley, L.D.; Harrison, T.M.; Leloup, P.H.; Ryerson, F.J.; Lovera, O.M.; Wang, J.-H. Direct dating of left-lateral deformation along the Red River shear zone, China and Vietnam. *J. Geophys. Res. Solid Earth* **2003**, *108*, 2127. [CrossRef]
55. Xu, Y.; Wang, C.Y.; Zhao, T. Using detrital zircons from river sands to constrain major tectono-thermal events of the Cathaysia Block, SE China. *J. Asian Earth Sci.* **2016**, *124*, 1–13. [CrossRef]
56. Jiang, X.Y.; Li, X.H.; Collins, W.J.; Huang, H.Q. U-Pb age and Hf-O isotopes of detrital zircons from Hainan Island: Implications for Mesozoic subduction models. *Lithos* **2015**, *239*, 60–70. [CrossRef]
57. Li, J.; Zhang, Y.; Zhao, G.; Johnston, S.T.; Dong, S.; Koppers, A.; Miggins, D.P.; Sun, H.; Wang, W.; Xin, Y. New insights into Phanerozoic tectonics of South China: Early Paleozoic sinistral and Triassic dextral transpression in the east Wuyishan and Chencai domains, NE Cathaysia. *Tectonics* **2017**, *36*, 819–853. [CrossRef]
58. Zhao, T.; Cawood, P.A.; Wang, K.; Zi, J.W.; Feng, Q.; Nguyen, Q.M.; Tran, D.M. Neoarchean and Paleoproterozoic K-rich granites in the Phan Si Pan Complex, north Vietnam: Constraints on the early crustal evolution of the Yangtze Block. *Precambrian Res.* **2019**, *332*, 105395. [CrossRef]
59. Kruskal, J.B. Multidimensional scaling by optimizing goodness of fit to a nonmetric hypothesis. *Psychometrika* **1964**, *29*, 1–27. [CrossRef]
60. Vermeesch, P. Multi-sample comparison of detrital age distributions. *Chem. Geol.* **2013**, *341*, 140–146. [CrossRef]
61. Wang, C.; Liang, X.; Foster, D.A.; Liang, X.; Zhang, L.; Su, M. Provenance and drainage evolution of the Red River revealed by Pb isotopic analysis of detrital K-feldspar. *Geophys. Res. Lett.* **2019**, *46*, 6415–6424. [CrossRef]

minerals

MDPI

Article

Sedimentology, Provenance and Radiometric Dating of the Silante Formation: Implications for the Cenozoic Evolution of the Western Andes of Ecuador

Cristian Vallejo [1,2,*], Santiago Almagor [1], Christian Romero [3], Jose L. Herrera [4,5], Vanessa Escobar [1], Richard A. Spikings [6], Wilfried Winkler [7] and Pieter Vermeesch [4]

1 Departamento de Geología, Facultad de Ingeniería en Geología y Petróleos, Escuela Politécnica Nacional, Quito 170517, Ecuador; santiago.almagor@geostrats.com (S.A.); vanessa.escobar@epn.edu.ec (V.E.)
2 Geostrat SA. Jorge Drom N39-188 y Jose Arizaga, Quito 170507, Ecuador
3 Instituto de Investigación Geológico Energético (IIGE), Quito 170518, Ecuador; wladimir.romero@geoenergia.gob.ec
4 Department of Earth Sciences, University College London, Gower Street, London WC1E 6BT, UK; josel.herrera@espoch.edu.ec (J.L.H.); p.vermeesch@ucl.ac.uk (P.V.)
5 GIDAC, Instituto de Investigaciones, Escuela Superior Politécnica de Chimborazo, Riobamba 060155, Ecuador
6 Department of Earth Sciences, University of Geneva, CH-1205 Geneva, Switzerland; Richard.Spikings@unige.ch
7 Department of Earth Sciences, ETH-Zürich, CH-8092 Zürich, Switzerland; wilfried.winkler@erdw.ethz.ch
* Correspondence: cristian.vallejov@epn.edu.ec; Tel.: +593-998726188

Received: 31 August 2020; Accepted: 10 October 2020; Published: 21 October 2020

Abstract: The Silante Formation is a thick series of continental deposits, exposed along a trench-parallel distance of approximately 300 km within the Western Cordillera of Ecuador. The origin, tectonic setting, age and stratigraphic relationships are poorly known, although these are key to understand the Cenozoic evolution of the Ecuadorian Andes. We present new sedimentological, stratigraphic, petrographic, radiometric and provenance data from the Silante Formation and underlying rocks. The detailed stratigraphic analysis shows that the Silante Formation unconformably overlies Paleocene submarine fan deposits of the Pilalo Formation, which was coeval with submarine tholeiitic volcanism. The lithofacies of the Silante Formation suggest that the sediments were deposited in a debris flow dominated alluvial fan. Provenance analysis including heavy mineral assemblages and detrital zircon U-Pb ages indicate that sediments of the Silante Formation were derived from the erosion of a continental, calc-alkaline volcanic arc, pointing to the Oligocene to Miocene San Juan de Lachas volcanic arc. Thermochronological data and regional correlations suggest that deposition of the Silante Formation was coeval with regional rock and surface uplift of the Andean margin that deposited alluvial fans in intermontane and back-arc domains.

Keywords: provenance analysis; Western Andes; Miocene; Silante Formation; Ecuador

1. Introduction

The Western Andes of Ecuador located between 1° N to 3° S along the active margin of South America (Figure 1) are composed of allochthonous oceanic blocks, which accreted in the late Cretaceous period [1,2]. The events of accretion are partly considered to influence orogenic processes, and the chemical composition of arc volcanism in the Northern Andes [3,4]. Previous studies of the Western Cordillera of Ecuador improved our understanding of the tectonic and stratigraphic evolution of this accretionary complex [5–11]. However, the Cenozoic sedimentary and tectonic evolution of the Western Andes of Ecuador remains poorly understood.

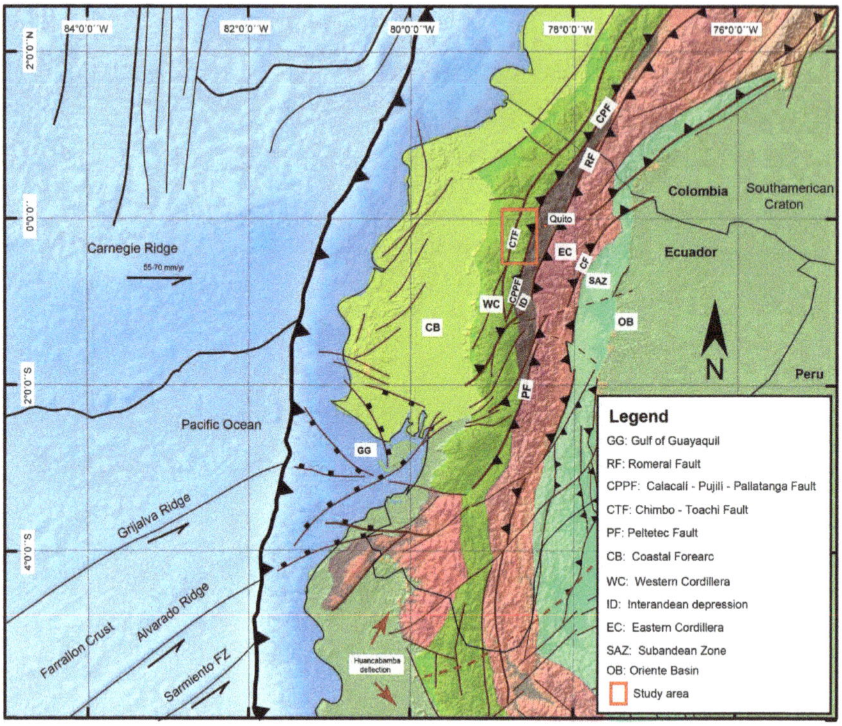

Figure 1. Lithotectonic map of Ecuador (modified from Luzieux et al. [1]) and location of the study area.

The Paleocene to Miocene period is particularly significant because several ore deposits formed over a large latitudinal range (0° to 3°30′ S) within the Western Cordillera during the Cenozoic (42–6 Ma; [12,13]). Thus, a reconstruction of the areal distribution and temporal framework of magmatism and its volcanic products will improve our understanding of the spatial distribution of mineral deposits.

The Silante Formation is a ~2000-m-thick series of red beds with intercalated conglomerates and sandstones rich in volcanic material [14]. It occupies a large area within the Western Cordillera, extending from 1° North to 1° South (Figure 2). The Silante Formation is intruded by mafic (diorite and andesite) dikes and sills [7,15]. Several authors have proposed different stratigraphic and geodynamic interpretations for the origin of these thick continental deposits [5–7,11–18]; however, the stratigraphic relationships of the Silante Formation remain unsolved, mainly because of the discontinuous rock exposures, and the absence of a coherent chronostratigraphic framework.

This study presents new sedimentological, stratigraphic, structural and geochronological data from several exposures of the Silante Formation and the underlying rocks. These data are used to determine the history of sedimentation and the tectonic regime. Radiometric dating and provenance studies including heavy minerals and single grain geochemistry are used to constrain the age, mineralogical composition and approximate location of the sediment source regions, which facilitates tectonic reconstructions and stratigraphic correlations [2]. This study addresses the geological evolution of a segment of the Western Cordillera of Ecuador and includes a detailed description of rock outcrops exposed along east to west road sections that traverse the Western Cordillera (Figure 2). These traverses are the Nono–Tandayapa, Calacalí–Nanegalito, Quito–Santo Domingo, Quito–Chiriboga and Otavalo–Selva Alegre road sections.

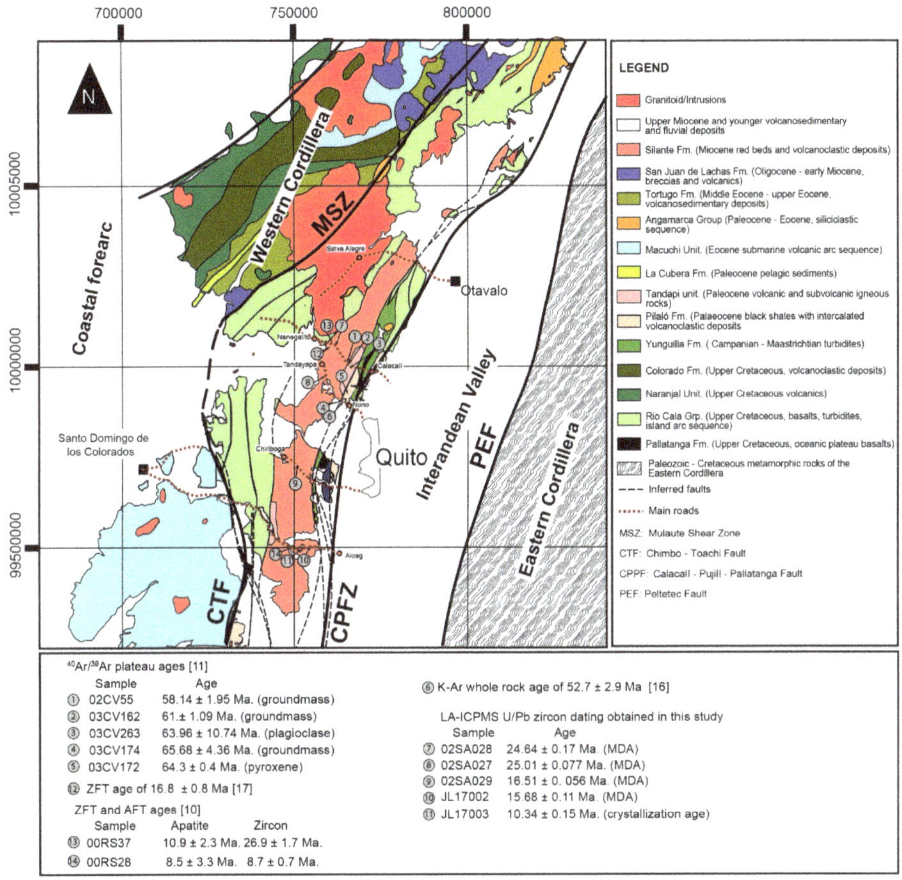

Figure 2. Geological map of the Western Cordillera from 1° North to 1° South, showing the main geological units (modified after [18]) and radiometric ages from the literature [10,11,16,17] and this study.

2. Regional Geology

Ecuador can be subdivided into five morphotectonic regions (Figure 1). (1) The coastal forearc is composed of mafic oceanic crust [1,3,19] and is covered by Paleogene to Neogene forearc deposits [2]. (2) The Western Cordillera (Figure 2) consists of mafic and intermediate extrusive and intrusive rocks that are tectonically juxtaposed with sedimentary rocks of Late Cretaceous to Miocene age [2,9,11]. (3) The Interandean Valley or Interandean Depression is located between the Western and Eastern cordilleras and is covered by thick Quaternary volcanic deposits [20], which are underlain by a crystalline basement composed of metamorphic and mafic rocks [20–22]. On its western flank, the Interandean Valley is bound by the Calacalí–Pujilí Fault [2,23]. This fault system (Figure 2) defines a suture between the South American continental margin and the accreted oceanic rocks [2,23]. (4) The Eastern Cordillera is formed by Paleozoic to Jurassic metamorphic rocks, and Mesozoic granitoids [10,24]. The Eastern Cordillera is separated from the Interandean Valley by the Peltetec Fault, which is the southward continuation of the Silvia-Pijao Fault of Colombia [22,25]. The east verging Cosanga Fault corresponds to the eastern limit of the Eastern Cordillera [24]. (5) The Oriente Basin is a Late Cretaceous–Quaternary retroarc foreland basin that developed on the South American plate margin in response to the weight of the adjacent Eastern Cordillera [26–28].

2.1. The Western Cordillera

The Western Cordillera of Ecuador consists of oceanic mafic rocks interpreted to have accreted to South America during the Late Cretaceous [2,19,29]. Sedimentary and volcano-sedimentary units overlying the allochthonous basement have a complex structural organization (Figure 2) due to the activity of NS striking faults [2,23]. The juxtaposition of turbidite successions and volcano-sedimentary rocks of similar lithologies albeit contrasting depositional ages has complicated stratigraphic correlations and tectonic reconstructions [2].

2.2. The Pallatanga Block and the Allochthonous Basement of Western Ecuador

The Pallatanga block corresponds to the allochthonous basement of the Western Cordillera of Ecuador and includes sedimentary and volcanic formations, which can be grouped into: (1) Basement rocks including basalts of the Pallatanga Formation and ultramafic rock of the San Juan complex (2) Late Cretaceous submarine basaltic lavas and volcaniclastic rocks of the Rio Cala arc; (3) Volcanic and subvolcanic rocks of the Tandapi unit of latest Maastrichtian to Paleocene age [2]; (4) Paleocene–Eocene submarine deposits of the Angamarca Group [23]; and (5) Oligocene–Miocene subaerial volcanic and volcaniclastic rocks of calc-alkaline affinity [2,18]. Oligocene–Miocene volcanism includes the San Juan de Lachas Formation in northern Ecuador [18,30], and the Saraguro Formation in southern Ecuador [31].

The Pallatanga Formation is the basement of the Western Cordillera and includes submarine basaltic lavas and dolerites. The basalts display flat primitive mantle- and chondrite-normalized REE patterns, very similar in chemical composition to basalts from the Caribbean Plateau [1,8], interpreted to have formed in an intraoceanic setting. Oceanic plateaus have thicknesses that are usually more than 10 km, and can exceed 30 km [32], which renders them difficult to subduct due to excessive positive buoyancy. Consequently, oceanic plateau fragments can be incorporated to the continental margin.

The San Juan ultramafic complex is exposed southwest of Quito and includes peridotites, dunites, and layered gabbros. REE geochemistry and isotopic data suggest that the San Juan complex represents the intrusive components of an oceanic plateau [2,33]. Radiometric ages obtained from gabbros of the San Juan complex include a zircon U-Pb age of 87.1 ± 1.7 Ma [11], which is considered to be the most accurate estimate of the crystallization age of the oceanic plateau basement of the Western Cordillera.

2.3. The Yunguilla Formation

The Yunguilla Formation [34], is exposed on the eastern flank of the Cordillera Occidental (Figure 2) and lithologically includes relatively thin beds ranging from 10 to 20 cm in thickness, which display a rhythmic stratification pattern of massive siltstones and fine-grained sandstones that alternate with mudstones. The sandstones contain quartz and display T_{bde} turbidite subdivisions, interpreted as intermediate to distal parts of a submarine fan [17]. Mapping of the Yunguilla Formation along the Western Cordillera is sparse and is mostly based on lithological characteristics, therefore it can be confused with Paleocene and Eocene turbidites of the Angamarca Group.

The presence of ammonites *Phylloceras* sp. and *Exiteloceras* sp. suggest a Late Campanian to Early Maastrichtian age [9]. Heavy mineral assemblages in the Yunguilla Formation include zircon, tourmaline, rutile, garnet and epidote with minor amounts of titanite, anatase and brookite, which indicate significant input of granitic and metamorphic detritus, derived from the reworking of older formations of the present-day Eastern Cordillera [2]. The Yunguilla Formation was deposited in a north–south oriented forearc basin along the South American continental margin [2] and has a tectonic contact with the underlying Pallatanga Formation [23].

2.4. The Tandapi Unit

The Tandapi unit [5] includes a sequence of volcanic rocks, tuffs and conglomerates exposed along the Alóag–Santo Domingo highway. Egüez [7] defined the Tandapi unit as a sequence of andesites and volcanic breccias in transitional contact with the overlying red beds of the Silante Formation.

Lithologically, the Tandapi unit corresponds to andesites and breccias, which have a greenish-grey stain, porphyritic texture and a fresh appearance [5]. The lavas are porphyritic andesites with plagioclase, hornblende and pyroxene within a pilotaxic or hyalopilitic matrix, formed by microliths of plagioclase and alteration minerals. Isotopic data obtained from andesites and basalts of the Tandapi unit in the Alóag–Santo Domingo road indicate that the volcanic rocks erupted through isotopically juvenile rocks [22,35].

Egüez [7] proposed a Paleocene to Eocene age for the Tandapi unit because it was considered to overlie the Upper Cretaceous Pilaton Formation. Vallejo et al. [11] correlated the Tandapi unit with igneous rocks in contact with red beds of the Silante Formation along the Nono–Tandayapa and Calacalí–Nanegalito sections (Figure 4). These authors reported $^{40}Ar/^{39}Ar$ plateau ages of 58.1 ± 1.95 Ma (groundmass), 61 ± 1.09 Ma (groundmass) and 63.96 ± 10.7 Ma (plagioclase), suggesting that the volcanic rocks of the Tandapi unit were deposited during the late Maastrichtian to early Paleocene.

2.5. The Angamarca Group and the Pilalo Formation

The Angamarca Group is a basin fill sequence composed of siliciclastic sedimentary rocks that include turbiditic sandstones, conglomerates and limestone intervals that were deposited from the Paleocene to Oligocene [17]. From base to top the Angamarca Group is subdivided into the Pilalo, Saquisilí, Apagua, Unacota and Rumi Cruz formations [17,22].

The Pilalo Formation was defined by Egüez and Bourgeois [36] as "Volcánicos Pilalo". It contains coarse-grained turbiditic sandstones, black shales, siltstones, reworked tuffs and matrix supported breccias with andesitic fragments. Egüez and Bourgeois [36] reported a lower member of the Pilalo Formation that includes volcanic rocks with green and red detritus, and intercalated lavas, whereas the upper part includes calcareous siltstones. The Pilalo Formation is overlain by Eocene limestones of the Unacota Fm., which is interpreted as a concordant contact [7,22]. Vallejo [22] reported the presence of the Pilalo Formation to the northwest of Quito along the Nono–Tandayapa road section. At this locality, Savoyat et al. [37] described the foraminiferal fauna *Epigonal Rzehakina* within two samples collected near the Alambí River, which indicates a Paleocene age [22]. Vallejo [22] reported a $^{40}Ar/^{39}Ar$ plateau ages of 34.81 ± 1.35 Ma from an andesitic dyke that crosscuts marine sedimentary rocks of the Pilalo Formation, and 65.68 ± 4.36 Ma (groundmass) from lavas at the top of the Pilalo Formation along the Nono–Tandayapa road section (Figure 2).

The Saquisilí Formation includes dark grey micaceous sandstones, siltstones and some calcareous strata [9,17] that were deposited as turbidites. The Saquisilí Formation is limited by faults to the East and West [17] and rests discordantly on pelagic chert of the Campanian-Maastrichtian Yunguilla Formation. Hughes and Bermudez [17] proposed an early to middle Paleocene age based on the presence of foraminifera microfossils collected close to the type locality.

The Apagua Formation overlies the Saquisili Formation consists of medium-grained sandstones, siltstones and shales forming turbiditic beds. The sandstones contain quartz, mafic minerals, lithic fragments and feldspar [2,22]. Foraminiferal fauna from this unit indicate a middle Paleocene to middle Eocene age [17].

The Rumi Cruz unit is the youngest part of the Angamarca Group and includes quartz-rich sandstones with cross-stratification, red mudstones and massive conglomerates, that were probably deposited in a fan delta system [17]. The lithofacies association suggest that the Angamarca Group represents an upward-shallowing progradational succession that records a shift from a submarine fan to fan delta environment [2,22].

2.6. The Oligocene to Miocene San Juan De Lachas Continental Arc

The San Juan de Lachas Formation integrates matrix supported breccias with intercalations of andesitic lavas and volcanoclastic deposits. It unconformably overlies the volcanoclastic rocks of the Late Cretaceous Pilatón Formation [18] and is exposed in the northern part of the Western Cordillera (Figure 2). The San Juan de Lachas Formation was extensively analysed by Van Thournout [38] who reported a K-Ar age (hornblende) of 32.6 Ma from an andesitic dike that crosscuts a sequence of lava and agglomerates of similar composition. Boland [18] reported two K-Ar ages (hornblende) of 19.8 ± 3 Ma and 36.3 ± 2 Ma of andesites rich in hornblende and plagioclase. Samples collected in the Guayllabamba River yield zircon fission track ages of 23.5 ± 1.5 Ma and 24.5 ± 3.1 Ma [18]. Vallejo [22] obtained an $^{40}Ar/^{39}Ar$ plateau age (hornblende) of 32.9 ± 1.2 Ma, from an andesite collected east of the towns of Jijón and Caamaño. Geochemical analysis of the andesitic lavas shows that they are calc-alkaline in composition, and hence probably erupted in a continental arc setting [18,22].

3. The Silante Formation and Its Stratigraphic Problem

The Silante Formation is a continental sedimentary sequence, which includes volcanic-rich sandstones, conglomerates and red mudstones [14]. The Silante Formation is exposed along the Western Cordillera of Ecuador, from the Alóag–Santo Domingo road northward to the Colombian border [2]. Most of the sedimentary rocks of the Silante Formation have experienced very little transport, as revealed by high angularity and poor sorting of crystal grains and the clasts [2,17]. The volcanic-rich sandstones have a purple color and contain crystalline fragments of plagioclase, pyroxene, hornblende, zircon, quartz and abundant lithic fragments. Clasts in conglomerates of the Silante Formation are lithologically similar to the Yunguilla, Rio Cala and Pallatanga formations, suggesting reworking of older sedimentary and volcanic rocks [2]. A three-meter sequence of yellow laminated shales, with the presence of well-preserved angiosperm leaves are reported by Boland et al. [18] in the Calacalí–Nanegalito section. The laminated shales were deposited in a terrestrial environment [18].

There are significant differences in the stratigraphic position, age and depositional environment attributed by different authors (Figure 3). Savoyat et al. [37] proposed a Paleocene age for the Silante Formation based on the presence of foraminifera *Gaudryina* aff. *laevigata Franke*, *Globotruncana* sp. and *Cibicides* sp., which were found in the Silante Formation and were presumably reworked from the Paleocene Pilalo Formation. Henderson [6] indicated the presence of lavas at the top of the Silante Formation and mentions that base and top criteria would indicate that the Silante Formation is covered by the Yunguilla Formation along the Nono–Tandayapa road section (Figure 3). However, the base and top criteria are not clearly described or shown with photographic or stratigraphic evidence. Kehrer and Van der Kaaden [5] proposed a Paleocene to early Eocene age for the Silante Formation and correlated it with Paleocene molasse type deposits of the Tiyuyacu Formation of eastern Ecuador.

Egüez [7] proposed that Silante Formation is transitionally overlying the Tandapi Unit along the Quito–Santo Domingo road, and was contemporary with submarine deposits of the Eocene Apagua Formation. This author also observed the presence of reworked detrital elements in the conglomerates of the Silante Formation, whose lithologies correspond to calc-alkaline lavas similar to the Tandapi unit. The transitional contact between the Tandapi unit and the Silante Formation proposed by Egüez [7] was questioned by Van Thournout [38], who reinterpreted the sequence of lavas and breccias of the Silante Formation and suggest they were deposited during the Oligocene.

The Silante Formation was redefined by Hughes and Bermudez [17] and Boland et al. [18]. These authors included within the Silante Formation andesites, dacites and volcanic breccia intercalations of calc-alkaline affinity, together with the continental red-bed sequence. The non-tectonic contact between the Silante Formation and the mapped Yunguilla Formation along the Calacalí–Nanegalito road was interpreted as a depositional hiatus (paraconformity), suggesting the Silante Formation is post-Maastrichtian (Figure 3).

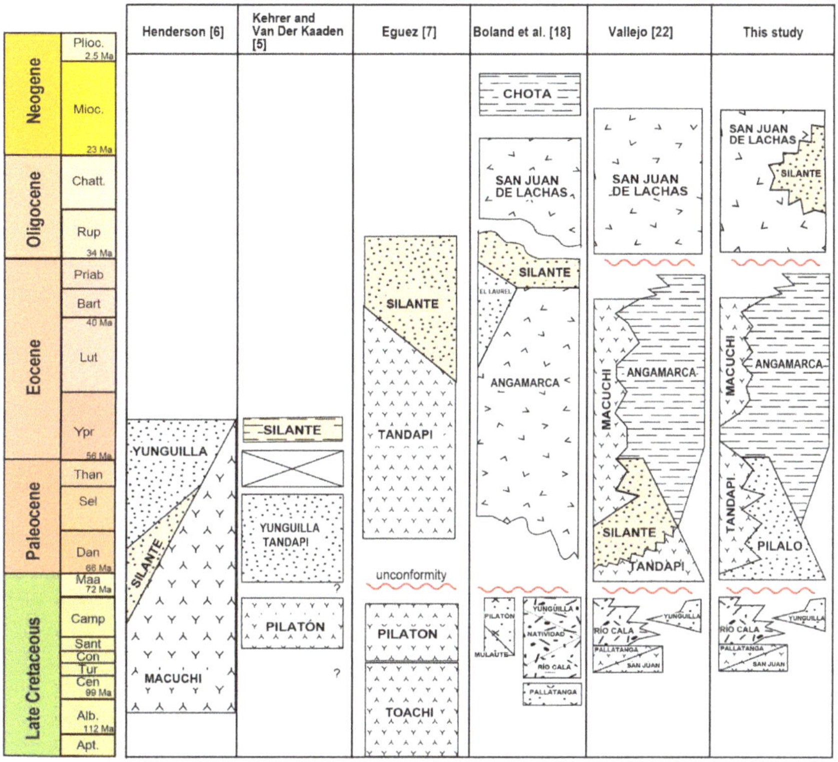

Figure 3. Stratigraphic position of the Silante Formation according to different authors [5–7,18,22]. Time scale from Cohen et al. [39].

Wallrabe-Adams [16] reported a K-Ar age of 52.7 ± 2.9 Ma (whole rock) from a lava collected along the Nono–Tandayapa road. The rock is interpreted to represent the top of the Silante Formation by the author. Hughes and Bermudez [17] assigned a depositional age of 16.8 ± 0.8 Ma, in which is a zircon fission track date obtained from sedimentary rocks collected along the Calacalí–Nanegalito road.

Vallejo [22] obtained $^{40}Ar/^{39}Ar$ plateau ages (groundmass and plagioclase) from igneous rocks mapped by Boland et al. [18] within the Silante Formation. Based on the ~65.68 to 58.1 Ma obtained from these rocks, this author proposed that the Silante Formation was deposited during the Maastrichtian to early Paleocene times (Figure 3).

4. Methodology

Fifteen sandstone samples were used for the study of heavy minerals (Table 1). The sandstones were treated following standard laboratory procedures, which included crushing, sieving and density separation [40,41] using sodium politungstate (density = 2.89 g/cm^3). The heavy minerals were mounted on glass slides using piperine (refractive index = 1.68).

Table 1. Sample locations, and analysis performed in this study.

Sample	Latitude	Longitude	Area	Formation	Lithology	Analysis			
						Heavy Minerals	Clinopyroxene Geochemistry	QFL Analysis	U-Pb Dating
00RS26	0°25′6.94″ S	78°47′27.77″ W	Aloag–Santo Domingo road	Silante	volcaniclastic sandstone		X		
JL17003	0°26′25.22″ S	78°44′8.25″ W	Aloag–Santo Domingo road	Silante	quartzdiorite				X
JL17002	0°26′18.81″ S	78°42′24.19″ W	Aloag–Santo Domingo road	Silante	volcaniclastic sandstone				X
01SA005	0°26′19.14″ S	78°42′23.48″ W	Aloag–Santo Domingo road	Silante	volcaniclastic sandstone	X		X	
01SA008	0°26′19.14″ S	78°42′23.48″ W	Aloag–Santo Domingo road	Silante	volcaniclastic sandstone			X	
02CV135	0°26′50.16″ S	78°42′45.58″ W	Aloag–Santo Domingo road	Silante	volcaniclastic sandstone			X	
00RS34	0°0′22.84″ N	78°34′48.29″ W	Calacali–Nanegalito road	Pilalo	volcaniclastic sandstone	X	X		
00RS35	0°0′22.52″ N	78°35′27.44″ W	Calacali–Nanegalito road	Pilalo	volcaniclastic sandstone		X		
02CV56	0°1′14.72″ N	78°36′44.37″ W	Calacali–Nanegalito road	Silante	volcaniclastic sandstone		X		
04SA049	0°1′19.56″ N	78°36′52.71″ W	Calacali–Nanegalito road	Silante	volcaniclastic sandstone	X		X	
02SA028	0°1′28.45″ N	78°39′1.23″ W	Calacali–Nanegalito road	Silante	volcaniclastic sandstone				X
04SA042	0°0′24.21″ N	78°35′28.54″ W	Calacali–Nanegalito road	Pilalo	volcaniclastic sandstone	X			
04SA053	0°0′14.16″ N	78°35′22.04″ W	Calacali–Nanegalito road	Pilalo	volcaniclastic sandstone	X		X	
04SA034	0°2′0.05″ N	78°40′51.36″ W	Calacali–Nanegalito road	Silante	volcaniclastic sandstone			X	
04SA051	0°1′29.39″ N	78°37′5.60″ W	Calacali–Nanegalito road	Silante	volcaniclastic sandstone			X	
00RS4	2°39′31.22″ S	79°26′47.30″ W	Cuenca–La Troncal road	Yunguilla	arkose sandstone	X			
00RS2	2°38′59.85″ S	79°26′55.05″ W	Cuenca–La Troncal road	Yunguilla	arkose sandstone	X			
WW3311	3°5′7.77″ S	79°0′29.94″ W	Cumbe	Yunguilla	arkose sandstone	X			
03CV172	0°1′49.69″ S	78°33′35.47″ W	Nono	Pilalo	andesite		X		
02SA027	0°1′55.20″ S	78°38′30.84″ W	Nono–Tandayapa road	Silante	volcaniclastic sandstone	X			X
05SA075	0°1′25.03″ S	78°38′43.42″ W	Nono–Tandayapa road	Silante	volcaniclastic sandstone	X			

Table 1. *Cont.*

Sample	Latitude	Longitude	Area	Formation	Lithology	Analysis			
						Heavy Minerals	Clinopyroxene Geochemistry	QFL Analysis	U-Pb Dating
02SA016	0°3′30.12″ S	78°36′41.58″ W	Nono–Tandayapa road	Pilalo	volcaniclastic sandstone	X			
02SA019	0°3′22.02″ S	78°36′56.84″ W	Nono–Tandayapa road	Pilalo	volcaniclastic sandstone	X			
05SA066	0°3′21.69″ S	78°36′55.48″ W	Nono–Tandayapa road	Pilalo	volcaniclastic sandstone			X	
05SA072	0°1′54.91″ S	78°38′30.55″ W	Nono–Tandayapa road	Silante	volcaniclastic sandstone			X	
05SA074	0°1′25.03″ S	78°38′43.42″ W	Nono–Tandayapa road	Silante	volcaniclastic sandstone			X	
02CV99	0°16′37.91″ N	78°27′40.76″ W	Otavalo–Selva Alegre road	Silante	volcaniclastic sandstone		X	X	
CV380	0°17′6.93″ N	78°29′58.32″ W	Otavalo–Selva Alegre road	Silante	volcaniclastic sandstone			X	
02SA029	0°16′42.49″ S	78°42′31.86″ W	Quito–Chiriboga road	Silante	volcaniclastic sandstone	X			X
06SA081	0°16′43.63″ S	78°42′59.28″ W	Quito–Chiriboga road	Silante	volcaniclastic sandstone	X		X	
02CV33	0°17′51.60″ S	78°38′56.46″ W	Quito–Chiriboga road	Pilalo	volcaniclastic sandstone	X			
06SA078	0°16′42.49″ S	78°42′31.86″ W	Quito–Chiriboga road	Silante	volcaniclastic sandstone			X	

X symbol in the table corresponds to the analysis performed to each sample.

Heavy mineral identification was made based on the optical properties of individual minerals, using a transmitted light Zeiss Primotech microscope. The mineral proportions were estimated by counting 300 detrital grains, following the procedure described by Mange and Maurer [40].

The provenance analysis of this study included the petrographic modal analysis of 14 sandstones, determining the relative proportions of detrital grains, which can be used to estimate the tectonics and paleogeography of the source regions [42,43]. The detrital components have been divided into feldspar (F), total quartz (Q), polycrystalline quartz (Qp), monocrystalline quartz (Qm), total lithic (Lt = L + Qp), lithic fragments (L), metamorphic lithic fragments (Lm), sedimentary lithic fragments (Ls), and volcanic lithic fragments (Lv). A total of 300 points were counted for each thin section using the Gazzi-Dickinson point-counting method [44,45]. The detrital components were plotted in QmFLt, QFL and LvLmLs ternary discrimination diagrams proposed by Dickinson et al. [46] and Dickinson [43]. Analyses were performed at the Petrology Laboratory of the Escuela Politécnica Nacional, Quito, Ecuador.

Single grain clinopyroxenes geochemistry was used to define the volcanic affinity of the volcaniclastic rocks of the Pilalo and Silante formations. Clinopyroxene grains were separated from the heavy mineral concentrates using a Frantz magnetic separator. Individual grains were mounted in an epoxy capsule and analyzed for major oxides and REE at ETH-Zürich [22]. Rock samples were mechanically and chemically disaggregated and processed according to standard heavy mineral separation techniques [28,40,41]. U-Pb ages of detrital zircons were obtained from four sandstones of the Silante Formation and from an intrusion that crosscuts this formation. Zircons were picked up from the non-magnetic fraction of the heavy minerals. Each sample was analyzed in a multi-selector laser-coupled plasma-mass spectrometer (LA-ICP-MS) at the University College London (UCL). All ages were obtained using a New Wave 193 nm aperture-imaged frequency-quintupled laser ablation system coupled to an Agilent 7700 quadrupole-based ICP-MS (Agilent Technologies, Santa Clara, CA, USA). Operating condition for zircon dating uses an energy density of ca 2.5 J/cm^2 and a repetition rate of 10 Hz. Repeated measurements of external zircon standard Plešovice [47]) and NIST 612 silicate glass [48] are used to correct for instrumental mass bias and depth-dependent inter-element fractionation of Pb, Th and U. 91500 zircon [49] was used as secondary age standard. Data were processed using GLITTER data reduction software (v. 4.4, Gemoc, Sydney, Australia). Final data processing, statistical processing and graphs development were done using the software IsoplotR (v. 3.3, UCL, London, UK) online mode [50].

Maximum depositional ages for the Silante Formation were calculated using the minimum age model of Galbraith [51], included with the radial plot functionality of RadialPlotter [52] and IsoplotR [50]. The radial plot is a graphical device that was also invented by Galbraith [53] with the aim to simultaneously visualize measurements and their uncertainties. The minimum age calculation algorithm of Galbraith [51] converges to a specific value with increasing sample size.

5. Results

5.1. Stratigraphy of the Nono–Tandayapa and Calacalí–Nanegalito Road Sections

The Nono–Tandayapa and Calacalí–Nanegalito road sections include almost continuous exposures of the Silante Formation and underlying series. In these sections Boland et al. [18] assumed a non-tectonic contact between the Yunguilla and Silante formations. This contact was interpreted as a possible depositional hiatus (discordance). However, biostratigraphic studies based on foraminiferal fauna collected in the Alambi River by Savoyat et al. [36] yielded a Paleocene age (Danian) for sedimentary rocks mapped as the Yunguilla Formation. Therefore, Vallejo [22] discarded the presence of the Yunguilla Formation and proposed that the Paleocene rocks exposed in this area are correlatable with the Paleocene Pilalo Formation rather than the Yunguilla Formation. In this study, we follow the stratigraphy proposed by Vallejo [22] and present new stratigraphic and sedimentological data from the Nono–Tandayapa and Calacalí–Nanegalito road sections (Figure 4).

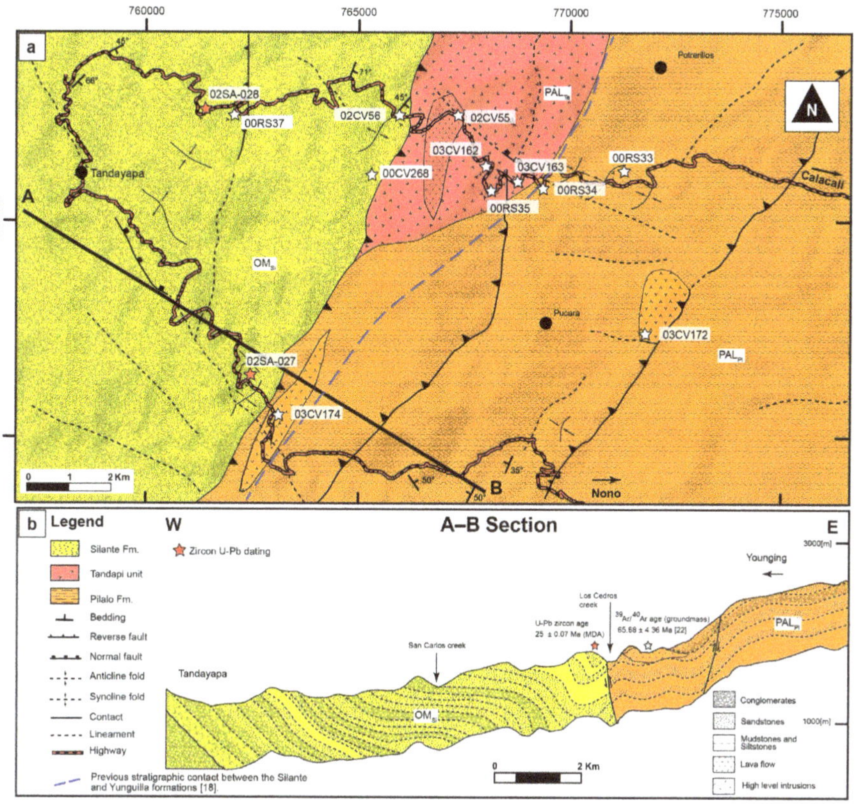

Figure 4. (**a**) Geological map of the area of study, and (**b**) schematic cross sections of the Nono–Tandayapa road section.

The western series along the Nono–Tandayapa road section corresponds to the Silante Formation (Figure 4), while the central series is an igneous sequence (Tandapi unit) that intrudes a submarine sedimentary sequence of ~1000 m in thickness, ascribed to the Pilalo Formation. In the analyzed road sections, there is a tectonic contact between the Silante Formation and the Tandapi unit, which is located between UTM 766182/2365 and 766106/2020 on the Calacalí–Nanegalito road (Figure 4). In the Nono–Tandayapa section the tectonic contact between the Pilalo Formation and the Silante Formation is located at los Cedros Creek (UTM 763032/9995645).

The Pilalo Formation is a coarsening-upwards submarine succession exposed at the eastern border of the studied section (Figure 5). The sedimentary rocks are folded and intercalated with lavas at the top. The intercalated andesitic to basaltic lavas yield a $^{39}Ar/^{40}Ar$ plateau age (groundmass) of 65.68 ± 4.36 Ma [2] that is coeval (within uncertainty) with the host sedimentary rocks.

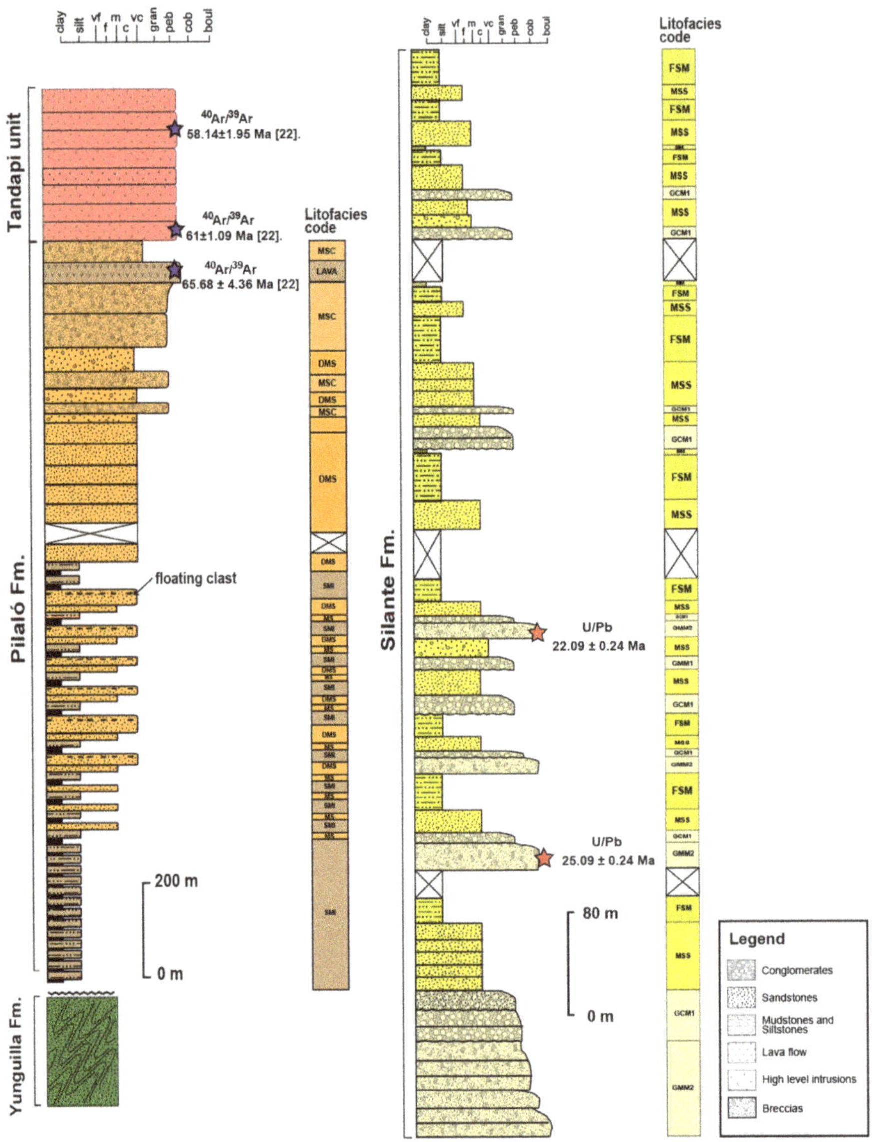

Figure 5. Composite stratigraphic column of the Nono–Tandayapa road section with lithofacies described for the Pilalo and Silante formations (lithofacies codes are discussed in text). Radiometric ages obtained by Vallejo [22] and this study are located with the proposed composite stratigraphic column of the study area.

The rocks at the top of the sedimentary sequence of the Pilalo Formation have a reddish color due to oxidation, and thus they were previously mapped as the Silante Formation [18]. However, they resemble turbidite beds falling out from highly concentrated flows with Bouma T_{ade} subdivisions.

Porphyry type high level intrusions mapped as part of the Tandapi unit crosscut the Pilalo Formation close to the contact with the Silante Formation (Figure 4). Vallejo [22] reported $^{39}Ar/^{40}Ar$

plateau ages of two samples collected in the Calacalí–Nanegalito road (Figure 2) that yielded ages of 61 ± 1.09 Ma (groundmass) and 63.96 ± 10.74 Ma (plagioclase).

Our structural, biostratigraphic and radiometric ages obtained along the Calacalí–Nanegalito and Nono–Tandayapa sections suggest that the previous stratigraphic contact between the Silante and Yunguilla Formation, as proposed by Boland [18], does not exist. Sedimentary rocks of the Yunguilla Formation were not identified in the present study area. Instead, the Pilalo Formation together with the high-level intrusion of the coeval Tandapi unit were thrusted onto the Silante Formation.

5.2. *Sedimentology of the Pilalo Formation*

The Pilalo Formation exposed along the Calacalí–Nanegalito and Nono–Tandayapa road sections is a thick sequence composed of sedimentary rocks deposited in a submarine environment. The bed-sets show an upward-coarsening and thickening trend, suggesting a progradational sequence stratigraphic pattern (Figure 5). In the Pilalo Formation, we identified four lithofacies, which are described in the following section. The acronym indicates the dominant lithology and the most important sedimentological features (Table 2).

Table 2. Lithofacies of the Pilalo and Silante formations.

Lithofacies	Description	Interpretation
	Pilalo Formation	
SMI	Intercalations of dark grey laminated siltstones and mudstones	Silty-muddy turbidites (Bouma T_{de} subdivisions) generated by low-density turbidity currents in the distal part of a submarine fan
MS	Massive sandstones with load cast, and volcanic clasts	Sandy turbidites generated by high-density turbidity current infilling channels in the middle part of a submarine fan
DMS	Massive sandstones with floating clasts of volcanic origin	Sandy debrites deposited by plastic flows in channels of the middle part of a submarine fan
MSC	Matrix-supported conglomerates with volcanic clasts	Conglomerates generated by debris flows. The common association with turbidites and sandy debrites suggests the upper part of a submarine fan
	Silante Formation	
MM	Reddish unstructured mudstones and siltstones with poorly developed parallel lamination.	Reddish color in mudstones suggest floodplain deposits deposited under oxidizing conditions
FMS	Reddish siltstones and fine-grained sandstones	Distal facies of alluvial fan margins in semi-arid continental environments
MSS	Structureless fine-grained sandstones	Sandy lithofacies deposited by a hyper-concentrated flow in the middle part of an alluvial fan
SPL	Reddish to yellowish sandstones with parallel lamination	Channels or sheet flood deposits deposited in the middle part of an alluvial fan
GMM1	Massive matrix-supported conglomerates	Deposits produced by debris flows in the upper part of an alluvial fan
GCM1	Massive clast-supported conglomerates	Deposits formed by debris flows in the upper part of an alluvial fan system
GMM2	Matrix-supported conglomerates, poorly sorted	Proximal facies of an alluvial fan depositional system, close to the alluvial fan head

5.2.1. Lithofacies SMI: Siltstones with Mudstone Intercalations

The lithofacies consists of a sequence of rhythmically bedded dark grey mudstone with siltstones (Figure 6a). The siltstones show parallel lamination and the beds range in thickness from 10 to 20 cm. Sporadic calcareous cement is also observed in this lithofacies. Some of these beds are rich in marine microfossil fauna, which are mostly foraminifera. This lithofacies dominates the lower part of the succession (Figure 5).

Figure 6. Photographs of representative lithofacies of the Pilalo Formation. (**a**) Lithofacies SMI, siltstone with mud intercalations; (**b**) lithofacies MS, structureless sandstones; (**c**) lithofacies DMS, massive sandstone with floating clasts; (**d**) lithofacies MSC, matrix supported conglomerates.

Parallel lamination could be a result of two main processes: (1) decelerating turbidity currents [54,55], or (2) bottom-current reworking [56,57]. The presence of parallel laminated siltstones with mudstone intercalations suggests silty-muddy turbidites (Bouma T_{de} subdivisions) that were generated by low-density turbidity currents in the distal part of a submarine fan [57].

5.2.2. Lithofacies MS: Massive Sandstones with Load Cast

Lithofacies MS consists of medium to coarse-grained sandstones composed of plagioclase, pyroxene and amphibole, which are accompanied by a few dark-colored siltstones, and green mudstone intrabasinal lithic fragments (Figure 6b). The bed thickness of this lithofacies varies between 0.5 to 1 m. The base of the sandstone beds displays load cast, and ball-and-pillow structures.

The presence of load cast and ball-and-pillow structures suggest rapid deposition of water-rich sediment by high-density turbidity currents (Bouma T_a subdivisions). Presumably, they represent fill deposits in channels in a middle fan environment [55,57]. The presence of plagioclase, pyroxene and amphibole suggests a volcanic source. The green color of the mudstones is a result of alteration of volcanic glass to chlorite.

5.2.3. Lithofacies DMS: Massive Sandstones with Floating Clasts

This lithofacies is characterized by medium to coarse-grained sandstone beds that are 0.3 to 0.8 m thick, with a massive structure. The sandstones are green and include plagioclase, pyroxene, chlorite and epidote crystals. It is common to find angular and sub-rounded volcanic and sedimentary floating clasts. The main characteristic of this lithofacies are intrabasinal floating clasts occurring at the top of the beds (Figure 6c).

Massive sands with floating clasts are either deposited by (1) high-density turbidity currents [58], or (2) sandy debris flows where the transport mechanism is a plastic flow and deposition is caused by mass freezing of sediments [59]. The abrupt freezing of the sediments impedes settling of the clasts towards the bottom of the stratum. Floating clasts in sandstones interspersed with siltstones and mudstones can be deposited by plastic flows defined as sandy debris flows [60]. This lithofacies represents infilling deposits of channels in a middle fan environment. The presence of plagioclase and pyroxene suggests a volcanic source. Epidote and chlorite formed via the alteration of mafic minerals and volcanic glass, respectively.

5.2.4. Lithofacies MSC: Matrix-Supported Conglomerates

Lithofacies MSC is characterized by the presence of green, matrix-supported conglomerates with subangular, partially oxidized andesitic clasts, with clast sizes that vary from pebbles to granules, and bed thickness that span between 1 to 2 m. The base of these deposits is erosive, forming a scour surface on top of the underlying beds (Figure 6d). The matrix of the conglomerates is composed of coarse-grained sandstone carrying plagioclase, amphibole, chlorite and epidote crystals.

The angularity of the clasts and the poorly sorted nature of the deposit reveals that they have not been transported over long distances. The structureless nature, together with the immaturity of clasts suggest a deposit generated by debris flows [61]. These flows are erosive and channelized, forming scoured lower contacts, and settled very rapidly, resulting in massive and ungraded beds [62]. The association with sandy debrites and turbidite beds (lithofacies DMS and MS) suggests that MSC lithofacies was formed within a submarine upper fan environment [60]. The mineralogy of the matrix and the clast lithology suggest a volcanic source.

5.3. Sedimentology of the Silante Formation

Alluvial fan sedimentary rocks of the Silante Formation are exposed along the Nono–Tandayapa and Calacalí–Nanegalito sections. We recognized 7 lithofacies in the Silante Formation, which has a thickness of approximately 900 m in this region (Figure 5).

5.3.1. Lithofacies MM: Mudstones

This lithofacies includes reddish couples of mudstones and siltstones (Figure 7a) with an approximate thickness that range from 1 to 3 cm. The mudstones are unstructured, and the siltstones include poorly developed parallel lamination. The beds have a tabular geometry and sharp basal contacts. This lithofacies occurs intercalated with the FMS lithofacies (see below).

The reddish color of the mudstones suggests subaerial deposition in a terrestrial environment under oxidizing conditions. Parallel lamination in the siltstones suggests they are floodplain deposits, which can develop in an inactive part of the alluvial fan system that occasionally receives sediment during flood events [63].

Figure 7. Photographs of representative lithofacies of the Silante Formation. (**a**) Lithofacies MM, mudstones and siltstone;s (**b**) lithofacies FMS, structureless sandstones; (**c**) lithofacies MSS, massive sandstone; (**d**) lithofacies SH, matrix supported conglomerates; (**e**) lithofacies GMM1, massive matrix supported conglomerates; (**f**) lithofacies GCM1, massive clast supported conglomerates; (**g**) lithofacies GMM2, matrix supported conglomerates; (**h**) representative outcrop of the Silante Formation and the spatial distribution of the lithofacies.

5.3.2. Lithofacies FMS: Massive Siltstones

The FMS lithofacies consists of reddish, unstructured beds that vary in grain size from silt to fine-grained sandstones (Figure 7b), with bed thicknesses that range from 0.3 to 0.8 m. Individual beds are tabular and associated with lithofacies MM.

The reddish siltstones and very fine-grained sandstones may represent distal facies of an alluvial fan in a semi-arid continental environment [64–66]. The presence of thin beds of fine-grained lithofacies suggests that they are distal facies of abandoned progradational lobes [64]. The association of silt and very fine sandstones (FMS) suggests deposition within quiet segments of alluvial fans, with sediments supplied during occasional flood events [67,68].

5.3.3. Lithofacies MSS: Massive Sandstones

The MSS lithofacies contains poorly sorted, ungraded fine sandstone beds with thicknesses up to 1 m and a tabular geometry (Figure 7c). The most abundant minerals within the sandstones are plagioclase, magnetite and mafic minerals, including pyroxene and amphibole. This lithofacies is usually associated with the FMS and MM lithofacies.

The unstructured sandstones were probably deposited from hyper-concentrated flows that were characterized by very high sediment to water ratios [69]. The lack of internal sedimentary structure in these poorly sorted sandstones suggest an abrupt deceleration and quick deposition of sediment with insufficient time to create bedforms [64,70]. This lithofacies commonly occurs within middle alluvial fans proximal to the mountain front, where the flow becomes unconfined, dewaters, and rapidly settles [71].

5.3.4. Lithofacies SPL: Sandstones with Parallel Lamination

The lithofacies SPL consists of reddish to yellow stained sandstones (Figure 7d), with grain sizes that vary from coarse to very coarse (0.5 to 2 mm). The sandstones have parallel lamination and are composed of plagioclase, pyroxenes, amphiboles and magnetite crystals.

The reddish color is considered to be a consequence of deposition in an oxidizing environment [72,73]. The parallel lamination of the fine to coarse-grained sandstones indicates a common transition to an upper-flow regime and deposition of planar bed flows in channels or sheet flood deposits [64,68]. The SPL Lithofacies was deposited in a middle alluvial fan area [68,70].

5.3.5. Lithofacies GMM1: Massive Matrix Supported Conglomerates

Lithofacies GMM1 comprises matrix supported conglomerates with thicknesses varying between 0.10 to 0.15 m. This lithofacies shows reddish and greyish colors with clast sizes ranging from granules to pebbles (Figure 7e). The matrix supported fabric is mainly composed of medium to coarse sand, which is rich in plagioclase and ferromagnesian minerals. Internally, the matrix-supported conglomerates are unstructured, while the andesitic clasts are sub-rounded.

Matrix supported conglomerates are generally produced by debris flows [74]. Due to the difficulty of transporting gravel clasts, these particles are concentrated at the base of a turbulent flow, forming a dense inertial layer. Lithofacies GMM1 was deposited in a subaerial environment within the upper part of an alluvial fan [68,73].

5.3.6. Lithofacies GCM1: Massive Clast Supported Conglomerates

The lithofacies GCM1 includes reddish, clast-supported conglomerate beds with thicknesses ranging from 0.6 to 2 m. The layers show moderate sorting, sharp bases and rounded clasts that range in size from granule to pebbles (2 to 32 mm), with high sphericity. The matrix is composed of fine to very fine particles, without a preferential orientation (Figure 8f). The clasts are dominantly andesites, with minor diorites.

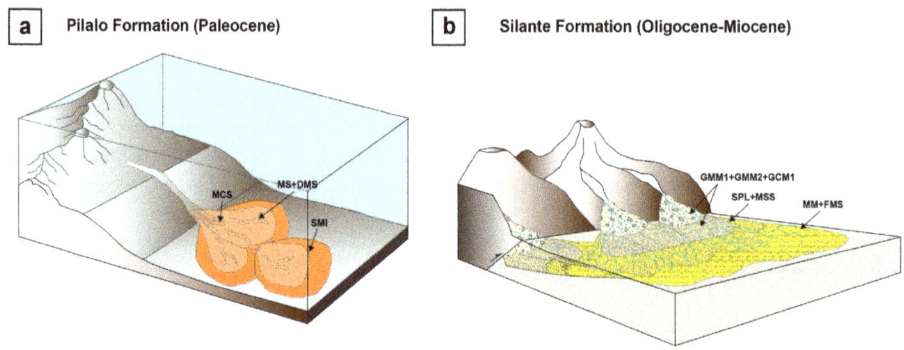

Figure 8. Depositional models for the Pilalo and Silante formations. (**a**) Submarine fan depositional model of the Pilalo Formation, (**b**) debris flow dominated alluvial fan model of the Silante Formation.

Clast supported conglomerates within alluvial fans are formed when clast-rich debris flows are deposited onto overlying beds that have a high permeability. The fine-grained fraction of the debris flow percolates downwards through the pore spaces of the underlying strata, producing a clast-supported conglomerate [68]. The GCM1 lithofacies infills channels in the upper part of an alluvial fan system [68].

5.3.7. Lithofacies GMM2: Matrix Supported Conglomerates

Lithofacies GMM2 includes poorly sorted dark green conglomerates with clast sizes that vary from granules to blocks, with edges varying from angular to sub-rounded. Most of the larger clasts with sub-rounded to subangular edges are andesites and basaltic andesites (Figure 7g). The smaller clasts with angular edges are similar to lithofacies FMS and MM. The matrix of this lithofacies is composed of fairly well-preserved crystals, with sizes that range from coarse to very coarse sand (0.5 to 2 mm), which is rich in plagioclase, pyroxenes and amphiboles. Inverse grading was observed at the base of the beds.

Poorly sorted and structureless conglomerates with angular clasts may represent the proximal facies of an alluvial fan depositional system [68,75,76], close to the alluvial fan head.

5.4. Depositional Environment of the Pilalo and Silante Formations

The lithofacies association of the Pilalo Formation suggests that the sediments were deposited within a submarine fan (Figure 8a). The distal fan deposits are intercalated with hemipelagic sediments, which corresponds to the transition of a distal depositional lobe to the basin plain [54–57]. The middle fan includes sandy channel deposits [68]. The coarse-grained sediments were deposited by debris flows in channels, which probably formed within the upper part of the fan [61,68]. In addition, the presence of andesitic clasts and abundant mafic minerals suggests a nearby volcanic arc.

Based on the observed lithofacies of the Silante Formation, we suggest that the sedimentary rocks formed in a debris flow dominated alluvial fan (Figure 8b). The upper fan consists of conglomeratic lithofacies deposited by debris flow process [67,68,74–76]. These deposits are mainly exposed in the westernmost part of the Silante Formation. In the Calacalí–Nanegalito section, these coarse deposits occur in greater proportions in the lower part of the stratigraphic column (Figure 5). The middle fan includes reddish sandy channels or sheet flood sediments that were deposited by water-laid processes [64,68–70,72]. The distal alluvial fan and flood plain deposits are represented by interfingering of siltstones and mudstones [65,68,73,75]. These fine-grained deposits are commonly observed in the eastern part of the Calacalí-Nanegalito and Nono-Tandayapa sections (Figure 4).

5.5. Provenance Analysis of the Pilalo and Silante Formations

Provenance analyses of clastic deposits of the Pilalo and Silante formations is used to refine the stratigraphic and tectonic interpretation by deciphering shifts in sediment source and source area location. Provenance shifts are associated with the evolution of the sedimentary basins, and hence the tectonic setting [2].

Here, we combine heavy mineral analyses, single grain geochemistry of clinopyroxenes and U-Pb detrital zircon dates. The U-Pb ages of detrital zircons can be used to estimate the maximum depositional age (MDA). Given that these units were deposited proximal to a magmatic arc, it is reasonable to suggest that the zircon U-Pb dates are close approximations of the time of deposition of the sedimentary rocks [77,78].

5.6. Heavy Mineral Data

The detrital assemblages of clastic sedimentary rocks reflect the mineralogical composition of the source regions that fed the catchment basin and its depositional history [2,79–82]. In this study, twelve sandstones of the Silante Formation and the underlying Pilalo Formation were selected for heavy mineral analyses (Figure 9). For comparison, the heavy mineral assemblages of three samples from the Campanian-Maastrichtian Yunguilla Formation are shown in Figure 9. Sample locations are presented in Table 1.

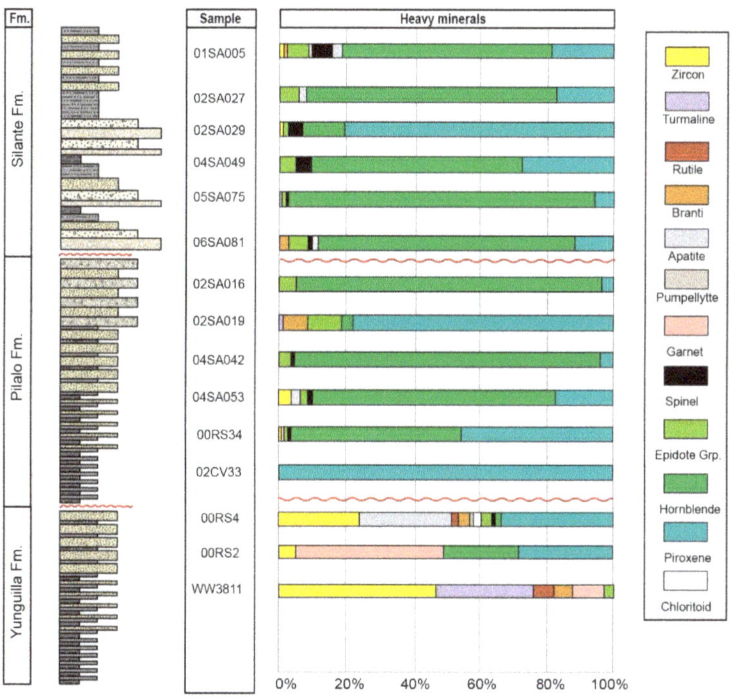

Figure 9. Composite stratigraphic column of the Yunguilla, Pilalo and Silante formations and heavy mineral frequencies (right). A clear change is observed in the composition of the Pilalo and Silante formations, which were derived from volcanic sources, whereas Campanian–Maastrichtian sediments of the Yunguilla Formation were shed from granitic and metamorphic sources that formed part of a continental plate, and now constitute the Eastern Cordillera.

All samples from the Pilalo Formation contain a high percentage of pyroxene (modal average of 26%), hornblende (average of 65%) and minerals of the epidote group (average ~4%). These samples yield a null ZTR (zircon-tourmaline-rutile assemblages) index with the exception of rocks 02SA019 and 04SA053, which give a low ZTR index due to a higher percentage of zircons (~3%) and tourmaline.

Samples from the Silante Formation contain a high amount of pyroxene (~27%), hornblende (~64%) and minerals of the epidote group (~3%). In general, samples of the Silante Formation have a very low ZTR index, and no metamorphic accessory minerals have been found. Rocks 01SA005 and 02SA029 contain euhedral zircon crystals (~1%), which are interpreted to have a volcanic origin. The high amounts of pyroxene, hornblende and apatite (~64%) suggests that the source area was strongly dominated by intermediate volcanic rocks [40].

For the Yunguilla Formation, the heavy minerals assemblages of sample WW3311 host a significant ZTR assemblage (82%), with minor amounts of brookite, anatase and titanite (~5%). Garnet and epidote represent ~13%. The heavy mineral in sample 00RS2 include pyroxenes and brown hornblendes (50%) that dominate over smaller amounts of garnet (42%) and zircon (6%). Sample 00RS4 of the Yunguilla Formation contains abundant ZTR group minerals (55%), while pyroxene and hornblende account for ~35%. The heavy minerals assemblages for the Yunguilla Formation suggest that sediments were derived from the erosion of a mixed granitic, metamorphic, and volcanic source.

Overall, the heavy mineral assemblages (Figure 9) reveal a strong volcanic input into the Pilalo and Silante formations, whereas the Yunguilla Formation was predominantly derived from the erosion of granitic rocks and/or metamorphic rocks of the Eastern Cordillera.

5.7. Clinopyroxene Single Grain Geochemistry of the Pilalo and Silante Formations

The chemical composition of clinopyroxene is directly related to the chemistry of their host lavas [81,82] and varies according to the magma type, and hence the tectonic setting [81–84]. Beccaluva et al. [84] proposed that the clinopyroxene compositional variability is mostly related to differences in the bulk chemistry of the host magmas, and is only partially due to physical conditions of crystallization, and magmatic fractionation.

We present major element compositions of clinopyroxenes from rocks of the Pilalo and the Silante formations. Samples 00RS34 and 00RS35 of the Pilalo Formation were collected in the Calacalí-Nanegalito road section, and andesite 03CV172 was collected to the north of the Nono locality (Figure 4). Three reddish volcanoclastic sandstones of the Silante Formation were sampled along the Alóag–Santo Domingo road (00RS26), the Otavalo–Selva Alegre road (02CV99), and from the Calacalí–Nanegalito road (02CV56). The full data of major element composition of clinopyroxene is available in the online supplementary material (Table S1).

All analysed clinopyroxenes lack zoning and were optically and chemically homogeneous. Major oxides geochemical data from the analysed clinopyroxenes yields diopsidic to augitic compositions.

To determine the magmatic affinity of the source of the detrital clinopyroxenes we used the discriminatory diagrams of Leterrier et al. [82] (Figure 11). The clinopyroxenes from sandstones of the Silante Formation were derived from a subalkaline volcanic source, although the discriminatory diagram of Al and Ti does not distinguish between a calc-alkaline or tholeiitic composition for these samples (Figure 10). However, the low Al and magnesium numbers (0.62 to 0.65) suggest that the source regions were composed of fractionated rocks, which generally supports a calc-alkaline affinity.

Detrital clinopyroxenes of the Silante Formation plot in the field of subalkaline basalts ($Ti < 0.025$ apfu), which formed within a subduction zone setting (Figure 11). However, a comparison of Al and Ti does not show a clear distinction between tholeiitic and calc-alkaline affinities. This can be due to the presence of clasts of the Pallatanga Formation and Rio Cala Group, which have a tholeiitic affinity [22].

Clinopyroxene from the Pilalo Formation gave a tholeiitic composition (Figure 11a,c) and probably formed in a subduction zone setting (Figure 11b).

In addition, we used clinopyroxene REE compositions to decipher the geochemical affinity of the magmatic source. The analysed samples include a lava within the Pilalo Formation (03CV172), and two sandstone samples of the Silante Formation (00RS26, and 02CV56).

LREE enrichments (relative to HREE) is regularly found in rocks formed in continental volcanic arcs, whereas a depletion of LREE, and flat REE patterns are more indicative of a primitive island arc, MORB and mantle-plume related, volcanic rocks [85]. In order to determine the composition of the clinopyroxene parent magmas, REE compositions of melts in equilibrium with these minerals were calculated using experimentally derived partition coefficients for clinopyroxenes crystallizing in basaltic rocks [86]. REE values of the analyzed samples were normalized using chondrite values published by Sun and McDonough [87]. The full data of REE composition of clinopyroxene is available in the online supplementary material (Table S2).

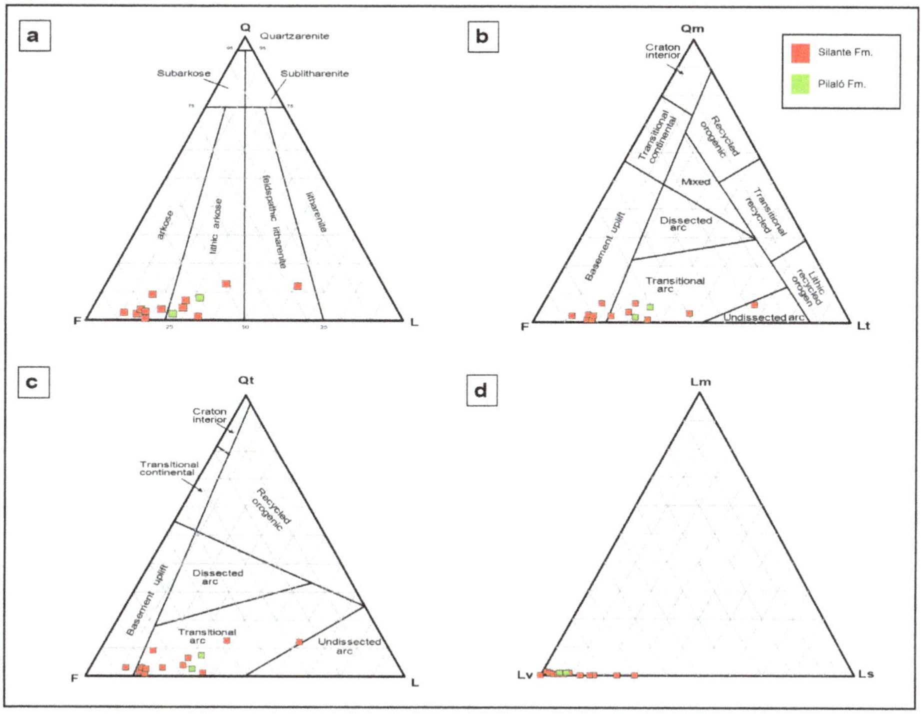

Figure 10. Sandstone ternary diagrams for the Pilalo and Silante formations. (**a**) Ternary diagram with discrimination fields using the terminology of Folk [88]. (**b**) Ternary discriminatory diagram of monocrystalline quartz (Qm), feldspar (F) and total lithics (Lt) after Dickinson [43]. (**c**) Ternary discriminatory diagram of quartz (Qt), feldspar (F) and lithics (L) after Dickinson [42]. (**d**) Ternary discriminatory diagram of metamorphic lithics (Lm), sedimentary lithics (Ls) and volcanic lithics (Lv) after Dickinson [43].

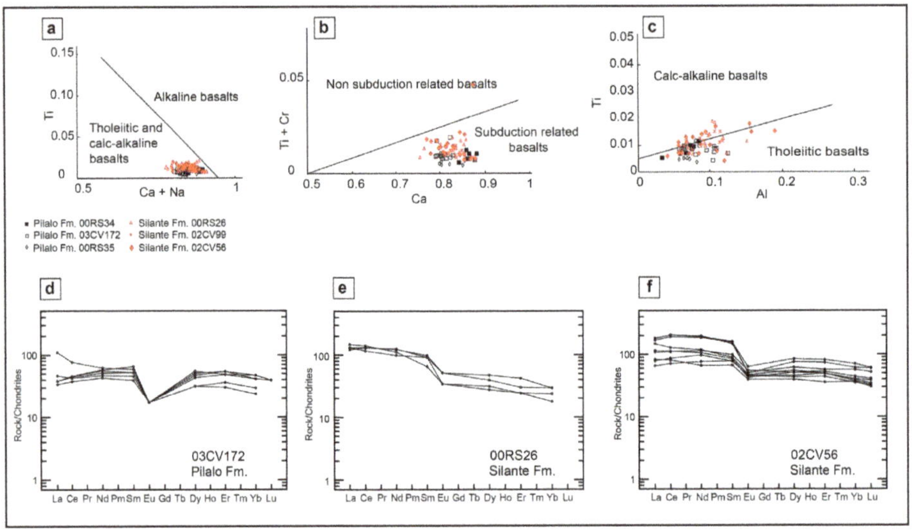

Figure 11. Geochemistry of detrital clinopyroxenes from the Silante and Pilalo formations. (**a**) Ti vs. Ca + Na discriminatory diagram from Leterrier et al. [82]. (**b**) Ti + Cr vs. Ca discriminatory diagram from Leterrier et al. [82]. (**c**) Ti vs. Al discriminatory diagram from Leterrier et al. [82]. (**d**) Chondrite-normalized REE plots of calculated melts in equilibrium with clinopyroxenes of a lava intercalated within the Pilalo Formation (sample 03CV172). (**e**) Chondrite-normalized REE plots of calculated melts in equilibrium with clinopyroxenes of the Silante Formation (sample 00RS26). (**f**) Chondrite-normalized REE plots of calculated melts in equilibrium with clinopyroxenes of the Silante Formation (sample 02CV56). Chondrite normalizing values after Sun and McDonough [87].

For sample 03CV172, the calculated melt in equilibrium with the clinopyroxenes yields a REE chondrite normalized profile that is generally flat (Figure 11), and the $(La/Yb)_N$ ratio is ~1, suggesting a primitive nature of the magmas in which the clinopyroxene crystallized. The calculated melt in equilibrium with clinopyroxenes extracted from sandstones of the Silante Fm. shows LREE enrichments up to 100 times chondritic values (Figure 11e,f). The $(La/Yb)_N$ ratios vary between 5.32 (00RS26) and 2.58 (02CV56), which are typical for arc rocks [86].

Summarizing, clinopyroxenes of the Silante Formation are compositionally distinct from clinopyroxenes extracted from the Pilalo Formation. The enrichment of LREE and low Ti values of detrital clinopyroxenes of the Silante Formation (Figure 11e,f) are probably the result of dehydration of subducted oceanic crust, which released fluids that have low concentrations of Ti [82] and high concentrations of REE. The LREE enrichment, together with the low to medium concentrations of Al suggests that the volcanic source rocks of the Silante Formation is more evolved than the volcanic source of the Pilalo Formation.

5.8. QFL Analysis (Quartz-Feldspar-Lithic Fragments)

Fourteen petrographic thin sections of sandstones of the Pilalo and Silante formations were analysed for modal sandstone composition. The relative proportions of detrital grains, including feldspar (F), quartz (Q), and lithic fragments (L), combined with grain size, provide a formal name for the sedimentary rock [41,88], and constrain the tectonic settings of the source regions [42]. According to the QFL classification diagram of Folk [88], sandstones of the Silante Formation are classified as arkoses to lithic arkoses, whereas samples of the Pilalo Formation are lithic arkoses (Figure 10a).

In the Qm-F-Lt discriminatory diagram (Figure 10b), samples of the Pilalo Formation plot in the transitional arc field. Sedimentary rocks of the Silante Formation contain a greater amount of feldspar grains and volcanic lithic fragments, and these rocks cluster in the transitional arc and basement fields (Figure 10b,c). The LmLvLs diagrams for both formations reveal high contents of volcanic lithics and small amounts of sedimentary lithoclasts (Figure 10d).

The relative proportions of quartz, feldspar, and lithic fragments in the Pilalo and Silante formations reveal provenance shifts that can be related to changes in their sediment source areas. Overall, the results imply that clastic sedimentary rocks in the Silante and Pilalo basins were supplied from the erosion of volcanic arcs.

5.9. U-Pb Ages of the Silante Formation

U-Pb zircon dates were obtained from detrital zircons from sedimentary rocks of the Silante Formation, including rocks 02SA027, 02SA028, 02SA029 and JL17002 (Table 1, Figure 2). We also present zircon U-Pb analyses of quartz diorite (JL17003), which intrudes sedimentary rocks of the Silante Formation to the southwest of Quito, along the Quito–Santo Domingo road (Figure 4). The full LA-ICP-MS U-Pb zircon age data is available in the online supplementary material (Table S3).

Sandstone 02SA027 shows a unimodal peak at ~25 Ma (Figure 12a), with an MDA age of 25.017 ± 0.07 Ma (2σ) (Oligocene). A single detrital zircon with an age of 553.9 ± 6.46 Ma was measured. This zircon can be derived from the Brazilian belt (500–700 Ma), formed during the assembly of Gondwana [41,89]. Furthermore, it is a common population in detrital zircons obtained within the Eocene Angamarca Group of western Ecuador [2]. In sample 02SA028 a unimodal peak (Figure 12b), with an MDA age of 24.64 ± 0.17 Ma (2σ) was obtained.

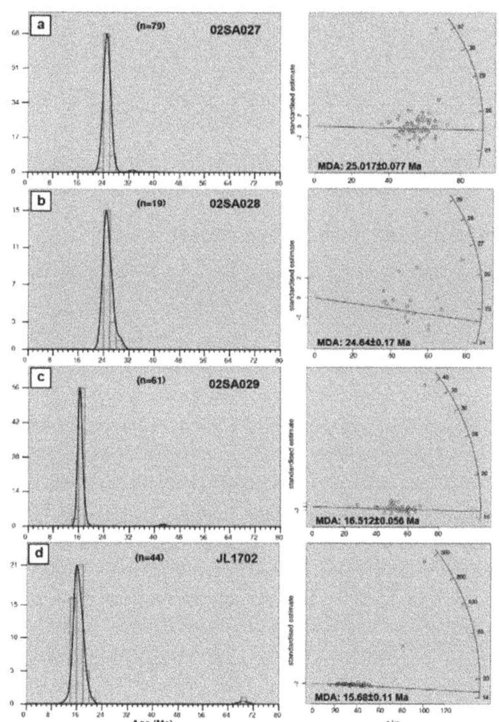

Figure 12. Frequency and probability density plots of detrital zircon U-Pb ages (left) and maximum depositional ages (MDA, right) from samples of the Silante Formation (**a**) sample 02SA027; (**b**) 02SA028; (**c**) 02SA029; (**d**) JL1702. The t/σ ratio on X-axis indicates the precision.

Sample 02SA029 also shows a unimodal peak at ~16 Ma (early Miocene). The MDA age obtained (Figure 12c) is 16.51 ± 0.056 Ma (2σ). A single detrital zircon revealed an age of 42.6 ± 0.6 Ma which can be correlated with the Macuchi volcanism of the Western Cordillera of Ecuador, dated at 42.62 ± 1.3 Ma with the U-Pb zircon dating method [90].

Sample JL17002 revealed a unimodal peak at ~16 Ma (early Miocene), and an MDA of 15.68 ± 0.11 (Figure 12d). In addition, two detrital zircons with ages of 68.9 ± 1.7 Ma, and 67.8 ± 2.6 Ma were detected. These Maastrichtian zircons are probably derived from the erosion of the Pilalo and Tandapi rocks.

Quartz diorite JL17003 intrudes sedimentary rocks of the Silante Formation and is exposed along the Alóag–Santo Domingo road section (Figure 2). The sample yields a weighted mean zircon U-Pb (LA-ICPMS) age of 10.31 ± 0.27 Ma (Figure 13), interpreted as crystallization age.

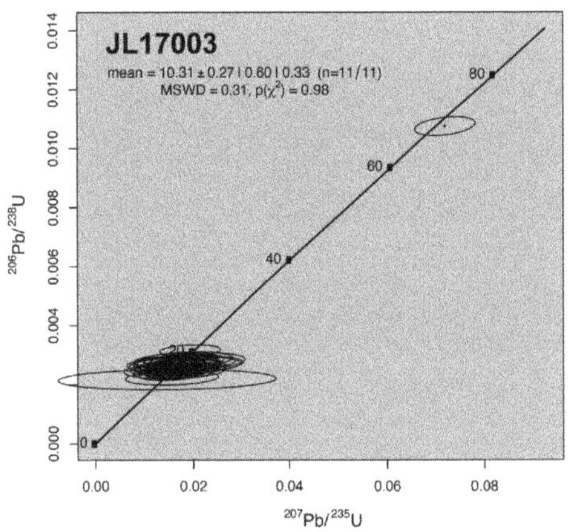

Figure 13. U-Pb LA-ICPMS zircon age of a quartz diorite intruding the Silante Formation (sample JL17003), plotted on a concordia diagram. Errors are given at the 2σ level.

6. Discussion: Paleogeographic and Paleotectonic Model

Sedimentological information and provenance analysis in the study area of the Western Cordillera of Ecuador show that turbidite sedimentation prevailed during the Paleocene, with the deposition of the Pilalo Formation. The main sedimentary source was a tholeiitic volcanic arc that can be related to the Tandapi arc reported by Vallejo et al. [2], which formed on top of an oceanic plateau basement. Therefore, the Paleocene sedimentary rocks described in this study do not correlate with the Campanian to Maastrichtian Yunguilla Formation. The Yunguilla Formation was also deposited in a submarine fan system, although provenance analysis indicates the sediments were eroded from a continental crust block, which is probably currently represented by the Eastern Cordillera.

The chronostratigraphic and sedimentological data suggest that the Silante Formation was deposited during the late Oligocene to Miocene period (~25–16 Ma) within a debris flow dominated alluvial fan system. The Silante Formation formed after a period of major rock uplift and erosion of the Western Cordillera. The Paleocene to Eocene submarine fan deposits of the Angamarca Group were probably partly eroded in the study area before deposition of the Silante Formation, and they are exposed several kilometers to the north and south of the study area (Figure 2).

Single grain geochemistry and petrographic analysis show that the Silante Formation received detrital material from a coeval calc-alkaline volcanic arc, which was presumably the San Juan de Lachas continental arc located in the Western Cordillera, to the northwest of the current exposures of the Silante Formation (Figure 2). A western location for the source of the Silante Formation is also suggested by the abundance of coarse-grained deposits to the west of the studied sections (Figure 4).

Petrographic analysis and detrital U-Pb zircon ages reveal a minor contribution of sediment to the Silante Formation from basement rocks, along with material from the erosion of the underlying Pilalo Formation, and the Eocene Macuchi submarine arc. Clinopyroxene compositions show that these were derived from calc-alkaline rocks, with a minor contribution from a tholeiitic volcanic source. The latter is consistent with the presence of clasts from the Pallatanga and Rio Cala volcanic rocks, suggesting the that basement was unroofed during sedimentation.

The Miocene period in Ecuador is generally characterized by the formation of large alluvial fan systems, including the Biblian alluvial fan in southern Ecuador in an intramontane setting [31]. The Arajuno Formation in the back-arc foreland basin also represents an alluvial fan system sourced from the Andes [91] and has been dated at ~17–13 Ma using U-Pb ages of detrital zircons [28].

(U-Th)/He and $^{40}Ar/^{39}Ar$, fission track data from the Andes of Ecuador [92] reveals rapid cooling and exhumation during 25–18 Ma, which correlates with a change in the vector of the subducting plate from ESE to E at 25 Ma, as a result of the breakup of the Farallon Plate [92–94], forming the Nazca and Cocos plates. Compressional events have also been dated between 25 and 15 Ma in the Andes of Bolivia and Peru [92–97]. Clearly, major plate rearrangements at 25 Ma affected the South American Plate to the north and south of the Huancabamba Deflection [92]. In addition, middle to late Miocene broadening of arc magmatism in the Eastern and Western Cordilleras [12,28] accounts for appearance of syndepositional age signatures within the Western Cordillera and the back-arc region [27,28].

Palinspastic constraints based on the results of this study are presented in Figure 14. During the Paleocene, the Pilalo Formation was deposited in a submarine fan depositional environment, with sediments sourced from the erosion of the tholeiitic Tandapi volcanic arc (Figure 14a). Igneous rocks associated with this volcanic arc include the Tandapi volcanic and subvolcanic rocks that are exposed in the study area with crystallization ages of ~65 to 61 Ma [11].

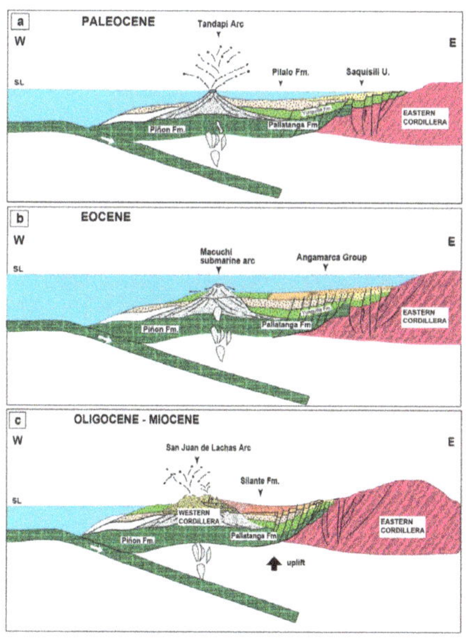

Figure 14. Tectonic evolution of the Western Cordillera of Ecuador during the Cenozoic. The initial uplift started at the end of the Cretaceous. (**a**) Paleocene paleogeography of the Pilalo Formation submarine fan with the coeval Tandapi arc; (**b**) Eocene Macuchi submarine arc and sedimentation of the Angamarca Group; (**c**) during the Oligocene to Middle Miocene there is a rapid uplift of the Eastern and Western Cordilleras, and the deposition of the Silante Formation in an intramontane basin setting. SL: sea level.

Submarine sedimentation continued during most of the Eocene period with the submarine fan of the Angamarca Group [22], depositing sediments that were mainly sourced from the Eastern Cordillera (Figure 14b).

A change from submarine to subaerial sedimentation occurred in broad areas of the Western Cordillera during the Oligocene [2], which coincides with a rapid rock uplift and exhumation of the Eastern and Western cordilleras [2]. The San Juan de Lachas continental arc was active during this period (Figure 14c) and extended into the Miocene [18], and supplied sediments to the Silante alluvial fan, which was deposited in an intramontane basin setting (Figure 14c).

The last deformation event recorded in rocks of the Silante Formation postdates the intrusion of the quartz diorite dated in this study, which yields a U-Pb crystallization age of 10.31 ± 0.27 Ma. This deformation phase may coincide with a latest Miocene-Pliocene phase of deformation, prior to the deposition of Quaternary volcanic ash of the Cangahua Formation [2,92,98].

7. Conclusions

Stratigraphic and sedimentological analyses of the Pilalo Formation, which is in fault contact with the Silante Formation, indicate that the sediments were deposited in the distal parts of a submarine fan. The Pilalo Formation was coeval with the latest Maastrichtian to Paleocene Tandapi volcanic arc, which formed on top of the accreted terranes of western Ecuador.

Radiometric dating indicates the Silante Formation in the study region was deposited during the late Oligocene to middle Miocene, subsequent to local erosion of the Eocene Angamarca Group.

Stratigraphic and sedimentologic analyses of the Silante Formation show that the sediments were deposited in a continental alluvial fan system that was dominated by debris flow processes and formed in an intramontane setting. Petrographic and provenance analyses of the same sediments suggest they were derived from the erosion of an andesitic continental volcanic arc, which was probably the Oligocene San Juan de Lachas arc.

Thermochronological data and regional correlations suggest that deposition of the Silante Formation was coeval with regional rock uplift and exhumation of the Andean margin, which gave rise to alluvial fans in intermontane and back-arc domains. This phase of regional deformation is temporally correlated with a phase of major plate reorganization during the breakup of the Farallon Plate.

Supplementary Materials: The following are available online at http://www.mdpi.com/2075-163X/10/10/929/s1, LA-ICP-MS U-Pb zircon age data and clinopyroxene geochemistry. Table S1. Chemical composition and structural formula of clinopyroxene phenocrysts. Table S2. REE elements composition of clinopyroxenes from the Western Cordillera. Table S3. LA-ICP-MS U–Pb dating data of detrital zircons sample.

Author Contributions: Conceptualization, C.V., S.A., V.E. and C.R.; methodology, C.V. and J.L.H.; software, V.E.; validation, R.A.S. and W.W.; formal analysis, P.V. All authors have read and agreed to the published version of the manuscript.

Funding: This work is part of the EPN project PIS-18-04. The authors would like to express their gratitude to the company Geostrat S.A. Ecuador for the partial financial support of this project.

Acknowledgments: We thank Bernardo Beate and Brian K. Horton for fruitful discussions about the geology of Ecuador and the Western Cordillera. Fieldwork benefited from the knowledge of Efrain Montenegro, Michel Toro and Arturo Egüez.

Conflicts of Interest: The authors declare no conflict of interest.

References

1. Luzieux, L.D.; Heller, F.; Spikings, R.; Vallejo, C.F.; Winkler, W. Origin and Cretaceous tectonic history of the coastal Ecuadorian forearc between 1° N and 3° S: Paleomagnetic, radiometric and fossil evidence. *Earth Planet. Sci. Lett.* **2006**, *249*, 400–414. [CrossRef]
2. Vallejo, C.; Spikings, R.A.; Horton, B.K.; Luzieux, L.; Romero, C.; Winkler, W. Late Cretaceous to Miocene stratigraphy and provenance of the coastal forearc and Western Cordillera of Ecuador: Evidence for accretion of a single oceanic plateau fragment. In *Andean Tectonics*, 1st ed.; Horton, B., Folguera, A., Eds.; Elsevier: Amsterdam, The Netherlands, 2019; pp. 209–236.

3. Feininger, T.; Bristow, C.R. Cretaceous and Palaeogene geologic history of Coastal Ecuador. *Geol. Rundsch.* **1980**, *69*, 40–44. [CrossRef]
4. Chiaradia, M. Adakite-like magmas from fractional crystallization and melting-assimilation of mafic lower crust (Eocene Macuchi arc, Western Cordillera, Ecuador). *Chem. Geol.* **2009**, *265*, 468–487. [CrossRef]
5. Kehrer, W.; Van Der Kaaden, G. Notes on the geology of Ecuador, with special reference to the Western Cordillera. *Geol. Jahrb.* **1979**, *35*, 5–57.
6. Henderson, W.G. Cretaceous to Eocene volcanic arc activity in the Andes of northern Ecuador. *J. Geol. Soc. Lond.* **1979**, *136*, 73–78.
7. Egüez, A. Evolution Cenozoique de la Cordillere Occidentale Septentrionale d'Equateur (0°15′ S–01°10′ S) les Mineralisations Associetees. Ph.D. Thesis, University of Pittsburgh Medical Center, Paris, France, 1986.
8. Kerr, A.C.; Aspden, J.A.; Tarney, J.; Pilatasig, L.F. The nature and provenance of accreted oceanic Blocks in western Ecuador: Geochemical and tectonic constraints. *J. Geol. Soc.* **2002**, *159*, 577–594. [CrossRef]
9. Jaillard, E.; Ordoñez, M.; Suarez, J.; Toro, J.; Iza, D.; Lugo, W. Stratigraphy of the late Cretaceous Paleogene deposits of the cordillera occidental of central Ecuador: Geodynamic implications. *J. S. Am. Earth Sci.* **2004**, *17*, 49–58. [CrossRef]
10. Spikings, R.A.; Winkler, W.; Hughes, R.A.; Handler, R. Thermochronology of Allochthonous Blocks in Ecuador: Unraveling the accretionary and post-accretionary history of the Northern Andes. *Tectonophysics* **2005**, *399*, 195–220. [CrossRef]
11. Vallejo, C.; Winkler, W.; Spikings, R.A.; Luzieux, L.; Heller, F.; Bussy, F. Mode and timing of terrane accretion in the forearc of the Andes in Ecuador. In *Backbone of the Americas: Shallow Subduction, Plateau Uplift, and Ridge and Terrane Collision*, 1st ed.; Kay, S., Ramos, V., Dickinson, W., Eds.; Geological Society of America Memoirs: Boulder, CO, USA, 2009; Volume 204, pp. 197–216.
12. Schütte, P.; Chiaradia, M.; Beate, B. Geodynamic controls on Tertiary arc magmatism in Ecuador: Constraints from U–Pb zircon geochronology of Oligocene–Miocene intrusions and regional age distribution trends. *Tectonophysics* **2010**, *489*, 159–176. [CrossRef]
13. Beate, B.; Monzier, M.; Spikings, R.; Cotton, J.; Silva, J.; Bourdon, E.; Eissen, J.P. Mio-Pliocene adakite generation related to flat subduction in southern Ecuador: The Quimsacocha volcanic center. *Earth Planet. Sci. Lett.* **2001**, *192*, 561–570. [CrossRef]
14. Baldock, J. *Geology of Ecuador: Explanatory Bulletin of the National Geological Map of the Republic of Ecuador*, 1st ed.; DGGM: Quito, Ecuador, 1982; p. 70.
15. Herrera, J.L. La-ICP-MS Zircon U-Pb Geochronology on the Ecuadorian Andes and Forearc, Constraints in Their Cretaceous-Paleogene Evolution: SILANTE Formation, CATACOCHA Unit, CAÑAVERAL Formation. Master's Thesis, University College, London, UK, 2018.
16. Wallrabbe-Adams, H.J. Petrology and geotectonic development of the Western Ecuadorian Andes: The Basic Igneous Complex. *Tectonophysics* **1990**, *185*, 163–182. [CrossRef]
17. Hughes, R.; Bermudez, R. *Geology of the Cordillera Occidental of Ecuador between 0°00′ and 1°00′ S. Proyecto de Desarrollo Minero y Control Ambiental, Programa de Información Cartográfica y Geológica; Report Number 4*, 1st ed.; CODIGEM-BGS: Quito, Ecuador, 1997; p. 75.
18. Boland, M.P.; Pilatasig, L.F.; Ibadango, C.E.; McCourt, W.J.; Aspden, J.A.; Hughes, R.A.; Beate, B. *Geology of the Western Cordillera between 0°–1° N, Proyecto de Desarrollo Minero y Control Ambiental, Programa de Informacion cartografica y Geológica, Informe No. 10*; CODIGEM-BGS: Quito, Ecuador, 2000; p. 72.
19. Jaillard, E.; Ordoñez, M.; Benitez, S.; Berrones, G.; Jimenez, N.; Montenegro, G.; Zambrano, I. Basin Develoment in an Accretionary, Oceanic-floored Fore-Arc Setting: Southern Coastal Ecuador During Late Cretaceous-Late Eocene time. *AAPG Mem.* **1995**, *62*, 615–631.
20. Winkler, W.; Villagómez, D.; Spikings, R.; Abegglen, P.; Tobler, S.; Egüez, A. The Chota basin and its significance for the inception and tectonic setting of the Inter-Andean Depression in Ecuador. *J. S. Am. Earth Sci.* **2005**, *19*, 5–19. [CrossRef]
21. Cotteccia, V.; Zezza, F. The Eocene basement of the inter-Andean corridor in the Latacunga Ambato trough (Ecuador). *Geol. Appl. Idrogeol.* **1969**, *4*, 43–46.
22. Vallejo, C.F. Evolution of the Western Cordillera in the Andes of Ecuador (Late Cretaceous Paleogene). Ph.D. Thesis, ETH Zürich, Zürich, Switzerland, 2007.
23. Hughes, R.A.; Pilatasig, L.F. Cretaceous and Tertiary Block accretion in the Cordillera Occidental of the Andes of Ecuador. *Tectonophysics* **2002**, *345*, 29–48. [CrossRef]

24. Litherland, M.; Aspden, J.; Jemielita, R.A. *The Metamorphic Belts of Ecuador. Overseas Memoir 11*; BGS: Keyworth, UK, 1994; p. 147.
25. Spikings, R.; Cochrane, R.; Villagomez, D.; Van der Lelij, R.; Vallejo, C.; Winkler, W.; Beate, B. The geological history of northwestern South America: From Pangaea to the early collision of the Caribbean Large Igneous Province (290–75 Ma). *Gond. Res.* **2015**, *27*, 95–139. [CrossRef]
26. Ruiz, G.M.H.; Seward, D.; Winkler, W. Detrital thermochronology—A new perspective on hinterland tectonics, an example from the Andean Amazon Basin, Ecuador. *Basin Res.* **2004**, *16*, 413–430. [CrossRef]
27. Gombojav, N.; Winkler, W. Recycling of Proterozoic crust in the Andean Amazon foreland of Ecuador: Implications for orogenic development of the northern Andes: Multiple recycling of Proterozoic crust in the northern Andes. *Terra Nova* **2008**, *20*, 22–31. [CrossRef]
28. Gutiérrez, E.G.; Horton, B.K.; Vallejo, C.; Jackson, L.J.; George, S.W. Provenance and geochronological insights into Late Cretaceous–Cenozoic foreland basin development in the Subandean Zone and Oriente Basin of Ecuador. In *Andean Tectonics*, 1st ed.; Horton, B., Folguera, A., Eds.; Elsevier: Amsterdam, The Netherlands, 2019; pp. 237–268. [CrossRef]
29. Vallejo, C.F.; Spikings, R.A.; Winkler, W.; Luzieux, L.; Chew, D.; Page, L. The early interaction between the Caribbean Plateau and the NW South American Plate. *Terra Nova* **2006**, *18*, 264–269. [CrossRef]
30. Van Thournot, F.; Hertogen, J.; Quevedo, L. Allochtonous Blocks in northern Ecuador. *Tectonophysics* **1992**, *205*, 205–222.
31. Hungerbühler, D.; Steinmann, M.; Winkler, W.; Seward, D.; Egüez, A.; Peterson, D.E.; Hammer, C. Neogene stratigraphy and Andean geodynamics of southern Ecuador. *Earth Sci. Rev.* **2002**, *57*, 75–124. [CrossRef]
32. Sinton, C.W.; Duncan, R.A.; Storey, M.; Lewis, J.; Estrada, J.J. An oceanic flood basalts province within the Caribbean plate. *Earth Planet. Sci. Lett.* **1998**, *155*, 221–235. [CrossRef]
33. Mamberti, M.; Lapierre, H.; Bosch, D.; Jaillard, E.; Hernandez, J.; Polve, M. The Early Cretaceous San Juan Plutonic Suite, Ecuador: A magma chamber in an oceanic plateau? *Can. J. Earth Sci.* **2004**, *41*, 1237–1258. [CrossRef]
34. Thalmann, H.E. Micropaleontology of upper Cretaceous and Paleocene in western Ecuador. *AAPG Bull.* **1946**, *30*, 337–347.
35. Cosma, L.; Lapierre, H.; Jaillard, E.; Laubacher, G.; Bosch, D.; Desmet, A.; Mamberti, M.; Gabriele, P. Petrographie et geochimie des unites magmatiques de la Cordillère Occidentale d'Equateur (0°30′ S): Implications tectoniques. *Bull. Soc. Geól. Fr.* **1998**, *169*, 739–751.
36. Egüez, A.; Bourgois, J. La Formacion Apagua, edad y posición estructural en la Cordillera Occidental del Ecuador. In *Cuarto Congreso Ecuatoriano de Geologia Minas y Petroleos*; Escuela Politecnica Nacional: Quito, Ecuador, 1986; pp. 161–178.
37. Savoyat, F.; Vernet, R.; Sigal, J.; Mosquera, C.; Granja, J.; Guevara, G. *Formaciones Sedimentarias de la Sierra Tectonica Andina del Ecuador*; Servicio Nacional de Geologia y Mineria—Institut Francais du Petrole; 1970; unpublished report.
38. Van Thournout, F. Stratigraphy, Magmatism and Tectonism in the Ecuadorian Northwestern Cordillera: Metallogenic and Geodynamic Implications. Ph.D. Thesis, Katholieke Universiteit Leuven, Leuven, Belgium, 1991.
39. Cohen, K.M.; Finney, S.C.; Gibbard, P.L.; Fan, J. The ICS International Chronostratigraphic Chart. *Episodes* **2013**, *36*, 199–204. [CrossRef] [PubMed]
40. Mange, M.; Maurer, H. *Heavy Minerals in Colour*, 1st ed.; Chapman and Hall: London, UK, 1992; p. 147.
41. Valarezo, M.; Vallejo, C.; Horton, B.K.; Gaibor, J.; Esteban, J.; Jackson, L.J.; Carrasco, H.; Winkler, W.; Bernal, C.; Beate, B. Sedimentological and provenance analysis of the Río Playas stratigraphic section: Implications for the evolution of the Alamor-Lancones Basin of southern Ecuador and northern Peru. *J. S. Am. Earth Sci.* **2019**, *94*, 1–18. [CrossRef]
42. Dickinson, W.R.; Suczek, C.A. Plate tectonics and sandstone compositions. *AAPG Bull.* **1979**, *63*, 2164–2182.
43. Dickinson, W.R. Interpreting provenance relations from detrital modes of sandstones. In *Provenance of Arenites*, 1st ed.; Zuffa, G., Ed.; Reidel Publishing Company: Dordrecht, The Netherlands, 1985; pp. 331–361.
44. Gazzi, P. Le arenarie del flysch sopracretaceo dell'Appennino modenese; correlazioni con il flysch di Monghidoro. *Miner. Petrogr. Acta* **1966**, *12*, 69–97.
45. Dickinson, W.R. Interpreting detrital modes of graywacke and arkose. *J. Sed. Petrol.* **1970**, *40*, 695–707.

46. Dickinson, W.R.; Beard, L.S.; Brackenridge, G.R.; Erjavec, J.L.; Fergusen, R.C.; Inman, K.F.; Knepp, R.A.; Lindenberg, F.A.; Ryberg, P.T. Provenance of North American Phanerozoic sandstones in relation to tectonic setting. *Geol. Soc. Am. Bull.* **1983**, *94*, 222–235. [CrossRef]
47. Sláma, J.; Kosler, J.; Condon, D.; Crowley, J.; Gerdes, A.; Hanchar, J.; Horstwood, M.; Morris, G.; Nasdala, L.; Norberg, N.; et al. Plesovice zircon—A new natural reference material for U-Pb and Hf isotopic microanalysis. *Chem. Geol.* **2008**, *249*, 1–35. [CrossRef]
48. Jochum, K.; Weis, U.; Stoll, B.; Kuzmin, D.; Yang, Q.; Raczek, I.; Jacob, D.; Stracke, A.; Birbaum, K.; Frick, D.; et al. Determination of Reference Values for NIST SRM 610617 Glasses Following ISO Guidelines. *Geostand. Geoanal. Res.* **2011**, *35*, 397–425. [CrossRef]
49. Wiedenbeck, M.; Hanchar, J.M.; Peck, W.H.; Sylvester, P.; Valley, J.; Whitehouse, M.; Kronz, A.; Morishita, Y.; Nasdala, L.; Fiebig, J.; et al. Further characterisation of the 91500 zircon crystal. *Geostand. Geoanal. Res.* **2004**, *28*, 9–39. [CrossRef]
50. Vermeesch, P. IsoplotR: A free and open toolbox for geochronology. *Geos. Front.* **2018**, *9*, 1479–1493. [CrossRef]
51. Galbraith, R.F. *Statistics for Fission Track Analysis*; Chapman and Hall/CRC: Boca Raton, FL, USA, 2005; p. 224.
52. Vermeesch, P. RadialPlotter: A Java application for fission track, luminescence and other radial plots. *Radiat. Meas.* **2009**, *44*, 409–410. [CrossRef]
53. Galbraith, R. Graphical display of estimates having differing standard errors. *Tectonometrics* **1988**, *30*, 271–281. [CrossRef]
54. Lowe, D.R. Sediment gravity flows: II. Depositional models with special reference to the deposits of high-density turbidity currents. *J. Sediment. Petrol.* **1982**, *52*, 279–297.
55. Tucker, M.E. *Sedimentology Rocks in the Field*, 3rd ed.; Wiley: England, UK, 2003; pp. 83–158.
56. Shanmugam, G. The Bouma Sequence and the turbidite mind set. *Earth Sci. Rev.* **1997**, *42*, 201–229. [CrossRef]
57. Shanmugam, G.; Shrivastava, S.K.; Das, B. Sandy Debrites and Tidalites of Pliocene Reservoir Sands in Upper-Slope Canyon Environments, Offshore Krishna-Godavari Basin (India): Implications. *J. Sediment. Res.* **2009**, *79*, 736–756. [CrossRef]
58. Postma, G. Classification for sediment gravity-flow deposits based on flow conditions during sedimentation. *Geology* **1986**, *4*, 291–294. [CrossRef]
59. Shanmugam, G. Glossary: A supplement to "Submarine fans: A critical retrospective (1950–2015)". *J. Palaeogeogr.* **2016**, *5*, 1–20.
60. Shanmugam, G. 50 years of the turbidite paradigm (1950s–1990s): Deepwater processes and facies models—A critical perspective. *Mar. Pet. Geol.* **2000**, *17*, 285–342. [CrossRef]
61. Song, S.R.; Lo, H.J. Lithofacies of volcanic rocks in the central Coastal Range, eastern Taiwan: Implications for island arc evolution. *J. Asian Earth Sci.* **2002**, *21*, 23–38. [CrossRef]
62. Sohn, Y.K. Depositional processes of submarine debris flows in the Miocene fan deltas, Pohang Basin, SE Korea with special reference to flow transformation. *J. Sediment. Res.* **2000**, *70*, 491–503. [CrossRef]
63. Hughes, D.A.; Lewin, J. A small-scale flood plain. *Sedimentology* **1982**, *29*, 891–895. [CrossRef]
64. Gao, L.; Wang, X.; Yi, S.; Vandenberghe, J.; Gibling, M.; Lu, H. Episodic Sedimentary Evolution of an Alluvial Fan (Huangshui Catchment, NE Tibetan Plateau). *Quaternary* **2018**, *1*, 16. [CrossRef]
65. Heward, A. Alluvial fan sequence and megasequence model: With examples from Westphalian D-Stephanian B coalfields, Northern Spain. In *Fluvial Sedimentology*, 1st ed.; Miall, A.D., Ed.; Dallas Geological Society: Dallas, TX, USA, 1977; pp. 669–702.
66. Nichols, G.J.; Fisher, J.A. Processes, facies and architecture of fluvial distributary system deposits. *Sediment. Geol.* **2007**, *195*, 75–90. [CrossRef]
67. Sohn, Y.K.; Rhee, C.W.; Kim, B.C. Debris Flow and Hyperconcentrated Flood Flow Deposits in an Alluvial Fan, Northwestern Part of the Cretaceous Yongdong Basin, Central Korea. *J. Geol.* **1999**, *107*, 111–132. [CrossRef]
68. Reineck, H.; Singh, I. *Depositional Sedimentary Environments*; Springer: New York, NY, USA, 1980; p. 549.
69. Mack, G.H.; Leeder, M.R. Climatic and tectonic controls on alluvial-fan and axial- fluvial sedimentation in the Plio Pleistocene Palomas half graben, southern Rio Grande rift. *J. Sediment. Res.* **1999**, *69*, 635–652. [CrossRef]
70. Bourquin, S.; Rigollet, C.; Bourges, P. High-resolution sequence stratigraphy of an alluvial fan delta environment: Stratigraphic and geodynamic implications. An example from the Keuper Chaunoy Sandstones, Paris Basin. *Sediment. Geol.* **1998**, *121*, 207–237. [CrossRef]

71. Li, S.; He, H.; Hao, R.; Chen, H.; Bie, H.; Liu, P. Depositional regimes and reservoir architecture characterization of alluvial fans of Karamay oilfield in Junggar basin, Western China. *J. Petr. Sci. Eng.* **2020**, *186*, 1–15. [CrossRef]
72. Van Houten, F.B. Iron and clay in tropical savanna alluvium: A contribution to the origin of red beds. *Bull. Geol. Soc. Am.* **1972**, *83*, 2761–2772. [CrossRef]
73. Leleu, S.; Ghienne, J.F.; Manatschal, G. Upper Cretaceous-Palaeocene basin-margin alluvial fans documenting interaction between tectonic and environmental processes (Provence, SE France). *Geol. Soc. Lond. Spec. Publ.* **2005**, *251*, 217–239. [CrossRef]
74. Neves, M.A.; Morales, N.; Saad, A.R. Facies analysis of tertiary alluvial fan deposits in the Jundiaí region, São Paulo, southeastern Brazil. *J. S. Am. Earth Sci.* **2005**, *19*, 513–524. [CrossRef]
75. Nichols, G. *Sedimentology and Stratigraphy*, 2nd ed.; Wiley-Blackwell: Hoboken, NJ, USA, 2009; pp. 129–150.
76. Ferrando, S.; Bernoulli, D.; Compagnoni, R. The Canavese zone (internal Western Alps): A distal margin of Adria. *Schweizerische Mineralogische und Petrographische Mitteilungen* **2004**, *84*, 237–256.
77. Perez, N.D.; Horton, B.K. Oligocene-Miocene deformational and depositional history of the Andean hinterland basin in the Northern Altiplano Plateau, Southern Peru. *Tectonics* **2014**, *33*, 1819–1847. [CrossRef]
78. Horton, B.K.; Anderson, V.J.; Caballero, V.; Saylor, J.E.; Nie, J.; Parra, M.; Mora, A. Application of detrital zircon U-Pb geochronology to surface and subsurface correlations of provenance, paleodrainage, and tectonics of the middle Magdalena Valley basin of Colombia. *Geosphere* **2015**, *11*, 1790–1811. [CrossRef]
79. Morton, A.C.; Hallsworth, C.R. Identifying provenance-specific features of detrital heavy mineral assemblages in sandstones. *Sediment. Geol.* **1994**, *90*, 241–256. [CrossRef]
80. Luzieux, L.D. Origin and Late Cretaceous-Tertiary Evolution of the Ecuadorian Forearc. Ph.D. Thesis, ETH Zürich, Zürich, Switzerland, 2007.
81. Nisbet, E.; Pearce, J. Clinopyroxene Composition in Mafic Lavas from Different Tectonic Settings. *Cont. Miner. Petr.* **1977**, *63*, 149–160. [CrossRef]
82. Leterrier, J.; Maury, R.; Thonon, P.; Girard, D.; Marchal, M. Clinopyroxene composition as a method of identification of the magmatic affinities of paleo-volcanic series. *Earth Planet. Sci. Lett.* **1982**, *59*, 139–154. [CrossRef]
83. Le Bas, M.J. The role of aluminium in igneous clinopyroxenes with relation to their parentage. *Am. J. Sci.* **1962**, *260*, 267–288. [CrossRef]
84. Beccaluva, L.; Macciotta, G.; Piccardo, G.B.; Zeda, O. Clinopyroxene composition of ophiolite basalts as petrogenetic indicator. *Chem. Geol.* **1989**, *77*, 165–182. [CrossRef]
85. Rollinson, H. *Using Geochemical Data: Evaluation, Presentation, Interpretation*, 1st ed.; Longman: Singapore, 1993; p. 352.
86. Hart, S.R.; Dunn, T. Experimental cpx/melt partitioning of 24 elements. *Contrib. Miner. Petrol.* **1993**, *113*, 1–18. [CrossRef]
87. Sun, S.S.; McDonough, W.F. Chemical and isotopic systematics of oceanic basalts: Implications for mantle compositionand processes. In *Magmatism in the Ocean Basins*; Saunders, A.D., Norry, M.J., Eds.; Geological Society: London, UK, 1989; Volume 42, pp. 313–345.
88. Folk, R.L.; Andrews, P.B.; Lewis, D.W. Detrital sedimentary rock classification and nomenclature for use in New Zealand. *J. Geol. Geophys.* **1970**, *13*, 937–968. [CrossRef]
89. Litherland, M.; Klinck, B.A.; Oconnor, E.A.; Pitfield, P.E.J. Andean-trending mobile belts in the Brazilian shield. *Nature* **1985**, *314*, 345–348. [CrossRef]
90. Vallejo, C.; Soria, F.; Tornos, F.; Naranjo, G.; Rosero, B.; Salazar, F.; Cochrane, R. Geology of El Domo deposit in central Ecuador: A VMS formed on top of an accreted margin. *Miner. Depos.* **2016**, *51*, 389–409. [CrossRef]
91. Christophoul, F.; Baby, P.; Dávila, C. Stratigraphic responses to a major tectonic event in a foreland basin: The Ecuadorian Oriente Basin from Eocene to Oligocene times. *Tectonophysics* **2002**, *345*, 281–298. [CrossRef]
92. Spikings, R.A.; Crowhurst, P.V.; Winkler, W.; Villagomez, D. Syn- and post-accretionary cooling history of the Ecuadorian Andes constrained by their in-situ and detrital thermochronometric record. *J. S. Am. Earth Sci.* **2010**, *30*, 121–133. [CrossRef]
93. Pilger, R.H. Cenozoic plate kinematics, subduction and magmatism: South American Andes. *J. Geol. Soc.* **1984**, *141*, 793–802. [CrossRef]
94. Pardo-Casas, F.; Molnar, P. Relative motion of the Nazca (Farallon) and South America plate since Late Cretaceous time. *Tectonics* **1987**, *6*, 233–248. [CrossRef]

95. Sébrier, M.; Lavenu, M.; Fornari, M.; Soulas, P. Tectonics and uplift in the central Andes (Peru, Bolivia and northern Chile) from Eocene to present. *Geodynamique* **1988**, *3*, 85–106.
96. Sempéré, T.; Herail, G.; Oller, J.; Bonhomme, G. Late Oligocene-early Miocene major tectonic crisis and related basins in Bolivia. *Geology* **1990**, *18*, 946–949. [CrossRef]
97. Laubacher, G.; Naeser, C.W. Fission-track dating of granitic rocks from the Eastern Cordillera of Peru: Evidence for Late Jurassic and Cenozoic cooling. *J. Geol. Soc. Lond.* **1994**, *151*, 473–483. [CrossRef]
98. Hall, M.; Calle, J. Geochronological control for the main Tectono-Magmatic events of Ecuador. *Earth Sci. Rev.* **1982**, *10*, 215–239. [CrossRef]

Publisher's Note: MDPI stays neutral with regard to jurisdictional claims in published maps and institutional affiliations.

minerals

Article

Provenance Analysis of the Paleogene Strata in the Northern Qaidam Basin, China: Evidences from Sediment Distribution, Heavy Mineral Assemblages and Detrital Zircon U-Pb Geochronology

Jianguo Yin [1], Shuai Zhang [2,*] and Zhixiong Wu [3]

[1] Key Laboratory of Petroleum Resources, Gansu Province/Northwest Institute of Eco-Environment and Resources, Chinese Academy of Sciences, Lanzhou 730000, China; yinjianguo@nieer.ac.cn

[2] Oil and Gas Survey, China Geological Survey, Beijing 100083, China

[3] Research Institute of Exploration and Development, PetroChina Qinghai Oilfield Company, Dunhuang 736202, China; wzx19842003@163.com

* Correspondence: zhangshuai870517@163.com

Received: 31 August 2020; Accepted: 24 September 2020; Published: 27 September 2020

Abstract: Using provenance analysis to build an accurate source-to-sink relationship is the key to infer mountain building scenarios around the Qaidam Basin, and also important to understanding the uplift and expansion of the Tibetan Plateau. However, some conflicting provenance inferences are caused by different interpretations for the prevalent existence of the late Paleozoic to early Mesozoic age group in detrital zircon U-Pb age spectra of the Paleogene strata at the northern Qaidam Basin, and these need to be resolved. In this article, an integrated study of sediment distribution, heavy mineral assemblages, and detrital zircon U-Pb geochronology is carried out to analyze provenance of the Paleogene strata at the northern Qaidam Basin. The decreasing trends of the net sand to gross thickness ratios and conglomerate percentages away from the Qilian Mountains and Altyn Tagh range to basin interior clearly support they are the provenance areas. Sedimentation of materials from the Altyn Tagh range is spatially confined to a small area in front of the mountains. A large sandy body with a uniform distribution of detrital zircon ages (containing a lot of the late Paleozoic to early Mesozoic zircon ages) and heavy mineral assemblages in the Xiaganchaigou Formation is supplied by the Qilian Mountains.

Keywords: provenance analysis; stratigraphic thicknesses; net sand to gross thickness ratio; conglomerate percentage; heavy mineral analysis; detrital zircon U-Pb geochronology; Intermontane basin

1. Introduction

Tectonic activities of the northwestern Tibetan Plateau provide significant evidence for understanding the formation and evolution of the Tibetan Plateau. As the largest Cenozoic intermontane basin in the northwestern Tibetan Plateau, the Qaidam Basin accumulates a thick and continuous sedimentary succession and thus preserves important information to infer tectonic activities of its surrounding orogenic belts. Using provenance analysis to construct a reasonable source-to-sink relationship is the prerequisite to interpret the information of the sedimentary archive.

Numerous approaches have been applied to provenance analysis in the Qaidam Basin, including clast composition of conglomerate [1–3], sandstone modal analysis [4–7], detrital mineral U-Pb geochronology [8–13], paleocurrent analysis [3,4,10–13], heavy mineral analysis [4,5,14–16], and element geochemistry [7,17], etc. However, some discrepancies still exist in previous results of provenance analysis. Bush et al. (2016) [4] thought that a long E-directed drainage system transported materials eroded from the eastern Kunlun Mountains to the Dahonggou area (the Dhg in Figure 1c) and formed the Lulehe Formation (Fm.) in this area. Wang et al. (2017) [13] also suggested that the eastern Kunlun Mountains was the dominant provenance of the Lulehe and Xiaganchaigou Fms. in the Dahonggou area. Song et al. (2019) [12] pointed out the Altyn Tagh range was the provenance of the upper Lulehe, Xiganchaigou, and lower Shangganchaigou Fms. in the Dahonggou area. Lu et al. (2018) [11] and Zhuang et al. (2011) [3] thought the Lulehe Fm. in the Dahonggou area had a proximal northerly source area. The conflicting provenance inferences were mainly based on paleocurrent measurement and detrital zircon U-Pb geochronology from outcrops. The former is easily influenced by the high sinuosity of rivers, while the latter is easily influenced by the high complexity of rock types and ever-changing catchment area in provenance. When the paleocurrent only represents flow direction of one specified river reach but the entire drainage system, determining the provenance by comparisons of detrital zircon U-Pb age spectra may be error-prone and need more caution. The proportion variation of the late Paleozoic to early Mesozoic age group in detrital zircon U-Pb age spectra of the Paleogene strata in the Dahonggou area is the important evidence applied to analyze provenance in the above-mentioned articles and also the main factor to cause the conflicting provenance inferences. The controversy of provenance analysis refers to different tectonic geomorphology and mountain building scenarios for the orogenic belts surrounding the Qaidam Basin and need to be resolved.

The Qaidam Basin has been an endorheic intermontane basin in the Cenozoic [11]. The trunk rivers formed by confluence of the tributaries in the mountainous catchment areas flow through apex or intersection points and then become distributive drainage systems in the basin [18]. Without intensive impaction of the axial rivers, the resulted sedimentary bodies commonly show downstream decrease in channel proportion, channel scale and grain size [19], etc. Therefore, the regional sediment distribution (e.g., stratigraphic thicknesses, net sand-to-gross thickness (NTG) ratios, and conglomerate percentages, etc.) can be used to infer the delivery pathways of sediments. The sediment delivery pathways obtained from the sediment distribution trends are more representive than paleocurrent data of sparse outcrops. Although the stratigraphic information from sparse outcrops at the basin margin is not enough to acquire the regional changes in trends, the decades-long petroleum exploration and exploitation activities of the PetroChina Qinghai Oilfield Company have accumulated large amounts of drilled well data and make regional analyses of sediment distribution possible.

In this study, we analyzed the regional changes in trends of stratigraphic thicknesses, NTG ratios and conglomerate percentages in the northern Qaidam Basin. Heavy mineral analysis and detrital zircon U-Pb geochronology of nine samples were also conducted to infer the provenance. These results show that the provenance of the Paleogene strata in the northern Qaidam Basin is from the neighbouring Altyn Tagh range and Qilian Mountains, and that materials from the Altyn Tagh range are not southwardly transported to the Dahonggou area.

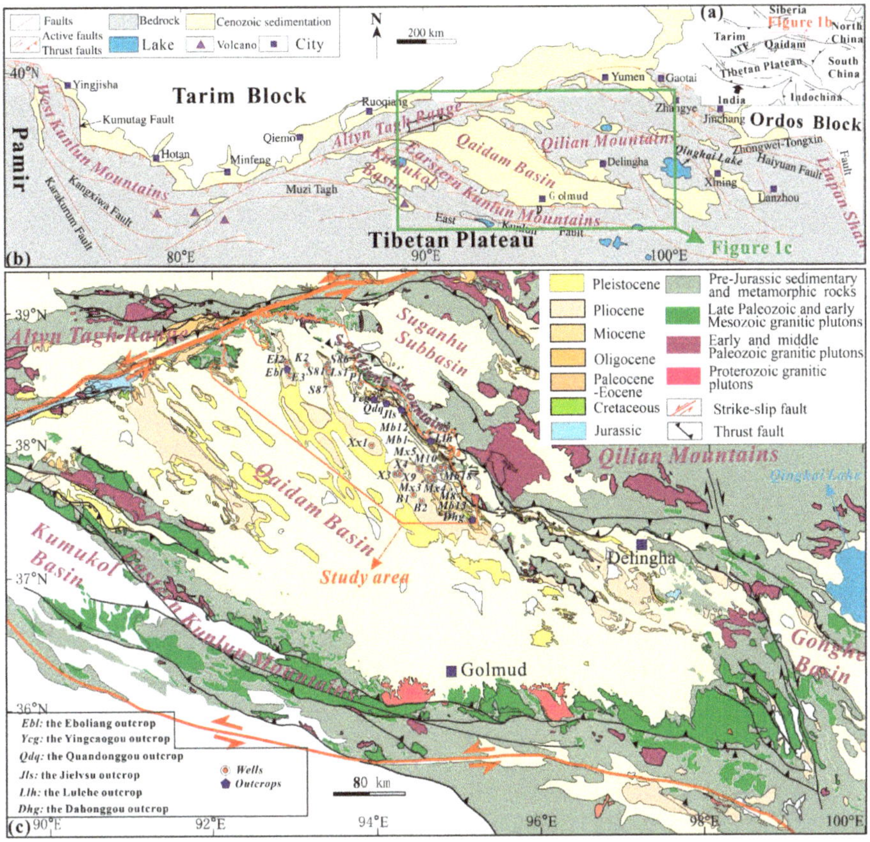

Figure 1. (**a**) Sketch map of the Tibetan Plateau and adjacent regions, showing the major faults. (**b**) Regional tectonic map around the Qaidam Basin located at the northeastern margin of the Tibetan Plateau (modified from Wang et al. 2015 [20]). (**c**) Detailed geological map of the study area in the northern Qaidam Basin (modified from Cheng et al. 2019a [21]).

2. Geological Background

The Qaidam Basin, which is an intermontane basin surrounded geographically by the Altyn Tagh Range to the northwest, the Qilian Mountains to the northeast, and the eastern Kunlun Mountains to the south (Figure 1), is the largest petroliferous basin in the Tibetan Plateau and infilled by thick Cenozoic strata atop the Jurassic, Cretaceous, or the basement.

2.1. Stratigraphy

These Cenozoic strata in the Qaidam Basin have been subdivided into seven stratigraphic units, including the Lulehe, Xiaganchaigou, Shangganchaigou, Xiayoushashan, Shangyoushashan, Shizigou, and Qigequan Fms. The forming ages of these strata have been previously delineated by ostracods, fossil mammals, lithostratigraphic correlations, and magnetostratigraphy [22–27]. The stratigraphic chronologies of the Cenozoic strata derived from the Dahonggou section by Wang et al. (2017) [13] is in conflict with other regional chronologies in the Qaidam Basin [28] and not adopted in this article. A more widely accepted division scheme of stratigraphic chronologies is that the Lulehe Fm. is 53.5–43.8 Ma [25,27,29], the Xiaganchaigou Fm. is 43.8–35.5 Ma [24,25,27], the Shanganchaigou Fm. is 35.5–22 Ma [22,24,25], the Xiayoushashan Fm. is 22–15.3 Ma [22–24], the Shangyoushashan Fm. is

15.3–8.1 Ma [23], the Shizigou Fm. is 8.1–2.5 Ma [23,24], and the Qigequan Fm. is 2.5–0 Ma [23]. The Paleogene strata, consisting of the Lulehe, Xiaganchaigou, and Shangganchaigou Fms., are easily distinguished from each another by the lithology in the northern margin outcrops of the Qaidam Basin [3,13]. This difference of lithology is the key foundation for tracing stratigraphic units from one location to another by subsurface data. The stratigraphic correlation of subsurface data is successfully built by the PetroChina Qinghai Oilfield Company.

2.2. Geochronological Characteristics of Potential Provenances of the Paleogene Strata in the Northern Qaidam Basin

In this article, detrital zircon U-Pb ages are used to infer the provenance evolution of the Paleogene strata in the northern Qaidam Basin. The zircon U-Pb age spectra of three potential provenances, namely the Qilian Mountains, the Altyn Tagh Range, and the eastern Kunlun Mountains, can be distinguished from one another is the prerequisite for applying this method. The characteristics of zircons sourced from the three provenance candidates are presented in Figure 2 [9].

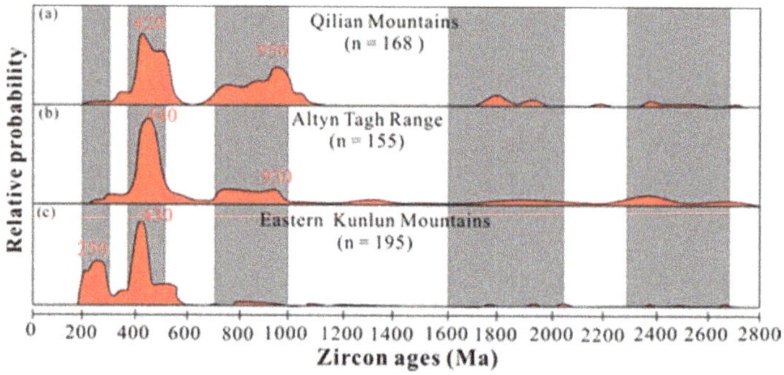

Figure 2. The relative probability plots of zircon U-Pb ages [9] from (**a**) basement rocks in the Qilian Mountains, (**b**) basement rocks in the Altyn Tagh Range and (**c**) basement rocks in the Eastern Kunlun.

The Qilian Mountains, located between the North Qilian Suture Zone and the North Qaidam ultrahigh-pressure metamorphic rocks (UHP Belt), mainly consists of a Precambrian crystalline basement overlain by thick and large-area Paleozoic sedimentary strata [30–33]. Additionally, forming-ages of granitoids outcropped in the Qilian Mountains adjacent to the northern Qaidam Basin margin are dominantly early Paleozoic [30,34]. In addition, a few late Paleozoic–early Mesozoic granitoids are exposed in the Saishiteng Mountains, northwestern part of the Qilian Mountains (Figure 1c) [13]. Relative probability plots of zircon U-Pb ages from basement rocks and magmatic intrusives of the Qilian Mountains are shown in Figure 2a [9] and indicate that zircons with early Paleozoic and Neoproterozoic ages are dominant. In addition, the Permian-Jurassic sedimentary rocks in the Qilian Mountains commonly have a major peak of late Paleozoic–early Mesozoic ages (200–300 Ma) in detrital zircon age spectra [35–37]. The late Paleozoic–early Mesozoic-aged zircons in the Permian- Jurassic sedimentary rocks are believed to originally be shed from the eastern Kunlun Mountains [4,35,36].

The basement rocks of the Altyn Tagh Range consist of the Archean–late Paleozoic magmatic rocks with zircon ages spanning from ~3.6 Ga to ~260 Ma [30,31]. In particular, the Early Paleozoic felsic rocks are widely distributed in the Altyn Tagh Range (Figure 2b) [31,37,38]. Additionally, several Permian plutonic terranes are exposed in the central part of the range (Figure 1c) [31] and make the late Paleozoic age as a major peak in detrital zircon U-Pb age spectra of the Cenozoic strata in the adjacent Eboliang outcop (Figure 7 in Cheng et al., 2016 [9]).

The late Paleozoic–early Mesozoic granitoids outcrop widely in the Eastern Kunlun Mountains, making 200–300 Ma one of two major age groups in the zircon U-Pb age spectrum of the East Kunlun

Shan (Figure 2c). The late Paleozoic–early Mesozoic granitoids record magmatic events related to the Paleo-Tethyan tectonic regime [39–41]. The early Paleozoic age group is the other major age group. The early Paleozoic tectono-magmatic events are comparable to the Qilian orogenic belt and constituent part of the Qilian-Kunlun Caledonian orogenic system [32,41]. Compared to the late Paleozoic–early Mesozoic and early Paleozoic ages, the Precambrian ages are rare [8,13] in the eastern Kunlun Mountains, although the Proterozoic rocks are exposed in both the eastern Kunlun Mountains and the Qilian Mountains [41].

3. Materials and Methods

In this study, well cuttings from 147 drilled wells were collected to calculate the stratigraphic thicknesses, NTG ratios and conglomerate percentages of the Lulehe, Xiaganchaigou and Shangganchaigou Fms. For each of these stratigraphic units, calculation only applies data from wells drilling through the corresponding unit. The wells with missing stratigraphic units are abandoned, too. Accordingly, 96, 123 and 122 in the 147 wells are finally available for the calculation of the Lulehe, Xiaganchaigou and Shangganchaigou Fms, respectively. The NTG ratio represents thickness ratio of conglomerates, sandstones and siltstones to stratigraphic unit in an individual drilled well. The conglomerate percentage represents ratio of the thickness of conglomerates, gravelly clastic rocks multiplied by a factor of 0.4 and gravel-bearing clastic rocks multiplied by a factor of 0.1 to the stratigraphic unit. According to the gravel contents in conglomerates (more than 30%), gravelly clastic rocks (between 5–30%) and gravel-bearing clastic rocks (less than 5%), the above two factors are chosen to transfer the thickness of gravelly clastic rocks and gravel-bearing clastic rocks to conglomerates.

Additionally, eleven sandstone samples collected from 8 cored wells were prepared for heavy mineral analysis and detrital zircon U-Pb dating (Table 1; Figure 1c). Separation, concentration, and identification of heavy minerals were performed at the Institute of Regional Geological Survey of Hebei Province, Langfang, Hebei Province, China by following the procedures outlined by Mange and Maurer (1992) [42] and Liu et al. (2013) [43]. Zircons (ca. 300 grains) were mounted in epoxy resin and then polished to obtain a smooth surface. Reflected and transmitted light as well as cathodoluminescence (CL) images were obtained to pick suitable dating targets, avoiding inclusion and cracks. Zircon U-Pb dating was carried out on Agilent 7200 ICP-MS combined with a Photon-machines Analyte Exite 193 nm laser ablation system at Lanzhou University (Lanzhou, China). In addition, ca. 80–100 effective zircon ages should be acquired to achieve statistically reliable detrital zircon U-Pb dating results. Zircon 91500, for which the preferred U-Pb isotopic ratios were found by Wiedenbeck et al. (1995) [44], was used as the external standard for U-Pb dating and was analyzed once every five analyses. The GLITTER 4.0 software was used to calculate the U-Pb isotope ratios and element contents. The U-Pb ages of detrital zircons younger than ca. 1000 Ma were calculated from $^{206}Pb/^{238}U$ ratios, whereas ages older than ca. 1000 Ma were calculated from $^{207}Pb/^{206}Pb$ ratios. Detrital zircon U-Pb ages with discordance degree >10% were excluded from the results. The probability density functions and histogram plots were plotted using the Isoplot 3.0 software [45]. A more detailed description of the sample separation methods and analytical procedures was given by Yuan et al. (2004) [46].

Table 1. Basic information of samples collected for heavy mineral analysis and detrital zircon U-Pb dating.

Samples	Depth (m)	Stratigraphic Units	Heavy Mineral Analysis	Zircon U-Pb Dating	Wells
B2u-s	3275	Xiaganchaigou Fm.	√	√	Well B2
B2d-s	3532	Xiaganchaigou Fm.	√	√	
Mx3-s	1240	Lulehe Fm.	√	√	Well Mx3
Mx4-s	1317	Lulehe Fm.	√	√	Well Mx4
Mx5u-s	1940	Xiaganchaigou Fm.	√	√	Well Mx5
Mx5d-s	2287	Lulehe Fm.	√	√	
X4-s	1806	Shangganchaigou Fm.	√	√	Well X4
Xx1u-s	4115	Xiaganchaigou Fm.	√	√	Well Xx1
Xx1d-s	4852	Lulehe Fm.	√	√	
Mb13-s	1450	Xiaganchaigou Fm.	√		Well Mb13
Mb18-s	1553	Xiaganchaigou Fm.	√		Well Mb18

4. Results

4.1. Sediment Distribution

Data of the stratigraphic thickness, NTG ratios and conglomerate percentages from drilled wells are listed in Supplementary Table S1. The drilled wells are presented in Supplementary Figure S1.

The thicknesses of the Lulehe Fm. are mainly less than 650 m (Figure 3a). A paleohigh is at southeastern margin of the study area, and some local sub-depocenters are near basin margin and show basinward thinning of thicknesses (Figure 3a). The NTG ratios of the Lulehe Fm. show basinward decrease (Figure 3b). The lowest NTG ratios are in the northwestern study area and far away from basin margin. The local sub-depocenters have relatively high NTG ratios and conglomerate percentages (Figure 3b,c). The conglomerate percentages of the Lulehe Fm. present the same trend as the NTG ratios (Figure 3c).

The thicknesses of the Xiaganchaigou Fm. are apparently increased and present basinward increasing trend (Figure 3d). The paleohigh at southeastern margin of the study area is covered by the Xiaganchaigou Fm., and a large depocenter is formed in the northwestern study area (Figure 3d). This depocenter has lowest NTG ratios (Figure 3e). The NTG ratios present decreasing trend from basin margin to the depocenter (Figure 3e). The conglomerate percentages of the Xiaganchaigou Fm. still present basinward decrease (Figure 3f). Compared to the Lulehe Fm., the conglomerate percentages of the Xiaganchaigou Fm. are lower, especially at eastern margin of the study area (Figure 3c,f). According to the isopach map and NTG ratios distribution, a large-scale sandy body is believed to be formed in southern study area (Figure 3d,e).

The thicknesses of the Shangganchaigou Fm. are obviously decreased and present basinward increasing trend (Figure 3g). The white parts in the study area are areas where the Shangganchaigou Fm. is completely removed after the later uplift (Figure 3g; Figure 11 in Cheng et al., 2016 [9]). There are two sub-depocenters formed in northwestern and central areas (Figure 3g). The NTG ratios present decreasing trend away from basin margin to the depocenters (Figure 3h). In the northern study area, the conglomerate percentages of the Shanganchaigou Fm. are larger than that of the Shanganchaigou Fm. (Figure 3f,i). Nevertheless, the former still maintains similar plane distribution characteristics of the latter (Figure 3f,i).

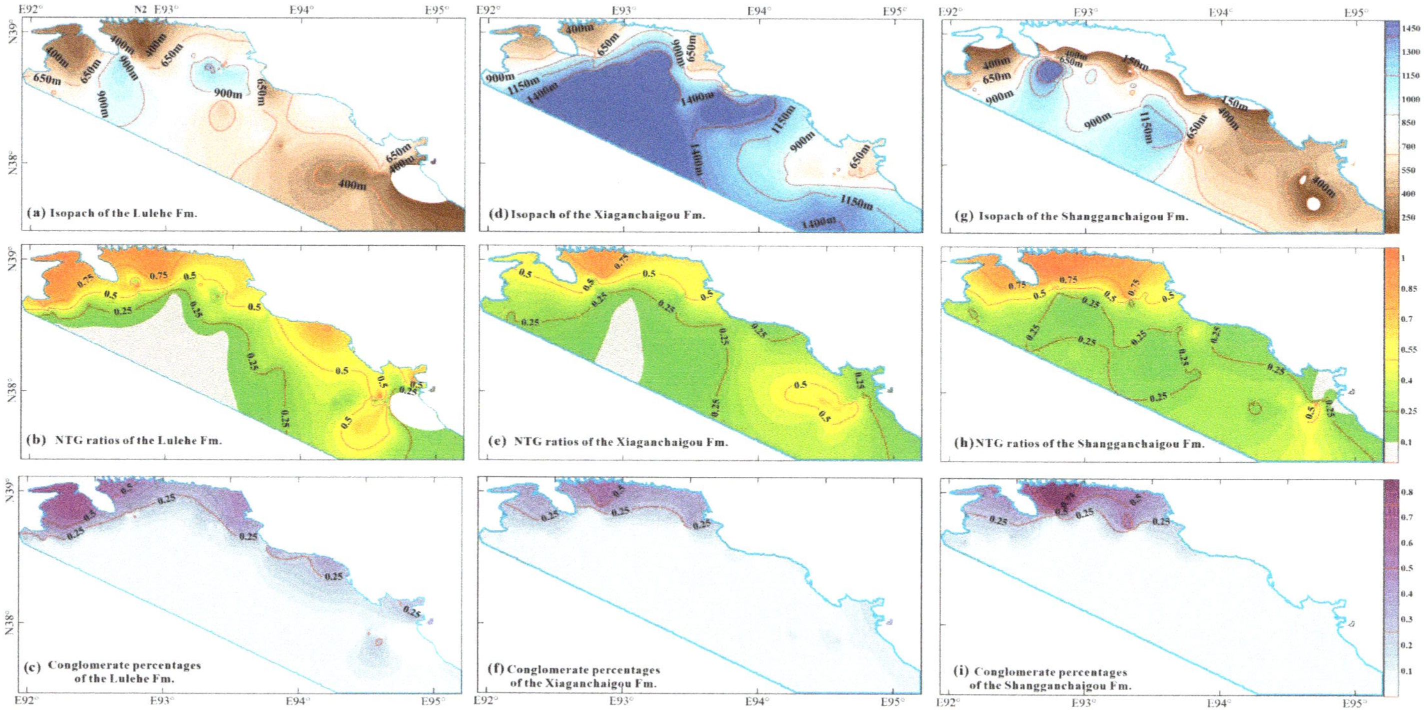

Figure 3. The plane distribution of stratigraphic thicknesses, NTG ratios and conglomerate percentages of the Paleogene strata. White area in study area represents area missing stratum. (**a**) Isopach of the Lulehe Fm. (**b**) NTG ratios of the Lulehe Fm. (**c**) Conglomerate percentages of the Lulehe Fm. (**d**) Isopach of the Xiaganchaigou Fm. (**e**) NTG ratios of the Xiaganchaigou Fm. (**f**) Conglomerate percentages of the Xiaganchaigou Fm. (**g**) Isopach of the Shangganchaigou Fm. (**h**) NTG ratios of the Shangganchaigou Fm. (**i**) Conglomerate percentages of the Shangganchaigou Fm.

4.2. Heavy Mineral Analysis

The heavy mineral analysis data are listed in Supplementary Table S2. Except for heavy mineral analysis data of wells B2, X4, Mx3, Mx4, Mx5, and Xx1, others are cited from Jian et al., 2013 [5]. Fe-oxide minerals (i.e., magnetite, hematite, and limonite) are excluded in heavy mineral analysis due to their susceptibility to hydraulic-sorting [47], so heavy mineral analysis data from Jian et al., 2013 [5] are renormalized and also listed in Supplementary Table S2. Drilled wells with heavy mineral assemblages are plotted to present the plane dissimilarities and similarities (Figure 4).

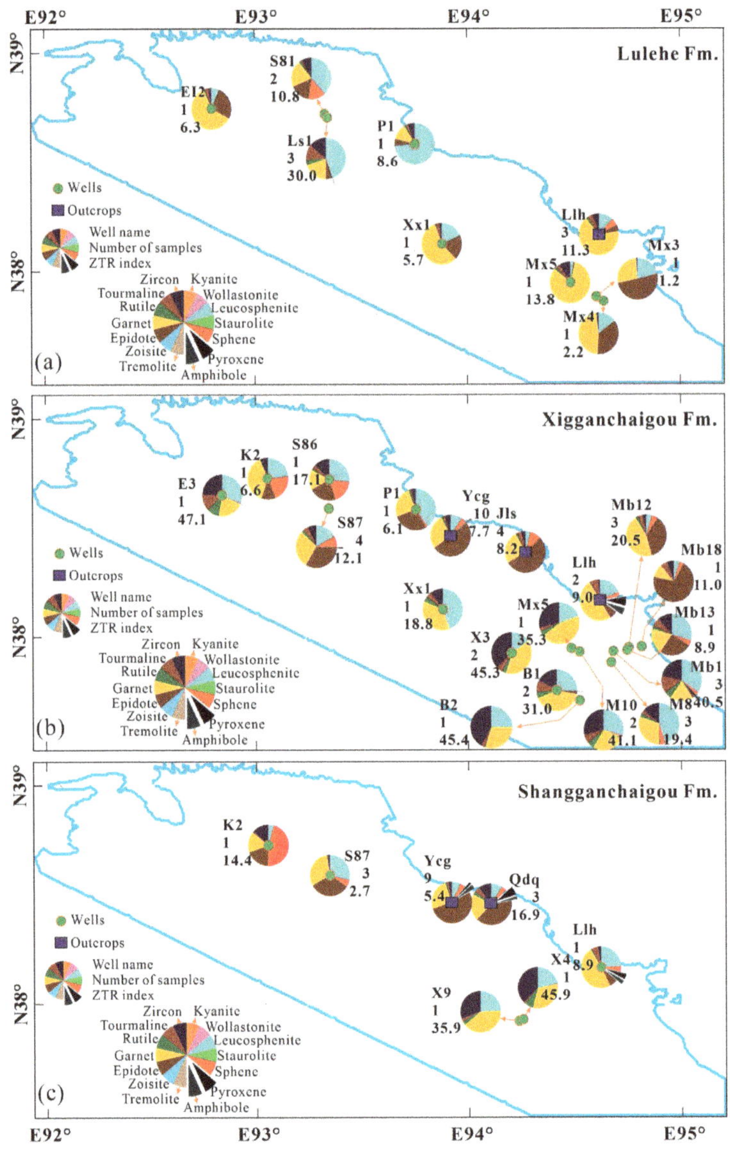

Figure 4. The characteristics of the heavy mineral assemblages of the Paleogene strata. (**a**) Lulehe Fm. (**b**) Xigganchaigou Fm. (**c**) Shangganchaigou Fm.

In the Lulehe Fm. (Figure 4a), the heavy mineral assemblages of Mx4 and Mx3 are similar, consist dominantly of epidote, garnet and leucosphenite, and display a very low abundance of ZTR minerals (zircon, tourmaline and rutile). Garnet is the dominant mineral species of Mx5 and Llh. S81 consists of ZTR minerals, garnet, epidote, sphene (titanite), and leucosphenite, and is different from near Ls1, which consists of ZTR minerals, garnet, and leucosphenite. P1 has the same heavy mineral types of Ls1 but is dominated by leucosphenite, accompanied by minor ZTR minerals, garnet and epidote.

In the Xiaganchaigou Fm. (Figure 4b), the heavy mineral assemblages of Xx1 and Mx5 are quite different from that of the Lulehe Fm., and consist dominantly of ZTR minerals, garnet, and leucosphenite. M1, M8, M10, B1, B2, and X3 have the same heavy mineral assemblages as Xx1 and Mx5. Ycg and Jls have similar heavy mineral assemblages composed dominantly of epidote and garnet. Epidote and garnet of P1 increase obviously. The heavy mineral assemblages of K2, S86 and S87 consist of ZTR minerals, garnet, epidote, sphene and leucosphenite, and are similar to S81 of the Lulehe Fm. Compared to Lulehe Fm., minor amphibole and pyroxene are showed in Llh. Ycg and Jls are similar and consist dominantly of epidote and garnet.

In the Shangganchaigou Fm. (Figure 4c), X9 and X4 (near X3) still consist dominantly of ZTR minerals, garnet and leucosphenite. Heavy mineral assemblages of Llh and S87 have no obviously change. Sphene of K2 increases obviously. Minor amphibole and pyroxene are showed in Ycg, which still consist dominantly of epidote and garnet. Qdq is very similar to Ycg.

4.3. Zircons Dating Results

The detrital zircon U-Pb dating results are listed in Supplementary Table S3. The Th/U ratios of the analyzed zircons are generally greater than 0.1 and indicative of the predominant magmatic origin (Figure 5) [48]. The detrital zircon U-Pb ages of the nine samples can be statistically subdivided into three groups: the Precambrian, spanning from ca. 2800 Ma to 550 Ma (peaks at ca. 800–900 Ma, ca. 1700–1800 Ma, and ca. 2300–2500 Ma), early to middle Paleozoic (peaks at ca. 410–450 Ma), and late Paleozoic to early Mesozoic (peaks at ca. 230–270 Ma) (Figures 5 and 6a–i). Except for sample Xx1d-s, the Precambrian zircons outnumber the late Paleozoic to Mesozoic zircons, and the peak height of the late Paleozoic to Mesozoic group increases with peak heights of the Precambrian group in detrital zircon age spectra (Figure 6a–i). The latter feature is even more obviously shown by the positive correlation between the "late Paleozoic to Mesozoic/early to middle Paleozoic zircons" ratio and the "Precambrian zircons/early to middle Paleozoic zircons" ratio (Figure 6j). Additionally, the above mentioned two features also can be observed in detrital zircon age spectra of the Cenozoic strata in the Dahonggou outcrop (Figure 11A in Bush et al., 2016 [4]).

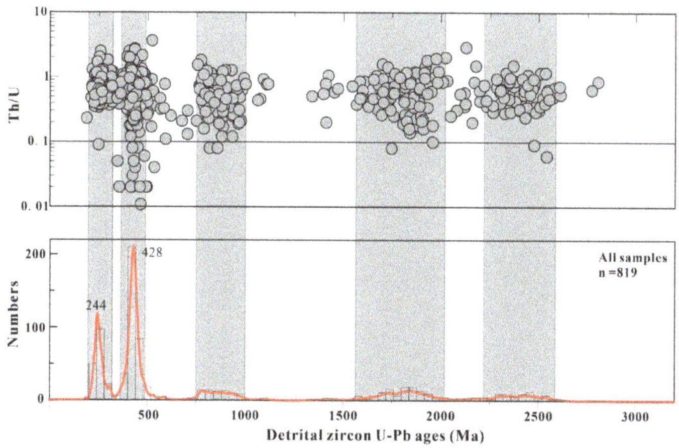

Figure 5. The Th/U ratios and detrital zircon ages of all analyzed samples.

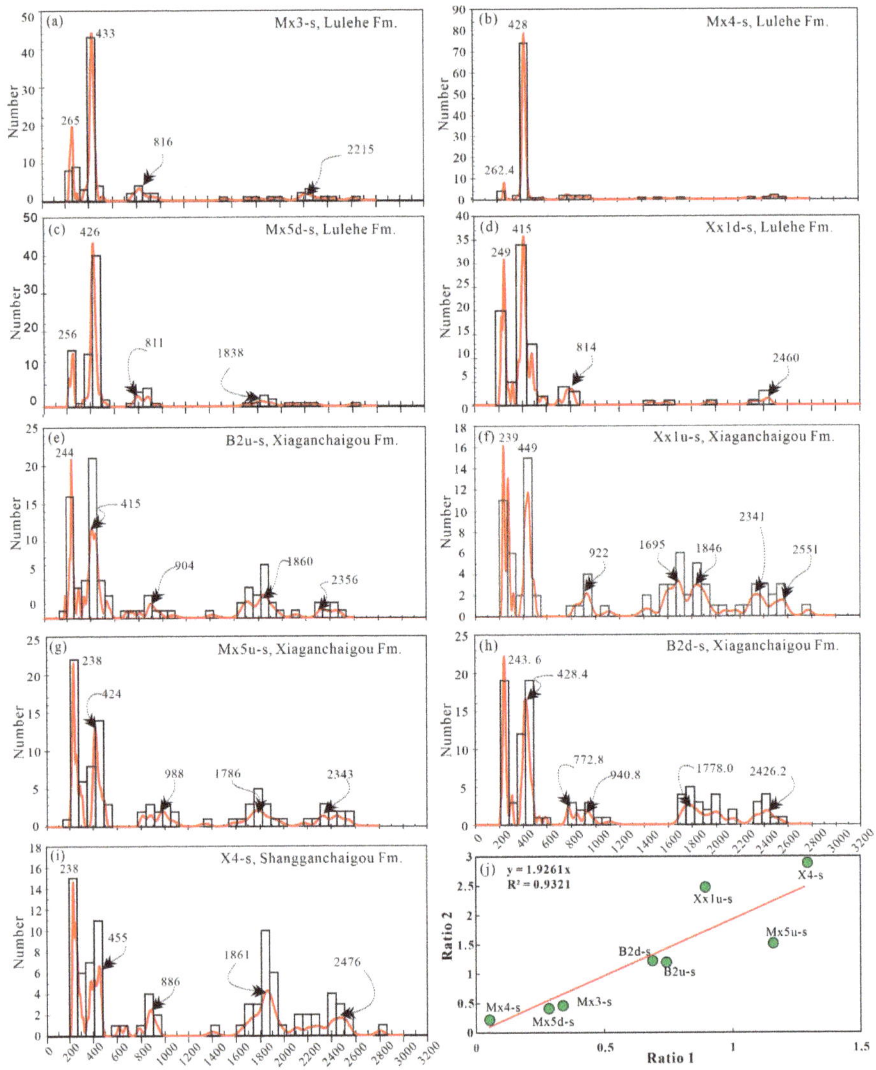

Figure 6. (**a–i**) Graphs of detrital zircon age spectra in study area. (**j**) Peaks height correlation between the Precambrian zircons and late Paleozoic to Mesozoic zircons. The ratio 1 represents "late Paleozoic to Mesozoic/early to middle Paleozoic zircons" ratio, and ratio 2 represents the "Precambrian zircons/early to middle Paleozoic zircons" ratio.

4.3.1. Samples from the Lulehe Fm.

Samples Mx5m-s, Mx3-s, and Mx4-s exhibit similar detrital zircon age spectra (Figure 6a–c), and the early to middle Paleozoic group is dominant in the spectra. The zircon ages in these three samples are distributed between 207 and 2608 Ma with a major ca. 430 Ma peak and minor peak at 260 Ma. The Precambrian group obtains three very low peaks at ca. 800 Ma, ca. 1800 Ma, and ca. 2200 Ma in these three samples.

The zircon ages in sample Xx1d-s are distributed between 211 and 3353 Ma. The zircon of 3353 Ma is obviously older than other zircons of all samples and is not shown in Figure 6d. The zircon ages in sample Xx1d-s mainly consist of early to middle Paleozoic group (peaks at 415 Ma) and late Paleozoic to Mesozoic group (peaks at 249 Ma) (Figure 6d). The Precambrian group only shows a low peak at 814 Ma and 2460 Ma (Figure 6d). This implies that the provenance of sample Xx1d-s differs from that of samples Mx5m-s, Mx3-s, and Mx4-s.

4.3.2. Samples from the Xiaganchaigou Fm.

Samples B2u-s, B2d-s, Mx5u-s, and Xx1u-s were collected from the Xiaganchaigou Fm. These four samples exhibit very similar detrital zircon age spectra (Figure 6e–h). The early to middle Paleozoic group and late Paleozoic to Mesozoic group are shown as two major peaks in detrital zircon age spectra (Figure 6e–h). The early to middle Paleozoic group peaks at ca. 410–450 Ma, and the late Paleozoic to Mesozoic group peaks at ca. 240 Ma. The Precambrian group exhibits three minor peaks at ca. 800–900 Ma, ca. 1700–1800 Ma, and ca. 2300–2500 Ma.

4.3.3. Samples from the Shangganchaigou Fm.

Sample X4-s was collected from the Shangganchaigou Fm. The detrital zircon age spectrum of the sample is the same as that of samples from the Xiaganchaigou Fm (Figure 6i).

5. Discussion

5.1. *Provenance Analysis of the Lulehe Fm.*

The decrease in trends of the NTG ratios and conglomerate percentages from the Altyn Tagh Range, Saishiteng Mountains and Qilian Mountains to basin interior imply that the Lulehe Fm. is sourced from these mountains (Figure 3b,c). Sandy bodies are mainly deposited in piedmont areas of these perveances and thus deny the existence of long axial rivers sourced from either the Altyn Tagh Range or the eastern Kunlun Mountains in this study area (Figure 3b). Widely deposited alluvial fans in front of these mountains, in which poorly sorted debris deposits (e.g., brownish-red matrix-supported conglomerates and muddy gravelly sandstones) are largely reserved, also support nearby provenances (Figure 7a) [49]. The plane distribution of sedimentary facies, presented by transition of the alluvial fans at basin margin to alluvial plain and lacustrine toward the basin interior, also suggests that sediments are delivered away from these mountains [49]. The published paleocurrent data observed from the Jls (southwest-directed flow [3]; named as Lake Mahai in Zhuang et al., 2011 [3]), Lulehe (southwest-directed flow [3]), and Dahonggou outcrops (southwest- or southeast-directed flow [4,12,50]) agree with delivery pathways of sediments from the Saishiteng Mountains and the Qilian Mountains. Isopach map shows that some local depocenters with relatively high NTG are developed close to basin margin and thin toward basin interior (Figure 3a). This may be ascribed to (1) local residue depressions caused by pre-Cenozoic tectonic inversion from extension to compression [51], (2) newly formed local depressions due to the Cenozoic tectonic events [52], and (3) clastic wedges of alluvial fans thinning away from sources [11].

According to the bedrock's outcrops in the Qilian Mountains adjacent to the northern Qaidam Basin, the detrital zircon ages eroded from the Qilian Mountains should consist dominantly of the early to middle Paleozoic ages (Figure 2a) [4,13,53]. The detrital zircon age spectra of samples Mx5m-s, Mx3-s, and Mx4-s are also dominated by the early to middle Paleozoic ages (Figure 6a–c). Thus, the zircons dating results support the provenance of the Lulehe Fm. of Wells Mx5, Mx3, and Mx4 is the Qilian Mountains. The difference of heavy mineral assemblages between Mx5, Mx3, and Mx4 can be explained by different catchment areas in the Qilian Mountains (Figure 4a). However, the detrital zircon age spectrum of sample Xx1d-s mainly consist of the early to middle Paleozoic ages and the nearly equivalent late Paleozoic to early Mesozoic ages (Figure 6d) and implies that sample Xx1d-s should have a different provenance. Though the Permian plutonic terranes are exposed in both the central part of the Altyn Tagh Range (Figure 1) [9,31] and the western Saishiteng Mountains (Figure 1c) [9,12], the distinctive heavy mineral assemblages of Xx1d-s, the NTG ratios, and conglomerate percentages tends to support provenance from the Saishiteng Mountains.

Thus, we can reasonably conclude that the Qilian Mountains, the Saishiteng Mountains and the Altyn Tagh Range are the provenances of the Lulehe Fm. in the study area (Figure 7a). The inference is compatible with the Paleocene-early Eocene uplift of these mountains by virtue of detrital zircon fission track evidence [20], detrital apatite fission track evidence [54,55], and structural analysis of seismic sections and outcrops [52,56]. In the paleogeographic map of the Lulehe Fm., provenances and widely deposited alluvial fans are shown (Figure 7a).

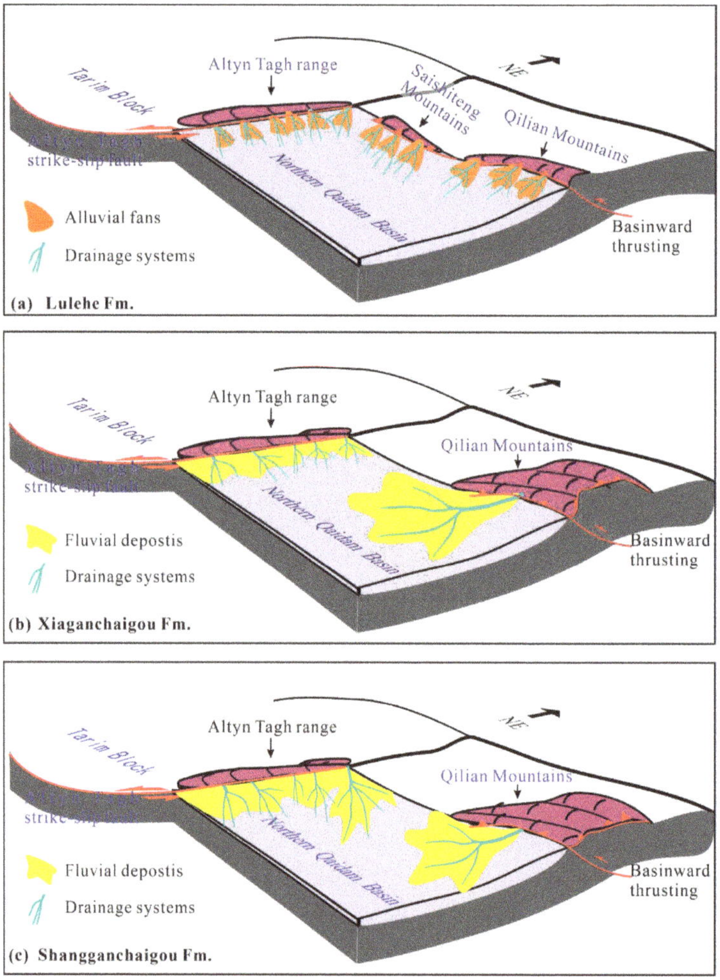

Figure 7. The paleogeographic maps of the Paleogene strata.

5.2. Provenance Analysis of the Xiaganchaigou Fm.

In the Xiaganchaigou Fm., a large depocenter is presented in the isopach map and located at central to northern study area (Figure 3d). Decreasing trends of the NTG ratios and conglomerate percentages from the Altyn Tagh Range and Qilian Mountains to the depocenter imply provenances from these mountains (Figure 3e,f). The low conglomerate percentages in front of the Qilian Mountains mean that the boundary between mountain and basin is further to the northeast (Figure 3f). The plane appearances of the NTG ratios and isopach map are very similar, and the depocenter has the lowest NTG ratios (Figure 3e,f). The features are related to the downstream allocation of sediments and roughly approximate to the published plane distribution of fluvial-lacustrine facies [49]. However, the NTG ratios show two distinctive differences, including that (1) the low NTG ratios in front of the Saishiteng Mountains disagree with a proximal source area and (2) a large scale sandy body is deposited in southern study area and sourced from the Qilian Mountains (Figure 3d,e and Figure 7b). The former can be interpreted by that the paleohigh which was located in the Saishiteng Mountains and supplied materials to the Lulehe Fm. was no longer existed in the depositional period of the Xiaganchaigou Fm. The same situation is experienced by the paleohigh close to the southeastern margin of the study area (Figure 3a,e). The latter can be further proved by the evidence from the heavy mineral assemblages and detrital zircon U-Pb dating, which present obvious differences in comparison with the Lulehe Fm (Figure 4b).

Wells in the area covered by the large scale sandy body (e.g., Xx1, X3, Mx5, M1, M8, M10, B1, and B2) have the same heavy mineral assemblages composed dominantly of ZTR minerals, garnet and leucosphenite, and are different from wells in front of the Altyn Tagh range (e.g., E3, K2, S86, and P1) (Figure 4b; Supplementary Figure S1; Figure 3d,e), suggesting they have the same provenance. Samples B2u-s, B2d-s, Mx5u-s, and Xx1u-s of the Xiaganchaigou Fm. collected from some of these wells exhibit highly similar detrital zircon age spectra characterized by two major peaks of the early to middle Paleozoic group and late Paleozoic to early Mesozoic group and some minor peaks of the Precambrian group, and also suggest they have the same provenance. Additionally, detrital zircon age spectra of the Xiaganchaigou Fm. from wells Mb18, Mb17 (Figure 3 in Wang et al., 2019 [57]) and the nearby Dahonggou outcrop have the same characteristics (Figure 11 in Bush et al., 2016 [4]). The large scale sandy body with uniform distribution of detrital zircon ages and heavy minerals assemblages that are sourced from the Qilian Mountains may implies a large catchment area is formed in the mountains. Such a catchment area is needed to provide enough sediments with uniformly mixed provenance signals [58].

The presence and absence of the late Paleozoic to early Mesozoic age group has been used as important evidence to infer provenance [4,12,13]. When the late Paleozoic to early Mesozoic zircons are presented as a major peak in age spectra of clastic rocks, the Qilian Mountains, in which the Permian–Triassic plutonic terranes are rarely exposed, is commonly excluded from the provenances in previous works [12,13]. However, the late Paleozoic and Mesozoic sedimentary strata are widely distributed in the Qilian Mountains and can supply considerable the late Paleozoic to early Mesozoic zircons [11,37]. Considering that the NTG ratios and conglomerate percentages distribution indicate a delivery path of sediments away from the Qilian Mountains (Figure 3e,f), the late Paleozoic to early Mesozoic zircons should be sourced from the recycled late Paleozoic and Mesozoic sedimentary strata in the Qilian Mountains. The age spectra of clastic rocks show that the late Paleozoic to Mesozoic group increases with the Precambrian group (Figure 6j), which is a significant constituent in age spectra of the Qilian Mountains [37], and further supports this view. In addition, the spectra of these samples are obviously different from those of the Cenozoic clastic rocks [9] and modern sediments [12] sourced from the Altyn Tagh range.

Thus, we can reasonably conclude that the Qilian Mountains and the Altyn Tagh Range are the provenances of the Lulehe Fm, and that materials eroded from the Altyn Tagh Range and the eastern Kunlun Mountains have not been delivered to the southern study area (Figure 7b). The inference is consistent with the previous works [4,5]. The southwest-directed paleoflow from the Dahonggou outcrop reveals a transverse dispersal away from the Qilian Mountains in the Xiagahnchaigou Fm. [4]. Jian et al. (2013) [5] concluded that the Altyn Tagh Range and Qilian Mountains were the potential provenance of the northern Qaidam Basin in the Paleogene and Neogene via the integrated analysis of framework petrography, heavy mineral analysis, and mineral chemistry. Materials of Wells B2, Mx5, X4, Mb17, Mb18, and Xx1 (located at depositional area A in Jian et al., 2013 [5]) are sourced from the Qilian Mountains. The materials from Altyn Tagh Range are only deposited at its front area [20] and blocked by the depocenter to transport onwards to the southern study area (Figure 3d, Figure 7b). The detrital apatite fission-track data from the JiuQuan Basin [59] and Qaidam Basin [54] both point to the rapid exhumation of the Qilian Mountains during the middle-late Eocene (~42–38 Ma). The apatite fission-track data of in situ bedrocks in the Qilian Mountains also reveal this rapid cooling event [60]. Zhuang et al. (2011) [3] propose the uplifted area in the Qilian Mountains experiences regional expansion during the depositional period of the Xiaganchaigou Fm. by detailed sedimentary analysis of outcrops. Intensive uplift and expansion of the uplifted area in the Qilian Mountains make formation of the large scale sandy body possible (Figure 7b).

5.3. Provenance Analysis of the Shangganchaigou Fm.

The NTG ratios present a decreasing trend away from the Altyn Tagh range and the Qilian Mountains to the two sub-depocenters at northwestern and central study areas and imply the Shangganchaigou Fm. is sourced from these mountains (Figure 3h,i; Figure 7c). Thus, the decreasing thicknesses are ascribed to the weaker tectonic activities of these mountains during deposition of the Shangganchaigou Fm. According to the distribution of NTG ratios and conglomerate percentages, the sedimentary bodies in front of the Altyn Tagh range and the Qilian Mountains exhibit progradation and retrogradation, respectively (Figure 3h,i; Figure 7c). This feature can be interpreted by interaction between deceasing accommodation and sediments supply under the control of weakened tectonic activities. Bao et al. (2017) [61] agreed that a relatively stable tectonic setting caused the low sediment flux in the Shangganchaigou Fm. He et al. (2018) [54] also considered that the Qilian Mountains did not experience rapid exhumation in the Shangganchaigou Fm. The heavy mineral assemblages of wells X9, X6, and X4 are also composed dominantly of ZTR minerals, garnet and leucosphenite, and suggest they are sourced from the Qilian Mountains (Figure 4c; Figure 7c). The good similarity of the detrital zircon age spectrum of sample X4-s with that of samples from the Xiaganchaigou Fm. further supports that the Qilian Mountains is still the provenance area (Figure 6i). The paleocurrent data observed from the Jls (northwest-directed flow [3]), Lulehe (northwest-directed flow [3]), and Dahonggou outcrops (southwest-directed flow [4,50]) suggest that the Qilian Mountains continuously supplied material into the Qaidam Basin in the Shangganchaigou Fm. Thus, the Shangganchaigou Fm. is still sourced from the Altyn Tagh range and Qilian Mountains and exhibits progradation and retrogradation in front of these mountains, respectively (Figure 7c).

6. Conclusions

In this article, the provenance of the Paleogene strata in the northern Qaidam Basin is inferred by virtue of evidence from sediment distribution, heavy mineral assemblages, and detrital zircon U-Pb dating. The main conclusions are as follows.

(1) The decreasing trends of the NTG ratios and conglomerate percentages away from the Qilian Mountains and Altyn Tagh range obviously support the fact that they are provenance areas of the Paleogene strata. The sandy bodies sourced from the Altyn Tagh range are spatially confined to a small area in front of the mountains during the Paleogene, and thus deny the existence of long drainage systems southwardly flowing from the Altyn Tagh range to the southern study area.

(2) Based on the sediment distribution in study area, the spatial dissimilarities and similarities exhibited by detrital zircon age spectra and heavy minerals assemblages can be used to further testify the provenance inference. Signals only from presence or absence of the spectral age group (i.e., the late Paleozoic to early Mesozoic age group) in detrital zircon age spectra are not enough to determine the provenance.

(3) In the Xiaganchaigou Fm., a large sandy body with uniform distribution of detrital zircon ages and heavy minerals assemblages is sourced from the Qilian Mountains, implying that a large catchment area formed in the mountains under the control of intensive tectonic activity. The recycled late Paleozoic and Mesozoic sedimentary strata in the Qilian Mountains supply considerable late Paleozoic to early Mesozoic zircons. In addition, we consider that variation in proportions of the

spectral age groups (e.g., the late Paleozoic to early Mesozoic age group) in detrital zircon U-Pb age spectra need to be cautiously applied in provenance analysis of the Qaidam baisn due to the complex rock types and intense tectonic activities of its surrounding mountains.

Supplementary Materials: The following are available online at http://www.mdpi.com/2075-163X/10/10/854/s1, Table S1: Data of the stratigraphic thickness, NTG ratios and conglomerate percentages, Table S2: Data of heavy mineral analysis, Table S3: Th/U ratios and detrital zircon U-Pb ages of the samples from the northern Qaidam Basin. Figure S1: The locations of drilled well.

Author Contributions: Conceptualization, J.Y. and S.Z.; methodology, J.Y. and S.Z.; validation, J.Y., S.Z., and Z.W.; formal analysis, J.Y., S.Z., and Z.W.; investigation, J.Y.; data curation, J.Y., S.Z., and Z.W.; writing—original draft preparation, J.Y. and S.Z.; writing—review and editing, J.Y., S.Z., and Z.W. All authors have read and agreed to the published version of the manuscript.

Funding: This research was funded by National geological survey project of China (Comprehensive survey of natural gas hydrate and petroleum resources in Domestic Permafrost Area, grant number: DD20190102).

Acknowledgments: The authors appreciate the prospective comments and suggestions from the reviewers and editors.

Conflicts of Interest: The authors declare no conflict of interest.

References

1. Cai, K.; Sun, M.; Yuan, C.; Xiao, W.; Zhao, G.; Long, X.; Wu, F. Carboniferous mantle-derived felsic intrusion in the Chinese Altai, NW China: Implications for geodynamic change of the accretionary orogenic belt. *Gondwana Res.* **2012**, *22*, 681–698. [CrossRef]
2. Wu, L.; Xiao, A.; Yang, S.; Wang, L.; Mao, L.; Wang, L.; Dong, Y.; Xu, B. Two-stage evolution of the Altyn Tagh Fault during the Cenozoic: New insight from provenance analysis of a geological section in NW Qaidam Basin, NW China. *Terra Nova* **2012**, *24*, 387–395. [CrossRef]
3. Zhuang, G.; Hourigan, J.K.; Ritts, B.D.; Kent-Corson, M.L. Cenozoic multiple-phase tectonic evolution of the northern Tibetan Plateau: Constraints from sedimentary records from Qaidam Basin, Hexi Corridor, and Subei basin, northwest China. *Am. J. Sci.* **2011**, *311*, 116–152. [CrossRef]
4. Bush, M.A.; Saylor, J.E.; Horton, B.K.; Nie, J. Growth of the Qaidam Basin during Cenozoic exhumation in the northern Tibetan Plateau: Inferences from depositional patterns and multiproxy detrital provenance signatures. *Lithosphere* **2016**, *8*, 58–82. [CrossRef]
5. Jian, X.; Guan, P.; Zhang, D.-W.; Zhang, W.; Feng, F.; Liu, R.-J.; Lin, S.-D. Provenance of Tertiary sandstone in the northern Qaidam Basin, northeastern Tibetan Plateau: Integration of framework petrography, heavy mineral analysis and mineral chemistry. *Sediment. Geol.* **2013**, *290*, 109–125. [CrossRef]
6. Rieser, A.B.; Neubauer, F.; Liu, Y.; Ge, X. Sandstone provenance of north-western sectors of the intracontinental Cenozoic Qaidam basin, western China: Tectonic vs. climatic control. *Sediment. Geol.* **2005**, *177*, 1–18. [CrossRef]
7. Sun, G.; Wang, M.; Guo, J.; Wang, Y.; Yang, Y. Geochemical Significance of Clay Minerals and Elements in Paleogene Sandstones in the Center of the Northern Margin of the Qaidam Basin, China. *Minerals* **2020**, *10*, 505. [CrossRef]
8. Cheng, F.; Fu, S.; Jolivet, M.; Zhang, C.; Guo, Z. Source to sink relation between the Eastern Kunlun Range and the Qaidam Basin, northern Tibetan Plateau, during the Cenozoic. *Geol. Soc. Am. Bull.* **2016**, *128*, 258–283. [CrossRef]
9. Cheng, F.; Jolivet, M.; Fu, S.; Zhang, C.; Zhang, Q.; Guo, Z. Large-scale displacement along the Altyn Tagh Fault (North Tibet) since its Eocene initiation: Insight from detrital zircon U-Pb geochronology and subsurface data. *Tectonophysics* **2016**, *677–678*, 261–279. [CrossRef]
10. Cheng, F.; Garzione, C.N.; Mitra, G.; Jolivet, M.; Guo, Z.; Lu, H.; Li, X.; Zhang, B.; Zhang, C.; Zhang, H.; et al. The interplay between climate and tectonics during the upward and outward growth of the Qilian Shan orogenic wedge, northern Tibetan Plateau. *Earth-Sci. Rev.* **2019**, *198*, 102945. [CrossRef]
11. Lu, H.; Ye, J.; Guo, L.; Pan, J.; Xiong, S.; Li, H. Towards a clarification of the provenance of Cenozoic sediments in the northern Qaidam Basin. *Lithosphere* **2018**, *11*, 252–272. [CrossRef]

12. Song, B.; Zhang, K.; Hou, Y.; Ji, J.; Wang, J.; Yang, Y.; Yang, T.; Wang, C.; Shen, T. New insights into the provenance of Cenozoic strata in the Qaidam Basin, northern Tibet: Constraints from combined U-Pb dating of detrital zircons in recent and ancient fluvial sediments. *Palaeogeogr. Palaeoclimatol. Palaeocol.* **2019**, *533*, 109254. [CrossRef]
13. Wang, W.; Zheng, W.; Zhang, P.; Li, Q.; Kirby, E.; Yuan, D.; Zheng, D.; Liu, C.; Wang, Z.; Zhang, H.; et al. Expansion of the Tibetan Plateau during the Neogene. *Nat. Commun.* **2017**, *8*, 15887. [CrossRef] [PubMed]
14. Li, L.; Guo, Z.; Guan, S.; Zhou, S.; Wang, M.; Fang, Y.; Zhang, C. Heavy mineral assemblage characteristics and the Cenozoic paleogeographic evolution in southwestern Qaidam Basin. *Sci. China Earth Sci.* **2015**, *58*, 859–875. [CrossRef]
15. Li, L.; Wu, C.; Yu, X. Cenozoic evolution of the Altyn Tagh and East Kunlun fault zones inferred from detrital garnet, tourmaline and rutile in southwestern Qaidam Basin (Northern Tibetan Plateau). *Basin Res.* **2018**, *30*, 35–58. [CrossRef]
16. Nie, J.; Ren, X.; Saylor, J.E.; Su, Q.; Horton, B.K.; Bush, M.A.; Chen, W.; Pfaff, K. Magnetic polarity stratigraphy, provenance, and paleoclimate analysis of Cenozoic strata in the Qaidam Basin, NE Tibetan Plateau. *Geol. Soc. Am. Bull.* **2019**, *132*, 310–320. [CrossRef]
17. Jian, X.; Guan, P.; Zhang, W.; Feng, F. Geochemistry of Mesozoic and Cenozoic sediments in the northern Qaidam Basin, northeastern Tibetan Plateau: Implications for provenance and weathering. *Chem. Geol.* **2013**, *360–361*, 74–88. [CrossRef]
18. Weissmann, G.S.; Hartley, A.J.; Nichols, G.J.; Scuderi, L.A.; Olson, M.; Buehler, H.; Banteah, R. Fluvial form in modern continental sedimentary basins: Distributive fluvial systems. *Geology* **2010**, *38*, 39–42. [CrossRef]
19. Owen, A.; Hartley, A.J.; Ebinghaus, A.; Weissmann, G.S.; Santos, M.G.M. Basin-scale predictive models of alluvial architecture: Constraints from the Palaeocene–Eocene, Bighorn Basin, Wyoming, USA. *Sedimentology* **2019**, *66*, 736–763. [CrossRef]
20. Wang, Y.; Zheng, J.; Zheng, Y.; Liu, X.; Sun, G. Paleocene-Early Eocene uplift of the Altyn Tagh Mountain: Evidence from detrital zircon fission track analysis and seismic sections in the northwestern Qaidam Basin. *J. Geophys. Res-Sol Ea.* **2015**, *120*. [CrossRef]
21. Cheng, F.; Garzione, C.N.; Jolivet, M.; Guo, Z.; Zhang, D.; Zhang, C.; Zhang, Q. Initial Deformation of the Northern Tibetan Plateau: Insights from Deposition of the Lulehe Formation in the Qaidam Basin. *Tectonics* **2019**, *38*, 741–766. [CrossRef]
22. Chang, H.; Li, L.; Qiang, X.; Garzione, C.N.; Pullen, A.; An, Z. Magnetostratigraphy of Cenozoic deposits in the western Qaidam Basin and its implication for the surface uplift of the northeastern margin of the Tibetan Plateau. *Earth Planet. Sci. Lett.* **2015**, *430*, 271–283. [CrossRef]
23. Fang, X.; Zhang, W.; Meng, Q.; Gao, J.; Wang, X.; King, J.; Song, C.; Dai, S.; Miao, Y. High-resolution magnetostratigraphy of the Neogene Huaitoutala section in the eastern Qaidam Basin on the NE Tibetan Plateau, Qinghai Province, China and its implication on tectonic uplift of the NE Tibetan Plateau. *Earth Planet. Sci. Lett.* **2007**, *258*, 293–306. [CrossRef]
24. Lu, H.; Xiong, S. Magnetostratigraphy of the Dahonggou section, northern Qaidam Basin and its bearing on Cenozoic tectonic evolution of the Qilian Shan and Altyn Tagh Fault. *Earth Planet. Sci. Lett.* **2009**, *288*, 539–550. [CrossRef]
25. Sun, Z.; Yang, Z.; Pei, J.; Ge, X.; Wang, X.; Yang, T.; Li, W.; Yuan, S. Magnetostratigraphy of Paleogene sediments from northern Qaidam Basin, China: Implications for tectonic uplift and block rotation in northern Tibetan plateau. *Earth Planet. Sci. Lett.* **2005**, *237*, 635–646. [CrossRef]
26. Yang, F.; Ma, Z.; Xu, T.; Ye, S. A tertiary paleomagnetic stratigraphic profile in Qaidam Basin. *Acta Pet. Sin.* **1992**, *13*, 97–101, (In Chinese with English abstract).
27. Zhang, W. The High Precise Cenozoic Magnetostratigraphy of the Qaidam Basin and Uplift of the Northern Tibetan Plateau. Ph.D. Thesis, Lanzhou University, LanZhou, China, 2006.
28. Song, B.; Spicer, R.A.; Zhang, K.; Ji, J.; Farnsworth, A.; Hughes, A.C.; Yang, Y.; Han, F.; Xu, Y.; Spicer, T.; et al. Qaidam Basin leaf fossils show northeastern Tibet was high, wet and cool in the early Oligocene. *Earth Planet. Sci. Lett.* **2020**, *537*, 116175. [CrossRef]
29. Ke, X.; Ji, J.; Zhang, K.; Kou, X.; Song, B.; Wang, C. Magnetostratigraphy and Anisotropy of Magnetic Susceptibility of the Lulehe Formation in the Northeastern Qaidam Basin. *Acta Geol. Sin.* **2013**, *87*, 576–587.
30. Gehrels, G.E.; Yin, A.; Wang, X.-F. Detrital-zircon geochronology of the northeastern Tibetan plateau. *Geol. Soc. Am. Bull.* **2003**, *115*, 881–896. [CrossRef]

31. Gehrels, G.E.; Yin, A.; Wang, X.-F. Magmatic history of the northeastern Tibetan Plateau. *J. Geophys. Res. Solid Earth* **2003**, *108*, 2423. [CrossRef]
32. Song, S.; Niu, Y.; Su, L.; Zhang, C.; Zhang, L. Continental orogenesis from ocean subduction, continent collision/subduction, to orogen collapse, and orogen recycling: The example of the North Qaidam UHPM belt, NW China. *Earth-Sci. Rev.* **2014**, *129*, 59–84. [CrossRef]
33. Yin, A.; Harrison, T.M. Geologic evolution of the Himalayan-Tibetan orogen. *Annu. Rev. Earth Pl. Sc.* **2000**, *28*, 211–280. [CrossRef]
34. Chen, X.; Gehrels, G.; Yin, A.; Li, L.; Jiang, R. Paleozoic and Mesozoic Basement Magmatisms of Eastern Qaidam Basin, Northern Qinghai-Tibet Plateau: LA-ICP-MS Zircon U-Pb Geochronology and its Geological Significance. *Acta Geol. Sin.* **2012**, *86*, 350–369.
35. Luo, K. *The Evolution of Carboniferous-Triassic Basin in Tuolai River on the Northern Margin of Center Qilian Block*; China University of Geosciences (Beijing): Beijing, China, 2015; (In Chinese with English Abstract).
36. Wu, L. *The Detrital-Zircon U-Pb Dating and Provenance Analysis for the Triassic Sandstone in Qilian Shan, Northeastern Margin of the Tibetan Plateau*; China University of Geosciences (Beijing): Beijing, China, 2013; (In Chinese with English Abstract).
37. Zhao, J.; Zeng, X.; Tian, J.; Hu, C.; Wang, D.; Yan, Z.; Wang, K.; Zhao, X. Provenance and paleogeography of the Jurassic Northwestern Qaidam Basin (NW China): Evidence from sedimentary records and detrital zircon geochronology. *J. Asian Earth Sci.* **2020**, *190*, 104060. [CrossRef]
38. Wang, C.; Liu, L.; Xiao, P.-X.; Cao, Y.-T.; Yu, H.-Y.; Meert, J.G.; Liang, W.-T. Geochemical and geochronologic constraints for Paleozoic magmatism related to the orogenic collapse in the Qimantagh–South Altyn region, northwestern China. *Lithos* **2014**, *202–203*, 1–20. [CrossRef]
39. Chen, X.; Gehrels, G.; Yin, A.; Zhou, Q.; Huang, P. Geochemical and Nd–Sr–Pb–O isotopic constrains on Permo–Triassic magmatism in eastern Qaidam Basin, northern Qinghai-Tibetan plateau: Implications for the evolution of the Paleo-Tethys. *J. Asian Earth Sci.* **2015**, *114*, 674–692. [CrossRef]
40. Dong, Y.; He, D.; Sun, S.; Liu, X.; Zhou, X.; Zhang, F.; Yang, Z.; Cheng, B.; Zhao, G.; Li, J. Subduction and accretionary tectonics of the East Kunlun orogen, western segment of the Central China Orogenic System. *Earth-Sci. Rev.* **2018**, *186*, 231–261. [CrossRef]
41. Mo, X.; Luo, Z.; Deng, J.-f.; Yu, X.; LIU, C.; Chen, H.; Yuan, W.; Liu, Y. Granitoids and Crustal Growth in the East-Kunlun Orogenic Belt (in Chinese with English abstract). *Acta Metall. Sin.* **2007**, *13*, 403–414.
42. Mange, M.A.; Maurer, H.F.W. *Heavy Minerals in Color*; Chapman and Hall: London, UK, 1992.
43. Liu, Y.; Hu, Z.; Li, M.; Gao, S. Applications of LA-ICP-MS in the elemental analyses of geological samples. *Chin. Sci. Bull.* **2013**, *58*, 3863–3878. [CrossRef]
44. Wiedenbeck, M.; Allé, P.; Corfu, F.; Griffin, W.L.; Meier, M.; Oberli, F.; Quadt, A.V.; Roddick, J.C.; Spiegel, W. Three natural zircon standards for U-Th-Pb, Lu-Hf, trace element and REE analyses. *Geostandard. Newslett.* **1995**, *19*, 1–23. [CrossRef]
45. Ludwig, K.R. User's Manual for Isoplot/Ex, Version 3.00, A Geochronological Toolkit for Microsoft Excel. *Berkeley Geochronol. Cent. Spec. Publ.* **2003**, *4*, 1–70.
46. Yuan, H.; Gao, S.; Liu, X.; Li, H.; Günther, D.; Wu, F. Accurate U-Pb Age and Trace Element Determinations of Zircon by Laser Ablation-Inductively Coupled Plasma-Mass Spectrometry. *Geostand. Geoanal. Res.* **2004**, *28*, 353–370. [CrossRef]
47. Liu, S.; Li, J.; Stockli, D.F.; Song, C.; Guo, B.; Stockli, L.D.; Ma, Z.; Li, X.; Peng, T. Reappraisal of Miocene eolian deposition in Tianshui Basin, China, based on an investigation of stratigraphy and provenance. *Geol. Soc. Am. Bull.* **2019**, *131*, 1312–1332. [CrossRef]
48. Belousova, E.A.; Griffin, W.L.; O'Reilly, S.Y.; Fisher, N.I. Igneous zircon: Trace element composition as an indicator of source rock type. *Contrib. Mineral. Petr.* **2002**, *143*, 602–622. [CrossRef]
49. Su, N.N.; Jin, Z.K.; Song, F.; Gu, J.F.; Zhang, W.D. Sedimentary facies of the Paleogene in Qaidam Basin. *J. China Univ. Pet.* **2014**, *38*, 1–9, (In Chinese with English abstract).
50. Ji, J.; Zhang, K.; Clift, P.D.; Zhuang, G.; Song, B.; Ke, X.; Xu, Y. High-resolution magnetostratigraphic study of the Paleogene-Neogene strata in the Northern Qaidam Basin: Implications for the growth of the Northeastern Tibetan Plateau. *Gondwana Res.* **2017**, *46*, 141–155. [CrossRef]
51. Fu, L.; Guan, P.; Zhao, W.; Wang, M.; Zhang, Y.; Lu, J. Heavy mineral feature and provenance analysis of Paleogene Lulehe Formation in Qaidam Basin. *Acta Petrol. Sin.* **2013**, *29*, 2867–2875, (In Chinese with English abstract).

52. Yin, A.; Dang, Y.-Q.; Zhang, M.; Chen, X.-H.; McRivette, M.W. Cenozoic tectonic evolution of the Qaidam basin and its surrounding regions (Part 3): Structural geology, sedimentation, and regional tectonic reconstruction. *Geol. Soc. Am. Bull.* **2008**, *120*, 847–876. [CrossRef]
53. Tang, W.; Zhang, Z.; Li, J.; Li, K.; Chen, Y.; Guo, Z. Late Paleozoic to Jurassic tectonic evolution of the Bogda area (northwest China): Evidence from detrital zircon U-Pb geochronology. *Tectonophysics* **2014**, *626*, 144–156. [CrossRef]
54. He, P.; Song, C.; Wang, Y.; Meng, Q.; Chen, L.; Yao, L.; Huang, R.; Feng, W.; Chen, S. Cenozoic deformation history of the Qilian Shan (northeastern Tibetan Plateau) constrained by detrital apatite fission-track thermochronology in the northeastern Qaidam Basin. *Tectonophysics* **2018**, *749*, 1–11. [CrossRef]
55. Jian, X.; Guan, P.; Zhang, W.; Liang, H.; Feng, F.; Fu, L. Late Cretaceous to early Eocene deformation in the northern Tibetan Plateau: Detrital apatite fission track evidence from northern Qaidam basin. *Gondwana Res.* **2018**, *60*, 94–104. [CrossRef]
56. Yin, A.; Dang, Y.-Q.; Wang, L.-C.; Jiang, W.-M.; Zhou, S.-P.; Chen, X.-H.; Gehrels, G.E.; McRivette, M.W. Cenozoic tectonic evolution of Qaidam basin and its surrounding regions (Part 1): The southern Qilian Shan-Nan Shan thrust belt and northern Qaidam basin. *Geol. Soc. Am. Bull.* **2008**, *120*, 813–846. [CrossRef]
57. Wang, Y.T.; Sun, G.Q.; Yang, Y.H.; Wang, M.; Ma, F.Q.; Zhang, C.J.; Zhao, J.; Shi, J.A. U-Pb geochronology analysis of detrital zircons in the Paleogene lower Xiaganchaigou formation in the northern Mahai area, northern margin of Qaidam Basin. *J. Lanzhou Univ.* **2019**, *55*, 141–148, 157, (In Chinese with English abstract).
58. Hartley, A.J.; Weissmann, G.S.; Nichols, G.J.; Warwick, G.L. Large Distributive Fluvial Systems: Characteristics, Distribution, and Controls on Development. *J. Sediment. Res.* **2010**, *80*, 167–183. [CrossRef]
59. He, P.; Song, C.; Wang, Y.; Chen, L.; Chang, P.; Wang, Q.; Ren, B. Cenozoic exhumation in the Qilian Shan, northeastern Tibetan Plateau: Evidence from detrital fission track thermochronology in the Jiuquan Basin. *J. Geophys. Res. Solid Earth* **2017**, *122*, 6910–6927.
60. Jolivet, M.; Brunel, M.; Seward, D.; Xu, Z.; Yang, J.; Roger, F.; Tapponnier, P.; Malavieille, J.; Arnaud, N.; Wu, C. Mesozoic and Cenozoic tectonics of the northern edge of the Tibetan plateau: Fission-track constraints. *Tectonophysics* **2001**, *343*, 111–134. [CrossRef]
61. Bao, J.; Wang, Y.; Song, C.; Feng, Y.; Hu, C.; Zhong, S.; Yang, J. Cenozoic sediment flux in the Qaidam Basin, northern Tibetan Plateau, and implications with regional tectonics and climate. *Glob. Planet. Chang.* **2017**, *155*, 56–69. [CrossRef]

minerals

MDPI

Article

Role of the Down-Bending Plate as a Detrital Source in Convergent Systems Revealed by U–Pb Dating of Zircon Grains: Insights from the Southern Andes and Western Italian Alps

Andrea Di Giulio [1], Chiara Amadori [1,*], Pierre Mueller [1] and Antonio Langone [2]

[1] Department of Earth and Environmental Sciences, University of Pavia, 27100 Pavia, Italy; andrea.digiulio@unipv.it (A.D.G.); pierre.mueller01@universitadipavia.it (P.M.)

[2] Institute of Geosciences and Earth Resources, National Research Council (C.N.R), 27100 Pavia, Italy; langone@crystal.unipv.it

* Correspondence: chiara.amadori@unipv.it

Received: 27 April 2020; Accepted: 13 July 2020; Published: 16 July 2020

Abstract: In convergent zones, several parts of the geodynamic system (e.g., continental margins, back-arc regions) can be deformed, uplifted, and eroded through time, each of them potentially delivering clastic sediments to neighboring basins. Tectonically driven events are mostly recorded in syntectonic clastic systems accumulated into different kinds of basins: trench, fore-arc, and back-arc basins in subduction zones and foredeep, thrust-top, and episutural basins in collisional settings. The most widely used tools for provenance analysis of synorogenic sediments and for unraveling the tectonic evolution of convergent zones are sandstone petrography and U–Pb dating of detrital zircon. In this paper, we present a comparison of previously published data discussing how these techniques are used to constrain provenance reconstructions and contribute to a better understanding of the tectonic evolution of (i) the Cretaceous transition from extensional to compressional regimes in the back-arc region of the southern Andean system; and (ii) the involvement of the passive European continental margin in the Western Alps subduction system during impending Alpine collision. In both cases, sediments delivered from the down-bending continental block are significantly involved. Our findings highlight its role as a detrital source, which is generally underestimated or even ignored in current tectonic models.

Keywords: subduction zones; U–Pb dating; sandstone petrography; provenance; peripheral bulge; tectonics

1. Introduction

Detrital petrology studies of sedimentary basin fill can yield insights into the tectonics that govern compressional settings. Convergent zones are characterized by tectonically active regions affected by exhumation, uplift, and erosion and are surrounded by synorogenic basins accommodating clastic sediments produced in these zones [1,2]. This leads to complex source-to-sink systems where a relatively continuous transfer of rock mass tries to keep the system in equilibrium (i.e., growing relief vs. erosion) [3,4]. Although orogenic prisms are generally thought to be the main (if not the unique) source for detrital sediments in convergent zones, other parts of geodynamic systems can be involved when exhumated [5,6]. Such conditions are met in the cases of the foreland sector in the back-arc area of Andean-type subduction zones or in that of the passive continental margin of a down-bending continental block arriving near a subduction front during the final steps of oceanic closure. Traditionally, the tectonic evolution of those source-to-sink systems is studied by the provenance analysis of clastic sediments accumulated in synorogenic basins, commonly used to decipher the history of eroded units from their erosional products. These studies include a variety of tools whose choice depends on a number of factors like the expected lithology of eroded rocks, the grain size of eroded sediment (gravel vs. sand vs. mud), and so forth. Among these tools, spanning from pebble clast petrology to mudstone and isotope geochemistry (i.e., Sm–Nd and Rb–Sr systems, [7]), the geo-thermochronology of specific types of sand-size clastic heavy minerals has rapidly developed and is now a standardized procedure (see [8] for an up-to-date review).

Nevertheless, we emphasize the importance of combining U–Pb dating of detrital zircons with sandstone petrography in order to (i) obtain a more complete picture of the parent rocks exposed in the source areas, including rock types lacking zircon crystals (e.g., carbonate rocks, basic and ultrabasic rocks, etc.); (ii) identify possible evidence of clastic grain recycling [9,10], which could potentially modify the interpretation of geochronological results; and (iii) analyze sediment composition and maturity, used as a tracer for source region relief and changes in climate of the source terrane (see [11] for an updated review).

Here, we present the results of two different studies in which U–Pb dating of detrital zircon grains, combined with other analytical approaches, has been employed to reconstruct the tectonic evolution of the convergent systems of the Southern Andes and the Western Italian Alps during some key and transient steps of their tectonic transformation. In particular, we describe the contribution of clastic zircon U–Pb geochronology for deciphering the signature with syntectonic sediments left by (i) the late Cretaceous transition from an extensional to a compressional retro-arc foreland basin of the Neuquén Basin in the Southern Andean region [12–15] and (ii) the Late Cretaceous final stages of subduction and related oceanic closure preceding the onset of continental collision in the Ligurian sector of the Western Italian Alps [16–18]. In both cases, results document the importance of the down-bending plate as a voluminous detrital source. Due to its low relief and the low degree of internal deformation, the provenance signal of the down-bending plate appears either underrepresented or absent to undetectable in the detrital record. However, when deduced, it may be linked to a flexural uplift response driven by the interplay between lithospheric properties of the plate (i.e., elastic thickness) [19] and the ensuing evolution of the drainage pattern [20].

2. Materials and Methods

In both presented case studies, the approach is based on provenance analyses of sandstone samples. Among the variety of tools available to address the scientific questions, we preferred to couple U–Pb radiometric age determination of detrital zircon grains with the standard point-counting analysis of sandstone framework grains.

U–Pb dating of detrital zircon grains has become a relatively standard approach for provenance study of clastic systems due to (i) the commonness of zircon crystals in parent rocks, and its resistance to diagenesis results in the relative abundance of zircon grains in clastic sedimentary rocks [21]; and (ii) its relatively low cost, allowing the collection of a statistically significant number of data (commonly >100) necessary for its application to clastic systems. In both presented case studies, U–Pb dating of detrital zircons was performed after the following standard procedure for sample preparation and heavy mineral separation. Sandstones were first crushed and then sieved between 300 and 63 microns. Zircon separates were prepared using heavy liquids and magnetic separation, then they were hand-picked under a binocular microscope, mounted into epoxy resin, and polished down to 0.25 microns to reveal the inner structures of zircon grains. This generalized sequence of steps is also suitable for apatite fission track analysis [22]. Prior to isotopic characterization, the internal structure of zircon was investigated under cathodoluminescence and the most suitable areas for the analysis were selected (see [12] for a detailed description of the morphology of the zircon grains). Rim and core ages [12,14] were collected to discriminate magmatic and metamorphic source rocks only during the deposition and evolution of the Neuquén group. For this reason, samples below the regional unconformity (i.e., Rayoso Formation at the base of the Neuquén Group) were not better investigated. U–Pb ratios in samples were determined with laser ablation inductively coupled plasma mass spectrometry (LA-MC-ICPMS) following the method described in [23,24].

In the Andean case, U–Pb geochronology was performed at Laboratorio de Geocronologia, Universitade de Brasilia (Brazil), following the method described in Matteini et al. [20] (Table S1), whereas U–Pb dating on detrital zircon from the Western Alps was performed at the LA-MC-ICPMS laboratory of the IGG-CNR of Pavia (Italy) following the method described in Tiepolo et al. [24] (Table S2).

When possible, 100 zircon grains of each sample were dated. As a whole, 325 (out of 512 analyzed) zircon grains for the Andean case and 356 (out of 432 zircon grain spots analyzed) for the Western Alps case were used. For the Andean case, only concordant or subconcordant (discordance smaller than ±3%) ages were considered, whereas for the Western Alps case, U–Pb ages with a discordance smaller than ±10% were considered as reliable. $^{206}Pb/^{238}U$ ages were used for grains younger than 1.2 Ga and $^{206}Pb/^{207}Pb$ for grains older than 1.2 Ga (see [12,14,17,18] and Tables S1 and S2).

Sandstone petrography was investigated through a standard thin-section point counting routine, following the Gazzi–Dickinson method [25,26] in order to minimize the possible bias introduced by different grain sizes of the studied samples [11,27]. According to the method, a double-point counting was performed on each sample, dealing with all rock constituents (framework grains, matrix, cements, pores; at least 250 framework grains counted) and fine-grained rock fragments (at least 200 fine-grained rock fragments counted) as proposed and described in detail by Cibin et al. [28] (Table S3). The goal of such double-point counting is to get a better picture of the rock types exposed in the source region and their tectonic meaning in terms of traditional provenance categories [29], including the occurrence of sedimentary rocks which could potentially deliver second-cycle zircon grains. As a whole, 33 thin sections were analyzed in the Andean case and 72 in the Western Alps case [14,17,18].

In the Southern Andes case study, thermochronological data from apatite grains were collected in order to provide a multiproxy dataset. Apatite fission track (AFT) and single-grain double-dating (AFT and U–Pb geochronology on the same grain) were performed to constrain the *T*–*t* path (temperature–time or cooling history) of rocks, both source and basin filling, with respect to a temperature range of 60–110 °C, called the Partial Annealing Zone (PAZ) [30,31]. Since fission tracks anneal primarily as a function of temperature, such that at temperatures higher than 110 ± 10 °C tracks are entirely annealed, temperatures cooler than 60 °C lead to minimal track annealing [32]. AFT analysis was performed on three samples where ~100 grains were analyzed for each detrital sample. Unfortunately, there were not enough confined lengths present in any given population to be statistically significant, so length data are not reported. For each sample, fission track grain–age distributions and populations were determined using both the Binomfit program [33] (which applies the binomial peak-fitting method [34]) and Density Plotter [35]. Calculated populations are reported by age and error (2σ). The apatite system is very sensitive to temperatures within the upper portion of the crust, less than about 4 km, and ages can either represent cooling events by exhumation of parent rocks or "geologically instantaneous" cooling following magmatism [36]. For this reason, 23 apatite AFT data Cretaceous in age (i.e., near syndepositional) were also coupled with apatite U–Pb dating. Apatite has also been employed in high-temperature thermochronology studies, which demonstrate that the U–Pb system in apatite has a closure temperature of about 450–550 °C [37,38]. LA-MC-ICPMS U–Pb dating of apatite is more challenging, as apatite typically contains lower U and Pb concentrations. In contrast to the well-documented polycyclic behavior of the stable heavy mineral zircon, apatite is unstable in acidic groundwater and weathering profiles, and it merely shows limited mechanical stability in sedimentary transport systems. Thus, it more likely represents first-cycle detritus; hence, U–Pb apatite dating would yield complementary information to U–Pb zircon provenance studies. Fission track and U–Pb dating are, therefore, two of the most useful and rapid techniques in apatite provenance studies. They yield complementary information, with the apatite fission track system indicating low-temperature exhumation ages and the U–Pb system yielding high-temperature cooling ages, which help constrain the timing of apatite crystallization [39,40]. Apatite U–Pb dating was conducted by laser ablation multicollector inductively coupled plasma mass spectrometry (LA-MC-ICPMS) at the Arizona LaserChron Center, following the methods used for zircon by Gehrels et al. [41,42] and Gehrels [43], modified for apatite by Thomson et al. [44].

3. Geological Framework and Results of the Southern Andes and Western Italian Alps Case Studies

3.1. The Case of the Retro-Arc Basin of Southern Andes between 34° and 40° S Latitude (Neuquén Basin, Argentina)

3.1.1. Geological Setting and Stratigraphy

The Neuquén Basin is a Mesozoic–Early Cenozoic foreland basin located in a retro-arc position with respect to the Andean volcanic arc between 34° and 40° S (Figure 1A). Due to its long and fairly continuous subsidence, the basin contains an almost complete record of Mesozoic Andean tectonics, erosion, and sedimentation. The geological evolution of the Neuquén Basin was largely controlled by changes in the tectonic regime along the western margin of South America during the Mesozoic and Cenozoic eras (see [12,14,15,45–47] for an up-to-date review). In particular, the Jurassic–Cretaceous evolution has been linked to changes in the subduction slab dip of the oceanic Nazca Plate under the eastern South American continental margin, and to changes in trench roll-back velocities [46], and is characterized by (i) a Jurassic–Early Cretaceous back-arc extensional phase linked to a steeply dipping subduction and (ii) a contractional retro-arc foreland basin phase linked to a temporary shallower dipping subduction due to a change in the absolute motion of South America. The sedimentary record preserved in the continental Cretaceous strata of the Neuquén Basin allows investigating the effects of that transition from an extensional back-arc to a contractional retro-arc foreland setting, notwithstanding that no significant facies changes occurred as depositional environments remained continental with fluvial deposits during the entire Cretaceous evolution (see [14] for a detailed description of depositional facies). In particular, the investigated effects concern (i) switches in source regions of the clastic detritus; (ii) exhumation patterns of source rocks; and (iii) stratigraphic architecture of the basin. Most Cretaceous sedimentary evolution is recorded by a thick continental clastic succession formed by the Bajada del Agrio and the Neuquén Groups, separated by a regional-scale Cenomanian unconformity (Figure 1B). This continental succession has been intensively studied in order to capture the tectonic signal linked to the evolution of that portion of the Andean system.

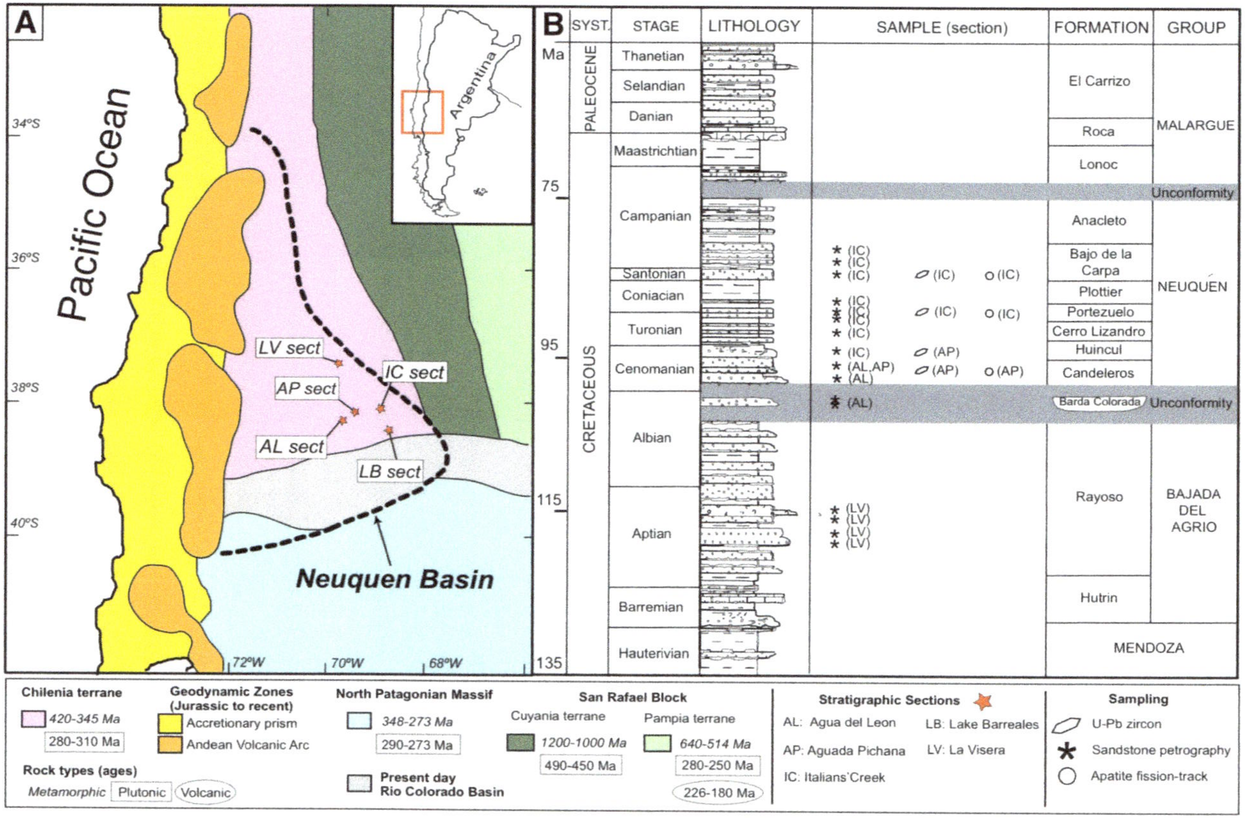

Figure 1. (**A**) Geological setting of the Neuquén Basin in the back-arc region of Southern Andes between 40° and 36° S latitude with locations of the sampled sections (modified after [48]). (**B**) Stratigraphy of the Cretaceous and lower Cenozoic filling of the Neuquén Basin with stratigraphic positions of the studied samples. Modified after [14], where also references on geochronological ages for different source terranes are reported.

3.1.2. The Detrital Record from Detrital Zircon U–Pb Dating, Sandstone Petrography, and Thermochronological Proxies

Sandstone provenance has been intensively studied in the last 10 years in the Andean retro-arc basins, frequently using U–Pb dating of zircon grains as the main or unique approach [13–16,47–51]; however, many of these geochronological studies concern the Meso-Cenozoic sedimentary sequences of the northern and southern sector of Andean mountain building [13,52–56]. Here, we particularly focus on the data reported in the regional studies by Di Giulio et al. [14,16]. These studies, performed on the Cretaceous part of the Neuquén Basin, reveal a distinct shift in sandstone provenance related to a change in source exhumation patterns of parent rocks as well as lag time across the basin-scale pre-Cenomanian unconformity that divides the Bajada del Agrio and the Neuquén Groups [13,14] (Figure 1B). Sandstone petrography of the same stratigraphic interval consistently shows a shift from a mixed to a volcanic arc signature across the Cenomanian unconformity that divides the Rajoso from the Candeleros Formation, and it shows a progressive return toward a mixed and then to cratonic signature along the Neuquén Group (Figure 2A). More specifically, sandstone detrital mode changes mostly reflect the variable abundance of fine-grained volcanic rock fragments in the studied formations, reaching a maximum in the lower Candeleros Formation and then rapidly decreasing up-section in the Neuquén Group. Consistently, U–Pb dating of zircon grains from clastic units across the unconformity in the Rayoso Formation and then along the Neuquén Group distinctly records an abrupt change of detrital age populations.

In the Neuquén Basin, from the Candeleros Formation, zircon grains age populations show a massive arrival of crystals with cooling ages equal or very close to the depositional age (few tens of millions to zero years of difference). This contribution is from syndepositional volcanics and/or juvenile volcanic material completely absent before (Rayoso Formation; Figure 2B,C). In the Rayoso Formation, the difference in time between its depositional age (Aptian–Albian) and the youngest detrital zircon population is ca. 100 Ma. Furthermore, the cumulative curve (Figure 2B) of the Rayoso Formation shows that most of the measured crystallization zircon ages are 250 Ma older than the depositional age as expected in tectonic settings characterized by erosion of old basement rocks.

Up-section (above Candeleros Formation), the progressive decrease of syndepositional volcanics and/or juvenile volcanic material is visible, as well as substitution by older zircon grains (Figure 2B,C). The picture is completed by geo-thermochronological data on apatite grains (U–Pb dating and fission tracks), that is, the double-dating method (Figure 3). Double-dating of apatite crystals was preliminarily used to detect possible synsedimentary volcanic contributions because AFT data is useful for unravelling the exhumation story of parent rocks within the upper few kilometers (i.e., their exhumation above the AFT annealing temperature). The results (Figure 3B) show that most of the analyzed apatite grains had a nonvolcanic origin, as their U–Pb and AFT ages were significantly different (delta age spanning from 150 to more than 300 Ma). Only one exception showed zero delta age between U–Pb and AFT ages, thus confirming its volcanic origin. For the nonvolcanic apatite grains, two broadly different age populations can be recognized and interpreted as two different source regions with respect to their tectonic history and lag time, defined as the difference between the AFT age and the depositional age of the rock. The sources are: (1) a rapidly exhuming region delivering apatite grains with AFT ages very close to the depositional age (very short lag time, ca. 10–20 Ma), shown by the single peak in Candeleros Formation (123.7 ± 4.7 Ma) and the youngest peak in Portezuelo and Bajo de la Carpa Formations (82.4 ± 5.5 and 94.7 ± 4.7 Ma, respectively), and (2) a sector delivering apatite grains with AFT ages several tens of Ma (>50 Ma) older than the depositional age of samples. This is inferred by the presence of a second older population in Portezuelo and Bajo de la Carpa Formations (158.3 ± 7.7 and 163 ± 9.5 Ma, respectively). Generally, a longer lag time testifies to the erosion of rocks that remained above the AFT annealing temperature for a considerably long time before arriving at the surface [36]; in addition, the proportion of the second source increases up-section (Figure 3A).

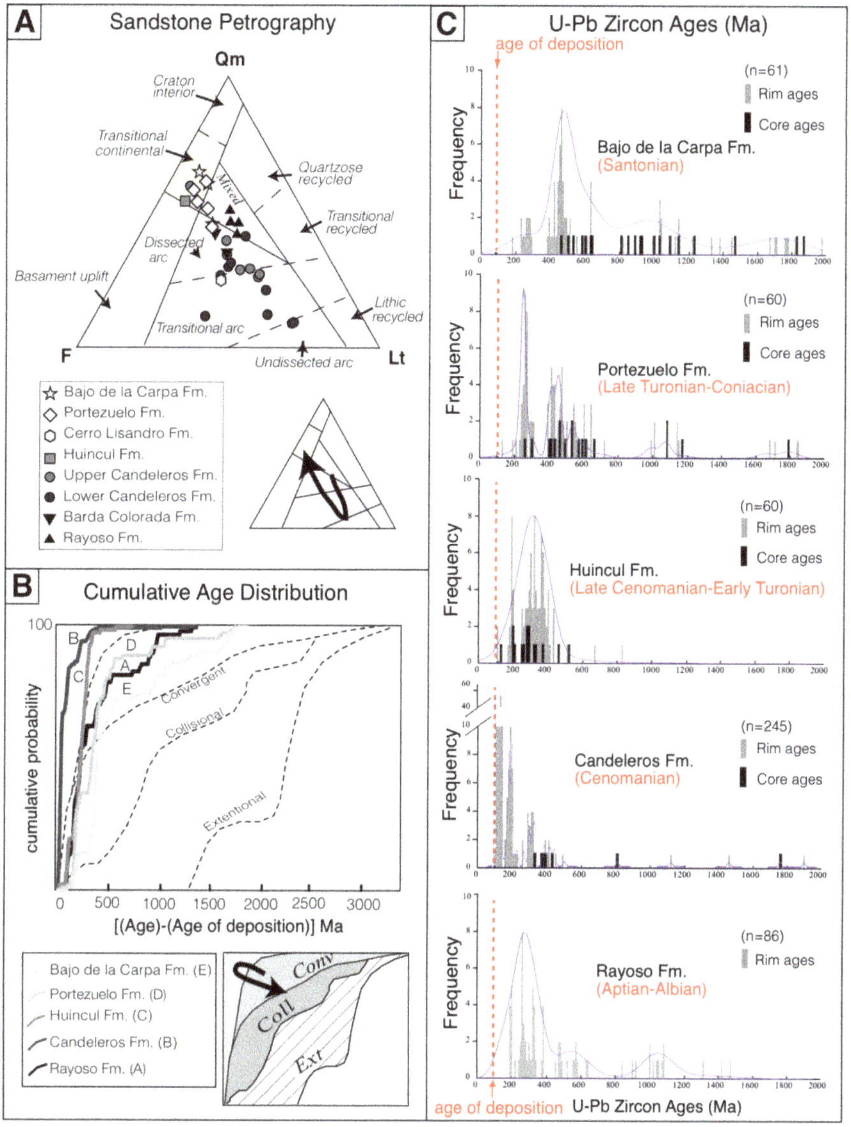

Figure 2. (**A**) Sandstone petrography in the Cretaceous Bajada del Agrio and Neuquén Groups of Neuquén Basin plotted in the QmFLt provenance diagram of Dickinson [29]. (**B**) Results from U–Pb dating of detrital zircons. Cumulative proportion curves showing the difference between the crystallization age for a detrital zircon and the depositional age of the sample in which it occurs. Solid lines indicate the studied samples; dashed lines bind the fields representing three tectonic settings as defined in [50]. (**C**) U–Pb ages (bins 10 Ma) shown from top to bottom according to their stratigraphic order (Table S1). Blue line is the kernel density estimate (KDE). Red text is the depositional age of the corresponding formations (see [14] and reference therein for the paleontological and geochronological data). For the stratigraphic position of studied samples, refer to Figure 1B. Modified from [14]. Stratigraphic positions of samples are reported in Figure 1B.

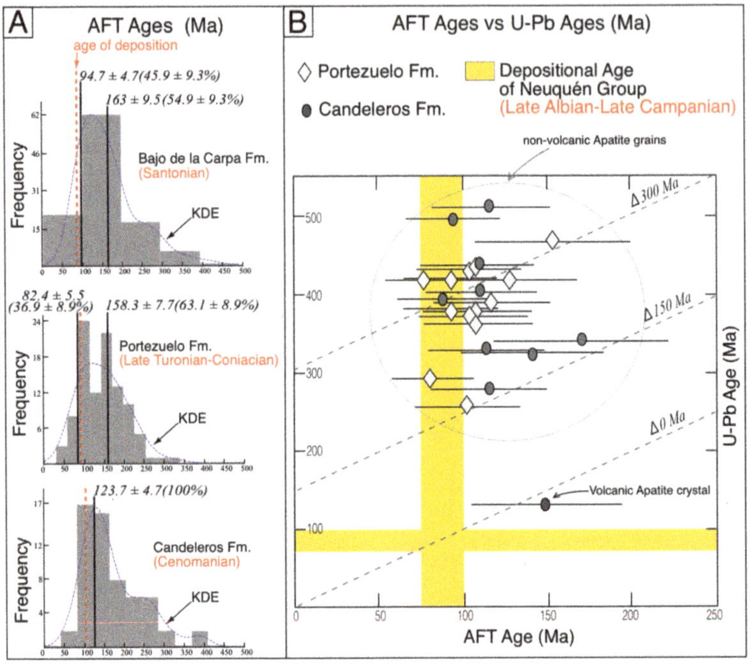

Figure 3. (**A**) Kernel density estimate (KDE, dashed blue line) and histogram distributions (grey rectangles) from apatite fission track data produced using Density Plotter [35] from Bajo de la Carpa, Portezuelo, and Candeleros Formations, shown in their stratigraphic order. Calculated populations are reported by age and error 2σ. (**B**) Results of double-dating on detrital apatite grains from the Neuquén Group; sample stratigraphic positions reported in Figure 1B. Note that no volcanic crystals (delta age = 0) are recognizable, with only one exception in the Candeleros Formation. Red text is the age of deposition (see [14] and reference therein for the paleontological and geochronological data).

3.1.3. Discussion on the Drainage Evolution and Related Provenance Changes in the Neuquén Basin

The collected multiproxy dataset allows tracing the tectonic evolution and the related rearrangement of the drainage pattern within the source-to-sink routing system formed by the Neuquén retro-arc basin and the Andean cordillera, between 36° and 40° S latitude from Albian to Santonian time. This evolution can be schematically summarized in the following main steps. During Albian time, the existence of a divergent drainage pattern in a back-arc basin setting is suggested by the lack of syndepositional, neovolcanic grains (Aptian–Albian in age) in the Rayoso Formation, as recorded by coupled zircon U–Pb ages, cumulative curve, and sandstone petrography (Figure 2).

In fact, at that time (>100 Ma), a volcanic arc was active and its overall detritus was likely transported westward, towards the pacific margin, with the only exception of ash layers. A volcanic detritus transported eastward would have needed to pass over the rift shoulders to finally end up in the basin; in this case, it would have been recorded by detrital zircon ages coeval with the sediments' depositional age. By contrast, the back-arc Neuquén Basin was mostly fed by the uplifted rift shoulders, mainly composed of Paleozoic–Proterozoic basement rocks and their Jurassic–Triassic covers shown by the highest peak of U–Pb zircon ages in the Rayoso Formation between ca. 200 and 400 Ma and minor peaks around 500–600 Ma and ca. 1.1 Ga (Figure 2C). At around 100 Ma, the region started to become affected by contractional deformation, possibly linked to the global-scale plate reorganization produced by the opening of the Southern Atlantic Ocean [57] (see also [15] for discussion). In the Neuquén Basin, this tectonic inversion triggered a rapid reorganization of the drainage pattern caused by the transition from a back-arc extensional setting to a retro-arc contractional basin. In the stratigraphic record, this transition is marked by the pre-Cenomanian regional-scale erosional unconformity between the Bajada del Agrio Group and the Neuquén Group occurring in an otherwise continuously subsiding basin. This erosion is thought to be the result of a weak but basin-scale uplift, due to large-scale continental folding (i.e., buckling) during the very beginning of contraction, before the full development of thrust tectonics and an adjacent foreland basin. After this uplift event, during Cenomanian–Turonian time (Candeleros, Huincul, and Portezuelo Formations), continental deposition restarted and both zircon U–Pb ages (highest peak at ca. 95–150 Ma in the Candeleros Formation) and sandstone petrography (volcanic lithic-rich) consistently recorded an important and abrupt shift, with poorly dissected syndepositional volcanic arc detritus (Figure 2A) arriving in the Neuquén Basin from the Cordillera together with a minor but ongoing contribution from Mesozoic (ca. 200 Ma detrital zircon age peak) and Paleozoic rocks exhumed during the Cretaceous (zircon ages of ca. 300–500 Ma) (Figure 2C). Paleocurrent data in the Neuquén Group show N–S to E–NE directions moving up-section [14] and allows confirming this reconstruction.

This evidence is interpreted as recording the beginning of contractional-related exhumation of the Cordillera, coupled with inversion of the drainage pattern from divergent to convergent and linked to transition of the Neuquén Basin from a back-arc to a retro-arc foreland basin. After this step, during the Late Cretaceous evolution, the integration of U–Pb zircon dating, sandstone petrography, and apatite double-dating results (showing no volcanic apatite grains deposited in the Neuquén group, except for only one crystal in the lower Candeleros Formation, Figure 3B) suggests a progressive increase of detrital input from old basement rocks. This included Proterozoic rocks with a multistage magmatic–metamorphic history (>500 Ma, rim-core ages, Figure 2C) exhumed during the Jurassic, together with a contribution from rocks that suffered Cretaceous exhumation with removal of the partial annealing zone during the Cenomanian (i.e., older AFT populations, Figure 3A), followed by increasing tectonic exhumation up to Santonian time. As proposed by [12] and [14], this multiproxy dataset indicates that a convergent drainage pattern developed into a contractional foreland setting during deposition of the Neuquén Group, with detrital material supplied by both the Andean Cordillera to the west and by the craton to the east, which was uplifted during eastward migration of the flexural peripheral bulge causing an increasing detrital contribution from the San Rafael continental block located in the foreland.

3.2. *The Case of the Western Ligurian Flysch (NW Italy) during Impending Collision*

3.2.1. Geological Setting and Stratigraphy

This case study addresses the detrital provenance of the terrigenous units included in the Cretaceous–Paleocene Western Ligurian Helminthoid Flysch Complex (WLF), which forms the uppermost part of the nappe pile of the Ligurian part of the Western Italian Alps, NW Italy (Figure 4A). The WLF represents the remnant of the accretionary wedge formed by the cover of the Piedmont–Ligurian ocean and its continental margin, scraped off along the Ligurian Alps transect of the Alpine subduction system in the framework of an intra-oceanic subduction zone [16,58–60] (Figure 4B–D). During the Late Eocene–Early Oligocene continental collision, the WLF was thrust over the European foreland and presently rests on the Mesozoic Dauphinois–Provençal succession to the west and southwest and on the Briançonnais units of the Ligurian Alps to the north [16,59,61–63]. The WFL comprises four main flysch units (from top to bottom of the complex): the San Remo–Monte Saccarello Unit, the Moglio–Testico Unit, the Borghetto d'Arroscia Unit, and the Colla Domenica–Leverone Unit (Figure 4B,C). The tectonically inverted chronostratigraphic organization of these units, divided by southward dipping thrusts, with the oldest unit resting on top of the nappe pile, documents the typical tectonic inversion of accretionary wedges [16,64]. Whilst the three lowermost and younger units underwent multiphase, ductile–brittle deformation, the oldest and structurally topmost San Remo–Monte Saccarello Unit is characterized by a rather simple and less intense structural overprint.

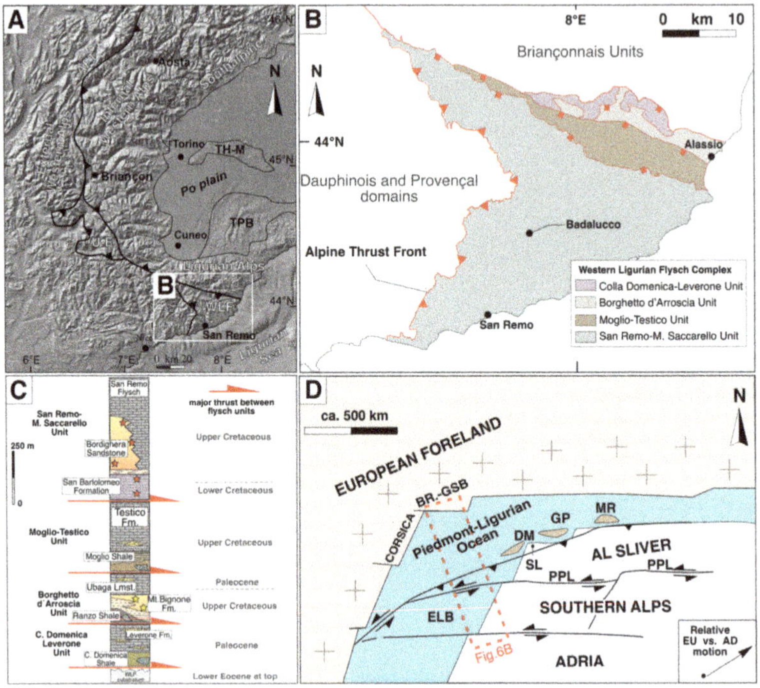

Figure 4. (**A**) Tectonic scheme of the Western Alps and location of Western Ligurian Flysch Complex (WLF); ATF: Alpine Thrust Front; U-E: Ubaye–Embrunais nappe; TH-M: Torino Hills–Monferrato arc; TPB: Tertiary Piedmont Basin. (**B**) Structural scheme of the WLF. (**C**) Stratigraphic framework of the WLF units and stratigraphic position of studied samples. (**D**) Paleogeographic reconstruction proposed for the Late Cretaceous setting of the Piedmont–Ligurian subduction zone along the Ligurian Alps transect, redrawn after [48]; BR-GSB: Briançonnais–Gran San Bernard domain; DM-GP-MR: Dora Maira–Gran Paradiso–Monte Rosa blocks; SL: Sesia–Lanzo zone; PPL: Paleo-Periadriatic Line; ELB: External Ligurian Basin; AL: Austro-Alpine continental sliver.

Two of the units of the WFL feature poorly dated upper Cretaceous, thick, sandstone-rich turbidite units [16] (Figure 4B): (i) The Borghetto d'Arroscia Unit contains quartz-rich turbidite fan deposits of the Monte Bignone Formation (sometimes called M. Bignone Quartzites in the regional literature) [65] interbedded with conglomeratic levels primarily comprising carbonate clasts [66], and divided by one of these conglomerate units into a lower and an upper quartzite member. (ii) The San Remo–Monte Saccarello Unit contains quartz-rich sandy turbidites in the uppermost part of the lowermost stratigraphic unit (San Bartolomeo Formation) that is stratigraphically overlain by a thick sand-rich turbidite fan unit, the Bordighera Sandstone [67]. The other WFL units are composed of shaly formations at the base (sometimes called basal complexes in the regional literature: Colla Domenica Shale, Ranzo Shale, Moglio Shale, lowermost part of the S. Bartolomeo Formation) and thick and monotonous calcareous-marly flysch successions (Leverone Formation, Ubaga Limestone, Testico Formation, San Remo Flysch) of the wide family of the Alpine Helminthoid Flysches.

3.2.2. The Detrital Record from Sandstone Petrography and U–Pb Dating of Detrital Zircon Grains

Provenance analyses of sandy siliciclastic turbiditic successions of both the San Remo–Monte Saccarello Unit and the Borghetto d´Arroscia Unit were performed by means of integrating U–Pb detrital zircon chronology and sandstone petrography (Tables S2 and S4). For petrographic investigations, a total of 40 samples were acquired from the terrigenous units of the San Remo–Monte Saccarello

Unit. Thirty sandstone samples were selected from two stratigraphic intervals of the Monte Bignone Formation in the Borghetto d´Arroscia unit: 27 samples from the Lower Quartzite member and 3 samples from the Upper Quartzites member. For detailed stratigraphic positions of samples processed for petrographic studies, the reader is referred to [14,15]. Samples for detrital zircon geochronology were collected from the same stratigraphic intervals (Figure 4C).

Based on sandstone modal analyses, the fine-grained turbidite sandstones of the San Bartolomeo Formation at the base of the San Remo Unit classify as quartz-rich sandstones (modal averages of $Qt_{69}F_{29}L_2$; Figure 5A). This relatively high compositional maturity [66] is accompanied by the well-sorted nature of the sandstone framework grains. Moderate to high roundness indicates high textural maturity. By contrast, the overlying coarse-grained Bordighera Sandstone turbidites represent typical arkoses, with roughly equal shares in quartz and feldspars (average modal composition: $Qt_{49}F_{48}L_3$), and typified by a poor sorting and angular to subangular framework grains. The high quartz proportions that typify the San Bartolomeo Formation samples suggest intense weathering of less stable feldspar grains during prolonged exposure on continental land masses and/or extended sedimentary recycling along lower-gradient areas [68]. In combination with the high textural maturity, according to traditional provenance categories [69], these observations indicate a stable craton to transitional continental provenance terrane. Further, this may include extensive shelfal areas where detrital material may have been subjected to additional textural maturation. Conversely, the abundance of unstable feldspar grains rules out such a provenance scenario for the younger coarse-grained Bordighera Sandstone [69]. Together with the markedly reduced textural composition, a first-cycle origin dominantly derived from an uplifting granitic basement block, with very minor contributions of low-grade metamorphic rocks and sedimentary rocks, may be inferred. Within both units, the dominance of monocrystalline over polycrystalline quartz is interpreted to indicate contributions of basement rocks [70,71].

Sandstones within the Borghetto d'Arroscia Unit exhibit variable degrees of sorting. Framework grains are typically subangular to rounded. The detritus generally contains high proportions of quartz, very low to low proportions of feldspar, and variable shares of rock fragments (mean $Qt_{83}F_5L_{12}$). Within this overall picture, the key observation is that modal percentages show significant variations, evolving from quartzose sandstones towards lithic and then to a lithic, subarkosic composition up-section (Figure 5A). The Lower Quartzite member sands feature high quartz proportions (average 86%), whereas quartz shares significantly decrease up-section towards the Upper Quartzite member (average 63%). On the contrary, feldspars account for relatively constant, low portions throughout the Lower Quartzite member (average 4%) but become more abundant in the Upper Quartzite unit (average 14%). A similar pattern can be observed in the shares of lithic fragments, which show an overall up-section increase (from 10% to 22% on average). In summary, a somewhat gradual change from quartz arenites to lithic arenites is recorded in the Lower Quartzite unit, delineated by an increase in the number of sedimentary and volcanic lithic fragments. By contrast, the Upper Quartzite member exhibits a further shift towards lithic subarkosic composition coupled with an increase in plutonic and metamorphic rock fragments. The high degree of compositional maturity characterizing the quartz-rich arenites of the Lower Quartzites in the Borghetto Unit implies intense sedimentary recycling [69], suggestive of temporary sediment storage along the source-to-sink pathway, similar to the interpretation of the provenance scenario obtained for the San Bartolomeo Formation in the San Remo-M. Saccarello unit. The compositional maturity could, hence, be explained by prolonged reworking in shallow marine environments along a passive margin shelf. The up-section increase of sedimentary lithic fragments in the M. Bignone Lower Quartzite member suggests progressive tectonic uplift and local denudation of a provenance terrane that, to a great extent, comprised a well-developed sedimentary cover sequence. Conversely, the increased feldspar and plutonic fragment portions in the M. Bignone Upper Quartzite member would indicate a further shift towards an increasing basement source [69].

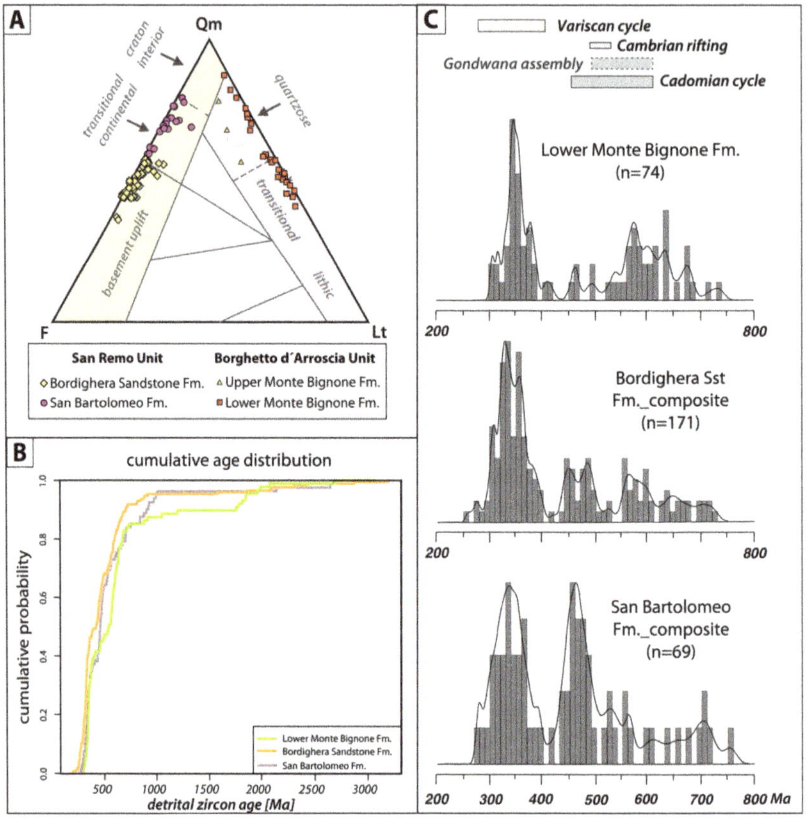

Figure 5. (**A**) Results from sandstone detrital framework analysis of the siliciclastic members of the WLF units plotted in the QmFLt provenance discrimination diagram of Dickinson [26] (Table S4). (**B**) Compilation of cumulative detrital zircon U–Pb age distributions of the investigated siliciclastic units of the WLF. (Table S2). (**C**) Probability density plots of the siliciclastic successions of the WLF covering the time range from 200 to 800 Ma; bins 10 Ma and discordance ±10%. Note that in this case, the depositional age (Late Cretaceous) is out of the plot area, as no zircon grains with a similar age occurred in studied samples.

The observed stratigraphic trends in sandstone modal composition in both the San Remo and Borghetto Units raise the question of whether they reflect changes in sediment provenance area (i.e., a provenance shift) or imply a change in tectonic regime affecting the same source area. U–Pb dating of detrital zircon grains was performed to solve this basic provenance question and further constrain source area reconstructions. A total of 356 zircon grains from six samples coming from the San Bartolomeo Formation (two samples), Bordighera Sandstone (three samples), and Mt. Bignone Formation (one sample) yielded concordant ages. Comparisons of the investigated detrital zircon signatures revealed marked similarities (Figure 5B,C; standard statistical evaluations of similarity between samples' detrital age distributions via KS-tests and MDS diagrams are provided in [17,18]). All detrital zircon age spectra were characterized by their primary Carboniferous age contributions around ca. 300–360 Ma. Significant populations of Ordovician ages around 450 Ma and 480 Ma were also common in all samples. Additional similarities were expressed in the shared broad age population ranging from ca. 535 to 660 Ma. Noteworthy, minor Upper Pennsylvanian-Guadalupian age peaks between ca. 270 Ma and 305 Ma were exclusive to samples of the San Bartolomeo Formation.

3.2.3. Discussion on the Erosion of the Passive Continental Margin during Oceanic Closure

The multiproxy sediment provenance study of the sandy units of the Western Ligurian Flysch Complex provides a better understanding of the precollisional evolution of the Piedmont–Ligurian ocean, and it specifically allows deciphering the role of the opposing continental margins as potential detrital sources during the final stages of oceanic closure. To this end, the distinction between the detritus delivered by the Adriatic margin (placed on the overriding plate) and that derived from the European passive margin (placed on the down-bending plate) is required.

The main populations of Carbonifeous detrital zircon U–Pb ages correspond to magmatic and metamorphic events that, in the Alpine region, occurred in the framework of the older Variscan orogeny [72]. The second most abundant age group that covers the time span between ca. 535 Ma and 660 Ma corresponds to geodynamic events associated with metamorphic and magmatic rocks that occurred in the course of the Cadomian orogeny [73]. Detrital ages between ca. 270 Ma and 305 Ma can be linked to post-Variscan magmatism associated with gravitational collapse of the thickened Variscan crust. With regard to the opposing continental margins that bounded the Piedmont Ligurian Ocean (i.e., the subducting paleo-Europe plate and the overriding Adriatic plate), pronounced late Variscan age distributions (270–330 Ma) provide fundamental insights into the provenance terrane, since granite emplacement primarily occurred along the paleo-European plate during that time span [72,74–76]; this conclusion is also supported by Multidimensional Scaling Analysis [18]. In fact, only older Variscan ages were found in Austroalpine and Carpathian basement units [77] (i.e., Adriatic plate).

These late Variscan age populations (Figure 5C) shared in all investigated clastic formations a similar provenance for all investigated WLF units. These results coupled with the detailed sedimentological description [17,18,63,67] of the clastic sequence allow inferring the principal sediment supply into the subduction zone staged from the Paleo-European continental margin (placed on the subducted plate), during its arrival close to the subduction zone facing the impending continental collision. Therefore, geochronological data provided by U–Pb detrital zircon ages helped to testify that (i) despite the observed compositional changes in sandstone modal composition, the terrigenous material was delivered by the same source terrane, and (ii) in the context of the Alpine subduction and age constraints on crustal growth stages recorded in the pre-Alpine basements [72,78], that terrane was likely placed on the passive continental margin of the subducting European plate. Nevertheless, in both investigated units, a substantial shift in tectonic stability of the source terrane was recorded first by the onset of coarse-clastic sedimentation, and then by its compositional evolution.

The development of a flexural bulge in response to the tectonic loading of the advancing Alpine accretionary wedge provides a feasible explanation for this reactivation and tectonic inversion of the lower plate margin. This would explain both the observed up-section trend in composition and the evolution of textural sediment maturity, which are interpreted to mirror the imminent arrival of the flexural bulge in the distal parts of the continental margin, triggering the reworking of quartz-rich shelf sands into the deep basin. More specifically, the up-section trend towards less mature, arkosic sands is thought to reflect the subsequent craton-ward migration (hinterland) of the flexural bulge, causing rapid unroofing of crystalline rocks and enhanced erosion rates. Increased sediment yield and sediment caliber might indicate the development of areas of high relief and/or reactivation of ancient fault scarps. The main implication from these observations is that the observed mineralogic up-section trend of decreasing sediment maturity of sediments delivered from the same source terrane can serve as a detrital tool for the inversion of the passive margin during its arrival near the subduction zone, and the forthcoming geodynamic transition from a subduction to a collisional scenario.

4. Overall Conclusions

Provenance of clastic sediments accommodated in syntectonic basins is commonly used as a tracer for reconstructing the evolution of convergent geodynamic settings. In these settings, the orogenic prism is commonly considered as the prevailing source for detrital material. By contrast, the examples from the Andean region and Western Ligurian Alps show the following:

- In the back-arc Andean region, a transition from an extensional to a contractional setting is recorded during late Albian–early Cenomanian time (Figure 6A) by detrital provenance, possibly linked to an overall plate reorganization that caused the decrease of subduction dip. The resulting flexure of the down-bending continental block in the back-arc Andean region triggered the erosion of the South American foreland craton, providing most of the detrital material to the majority of Upper Cretaceous continental successions accommodated in the back-arc Neuquén Basin during its contractional evolution.
- In such a back-arc contractional setting, the scarceness of syndepositional volcanic zircon crystals, combined with the contrasting exhumation rate of the source rocks revealed by double apatite dating, provide clues for discriminating sediments coming from the Cordillera volcanic arc/orogenic belt from sediments staged from the continental foreland.
- In the Ligurian transect of the Western Alps (Figure 6B), an intra-oceanic subduction was supposed due to the overall lack of terrigenous material arriving in the basin during most of the Cretaceous. In such a context, the Late Cretaceous arrival of voluminous sand-size terrigenous detritus remains enigmatic.
- Coupled U–Pb dating of clastic zircon grains and sandstone petrography support the European continental margin placed on the subducting plate as the source of that material, and its arrival close to the subduction zone is thought to announce the impending collision. Within this transient phase, decreasing sandstone maturity through time is thought to record the arrival of the flexural bulge on the passive continental margin, starting from a shelf region and then moving toward the hinterland. This passage first triggered the reworking of mature quartz-rich shelf sands, and afterwards it caused basement erosion and deposition in the subduction zone of basement-sourced arkoses.

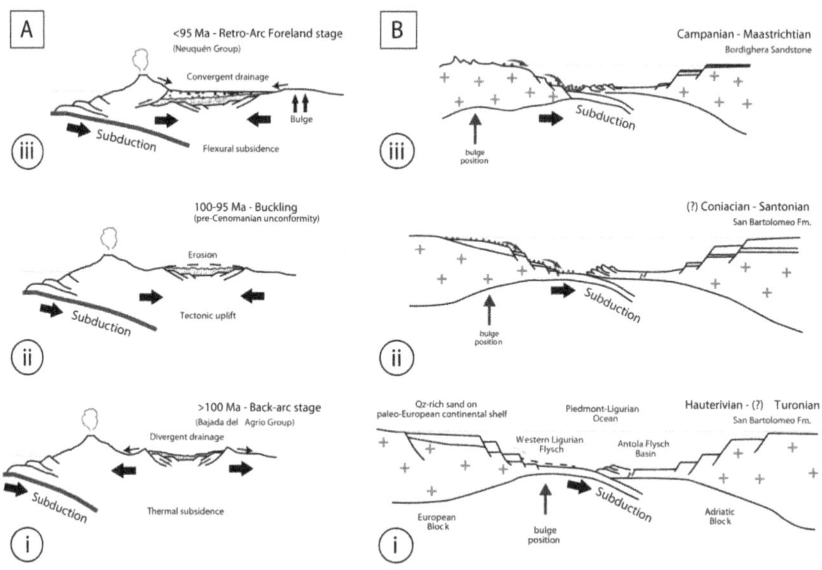

Figure 6. Tectonic model developed for the late Cretaceous evolution of (**A**) the Neuquén Basin in the retro-arc of Southern Andes, and (**B**) the Piedmont–Ligurian subduction zone along the Ligurian transect of Western Italian Alps. Note, in both, the major role of detrital source played by the continental down-bending block, uplifted and eroded by the passage of the peripheral bulge.

In summary, the findings derived from detrital studies of the south Andean system and the Ligurian transect of the Italian Alps system point out that even the down-bending block can be an important source, and sometimes even the only source, of terrigenous sediments in certain steps of the tectonic evolution of convergent systems. In the case of intra-oceanic subductions, that contribution can be considered a way to trace the final stages of oceanic closure and impending collision.

Supplementary Materials: The following are available online at http://www.mdpi.com/2075-163X/10/7/632/s1, Table S1: Geo-thermochronology Neuquén, Table S2: Geochronology Ligurian Alps, Table S3: Petrography Neuquén, Table S4: Petrography Ligurian Alps.

Author Contributions: A.D.G. and C.A. conceived the scientific question; P.M. and A.L. performed U–Pb zircon dating; A.D.G. and P.M. performed sandstone petrographic analyses; A.D.G., C.A., and P.M. provided figures/manuscript preparation. All authors contributed to the design of the manuscript and its final revision. All authors have read and agreed to the published version of the manuscript.

Funding: This research received no external funding.

Acknowledgments: Wilfried Winkler and Albrecht von Quadt are kindly acknowledged for inviting us to contribute to this Special Issue. All authors are indebted to two anonymous reviewers and Pereira, M.F. for their detailed and constructive revisions that led to significant improvements in the quality of the paper. We also want to thank the many colleagues that contributed to our Andean and Ligurian Alps research through the years; among them, Marco Patacci (Leeds, UK) and Victor Ramos (Buenos Aires, Argentina) are particularly acknowledged.

Conflicts of Interest: The authors declare no conflict of interest.

References

1. DeCelles, P.G.; Giles, K.A. Foreland basin systems. *Basin Res.* **1996**, *8*, 105–123. [CrossRef]
2. Garzanti, E. The Himalayan Foreland Basin from collision onset to the present: A sedimentary–petrology perspective. *Geol. Soc. Lond. Spec. Publ.* **2019**, *483*. [CrossRef]
3. Dahlen, F.A. Critical taper model of fold-and-thrust belts and accretionary wedges. *Ann. Rev. Earth Planet. Sci.* **1990**, *18*, 55–90. [CrossRef]
4. Malevieille, J. Impact of erosion, sedimentation, and structural heritage on the structure and kinematics of orogenic wedges: Analog model and case studies. *GSA Today* **2010**, *20*. [CrossRef]
5. Pereira, M.F.; Chichorro, M.; Johnston, S.T.; Gutiérrez-Alonso, G.; Silva, J.B.; Linnemann, U.; Hofmann, M.; Drost, K. The missing Rheic Ocean magmatic arcs: Provenance analysis of Late Paleozoic sedimentary clastic rocks of SW Iberia. *Gondwana Res.* **2012**, *22*, 882–891. [CrossRef]
6. Pereira, M.F.; Gutíerrez-Alonso, G.; Murphy, J.B.; Drost, K.; Gama, C.; Silva, J.B. Birth and demise of the Rheic Ocean magmatic arc(s): Combined U-Pb and Hf isotope analyses in detrital zircon from SW Iberia siliciclastic strata. *Lithos* **2017**, *278–281*, 383–399. [CrossRef]
7. Pereira, M.F.; Gama, C.; Dias da Silva, Í.; Fuenlabrada, J.M.; Silva, J.B.; Medina, J. Isotope geochemistry evidence for Laurussian-type sources of South-Portuguese Zone Carboniferous turbidites (Variscan orogeny). In *Pannotia to Pangea: Neoproterozoic and Paleozoic Orogenic Cycles in the Circum-North Atlantic Region*; Murphy, J.B., Ed.; Geological Society of London, Special Publication: London, UK, 2020. [CrossRef]
8. Malusà, M.; Fitzgerald, P.G. The geologic interpretation of the detrital thermochronology record within a stratigraphic framework, with examples from the European Alps, Taiwan and the Himalayas. *Earth Sci. Rev.* **2020**, *201*. [CrossRef]
9. Zuffa, G.G. Optical analyses of arenites: Influence of methodology on compositional results. In *Provenance of Arenites*; Series C; NATO Advanced Studies Institutes, Mathematical and Physical Sciences: Dordrecht, The Netherlands; Boston, MA, USA, 1985; pp. 333–361.
10. Di Giulio, A.; Ceriani, A.; Ghia, E.; Zucca, F. Composition of modern stream sands derived from sedimentary source rocks in a temperate climate (northern Apennines, Italy). *Sediment. Geol.* **2003**, *158*, 145–161. [CrossRef]
11. Garzanti, E. The Maturity Myth In Sedimentology and Provenance Analysis. *J. Sediment. Res.* **2017**, *87*, 353–365. [CrossRef]
12. Di Giulio, A.; Ronchi, A.; Sanfilippo, A.; Tiepolo, M.; Pimentel, M.; Ramos, V.A. Detrital zircon provenance from the Neuquén Basin (south-central Andes): Cretaceous geodynamic evolution and sedimentary response in a retroarc-foreland basin. *Geology* **2012**, *40*, 559–562. [CrossRef]
13. Horton, B.K. Sedimentary record of Andean mountain building. *Earth Sci. Rev.* **2018**, *178*, 270–309. [CrossRef]

14. Di Giulio, A.; Ronchi, A.; Sanfilippo, A.; Balgord, E.A.; Carrapa, B.; Ramos, V.A. Cretaceous evolution of the Andean margin between 36 °S and 40 °S latitude through a multiproxy provenence analysis of Neuquen Basin strata (Argentina). *Basin Res.* **2017**, *29*, 284–304. [CrossRef]
15. Fennell, L.M.; Iannelli, S.B.; Encinas, A.; Naipauer, M.; Valencia, V.; Folguera, A. Alternating contraction and extension in the Southern Central Andes (35° S–37° S). *Am. J. Sci.* **2020**, *319*, 381–429. [CrossRef]
16. Di Giulio, A. The evolution of Western Ligurian Flysch Units and the role of mud diapirism in ancient accretionary prisms (Maritime Alps, Northwestern Italy). *Geol. Rundsch.* **1992**, *81*, 655–668. [CrossRef]
17. Mueller, P.; Langone, A.; Patacci, M.; Di Giulio, A. Detrital signatures of impending collision: The deep-water record of the Upper Cretaceous Bordighera Sandstone and its basal complex (Ligurian Alps, Italy). *Sediment. Geol.* **2018**, *377*, 147–161. [CrossRef]
18. Mueller, P.; Langone, A.; Patacci, M.; Di Giulio, A. Towards a Southern European Tethyan Palaeomargin provenance signature: Sandstone detrital modes and detrital zircon U–Pb age distribution of the Upper Cretaceous-Paleocene Monte Bignone Sandstones (Ligurian Alps, NW Italy). *Int. J. Earth Sci.* **2019**, *109*, 201–220. [CrossRef]
19. Burov, E.B.; Diament, M. The effective elastic thickness (te) of the continental lithosphere: What does it really mean? *J. Geophys. Res.* **1995**, *100*, 3905–3927. [CrossRef]
20. Garcia-Castellanos, D. Interplay between lithospheric flexure and river transport in foreland basins. *Basin Res.* **2002**, *14*, 89–104. [CrossRef]
21. Hay, D.C.; Dempster, T.J. Zircon alteration, formatin and preservation in sandstones. *Sedimentology* **2009**, *56*, 2175–2191. [CrossRef]
22. Kohn, B.; Chung, L.; Gleadow, A. Fission-Track Analysis: Field Collection, Sample Preparation and Data Acquisition. In *Fission-Track Thermochronology and its Application to Geology*; Malusà, M.G., Fitzgerald, P.G., Eds.; Springer International Publishing: Cham, Switzerland, 2018; pp. 25–48. [CrossRef]
23. Matteini, M.; Junges, S.L.; Dantas, E.L.; Pimentel, M.M.; Bühn, B. In situ zircon U–Pb and Lu–Hf isotope systematic on magmatic rocks: Insights on the crustal evolution of the proterozoic Goiás Magmatic Arc, Brasília belt, Central Brazil. *Gondwana Res.* **2010**, *17*, 1–12. [CrossRef]
24. Tiepolo, M. In situ Pb geochronology of zircons with laser ablation-inductively coupled plasma-sector field mass spectrometry. *Chem. Geol.* **2003**, *199*, 159–177. [CrossRef]
25. Gazzi, P. Sulla determinazione microscopica della composizione mineralogical delle rocce, in particolare delle arenarie e delle sabbie. *Mineral. Petrogr. Acta* **1966**, *12*, 61–68.
26. Dickinson, W.R. Interpeting detrital modes of graywackes and arkoses. *J. Sed. Petrol.* **1970**, *40*, 695–707. [CrossRef]
27. Ingersoll, R.V.; Bullard, T.F.; Ford, R.D.; Grimm, J.P.; Pickle, J.D.; Sares, S.W. The effect of grain size on detrital modes: A test of the Gazzi-Dickinson point-counting method. *J. Sediment. Petrol.* **1984**, *54*, 103–116. [CrossRef]
28. Cibin, U.; Di Giulio, A.; Martelli, L. Oligocene-Early Miocene evolution of the Northern Apennines (northwestern Italy) traced through provenance of piggy-back basin fill successions. In *Tracing Tectonic Deformation Using the Sedimentary Record*; McCann, T., Saintot, A., Eds.; The Geologoical Society: London, UK, 2003; Volume 208, pp. 269–287.
29. Dickinson, W.R. Interpreting provenance relations from detrital modes of sandstones. In *Provenance of Arenites*; Zuffa, G.G., Ed.; NATO Advanced Studies Institutes, Mathematical and Physical Sciences: Dordrecht, The Netherlands; Boston, MA, USA, 1985; pp. 165–189.
30. Gleadow, A.J.W.; Fitzgerald, P.G. Uplift history and structure of the Transantarctic Mountains: New evidence from fission track dating of basement apatites in the Dry Valleys area, southern Victoria Land. *Earth Planet. Sci. Lett.* **1987**, *82*, 1–14. [CrossRef]
31. Gallagher, K.; Brown, R.; Johnson, C. Fission track analysis and its applications to geological problems. *Annu. Rev. Earth Planet. Sci.* **1998**, *26*, 519–572. [CrossRef]
32. Green, P.F.; Duddy, I.R.; Gleadow, A.J.W.; Tingate, P.R.; Laslett, G.M. Fission-track annealing in apatite: Track length measurements and the form of the Arrhenius plot. *Nucl. Tracks Radiat. Meas.* **1985**, *10*, 323–328. [CrossRef]
33. Brandon, M.T. Decomposition of mixed grain age distributions using Binomfit. *Track* **2002**, *24*, 13–18.
34. Galbraith, R.F.; Green, P.F. Estimating the component ages in a finite mixture. *Nucl. Tracks Radiat. Meas.* **1990**, *17*, 197–206. [CrossRef]

35. Vermeesch, P. On the visualisation of detrital age distributions. *Chem. Geol.* **2012**, *312–313*, 190–194. [CrossRef]
36. Carrapa, B.; DeCelles, P.G.; Reiners, P.; Gerhels, G. Apatite triple dating and white mica 40Ar/39Ar thermochronology of syn-tectonic detritus in the Central Andes: A multi-phase tectono-thermal history. *Geology* **2009**, *37*, 407–410. [CrossRef]
37. Chamberlain, K.R.; Bowring, S.A. Apatite-feldspar U–Pb thermochronometer: A reliable, mid-range (450 °C), diffusion-controlled system. *Chem. Geol.* **2000**, *172*, 173–200. [CrossRef]
38. Schoene, B.; Bowring, S.A. Determining accurate temperature-time paths from U–Pb thermochronology: An example from the Kaapvaal craton, southern Africa. *Geochim. Cosmochim. Acta* **2007**, *71*, 165–185. [CrossRef]
39. Chew, D.M.; Donelick, R.A. *Apatite Fission Track And U–Pb Dating by La–Icp–Ms*; Mineralogical Association of Canada Short Course: St. John's, NL, Canada, 2012; Volume 42, pp. 219–247.
40. Chew, D.M.; Sylvester, P.J.; Tubrett, M.N. U–Pb And Th-Pb dating of Apatite By LA–ICP–MS. *Chem. Geol.* **2011**, *280*, 200–216. [CrossRef]
41. Gehrels, G.E.; Valencia, V.; Pullen, A. Detrital zircon geochronology by Laser-Ablation Multicollector ICPMS at the Arizona LaserChron Center. In *Geochronology: Emerging Opportunities, Paleontology Society Short Course: Paleontology Society Papers*; Loszewski, T., Huff, W., Eds.; Cambridge University Press: Cambridge, UK, 2006; Volume 11, p. 10.
42. Gehrels, G.E.; Valencia, V.A.; Joaquin, R. Enhanced precision, accuracy, and efficiency, and spatial resolution of U-Pb ages by laser ablation-multicollector-inductively coupled plasmamass spectrometry. *Geochem. Geophys. Geosyst.* **2008**, *9*, 1–13. [CrossRef]
43. Gehrels, G.E. Detrital Zircon U-Pb Geochronology Applied to Tectonics. *Annu. Rev. Earth Planet. Sci.* **2014**, *42*, 127–149. [CrossRef]
44. Thomson, S.N.; Gehrels, G.E.; Ruiz, J.; Buchwaldt, R. Routine low–damage apatite U–Pb dating using laser ablation–multicollector–ICPMS. *Geochem. Geophys. Geosyst.* **2012**, *13*. [CrossRef]
45. Ramos, V.A.; Kay, S.M. Evolution of the Andean margin: A tectonic and magmatic view from the Andes to the Neuquén Basin (35°S–39° S lat). In *Overview on the Tectonic Evolution of the Southern Central Andes of Mendoza and Neuquén (35° S–39° S Latitude)*; Special Paper n. 40; Kay, S.M., Ramos, V.A., Eds.; The Geological Society of America: Boulder, CO, USA, 2006; pp. 1–17.
46. Ramos, V.A. The tectonic regime along the Andes: Present settings as a key for the Mesozoic regimes. *Geol. J.* **2010**, *45*, 2–25. [CrossRef]
47. Tunik, M.; Folguera, A.; Naipauer, M.; Pimentel, M.; Ramos, V.A. Early uplift and orogenic deformation in the Neuquén Basin: Constraints on the Andean uplift from U–Pb and Hf isotopic data of detrital zircons. *Tectonophysics* **2010**, *489*, 258–273. [CrossRef]
48. Balgord, E.A.; Carrapa, B. Basin evolution of upper Cretaceous-lower cenozoic strata in the Malargüe fold-and-thrust belt: Northern Neuquén Basin, Argentina. *Basin Res.* **2016**, *28*, 183–206. [CrossRef]
49. Naipauer, M.; Ramos, V.A. Changes in Source Areas at Neuquén Basin: Mesozoic Evolution and Tectonic setting based on U-Pb ages on zircons. In *Growth of the Southern Andes*; Folguera, A., Naipauer, M., Sagripanti, L., Ghiglione, M.C., Orts, D.L., Giambiagi, L., Eds.; Springer: New York, NY, USA, 2016. [CrossRef]
50. Cawood, P.A.; Hawkesworth, C.J.; Dhuime, B. Detrital zircon record and tectonic setting. *Geology* **2012**, *40*, 875–878. [CrossRef]
51. Fennell, L.M.; Folguera, A.; Naipauer, M. Cretaceous deformation of the southern Central Andes: Synorogenic growth strata in the NeuquénGroup (35° S 30–37° S). *Basin Res.* **2017**, *29* (Suppl. 1), 51–72. [CrossRef]
52. DeCelles, P.G.; Carrapa, B.; Gehrels, G.E. Detrital zircon UePb ages provide provenance and chronostratigraphic information from Eocene synorogenic deposits in northwestern Argentina. *Geology* **2007**, *35*, 323–326. [CrossRef]
53. Martin-Gombojav, N.; Winkler, W. Recycling of Proterozoic crust in the Andean Amazon foreland of Ecuador: Implications for orogenic development of the northern Andes. *Terra Nova* **2008**, *20*, 22–31. [CrossRef]
54. Sagripanti, L.; Bottesi, G.; Naipauer, M.; Folguera, A.; Ramos, V.A. U/Pb ages on detrital zircons in the southern central Andea Neogene foreland (36e37S): Constraints on Andean exhumation. *J. S. Am. Earth Sci.* **2011**, *32*, 555–566. [CrossRef]

55. Naipauer, M.; Tapia, F.; Mescua, J.; Farías, M.; Pimentel, M.M.; Ramos, V.A. Detrital and volcanic zircon U-Pb ages from southern Mendoza (Argentina): An insight on the source regions in the northern part of the Neuquén Basin. *J. South Am. Earth Sci.* **2015**, 1–18. [CrossRef]
56. Encinas, A.; Folguera, A.; Riffo, R.; Molina, P.; Fernández Paz, L.; Litvak, V.D.; Colwyn, D.A.; Valencia, V.A.; Carrasco, M. Cenozoic basin evolution of the Central Patagonian Andes: Evidence from geochronology, stratigraphy, and geochemistry. *Geosci. Front.* **2019**, *10*, 1139–1165. [CrossRef]
57. Borghi, P.; Fennell, L.; Omil, R.G.; Naipauer, M.; Acevedo, E.; Folguera, A. The Neuquén group: The reconstruction of a Late Cretaceous foreland basin in the southern Central Andes (35–37° S). *Tectonophysics* **2019**, *767*, 228177. [CrossRef]
58. Lanteaume, M. Considérations paléogéographiques sur la patrie supposée des nappes de flysch a Helminthoides des Alpes et des Apennins. *Bull. Soc. Géol. France* **1962**, *7*, 627–643. [CrossRef]
59. Vanossi, M.; Cortesogno, L.; Galbiati, B.; Messiga, B.; Piccardo, G.; Vannucci, R. Geologia delle Alpi liguri: Dati, problemi, ipotesi. *Mem. Soc. Geol. Ital.* **1986**, *28*, 5–75.
60. Di Giulio, A. Eo-Alpine geodynamics: An integrated approach. *Boll. Soc. Geol. It.* **1996**, *115*, 649–671.
61. Seno, S.; Dallagiovanna, G.; Vanossi, M. Pre-Piedmont and Piedmont-Ligurian nappes in the central sector of the Ligurian Alps: A possible pathway for their superposition on to the inner Briançonnais units. *Boll. Soc. Geol. Ital.* **2005**, *124*, 455–464.
62. Maino, M.; Casini, L.; Ceriani, A.; Decarlis, A.; Di Giulio, A.; Seno, S.; Setti, M.; Stuart, F. Dating shallow thrusts with zircon (U-Th)/He thermochronometry—The shear heating connection. *Geology* **2015**, *43*, 495–498. [CrossRef]
63. Mueller, P.; Maino, M.; Seno, S. Progressive Deformation Patterns from an Accretionary Prism (Helminthoid Flysch, Ligurian Alps, Italy). *Geosciences* **2020**, *10*, 26. [CrossRef]
64. Gasinski, A.; Slaczka, A.; Winkler, W. Tectono-sedimentary evolution of the Upper Prealpine nappe (Switzerland and France): Nappe formation by Late Cretaceous-Paleogene accretion. *Geodin. Acta* **1997**, *10*, 137–157. [CrossRef]
65. Galbiati, B. L'Unità di Borghetto d'Arroscia-Alassio. *Mem. Soc. Geol. It.* **1984**, *28*, 181–210.
66. Basu, A. Reading provenance from detrital quartz. In *Provenance of Arenites*; Zuffa, G.G., Ed.; Springer: Dordrecht, The Netherlands, 1985; pp. 231–247.
67. Mueller, P.; Patacci, M.; Di Giulio, A. Hybrid event beds in the proximal to distal extensive lobe domain of the coarse-grained and sand-rich Bordighera turbidite system (NW Italy). *Mar. Pet. Geol.* **2017**, *86*, 908–931. [CrossRef]
68. Dickinson, W.R.; Suczek, C. Plate tectonics and sandstone compositions. *Am. Assoc. Pet. Geol. Bull.* **1979**, *63*, 2164–2182. [CrossRef]
69. Dickinson, W.R.; Beard, L.S.; Brakenridge, G.R.; Erjavec, J.L.; Ferguson, R.C.; Inman, K.F.; Knepp, R.A.; Lindberg, F.A.; Ryberg, P.T. Provenance of North American Phanerozoic sandstones in relation to tectonic setting. *Geol. Soc. Am. Bull.* **1983**, *94*, 222–235. [CrossRef]
70. Palomares, M.; Arribas, J. Modern stream sands from compound crystalline sources: Composition and sand generation index. *Geol. Soc. Am. Spec. Pap.* **1993**, *284*, 313–322. [CrossRef]
71. Datta, B. Provenance, tectonics and palaeoclimate of Proterozoic Chandarpur sandstones, Chattisgarh Basin: A petrographic view. *J. Earth Sci.* **2005**, *114*, 227–245. [CrossRef]
72. Ballèvre, M.; Manzotti, P.; Dal Piaz, G.V. Pre-Alpine (Variscan) inheritance: A key for the location of the future Valaisan Basin (Western Alps). *Tectonics* **2018**, *37*, 786–817. [CrossRef]
73. Linnemann, U.; Pereira, F.; Jeffries, T.E.; Drost, K.; Gerdes, A. The Cadomian Orogeny and the opening of the Rheic Ocean: The diachrony of geotectonic processes constrained by LA-ICP-MS U–Pb zircon dating (Ossa-Morena and Saxo-Thuringian Zones, Iberian and Bohemian Massifs). *Tectonophysics* **2008**, *461*, 21–43. [CrossRef]
74. Casini, L.; Cuccuru, S.; Maino, M.; Oggiano, G.; Tiepolo, M. Emplacement of the Arzachena Pluton (Corsica–Sardinia Batholith) and the geodynamics of incoming Pangaea. *Tectonophysics* **2012**, *544–545*, 31–49. [CrossRef]
75. Fornelli, A.; Micheletti, F.; Piccarreta, G. Late-Proterozoic to Paleozoic history of the peri-Gondwana Calabria–Peloritani Terrane inferred from a review of zircon chronology. *Springerplus* **2016**, *5*, 1–19. [CrossRef] [PubMed]

76. Manzotti, P.; Ballèvre, M.; Poijol, M. Detrital zircon geochronology in the Dora–Maira and Zone Houillère: A record of sediment travel paths in the Carboniferous. *Terra Nova* **2016**, *28*, 279–288. [CrossRef]
77. Haas, I.; Eichinger, S.; Haller, D.; Fritz, H.; Nievoll, J.; Mandl, M.; Hippler, D.; Hauzenberger, C. Gondwana fragments in the Eastern Alps: A travel story from U/Pb zircon data. *Gondwana Res.* **2020**, *77*, 204–222. [CrossRef]
78. Von Raumer, J.F.; Stampfli, G.M.; Bussy, F. Gondwana-derived microcontinents—The constituents of the Variscan and Alpine collisional orogens. *Tectonophysics* **2003**, *365*, 7–22. [CrossRef]

minerals

Article

Complex Characteristic of Zircon from Granitoids of the Verkhneurmiysky Massif (Amur Region)

Maria M. Machevariani [1], Alexey V. Alekseenko [2,*] and Jaume Bech [3]

1 Department of Mineralogy, Crystallography, and Petrography, Saint Petersburg Mining University, 199106 Saint Petersburg, Russia; wmdmaria@gmail.com
2 Department of Geoecology, Saint Petersburg Mining University, 199106 Saint Petersburg, Russia
3 Faculty of Biology, University of Barcelona, 08002 Barcelona, Spain; jaumebechborras@gmail.com
* Correspondence: al.vl.alekseenko@gmail.com

Abstract: The study presents a complex characteristic of zircon from the Verkhneurmiysky intrusive series with Li-F granites. A wide range of morphological and chemical properties of zircon allowed us to obtain new information on the formation and alteration of zircon from biotite and zinnwaldite granitoids and to determine its features, which contribute to the correct definition of Li-F granites formed directly before the tin mineralization. The reviled trends of zircon morphology and composition evolution in the Verkhneurmiysky granites series are: the high-temperature morphotypes are followed by low-temperature ones with more complicated internal structure with secondary alteration zones, mineral inclusions, pores, and cracks; the increasing concentration of volatile (H_2O, F), large ion lithophile (Cs, Sr), high field strength (Hf, Nb) and rare-earth elements with decreasing crystallization temperatures and the determining role of the fluid phase (predominantly, F) in the trace element accumulation. The composition of zircon cores in biotite and zinnwaldite granites is very similar. However, the zircon rims from zinnwaldite granites are much more enriched in trace elements compared to those from biotite granites. The first study of zircon from the Verkhneurmiysky granitoids provides new data on the formation and alteration conditions of granitoids, including zinnwaldite ones.

Keywords: zircon crystal morphology; zircon textures; zircon trace elements; alteration of zircon; REE in zircon; rare metal granites; Li-F granites

Citation: Machevariani, M.M.; Alekseenko, A.V.; Bech, J. Complex Characteristic of Zircon from Granitoids of the Verkhneurmiysky Massif (Amur Region). *Minerals* **2021**, *11*, 86. https://doi.org/10.3390/min11010086

Received: 3 November 2020
Accepted: 12 January 2021
Published: 16 January 2021

1. Introduction

Granitoid plutons are widespread in the Badzhal and Myao-Chan ridges of the Amur Region and play an important role in the localization of the Badzhal-Komsomolsky tin ore district. At the same time, rare metal Li-F granites formed shortly before tin mineralization were found only in 1987–1990, when an expedition of the Leningrad Mining Institute conducted a special mapping of the territory of the Verkhneurmiysky ore cluster.

In recent years, rare-metal granites were determined only by their chemical composition, without taking into account their geological, mineralogical, and petrographic features. Due to the statistically insignificant differences in the composition of biotite and zinnwaldite granites, this approach leads to errors in geological mapping and mineral prospecting and exploration.

The comprehensive cutting edge study of accessory mineralization, including zircon, which is the most important mineral indicator of petrogenesis, effective geochronometer, and geothermometer, should contribute to the solution to problems of correct granitoid differentiation in the Amur region, the identification of rare-metal granites, and the study of their formation conditions.

The purpose of this study was to identify typomorphic features of accessory zircon from the granites of the Verkhneurmiysky massif. This set of characteristics should improve the scientific and methodological basis for the identification of rare-metal Li-F granite within the granitoid plutons. To solve these problems, we studied the composition, morphology, and internal structure of accessory zircon from biotite leucogranites and zinwalditic Li-F granites of the Verkhneurmiysky massif.

2. Geological Settings of the Verkhneurmiysky Massif and Characteristics of Granitoids

2.1. Geological Structure of the Verkhneurmiysky Massif

The Badzhal tin ore district is a part of the Sikhote-Alin-North-Sakhalin Lower Cretaceous orogenic belt, which extended for about 1500 km with a northeastern strike and a width of 600 km from the southern borders of Primorye to the Lower Priamurye and the northern tip of Sakhalin Island. The belt formation began in the Early Cretaceous, Neocomian stage and continued until the late Albian Stage within a transform margin.

From 1950 to 1970, the territory of Badzhal Ridge was studied in detail by the state geological survey for prospecting works. Geological maps at scales of 1:200,000 and 1:50,000 [1] were drawn. Gravimetric survey at scales of 1:100,000 resulted in the establishment of the largest gravimetric minimum in the Amur Region due to the presence of a hidden granite batholith, which includes the Verkhneurmiysky massif.

Late Cretaceous intrusions of biotite granites and leucogranites of the Badzhal complex, accompanied, especially in the east, by small massifs of monzonites, diorites, granodiorites, and granites of the Silinsky complex, significantly prevail in the study area. The general trend in the magmatic evolution is the increasing role of rhyolites and comagmatic granites, increasing alkalinity and potassium content in granitoids [2].

In the 1970s and 1980s, geological exploration was carried out within the Badzhal ore district, which resulted in the discovery of the major Pravourmiyskoye tungsten–copper deposit (balance reserves of 141.5 thousand tonnes of tin, 12.3 thousand tonnes of WO_3, and 40.1 tonnes of copper) and some smaller tin-bearing bodies [2]. Ore occurrences with Sn, W, Mo, and polymetallic deposits are grouped in three local ore clusters: Verkhneurmiysky, Verkhnebadzhalsky, and Hogdu–Lyanchlinsky, connected with extrusive dome structures, which siliceous intrusive and subvolcanic rocks of Badzhal complex in its core zone.

The Verkhneurmiysky tin ore deposit is the largest and most promising object in the southwestern part of the Badzhalsky region. It is notable for its complex geological structure and significant scale of metasomatic alterations. The geological and structural position of the ore node is determined by its confinement to a large Verkhneurmiysky granite massif and sublatitudinal structures [3,4].

The two main ore types are: cassiterite–quartz and cassiterite–silicate assemblages. The age of the Verkhneurmiysky massif is estimated from 83.4 $\pm$ 3.2 Ma to 98 $\pm$ 4 [2].

The rare metal granites immediately preceding tin mineralization were first found in 1987–1990 [2]. In 1993, a report was published giving a brief characterization of these granites in the Verkhneurmiysky massif, which are classified by petrochemical, geochemical, and mineralogical features as subalkaline rare-metal Li-F leucogranites [5].

Some researchers [6], point to the evolution in the formation of granites of the Verkhneurmiysky massif, which led to the intrusion of more and more evolved granite magmas with the final formation of Li-F granites, which are spatially and genetically associated with the Pravourmiyskoye tin deposit.

They also consider slab window as the most probable mechanism for magmatic rock formation of the Badzhal zone [7], and Li-F granites fluid sources that caused metasomatic alteration of rhyolite with the formation of ore-bearing quartz–topaz–siderophyllite greisens [6].

While the fact of spatial and genetic connection of the Pravourmiyskoye tin ore deposit with the formation of the Li-F granite at the final stage is not in doubt, the question of their genesis remains controversial.

Post-orogenic granitoids of the Verkhneurmiysky pluton are divided into the following types: (1) widespread biotite granites and leucogranites, which form major bodies and are united within the Badzhal complex; (2) rare subalkaline biotite monzodiorites and monzogranites of the Silinsky complex, composing poorly eroded dyke belts and stocks [2,8,9]; (3) rare stocks and dykes of subalkaline rare-metal Li-F granites of the Pravourmiysky complex [10].

The Pravourmiysky complex is located in the course of the Irungda-Makit River at the intersection of northwestern Orokotskaya and latitudinal Pravourmiyskaya disjunctive zones (Figure 1, red circle). The exposure of rare-metal granites composes less than 1% of the Badzhal zone.

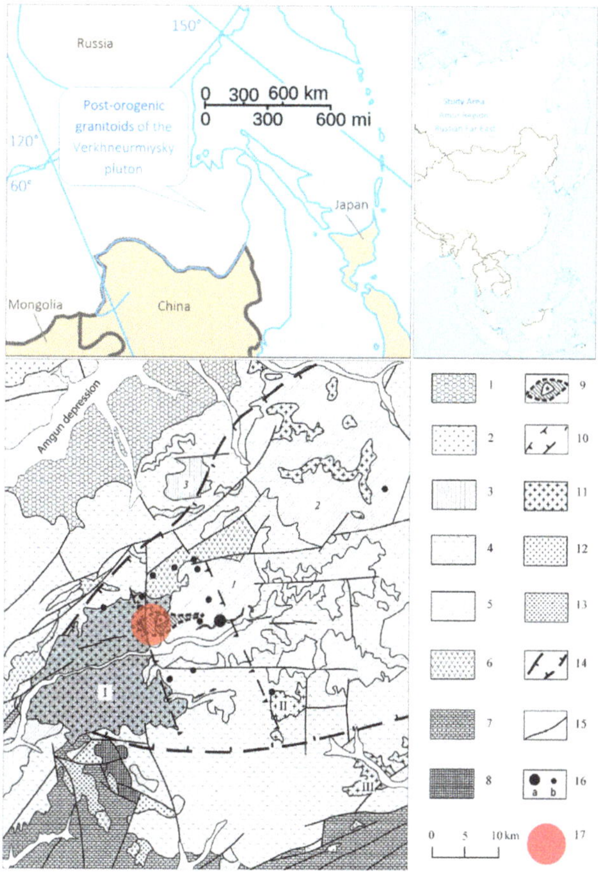

Figure 1. Upper part: Study area, based on the free blank map, *commons.wikimedia.org*. Lower part: Geological position of Li-F granites in Badzhal region (the scheme is based on the State Geological Map—1000/3, GGK-2009) [2]. 1—Oligocene–Miocene conglomerates, sandstones, argillites of the Verkhneamgunskaya Depression; 2–5—Upper Cretaceous volcanites: 2—Ignimbrites of trachyrhyolites, trachyrhyodacites, 3—Subvolcanic trachyandesites, quartz–monzonite porphyries, 4—Ignimbrites and rhyolites tuff, rhyodacites, 5—Subvolcanic rhyolites, rhyodacites; 6—Lower Cretaceous andesites, andesidacites and their tuffs; 7—Devonian–Perm terrigenous and clayey rocks; 8—Lower Proterozoic gneisses, amphibolites, quartzites; 9–11—Upper Cretaceous intrusive formations: 9—Areal of rare-metal granite magmatism (Pravourmiysky complex), 10—Orokotskaya zone of monzonitoid dyke magmatism (Silinsky complex), 11—Biotite, biotite-hornblende granites, leucogranites (Badzhal complex); 12—Lower and Upper Cretaceous granodiorites, quartz diorites (Laksky complex); 13—Late Paleozoic granites and granodiorites; 14—Boundaries of local minimum gravity of the second order; 15—Faults; 16—Large deposits (**a**); and ore manifestations (**b**); 17—Sampling area. Plutons: I—Verkhneurmiysky, II—Synchuginsky, III—Yarapsky. Extrusions: 1—Urmiysky, 2—Gerbinsky, 3—Kurkaltinsky.

2.2. Mineral and Chemical Composition of Granitoids of the Verkhneurmiysky Massif

Here we give a brief geological, petrographic, and geochemical characteristics of biotite granites of the Badzhal complex and zinwalditic granites of the Pravourmiysky complex.

Leucogranites of the Badzhal complex form middle-sized hypabyssal intrusive bodies, which stretch along with the magma-controlling faults. Crystallization of granites of the Badzhal complex lasted several billion years (crystallization started 94–95 billion years ago and completed ~92–93 billion years ago) and ended with the intrusion of porphyraceous granite dikes within the Verkhneurmiysky ore cluster.

Leucogranites are characterized by high concentrations of SiO_2 (70–75%), increased alkalinity ($\Sigma(K_2O + Na_2O)$) = 8–9%, K > Na, and alumina, but do not contain minerals with high alumina content. These granites belong to the type of leucogranite–alaskaite formation of tin provinces and are characterized by elevated concentrations of Li, Rb, Cs, Y, Yb, W, Mo, Sn, and lower concentrations of Ba and Sr. At the same time, the concentrations of Rb, Cs, W, Nb, Y, Yb, and Sc in the granites of later phases increase successively, while the Ba and Sr concentrations decrease. Consequently, the fractionation degree of granitoid systems (Rb/K, Rb/Ba, Rb/Sr, Cs/K, and K/Ba) increases too. We also revealed an increase in the Li, Sn, and F content and a decrease in the Mo concentration [2].

Post-magmatic alterations are widespread in the Badzhal leucogranites, including those acquired under the influence of late zinwalditic rare-metal granites. These alterations make leucogranites and zinwalditic granites quite similar in mineralogical and geochemical composition [2,11].

Li-F granites of the Verkhneurmiysky massif are the youngest intrusive complexes completing the Late Cretaceous rare-metal granite intrusive series [2,12]. According to the tectonic, geological, and structural conditions, as well as petrographic, mineralogical, and geochemical features—these granites belong to the Li-F type of subalkaline leucogranite formation of rare-metal granites. Li-F granites of Verkhneurmiysky massif are zinwalditic tantalum–tungsten–niobium-bearing granites with Nb-wolframite, W-ixiolite, and samarskite, specific for the in post-magmatic tungsten–tin deposits [13,14].

Li-F granites of the Verkhneurmiysk massif are characterized by a massive or eutaxitic texture, light gray or white color, and the same crystallinity as the host medium-grained leucogranite. Macroscopically, leucogranites are characterized by a lighter color, absence of porphyraceous phenocrysts, and chain aggregation of quartz. Rock-forming minerals are K-feldspar (33–43 vol.%), albite (27–38 vol.%), quartz (32–36 vol.%), and zinnwaldite (3–5 vol.%). Characteristic features of the Verkhneurmiysky Li-F granites are zinnwaldite nodules (up to 0.6 mm) with inclusions of fluorite and plagioclase with incorporated topaz crystals. The rounded pea-shaped quartz forms chains and snowball structures. Mica of the zinnwaldite series was defined as a Li-bearing siderophyllite that differs from the biotite of leucogranites by the increased content of Al, Li, Mn, and lower concentrations of Mg, Fe, and Ti [2].

The composition of the main accessory minerals of zinnwaldite granites differs significantly from biotite ones. The zinnwaldite granites contain topaz, fluorite, fluocerite, ixiolite-(W), fergusonite-(Y), ferberite, while apatite, allanite-(Ce), and anatase are absent. In addition, zinnwaldite granites have higher contents of zircon, monazite-(Ce), xenotime-(Y), thorite and decreased fraction of ilmenite and rutile [15]. One of the main typomorphic features of zinnwaldite granites is the presence of rare-earth and rare-metal-bearing minerals. They are characterized by the stable presence of W (fergusonite-(Y), calciosamarskite, samarskite-(Yb), eschinite-(Y), strüverite, ilmenite, liandratite), the predominance of Nb over Ta, a significant role of As, Bi, Th, and U, and low content of Sn. Thus, the feature of Li-F granites is the presence of W and Y-rare-earth elements (REE)-Nb-bearing mineralization [10].

Thus, we can conclude that the detection of ore-bearing granites within the Verkhneurmiysky massif is a promising task for modern geological exploration. The exploration for rare-metal mineralization in this region should consider not only the spatial and genetic relationship of Li-F granites with plutons of leucogranites and small intrusions of monzonites, but also all identified mineralogical and petrographic features of Li-F granites, including the detailed characteristics of the accessory minerals and, first of all—zircon. The comprehensive study of granitoids of the Verkhneurmiysky massif, described in the works of V.A. Alekseev, Yu.B. Marin, et al. [2,12], allow identifying the initial and final members of the Verkhneurmiyskaya granitoid series: biotite leucogranites of the Badzhal complex and zinnwaldite granites of the Pravourmiysky complex [2].

To clarify the geochemical types of biotite and zinnwaldite granite samples under study (Figure 2) and to compare them with the other granites of a similar genesis—we performed a bulk chemical analysis (XRF for petrogenic elements, ICP-MS for an extended set of elements).

(**a**) (**b**)

Figure 2. Chip samples of Li-F granites from Verkhneurmiysky massif: (**a**) medium-grained zinnwaldite granite; (**b**) medium-grained biotite granite.

The table of contents of granites from the Verkhneurmiysky massif, Severny massif (Chukotka), as well as zinnwaldite and protolithionitic granites from Erzgebirge (Germany) (Table 1) and their position on the Frost diagram [16] (Figure 3a) demonstrate that studied Verkhneurmiysky's granites are subalkaline and all of the compared rocks are rather similar in terms of the ratio of major oxides. Moreover, not only biotite and zinnwaldite granites compositions are close, but also the compositions of granites of the Far East, Chukotka, and the Erzgebirge in general.

The content of trace elements (Figure 3b) in biotite leucogranites reflects a rare metal melt nature, however, less contrasting than in the zinnwaldite Li-F granites of both Far Eastern massifs. Elevated concentrations of Li, F, Rb, Cs, Sn, W, Nb, Ta, Y, REE, Th, and other granitophile elements were also observed [2,10], with the accumulation of volatile (F), rare alkaline (Li, Rb) and some high field strength elements (Nb, Ta) in the zinnwaldite granites of the Erzgebirge. Some elements (Ba, Sr, Zr) in biotite granites have concentrations below the relative abundance of the elements and predictably increase in Li-F granites.

Comparison of the elemental composition of the studied granitoids is consistent with the geochemical evolution of the Far Eastern granitoid complexes [2], expressed in an increase in alumina content, accumulation of rare elements, reducing the values of K/Rb (Bt-Gr-0.67, Znw-Gr-0.64), Nb/Ta (Bt-Gr-7.7, Znw-Gr-5.1), a sharp increase in Rb/Sr (Bt-Gr-15.3, Znw-Gr-22.4), and a deepening of Eu anomaly on the course from biotite to zinnwaldite granites.

Table 1. The composition (single measurements) of granites of the Verkhneurmiysky massif (Amur region), Severny massif (Chukotka), and Erzgebirge (Germany, Czech Republic) according to published data [2,17] and the samples analyzed in this paper (*); Bt—Biotite granite, Znw—Zinnwaldite granite, Pr—Protolithionitic granite.

	Verkhneurmiysky Massif				Severny Massif		Erzgebirge	
	Bt *	Znw *	Bt	Znw	Bt	Znw	Pr	Znw
SiO_2	75.71	76.95	76.38	76.63	75.46	77.59	75.53	72.22
TiO_2	0.07	0.07	0.13	0.03	0.13	0.06	0.07	0.01
Al_2O_3	13.03	12.18	12.48	13.08	11.72	10.9	12.54	15.92
Fe_2O_3	1.59	1.93	0.27	0.15	1.68	1.49	0.6	0.2
FeO	–	–	1.58	1.09	–	–	0.78	0.45
MnO	0.019	0.027	0.01	0.02	0.02	0.04	0.05	0.09
MgO	0.07	0.06	0.11	0.05	0.17	0.05	0.07	0.09
CaO	0.52	0.51	0.53	0.23	0.68	0.27	0.64	0.38
Na_2O	3.53	3.3	3.53	4.24	3.84	3.45	3.49	4.83
K_2O	4.86	4.45	4.55	5.48	4.96	5.22	4.73	2.39
P_2O_5	0.03	0.02	0.02	0.02	0.07	0.03	0.011	0.015
LOI	0.13	0.18	0.74	0.11	1.26	0.54	0.87	1.95
B	–	–	38	35	9	10	–	–
F	–	–	2900	3557	560	2000	5100	7900
Li	–	–	160	363	95	364	98	384
Rb	322	314	594	750	290	860	802	1440
Cs	8.9	13.5	35	41	30	51	–	–
Sr	21	14	25	6	152	7	13	72
Ba	56	50	293	45	280	8	–	–
Zr	93	94	102	39	86	157	124	44
Sn	2	4	18	8	3	8	–	–
Pb	33	32	39	48	30	41	–	–
Bi	<0.1	0.1	4	13	2	7	–	–
Nb	15.8	14.3	46	54	13	20	52	109
Ta	2.06	2.8	9	14	3	10	7	52
W	<0.5	1.5	26	53	7	30	12	13
Th	32.1	32.6	19	12	8	29	58	15
Y	30.8	75	100	64	33	29	103	7.5
Ce	67.4	48.2	80	68	70	6	83	13.5

The content of oxides is given in wt.%, trace elements in ppm. Hereinafter, a dash (–) indicates cases when an analysis on the particular element is not performed.

(a)

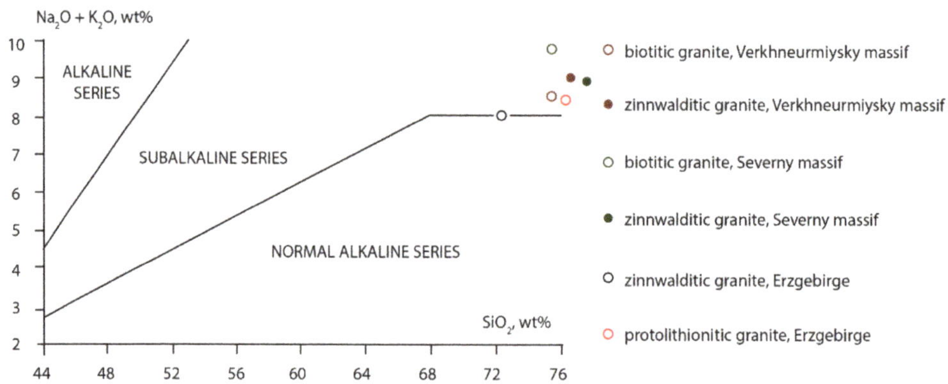

Figure 3. *Cont.*

(**b**)

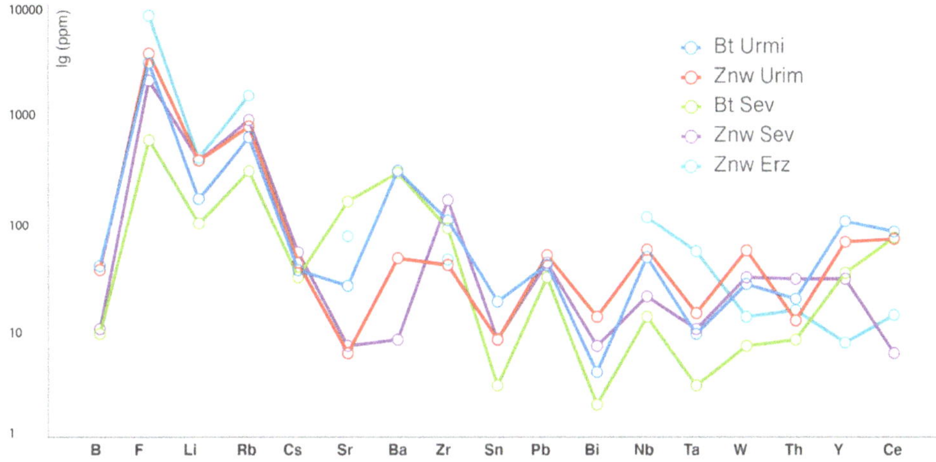

Figure 3. (**a**) $(Na_2O + K_2O)/SiO_2$ values for Verkhneurmiysky granitoids on the Frost diagram [16]; (**b**) Spectra of trace elements in biotite (Bt) and zinnwaldite (Znw) granites of Verkhneurmiysky massif (Urmi), Severny massif (Sev), and Erzgebirge (Erz).

3. Materials and Methods

This work is based on the samples collected by the staff of the Mineralogical Department (Saint Petersburg Mining University) during the fieldwork from 1987 to 1990, headed by Yu.B. Marin and focused on the special mapping of the Verkhneurmiysky ore cluster. We studied 13 bulk samples and the corresponding thin sections, zircon single fraction from 8 samples of biotite granites (more than 555 grains), and 5 samples of zinnwaldite Li-F granites (460 grains). The studies were conducted in the laboratories of the Mining University [18–20], the Russian Geological Research Institute (VSEGEI), the Institute of Precambrian Geology and Geochronology, the Russian Academy of Sciences (Saint Petersburg, Russia), and the Institute of Mineralogy, TU Bergakademie Freiberg (Freiberg, Germany) (Table 2).

The zircon extraction was made according to the following method: crushing (-0.5; -0.4; -0.315; -0.25) $\rightarrow$ elutriation, drying $\rightarrow$ electromagnetic separation $\rightarrow$ extraction of non-electromagnetic fraction $\rightarrow$ $CHBr_3$ bromoform (density of 2.899 g/cm^3 at 15 °C) $\rightarrow$ heavy fraction (accessory minerals) $\rightarrow$ methylene iodide CH_2J_2 (density 3.28 g/cm^3) $\rightarrow$ heavy fraction (zircon, monazite) $\rightarrow$ electromagnetic separation $\rightarrow$ extraction of non-electromagnetic zircon-bearing fraction. The final stage of sample preparation was the zircon handpicking from the non-electromagnetic heavy fraction of each sample. We used JSM-6460LV (Oxford) to get approximately 80 back-scattered electrones (BSE) images of zircon grains and to perform a wavelength-dispersive X-ray (WDX) analysis of zircon composition in thin sections of biotite granite (35 grains/66 points) and zinnwaldite granite (40 grains/167 points).

To study the morphology and surfaces of zircon faces, 118 grains were applied to electrically conductive tape with subsequent conductive film deposition. They were studied in the laboratory of the Geological Institute of the TU Bergakademie Freiberg on a scanning electron microscope JEOL JSM-7001F. The SE images were obtained with the following parameters: accelerating voltage—15 kV, probe current—20 pA, resolution—3.0 nm; BSE imaging was carried out under accelerating voltage of 20 kV and probe current of 14 pA. The analyzed crystals were cleaned, impregnated into epoxy resin, and polished to study grain inner structure in BSE and CL.

Table 2. The list of analyzed granite samples and extracted zircons, indicating the methods used and the corresponding number of measurements: OM—optical microscopy of granites in thin sections and extracted zircon grains, SEM—scanning electron microscopy (JSM-6460LV, Saint Petersburg, Russia) includes BSE imaging, WDX analysis of zircon and neighboring minerals in thin sections; SEM *—scanning electron microscopy (JEOL JSM-7001F, Freiberg, Germany) includes SE imaging of extracted zircon grains, applied to electrically conductive tape, to study the surface of crystal faces and BSE imaging of zircon grains, impregnated into epoxy resin and polished, to study the inner structure; SIMS—secondary ion mass spectrometry (Cameca IMS-4f, Yaroslavl, Russia) of zircon grains, impregnated into epoxy resin, to measure the trace element content (accompanied with BSE and cathodoluminescence (CL) imaging of zircon inner structure); Raman spectroscopy was applied to measure the degree of zircon crystallinity (Renishaw InVia Raman spectrometer, Saint Petersburg, Russia); XRF analysis of granite samples to determine the content of petrogenic elements and ***ICP-MS*** analysis for the petrogenic elements, as well as for a wide range of trace elements (Actlabs, Ancaster, ON, Canada).

Sample No.	Granite Type	Number of Zircon Grains	Analysis
60136	Biotite granite, Badzhal complex	130	OM; SEM (28 analysis); SEM * (39 zircon grains); XRF, ICP-MS
82053		60	OM
85200		75	OM
85203		75	OM
97029		90	OM
60070		50	OM; SEM (38 analysis); SEM * (19 grains); SIMS (30 analysis/16 grains); Raman (20 spectra/10 grains)
60186		35	OM
60108		40	OM
60063	Zinnwaldite granite, Pravourmiysky complex	35	OM
60205		200	OM, SEM (100 analysis); XRF, ICP-MS
82073		65	OM; SEM (67 analysis), SEM * (38 grains);
04001		70	OM; SEM * (24 grains);
82240		90	OM; SIMS (37 analysis/25 grains); Raman (16 spectra/8 grains)

The analysis of the trace elements content in zircon (67 analytical points/41 grains) was carried out on the Cameca IMS-4f ion microprobe (Valiev Institute of Physics and Technology of RAS, Yaroslavl, Russia) using standard methods [21,22]. The ion beam diameter was not more than 15–20 μm, the relative error for the majority of elements did not exceed 15%, and the detection limit is 10 ppb on average. We obtained data on the content of 11 lanthanides, Li, P, Ca, Ti, Sr, Y, Nb, Cs, Ba, Hf, Th U, F, and H_2O, and calculated the most important geochemical parameters, including Th/U ratio, Eu- and Ce-anomalies, ΣREE, ΣLREE, ΣHREE, and chondrite-normalized [23] LuN/LaN, LuN/GdN, and SmN/LaN ratios.

A non-destructive Raman spectroscopy technique was applied to measure the degree of crystallinity of zircon. Raman spectra were collected by means of the Renishaw InVia Raman spectrometer installed in the Department of Mineralogy, Saint Petersburg Mining University, using an excitation wavelength of 785 nm (diode NIR laser) and a thermoelectrically cooled charge-coupled device (CCD) detector equipped with a 1200 L/mm grating. The laser output power of 300 mW was reduced to 0.1% of the standard value, which is non-destructive to samples of interest. In each experiment, five scans were collected and averaged. The typical acquisition time was 20 s. The theoretical diffraction-limited confocal-spot diameter of the laser beam at the sample surface was approximately 1.5 mm with a Leica 50× long-working-distance objective and a 10× ocular. Repeated acquisitions of the crystals using the highest magnification (50×) were accumulated to improve the signal-to-noise ratio in the spectra. The Raman spectra of zircon from biotite granite (20 spectra/10 grains) and zinnwaldite granite (16 spectra/8 grains) were obtained from the analytical points of the preceding ion microprobe analysis of trace elements content. The spectra were recorded in a narrow range from 950 to 1050 cm^{-1}, containing ν_1 (975 cm^{-1}) and ν_3 (1008 cm^{-1}) bands, which are responsible for symmetric and anti-symmetric stretching of the SiO_4 group, respectively [24,25]. Spectra were calibrated using the 520.5 cm^{-1} line of a silicon wafer. Spectral manipulations such as smoothing (without baseline correction) and peak parameters recording were performed using the software package Wire 2.0.

No indications of surface damage and change of color by the laser radiation were observed during the sample checking after Raman measurements.

The bulk composition of granites was determined, using XRF analysis—for the petrogenic elements, and ICP-MS analysis for the petrogenic elements, as well as for a wide range of trace elements (Sc, Be, V, Cr, Co, Ni, Cu, Zn, Ga, Ge, As, Rb, Sr, Y, Zr, Nb, Mo, Ag, In, Sn, Sb, Cs, Ba, La, Ce, Pr, Nd, Sm, Eu, Gd, Tb, Dy, Ho, Er, Tm, Yb, Lu, Hf, Ta, W, Ti, Pb, Bi, Th, and U), in the Actlabs laboratory (Ancaster, ON, Canada).

4. Results and Discussion

4.1. Zircon Mineral Associations

The set of Li-F granite accessory minerals and their determining role in the prospecting works are briefly discussed in Section 2.2. The geochemical evolution of the granitoid series of the Far East can also be traced in the evolution of the composition of typomorphic accessory minerals: zircon, ixiolite-(W), ferberite-(Nb), allanite-(Y), and chernovite-(Y) [2]. An example of an indicator accessory mineral coexisting with zircon is allanite, which changes from Ce and alumina-bearing type in the complex of early biotite leucogranites to Y-bearing ferriferous allanites in zinnwaldite granites (Figure 4) [2,26].

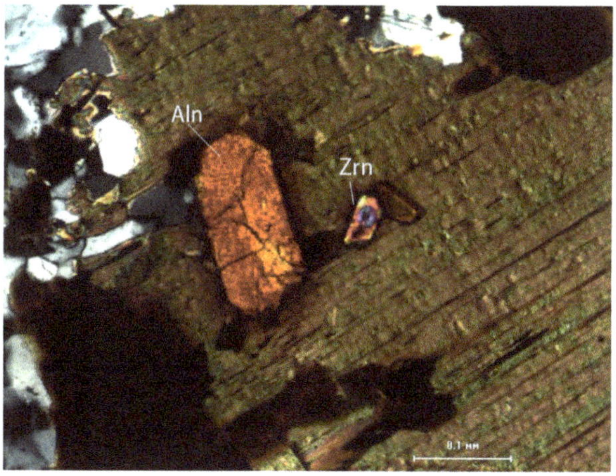

Figure 4. Accessory allanite (Aln) and zircon (Zrn) in zinnwaldite granite (sample 04001), nicols crossed.

The characteristic feature of slightly altered zircon from biotite granites (Zrn-Bt), usually located at the grain boundaries of rock-forming quartz and biotite crystals, is the formation of accretions and intergrowth with equal-sized and relatively small grains of xenotime. Xenotime forms short-prismatic crystals, euhedral grains and solid masses, with an average formula ($Y_{0.71–0.76}Gd_{0.02–0.03}Dy_{0.06–0.08}Er_{0.04}Yb_{0.03–0.07}$) [$PO_4$] (Figure 5a–c). In addition, zircon grains contain rare inclusions of table apatite crystals (Figure 5c).

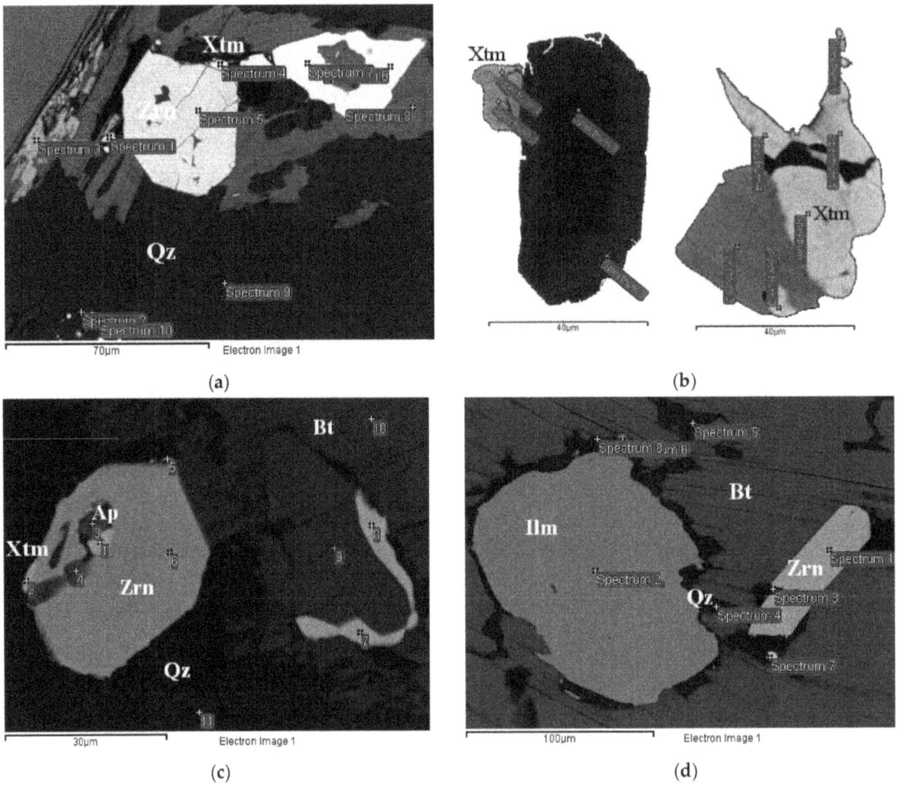

Figure 5. Associations of accessory zircon in biotite granite, BSE images: (**a**,**b**) zircon and xenotime intergrowth; (**c**) xenotime and apatite grains in zircon; (**d**) long-prismatic zircon and ilmenite in biotite.

There is a close association of long-prismatic zircon with tabular grains of ilmenite with quartz rims enclosed in biotite crystals. The characteristic feature of the titanium and iron oxide phase is the constant impurity of Mn (4.16–10.32%) (Figure 5d).

Zircon from zinnwaldite granites Zrn-Znw forms extremely small grains (30–50 µm) with varying alteration degrees and is located on the grain boundaries of anhedral microcline (Figure 6b,c), prismatic albite (Figure 6a–d), pea-shaped quartz (Figure 6c), as well as in small zinnwaldite nodules (Figure 6b) and inside zinnwaldite crystals (Figure 6e). Often there are intergrowths of anhedral spongy zircon with monazite of varying composition $(Ce_{0.46–0.47}La_{0.18–0.19}Nd_{0.16–0.17}Th_{0.04–0.12}Sm_{0.01–0.02})[PO_4]$ with the same grain size and a lower degree of alteration (Figure 6f).

Mica shows plastic deformations (Figure 6f), which are consistent with the concept of early syngenetic protoclase of rock-forming and accessory minerals of rare-metal granites associated with increased fluid pressure [10,27]. Zircon and xenotime occur together, both as the intergrowth of weakly altered grains of both minerals and as xenotime microinclusions in strongly altered zircon grains formed in the recrystallization and/or dissolution–precipitation processes (Figure 7a,b).

Xenotime crystals included in zircon show partial substitution with chernovite that forms an almost continuous xenotime–chernovite isomorphic series (Figure 7c). The chernovite formation is associated with the autometasomatic alterations in rare-metal granites which took place under the As-bearing fluids, characteristic to the rare-metal magma, enriched with both lithophile (W, Nb, Y, REE, Th, etc.) and chalcophile elements

(As, S, Sn, Cu, Bi, etc.) [14,28]. The appearance of the As-phases in Zrn-Znw is consistent with the sharply increased content of As in zinnwaldite granites.

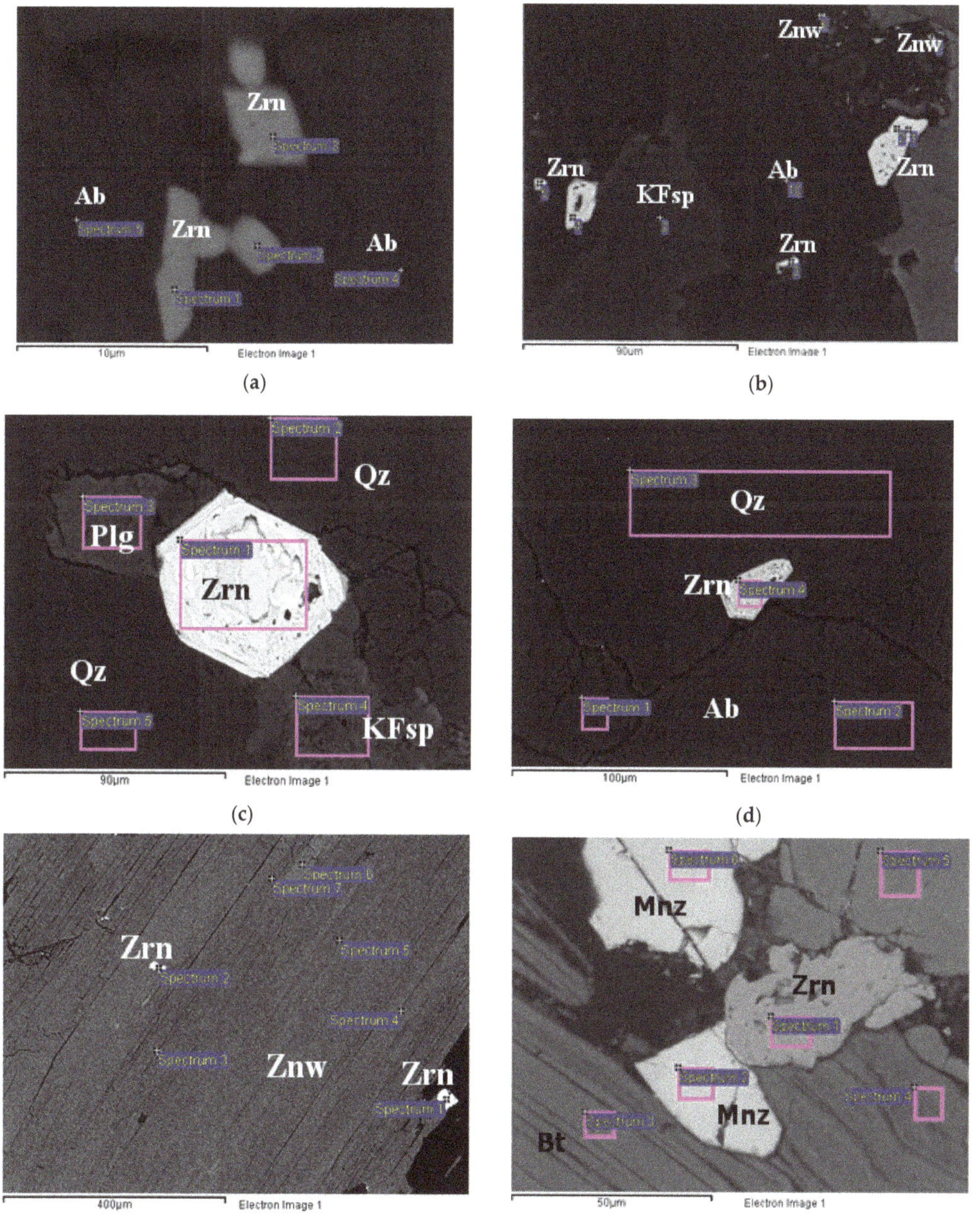

Figure 6. Accessociation of zircon in zinnwaldite granite, BSE images: (**a**) anhedral zircon grains in albite; (**b**) spongy zircon in zinnwaldite nodules; (**c**,**d**) euhedral zircon grains at the borders of rock-forming minerals; (**e**) zircon in zinnwaldite; (**f**) anhedral zircon intergrowth with monazite at the border of deformed biotite crystal.

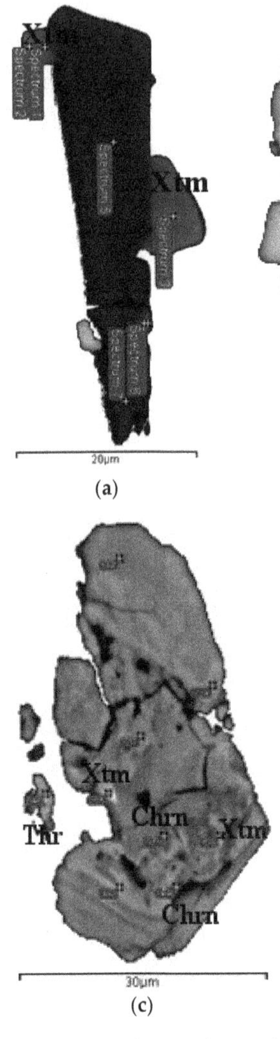

Figure 7. Associations of zircon from zinnwaldite granites with phosphate phases: (**a**) intergrowth of slightly altered long-prism zircon and xenotime; (**b**) xenotime phases in recrystallized zones of spongy zircon; (**c**) xenotime and chernovite phases in significantly altered zircon.

4.2. Zircon Morphology

Zircon morphology can record genetic information about changes in the crystallization parameters, which makes it reasonable to study morphological and anatomical features of both microcrystals and intergrowth of zircon.

Primary microscopic study of 13 samples (60–200 grains/samples) allowed identifying at least two types of zircon grains in all samples:

- The first type is represented by the relatively short prisms, brown and honey-colored, semi-transparent idiomorphic crystals. Some grains have rims, which are cleaner, lighter, and have a lower refractive index. Such rims may indicate postmagmatic changes or at least a significant interruption in zircon growth and a change in physical and chemical growth conditions. Almost all grains have metamict cores and partial amorphization in the near-core regions (Figure 8a,b);

- The second type includes virtually colorless, long-prismatic grains without metamict cores; however, mineral and melt inclusions are present in considerable quantities (Figure 8c,d).

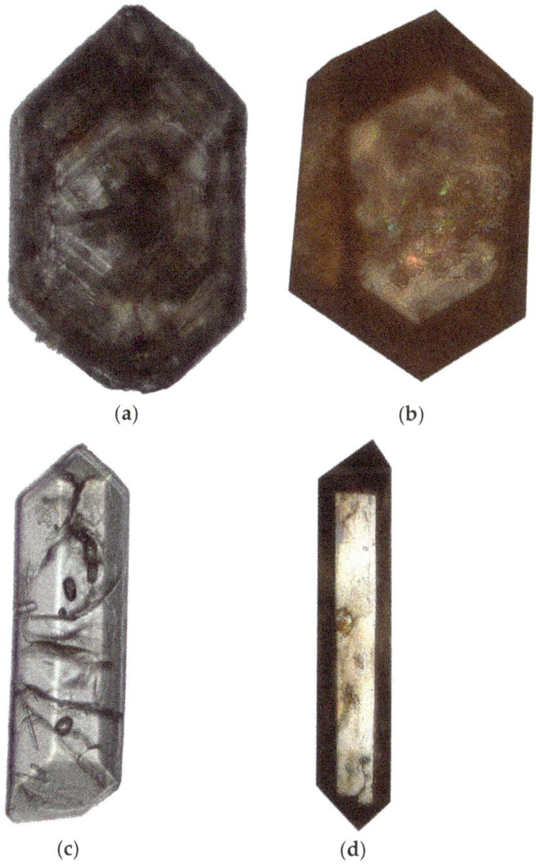

Figure 8. (**a**,**b**) Short-prismatic, honey-brown euhedral zircon grain of the first type; (**c**,**d**) Colorless, long-prismatic zircon grain of the second type.

The division into long- and short-prismatic zircon varieties is based on the median elongation value of 1.92 for biotite granites, and a median elongation value of 1.55 for zinnwaldite granites.

This approach partially solves the question of adequate interpretation of morphotype distribution in the Pupin diagram [29], since the entire zircon population, can record only the general trend of all morphotypes predominance, while the division into varieties records the stages of zircon evolution even within a single sample. The result of a statistically valid division of the zircon sample into short- and long-prismatic varieties is the presence of two different evolutionary trends. The trend of the long-prismatic zircon coincides with the general trend of the entire sample, while the evolution curve of the short-prismatic variety is directed towards the lower lines of the Pupin diagrams (Figure 9).

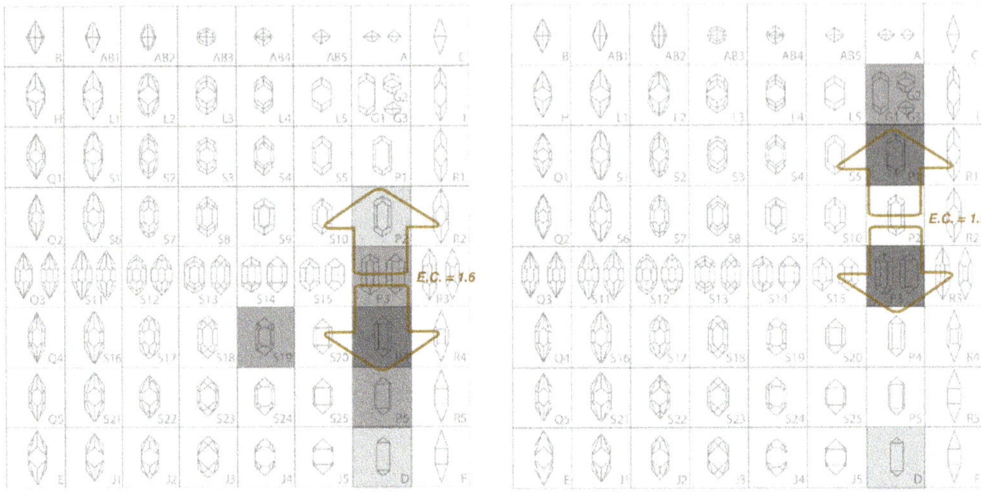

(**a**) (**b**)

Figure 9. Pupin diagram for: (**a**) Zrn-Bt and (**b**) Zrn-Znw. The tone gradient from black to light grey indicates the degree of abundance (from maximum to minimum, respectively) of zircon morphotypes in each sample. The contour arrows show that short-prismatic zircon tends to the upper field of the diagram, and long-prismatic zircon—to the lower one.

The two distinct zircon morphotypes are taken as evidence of the fluid phase evaporation from the melt [30], the formation of two zircon varieties within a single sample may be related to the F-containing fluids, which separated during emanation differentiation and affected already crystallized rocks. The scattering of the Zrn-Bt morphotypes along the horizontal axis demonstrates breaking points in its evolution and characterizes the heterogeneity of the melt. The relatively low intended temperature of Zrn-Znw formation indicates the fluid saturation of the melt and indirectly confirms the transfer of lithium and tin in the composition of F-complexes. Thus, the most abundant Zrn-Bt and Zrn-Znw morphotypes are P4 and P1/P3, respectively (Figure 9).

4.3. SE and BSE Zircon Images

The electron microscopy study of zircon confirmed the presence of short- and long-prismatic varieties in both studied granitoids and allowed the tracing of some genetic features of the formation processes.

Zircon from biotite granites (Zrn-Bt) is enriched with impurities, shows partially or completely altered cores of short-prismatic crystals with preserved oscillatory zoning, saturated with micropores and microinclusions of coffinite–thorite, monazite–cheralite phases and dark (in BSE) rims around the central zones (Figure 10a,b).

The formation of such a spongy texture with an abundance of mineral inclusions is a consequence of the dissolution–precipitation of zircon enriched with impurities under the influence of fluids, seeping through the numerous fissures formed during crystal growth [31–35]. In contrast, in the long-prismatic zircon Zrn-Bt, the cores have less impurities and have an elongated habit with reduced dipyramids and weak zoning (Figure 10c,d). Both types of Zrn-Bt have similar corrosion of the grain edges, presence of thin zones with wavy edges, and concentric heterometry-induced cracks that intersect the primary growth zoning.

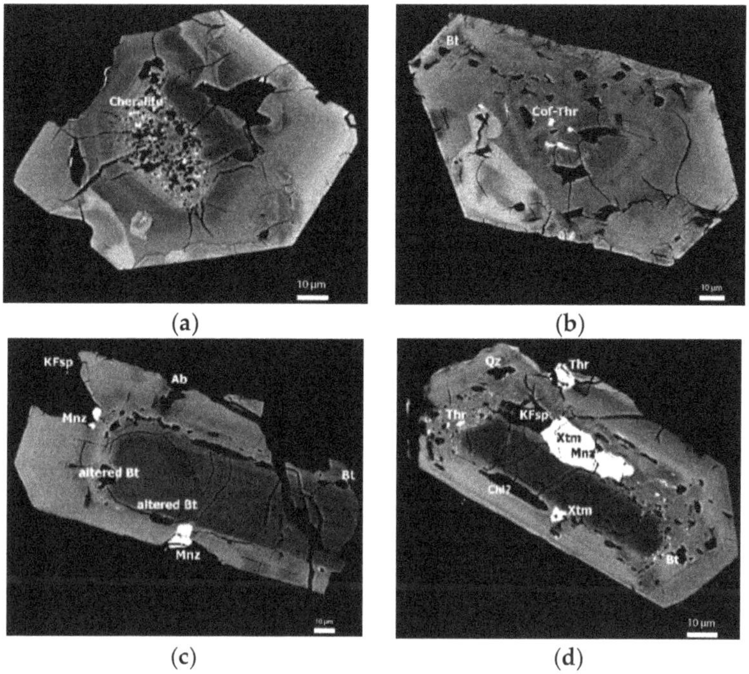

Figure 10. BSE images of short-prismatic (elongation coefficient < 1.5) long-prismatic (elongation coefficient > 1.5) varieties of Zrn-Bt: (**a**) the completely recrystallized central part, the micro inclusions of the monazite–cheralite phases, wavy boundaries of the altered zones cross the initial fine zoning; (**b**) the completely recrystallized central domain and rim form a common area crossing the primary fine zoning, a concentric system of fractures elongated from core to rim; (**c**) zone of secondary alterations around impurity depleted core, monazite inclusions, depleting zircon growth; (**d**) cracked core with weakly zoning surrounded with wide (11–15 µm) porous zone of secondary alterations with inclusions of xenotime and monazite, potassium feldspar, thorite, biotite, and quartz.

Morphological and inner texture features of zircon in biotite granites allowed assuming some stages of mineral formation and transformation, namely: alkalinity increase in the crystallization medium (dipyramid facets growth) at the rims growth and active influence of the fluid phase at the final magmatic stages of the massif formation (similar nature of secondary alterations in the short and long-prismatic variety of zircon Zrn-Bt). Zircon Zrn-Znw from Li-F zinnwaldite granites does not form parallel intergrowth but is characterized by a complex internal structure and intergrowth with micro-spherulitic aggregates of the chernovite Y[AsO_4]—xenotime Y[PO_4] series (Figure 11).

It should be noted that the formation of similar chernovite aggregates in Li-F granites of Zinnwald is connected with the arsenic supply at the oxidation stage [36], and in A-type granites (Zinnowitz-Rejkovo) post-magmatic fluids are considered as a unique source of arsenic [37]. Thus, the nature of such intergrowth is a strong argument for autometasomatic alterations in rare-metal Li-F granites [10,28].

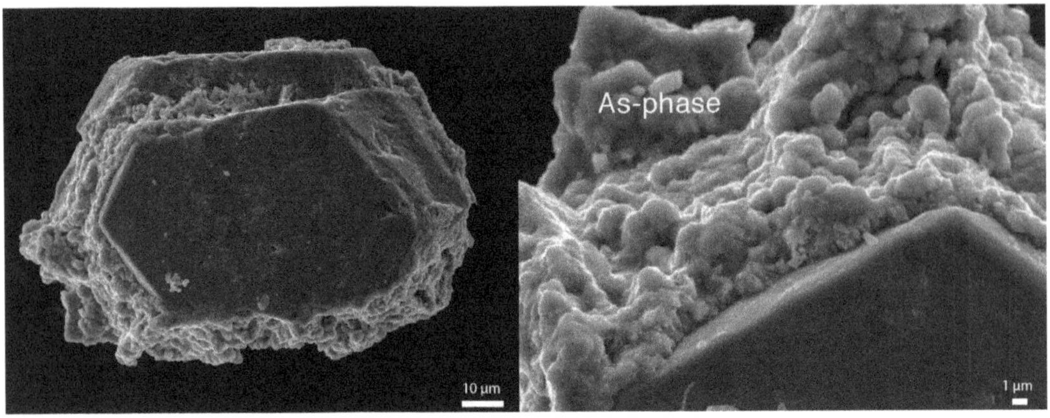

Figure 11. SE image of Zrn-Znw covered with micro spherolite aggregates of chernovite Y[AsO_4]—xenotime Y[PO_4] phases.

Crystals of the long-prismatic Zrn-Znw variety are characterized by spotty zoning (Figure 12b) with rare remains of the primary oscillatory one, obscured by an uneven distribution of secondary blocks of trace elements concentration/depletion [2,38]; increased rims brightness in BSE images, indicating enrichment with heavy elements; increased amount and irregular distribution of cracks. Postmagmatic alterations of long-prismatic Zrn-Znw under the influence of the fluid phase are evidenced by superimposed wavy zoning and chains, formed after gas–liquid inclusions decrepitation, and crossing the primary growth zoning (Figure 12a).

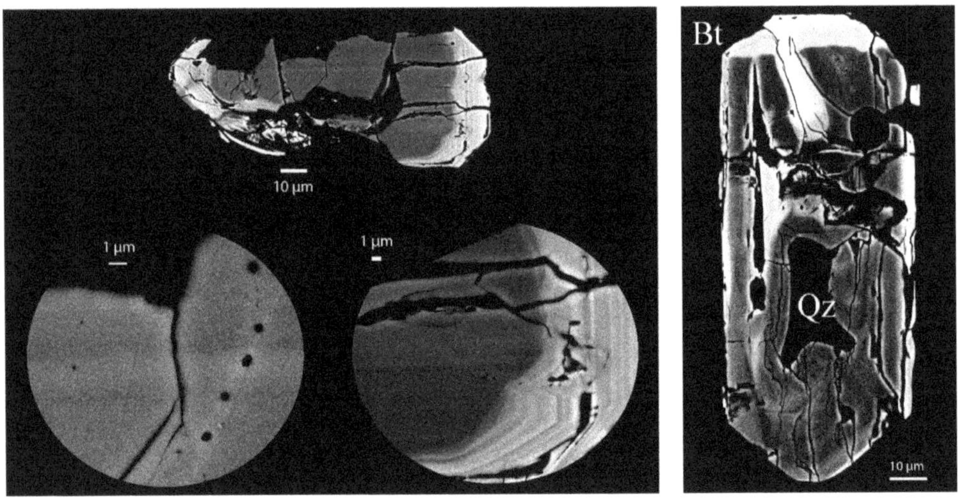

Figure 12. (**a**) Long-prismatic Zrn-Znw crystal with a chain of open gas–liquid inclusions (left) and primary oscillatory and superimposed wave-shaped zoning (right); (**b**) BSE image of the long-prismatic Zrn-Znw crystals demonstrating the spotty zoning and inclusions of rock-forming biotite (Bt) and quartz (Qz).

Short-prismatic Zrn-Znw variety shows the most complex inner structure: the core/rim boundaries and the primary growth zoning are obscured and give way to the secondary wavy zones and homogeneous blocks with soft edges; pores of different sizes are widespread and concentrate numerous inclusions of thorite, xenotime, chernovite, and monazite.

Distinctive features of the short-prismatic Zrn-Znw are the maximum content of heavy trace elements (relatively bright BSE images), as well as the secondary wavy zoning, porosity, and abundance of inclusions, which occupy a significantly larger grain area (in the section) than similar alteration zones in zircon from biotite granites. Secondary alterations can be seen on the surface of Zrn-Znw faces as the abundance of caverns and dissolution relief.

Thus, the contrast of morphological features of both zircon types from the initial (Zrn-Bt) and final (Zrn-Znw) granite series is an important additional tool to reveal rare-metal Li-F granites within the Verkhneurmiysky massif. The oscillatory zoning of Zrn-Bt reflects the long path of water-poor magma evolution, which is able to travel long distances; the spotty zoning and increased porosity of Zrn-Znw indicate less stable growth conditions, increased activity of the fluid phase, and lattice defects due to secondary alterations and formation of mineral inclusions (Figure 13).

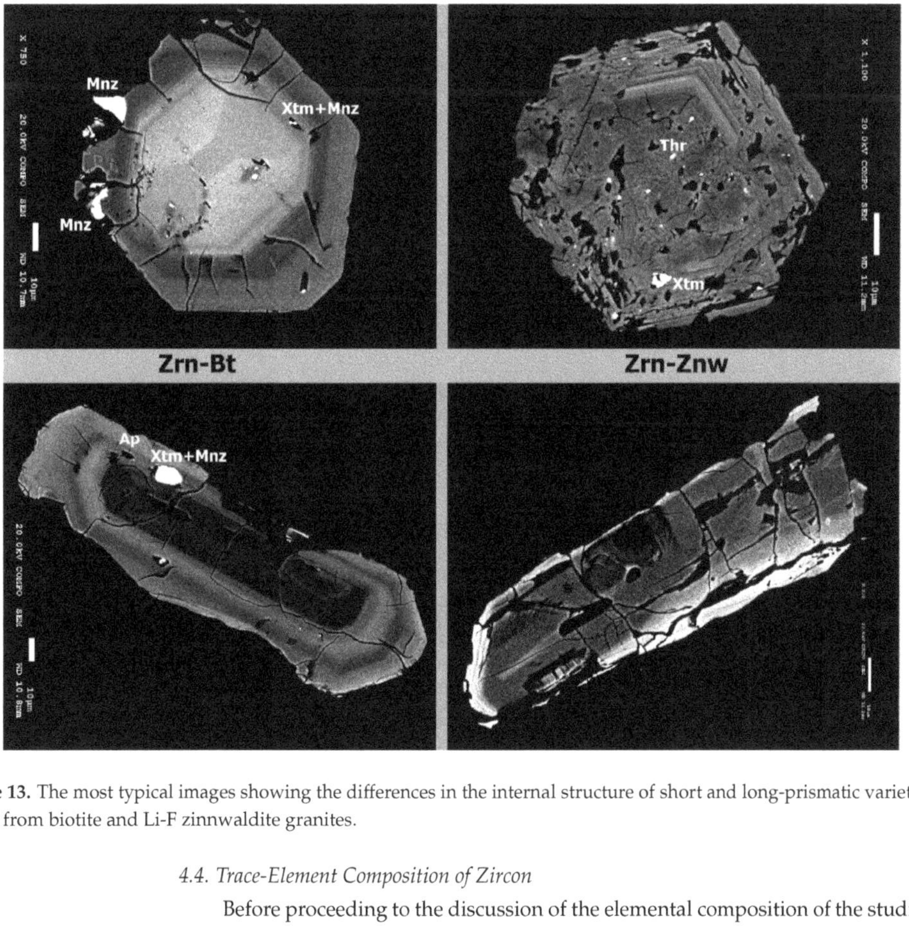

Figure 13. The most typical images showing the differences in the internal structure of short and long-prismatic varieties of zircon from biotite and Li-F zinnwaldite granites.

4.4. Trace-Element Composition of Zircon

Before proceeding to the discussion of the elemental composition of the studied zircon, it is necessary to make a brief comparison with zircon from granites of similar genesis from different geographical regions, for example: biotite and Li-F granites of the Severny massif

(Chukotka) [2], Li-F granites of Zinnwald deposit (Germany) [39,40] and zinnwaldite granites of Mole massif (Australia) [41] (Tables 3 and 4).

Table 3. Composition of zircon (ppm) from biotite leucogranites of the Verkhneurmisky massif, Amur region (median and quartiles calculated from ion probe data), and the Severny massif, Chukotka [2].

Element	Zircon from Biotite Granites											
	Verkhneurmisky Massif Amur Region						Severny Massif Chukotka					
	Core			Rim								
	Quartile 50%	Quartile 75%	Quartile 25%	Quartile 50%	Quartile 75%	Quartile 25%						
Li	1.02	1.91	0.59	1.82	2.06	1.59	0.1	0.1	0.05	0.57	0.1	2.38
P	421	706	289	893	1077	778	–	–	–	–	–	–
Ca	11.1	45.6	8.05	3.94	7.49	2.29	0.81	0.72	0.55	132	0.38	5.99
Ti	36.6	79.5	24.6	7.10	22.5	3.34	10.7	9.04	21.3	5.63	17.5	7.46
Sr	1.32	1.79	0.97	2.04	2.21	1.83	0.63	0.67	0.6	1.09	0.72	1.27
Y	3084	5327	2001	6469	7449	5718	504	516	912	1946	637	2225
Nb	37.9	65.4	25.2	110	135	94.1	26.7	24.7	22	43.5	20.6	62.7
Cs	0.89	1.47	0.66	1.88	1.98	1.74	–	–	–	–	–	–
Ba	4.05	4.83	1.69	1.56	2.47	1.41	1.47	1.21	1.06	1.78	1.39	1.55
Hf	9074	11,403	7523	11,856	13,484	10,437	7191	7547	6471	8281	7136	8593
Th	618	1974	275	1973	2378	1497	62.1	70.2	64.1	583	68.5	658
U	1673	2428	859	8437	8783	6768	227	280	190	2534	230	3582
La	1.59	3.46	0.27	0.50	0.86	0.13	0.03	0.05	0.09	2.59	0.1	1.15
Ce	26.3	48.9	16.9	26.9	30.4	23.1	7.61	7.04	4.84	21.3	5.38	17.8
Pr	1.31	3.21	0.43	0.47	0.80	0.14	0.06	0.03	0.1	1.38	0.08	0.6
Nd	8.69	23.8	3.72	4.14	6.05	2.71	0.86	0.79	1.54	8.04	1.15	4.06
Sm	12.5	27.2	6.95	11.5	14.1	8.85	1.66	1.67	3.51	6.58	2.03	5.17
Eu	0.25	0.46	0.17	0.13	0.16	0.04	0.02	0.03	0.13	0.01	0.02	0.02
Gd	59.6	81.7	39.7	76.7	95.5	50.9	8.54	8.39	19.4	27.2	10.9	27.3
Dy	257	284	200	461	552	335	36.5	36.7	70.4	136	46.8	159
Er	510	635	328	1118	1299	897	75.5	82.2	140	302	98.1	371
Yb	945	1140	523	2274	2384	1884	138	150	234	570	173	769
Lu	140	168	77.3	337	351	293	21	22.9	36.7	85.3	27.5	113
H_2O	943	1500	723	1270	1568	1110	–	–	–	–	–	–
F	22.0	32.3	13.0	11.4	15.0	9.76	–	–	–	–	–	–
Th/U	0.38	0.47	0.31	0.25	0.28	0.19	0.27	0.25	0.34	0.23	0.3	0.18
Eu/Eu*	0.03	0.04	0.02	0.01	0.02	0.01	0.02	0.03	0.05	0.00	0.01	0.01
Ce/Ce*	4.91	11.1	3.98	11.5	37.3	9.86	41.4	42.5	12.8	2.37	14.2	5.2
ΣLREE	37.3	81.0	19.9	30.9	36.6	28.8	9	8	7	33	7	24
ΣHREE	1886	2271	1243	4280	4573	3401	280	301	500	1120	357	1440

Table 4. Composition of zircon (ppm) from Li-F granites of the Verkhneurmisky massif, Amur region (median and quartiles calculated from ion probe data), and the Severny massif, Chukotka [2], Mole massif (Australia) [41] and Zinnwald Massif (Germany) [40].

	Zircon from Li-F Granites											
	Verkhneurmisky Massif											
Element	Core			Rim			Severny Massif			Mole Massif	Zinnwald Deposit	Erzgebirge
	Quartile 50%	Quartile 75%	Quartile 25%	Quartile 50%	Quartile 75%	Quartile 25%						Mean Value
Li	0.62	0.82	0.29	3.38	9.33	2.41	5.45	14.1	0.57	93	–	–
P	466	473	288	1137	1499	944	–	–	–	–	3392	–
Ca	16.7	33.3	6.07	126	311	72.4	0.19	7.07	173	–	7218	194.9
Ti	6.80	11.4	4.66	36.3	97.1	9.59	25.7	11.2	7.57	–	–	17.79
Sr	1.31	1.55	1.05	6.63	13.2	3.91	0.57	1.08	0.56	–	–	1.19
Y	3105	3371	2168	6663	10,682	4240	659	572	1478	7430	2225	2864.6
Nb	49.8	89.3	25.3	647	962	340	107	64.8	36.7	72.0	–	17.23
Cs	1.98	11.59	0.50	9.35	52.59	4.45	–	–	–	14.0	–	–
Ba	1.98	4.02	1.10	7.09	15.2	2.41	1.59	2.42	1.32	–	–	1.21
Hf	7523	8167	6913	15,023	17,883	12,567	6271	7360	7260	15,400	36,400	8833.4
Th	421	691	291	1844	3086	1305	30.5	50.1	369	2515	2293	297.5
U	1052	1722	909	8395	13,374	6901	56	156	1503	8192	6972	1413.2
La	0.99	1.92	0.34	8.28	12.0	2.70	0.11	0.41	3.12	22.7	–	1.87
Ce	15.5	17.6	11.0	64.4	92.5	44.0	33.6	28.6	27.5	111	–	18.15
Pr	0.77	1.74	0.28	4.85	7.72	1.80	0.21	0.64	1.84	13.0	85	1.15
Nd	7.04	13.10	3.32	29.36	42.71	12.11	2.08	4.78	9.29	70.0	–	10.63
Sm	15.53	19.51	8.07	30.05	50.15	20.29	3.73	4.48	7.21	48.9	345	14.27
Eu	0.16	0.36	0.14	0.29	0.49	0.12	0.32	0.5	0.01	1.30	–	1.29

Table 4. *Cont.*

Element	Zircon from Li-F Granites											
	Verkhneurmisky Massif						Severny Massif			Mole Massif	Zinnwald Deposit	Erzgebirge
	Core			Rim								
	Quartile 50%	Quartile 75%	Quartile 25%	Quartile 50%	Quartile 75%	Quartile 25%						Mean Value
Gd	54.7	69.9	45.3	102	145	65.2	15.0	11.9	25.8	143	–	63.44
Dy	218	312	163	541	824	382	49.7	39.1	110	713	1505	255.1
Er	456	566	320	1361	2011	1001	92.7	84.2	224	1227	2838	467.2
Yb	835	947	622	3661	4678	2495	168	194	398	2742	10,315	745.1
Lu	124	146	115	515	668	379	25.2	30.9	59.2	475	–	112.5
H_2O	536	853	418	2294	3974	1323	–	–	–	–	–	–
F	66.4	87.3	20.5	186	345	108	–	–	–	–	8400	–
Th/U	0.40	0.51	0.37	0.22	0.30	0.16	0.54	0.32	0.25	0.31	0.33	0.21
Eu/Eu*	0.02	0.04	0.02	0.02	0.02	0.01	0.13	0.21	0.01	0.05	–	0.13
Ce/Ce*	5.63	9.13	3.68	2.32	5.26	1.79	53.7	13.4	2.78	1.60	–	2.99
ΣLREE	23.10	42.38	14.10	99.52	150.93	67.95	36	34	42	217	85	47.36
ΣHREE	1707	2058	1250	6485	8740	4390	350	360	817	5907	14,658	1643.38

Analyzing the published data, we can say that, in general, the level of trace elements content in zircon from granitoids is relatively low: Hf (0.39–3.98%), Y (0.1–0.5%), REE, P, U, Th (up to 0.5%), and ~0.0 n% for the rest [42]. An exception is a zircon from the Li-F granites from Europe (Bohemia; Carpathians), Australia (New England batholith), Asia (Transbaikalia), Africa (Arabian desert), and North America (Guadalupe Mountains), with concentrations of rare elements reaching 10 wt.% or more [2,43].

The studied zircon from the Verkhneurmiysky granites (both biotite leucogranites and zinnwaldite granites) demonstrates the most similar composition with zircon from Chukotka granites (Severny massif), as well as with zircon from rare-metal topaz-bearing granites of the Mole massif (Australia) [41], Zinnwald Massif (Germany) [40], and Erzgebirge (Germany) [44] where REE concentrations almost reach the level of hydrothermal mineralization [41,45].

The distinctive feature of zircon from Verkhneurmiysky granites is the increased content of Hf, reaching the level of zircon from the granite of the Erzgebirge (Germany) and Mole massif (Australia). Hf content is significantly increased in the altered rims of the studied zircon, reflecting the process of the melt differentiation and hydrothermal alterations, and almost reach the anomalous values typical of zircon from Zinnwald granite (Figure 14a). Compared to zircon of the Severny Massif, the Verkhneurmiysky zircon shows approximately equal levels of actinoids (Th, U) and high-field-strength (HFS) elements (Ti, Nb) and sharply increased levels of both groups of elements in the altered zircon rims in Zrn-Znw (Figure 14b).

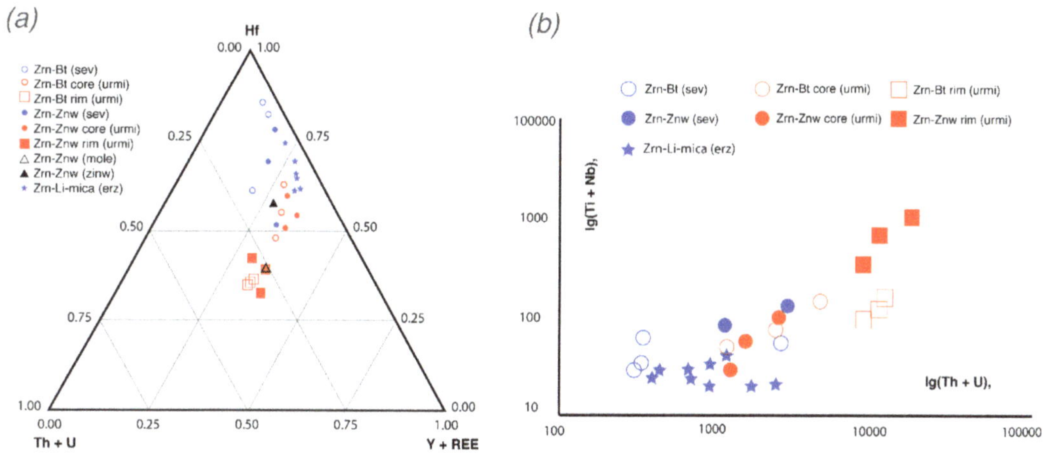

Figure 14. Discrimination charts of zircon composition from biotite (Bt) and zinnwaldite (Znw) granites of Verkhneurmisky (*Urmi*), Severny (*Sev*), Mole (*Sev*), Zinwald (*Zinw*), and Erzgebirge (*Erz*) massifs in (**a**) (Th + U)—Hf—(Y + REE) and (**b**) (Th + U)—(Ti + Nb) coordinates.

The electron-microscopic studies have already shown that all zircon types from Verkhneurmiysky granites show anomalously high concentrations of U, Th, REE, and Y. According to many researchers [46–48], the probability of finding these elements in the structure of zircon as an isomorphic impurity is small; whereas they rather form their own minerals enclosed in the zircon matrix. From this point of view, anomalous U, Th, REE, and Y concentrations associated with their own minerals (uraninite, coffinite, thorite, monazite, allanite, etc.), either captured during crystal growth or formed in the process of zircon secondary alterations, are quite understandable. Such anomalous values were not considered in the composition analysis of zircon under study.

The ion microprobe analysis of zircon also allows limiting the samples by the grain alteration degree and titanium content for the correct application of the Ti-in-zircon thermometer [49–51]. Thus, the median values of Ti-in-zircon temperatures, calculated for the unaltered zircons with a Ti content not exceeding 30 ppm, are 831 °C and 710 °C for the Zrn-Bt and Zrn-Znw, respectively. Zircon saturation thermometry [52–54], based on the granite's composition (Table 1) and calculated with M = (Na + K + 2·Ca)/(Al·Si), provides the values of 761 °C and 685 °C for the Zrn-Bt and Zrn-Znw, respectively. Despite the discrepancy between the absolute temperature values, obtained by different methods, the fact that the temperature of zircon formation decreases with time seems to be reliable.

Even in the case of pronounced contrast of morphology, microtexture, and zoning of zircon, the convergence of these features complicates the ambiguity of zircon typing. The relative reliability of Zrn-Znw type definition is achieved only when we consider the element composition. According to the results of the study on the zircon internal structure, the dataset of zircons in situ analysis was divided into the "core" and "rim" groups. The need for such separation was confirmed on the factor score diagrams for each sample, where the points corresponding to cores and rims are clearly split up (Figure 15).

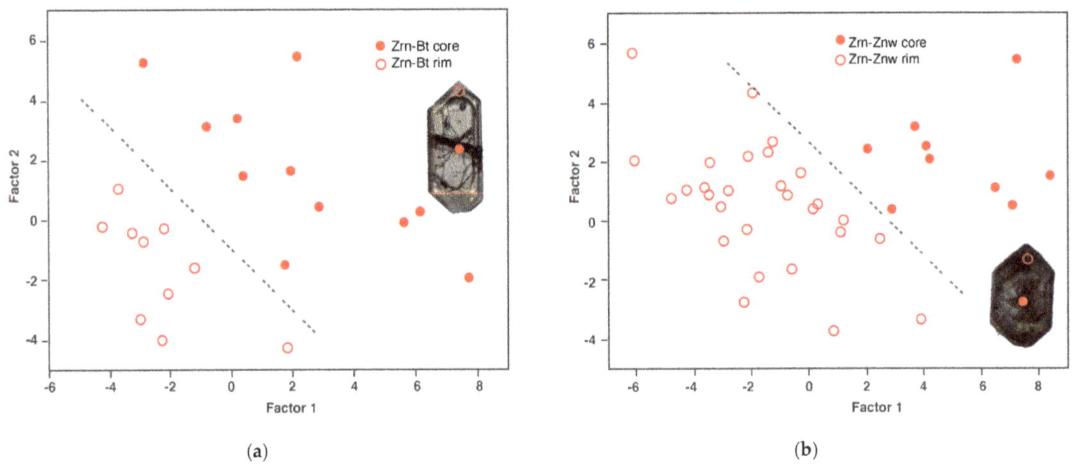

Figure 15. Factor score diagrams showing two fields of analytical points corresponding to the cores and altered rims of zircon from: biotite (**a**) and zinnwaldite (**b**) granites.

Thus, it is reasonable to consider the relationships between cores and rims within each zircon type, assuming that the magmatic history of zircon formation should be reflected in compositional and structural changes of the unaltered central domains of grains, while the intensity of the secondary alterations affects the composition of rims. The accumulation of U, Th, Nb, HREE, Cs, P, Y, Li, Sr, H_2O, and Hf (in descending order of accumulation intensity) and decrease in LREE, F, Ba, Ca, and Ti concentrations (Figure 16a,c) are observed in the rims of zircon from biotite granite. Zircon from zinnwaldite granite demonstrates a general increase in the concentrations of all trace elements (Figure 16a,c).

It is interesting to note that there are no significant variations in the composition of the central unaltered domains in zircon from different granites, which suggests similar physical and chemical parameters of zircon crystallization during the formation of the Verkhneurmiysky granitoid series (Figure 16a). Zircon from zinnwaldite granites has only initially slightly increased content of F, Cs, and Nb, emphasized by increasing zircon/melt distribution coefficients in the transition from biotite to zinnwaldite granites, which is likely a reflection of the increased degree of emanation differentiation in the final stages of granite magmatism. Additionally, cores are characterized by the REE spectra typical for magmatic zircon—the smooth growth of chondrite-normalized REE concentrations with an increase in atomic number, disturbed by Ce- and Eu-anomalies. Zircon from zinnwaldite granite differs in a slightly lower content of LREE and a bit weaker Ce-anomaly

(Figure 16c,d). Positive Ce-anomaly is controlled by oxygen fugacity and is explained by the fact that Ce^{4+} is incorporated in the structure of zircon according to the isovalent isomorphism. Moreover, the radius of Ce^{4+} ion is close to HREE ones, so the calculated value of zircon/melt distribution coefficient for Ce^{4+} is several orders of magnitude higher than for Ce^{3+}. The depth of the Eu-anomaly depends on the Eu^{2+}/Eu^{3+} ratio, and its manifestation in zircon REE spectra is caused by early processes of acid melts appearance as a result of crystallization differentiation. However, in this case, it seems more likely to associate the variations of Eu-anomaly in the melt not only with the participation of feldspars in the crystallization process but also with the fluid differentiation.

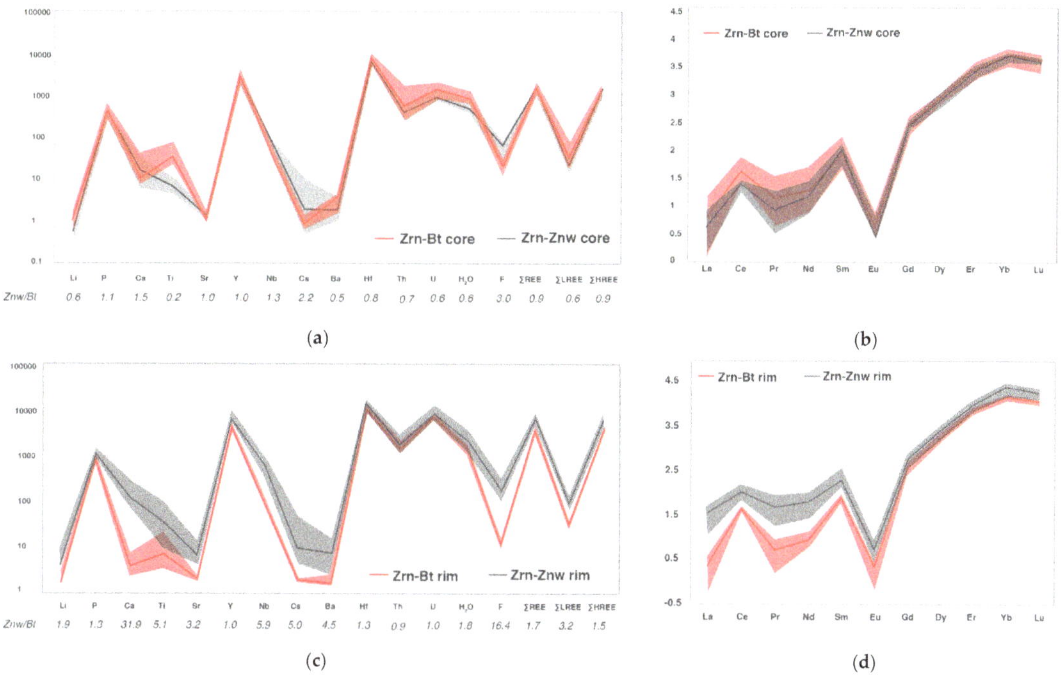

Figure 16. Contents of trace elements and REE spectra in (**a**,**b**) cores and (**c**,**d**) rims of zircon grains from biotite Zrn-Bt and zinnwaldite Zrn-Znw granites; solid lines—median values, dashed lines—upper/lower quartile boundaries.

Against the similar composition of cores in zircon of different types, the content of trace elements in the altered crystals domains experiences a sharp jump in the transition from Zrn-Bt to Zrn-Znw, which can be caused by the active influence of fluid phase accompanying the intrusion of the zinnwaldite granites. This assumption is also supported by the widespread secondary alterations in zircon Zrn-Znw, such as a decrease in the degree of grains crystallinity and the development of spongy texture, saturated with pores and inclusions. It should be noted that the set of elements (P, Y, Th, and U) corresponding to the composition of common mineral inclusions (xenotime, coffinite, and thorite) has an approximately equal level of concentrations in the rims of all zircons under study. This suggests that the evolution of zircon composition is regulated not only and not so much by the parameters of the crystallization medium as by the isomorphic capacity of the mineral lattice. REE spectra illustrate sharp enrichment of altered Zrn-Znw domains with rare earth elements, with the flattering on the left side of the spectrum (LREE), reduction of Ce-anomaly, and decrease in the ΣHREE/ΣLREE ratio. This may be the result of a destructive fluid effect on the zircon lattice that facilitates the isomorphism of LREE [55] (Figure 16d).

The SmN/LaN ratio together with the Ce-anomaly (Ce/Ce*) value allows tracing the distribution of points corresponding to zircon from biotite leucogranites and zinnwaldite granites between the fields of magmatic and hydrothermal zircon [56]. The density of Zrn-Znw sampling points distribution shows maximum near the field of hydrothermal zircon. Consequently, a relative enrichment of Ce^{3+} is observed for Zrn-Znw, which is typical for secondary altered zircons [57]. The separation of Zrn-Bt and Zrn-Znw on the diagram in SmN/LaN-Ce/Ce* coordinates illustrates the full post-magmatic processing of magmatic zircon from zinnwaldite granites and serves as its additional characteristic.

Statistical analysis of zircon composition data included the results of preliminary correlation analysis of SIMS data, which revealed a strong correlation between such groups of elements as Cs, Sr, and P; Ca, Ba, and Li in zircon from biotite granites, and the absence of such correlations, along with the shift of the negative correlation between Ca and Sr, on the strong positive one in zircon from zinnwaldite granites. Zrn-Znw is characterized by a strong positive correlation between halogens (F), LIL (Ba, Cs), LREE, and H_2O, which is explained by the formation of late magmatic Zrn-Znw grains with active gas evaporation from the fluid-saturated melt and imperfection of Zrn-Znw structure caused by fluid action and facilitating the entrance of incompatible elements into the zircon structure. Similar dependencies in zircon Zrn-Bt are very weak and are due to the contribution of rare-metal rims, which formed around the early magmatic zircon during the intrusion of the Li-F granites. The principal component analysis allowed interpreting two significant factors overlapping 73% of the total dispersion (Figure 17a): the first factor is the formation of rims enriched with impurities under the temperature decrease; the second factor reflects the process of REE fractionation, and the formation of mineral inclusions (xenotime, thorite, coffinite) in the altered zircon domains. The reliability of interpretation of the first factor is confirmed by the factor score diagrams, where the points corresponding to cores and rims are clearly split up (Figure 17b).

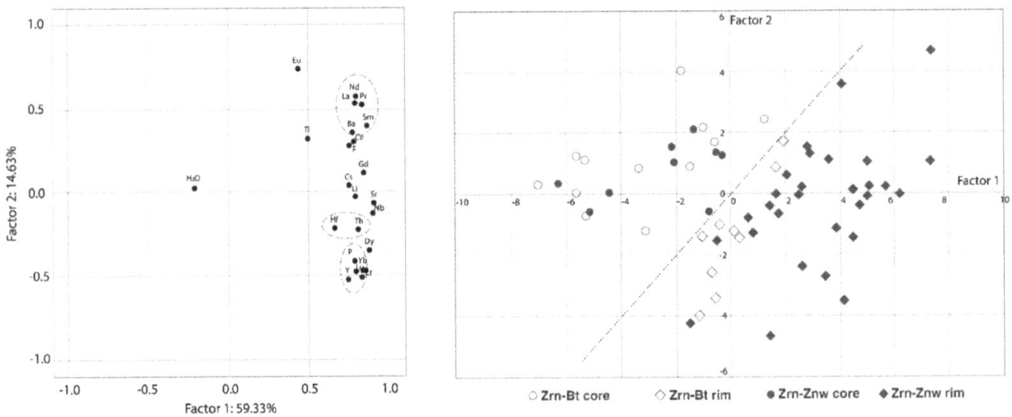

(a) (b)

Figure 17. (**a**) Factor loadings and (**b**) factor scores graphs plotted for the general sample of zircon from the Verkhneurmiysky granitoids.

Thus, we identified the main trends in the composition evolution of zircon from the Verkhneurmiysky intrusive series with Li-F granites, which are determined by both the degree of melt differentiation and the intensity of post-magmatic processes, as well as the special features of the zircon lattice: accumulation of HFS (Hf, Nb) and rare-earth (REE, Y) elements at a decrease in the Th/U ratio, which sensitively records the degree of zircon recrystallization and positively correlates with Eu-anomaly (Figure 18a); an abrupt increase in the concentration of volatile (H_2O, F) and rare alkalis (Li, Cs) at the formation of Li-F granites, which directly affected the alteration of zircon from biotite leucogranites (Figure 18b).

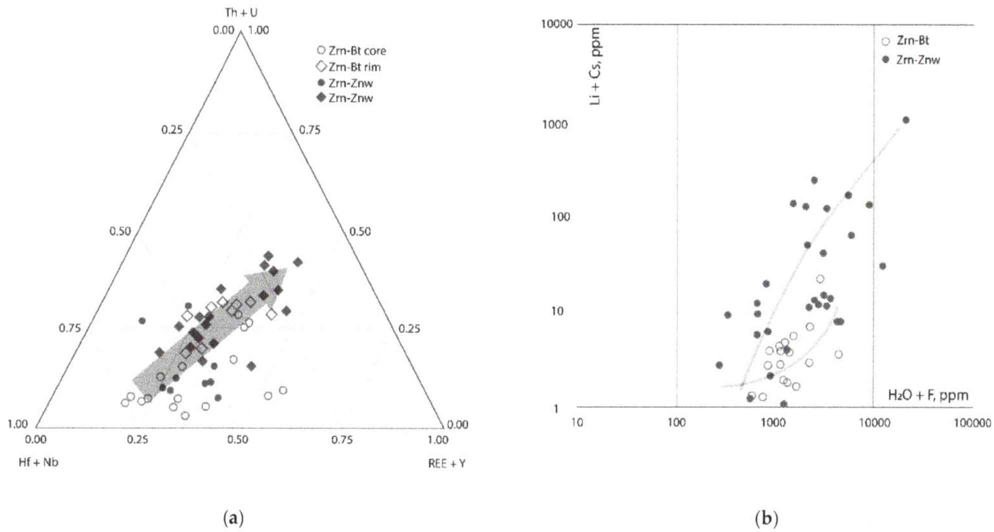

Figure 18. (**a**) Trend of accumulation of HFS and rare-earth elements with an increase in the gradient of uranium concentration; (**b**) abrupt increase in the concentration of volatile and rare alkalis at the stage of Li-F granites formation, which had a direct impact on the alteration of zircon from biotite leucogranites.

4.5. Raman Spectroscopy of Zircon

The set of rather conformal spectra and wavenumbers of Raman modes for both zircon types under study agree with other studies [24,25,58,59] and characterized by the left side wave-shaped curves elevation with the extremum value in the range of 200–350 cm^{-1}, overlapping an external translational (202, 225, 355 cm^{-1}) modes of SiO_4 tetrahedra (Figure 19).

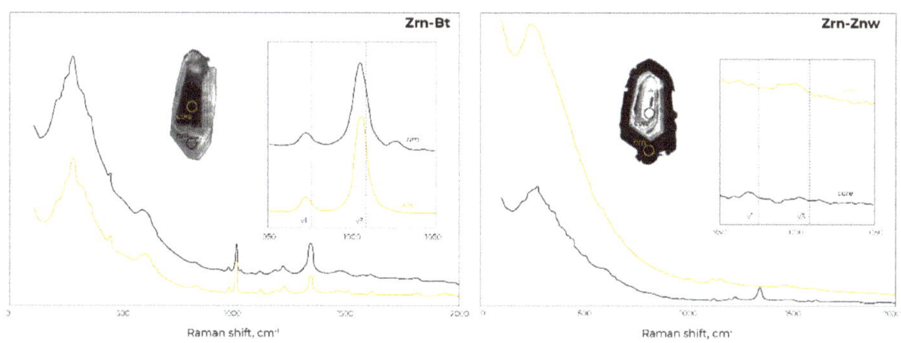

Figure 19. Raman spectra of zircon from biotite (Zrn-Bt) and zinnwaldite (Zrn-Znw) granites at the certain points corresponding to the zircon ion microprobe analytical points; the dotted red lines indicate the position of the characteristic peaks of SiO_4 oscillations in crystalline zircon: ν_1—975 cm^{-1}, ν_3—1008 cm^{-1}.

The spectra of the Zrn-Znw without distinction of core/rim crystal zones are characterized by whole peaks reduction grading into the wave-shaped curves. Whereas in passing from core to rim in the vast majority of Zrn-Bt grains, there is a gradual shift in the red region for the $\nu_1(SiO_4)$ and $\nu_3(SiO_4)$ peaks, accompanied by peak tailing up to complete disappearance of the peaks and the emergence of so-called amorphous halo, typical for zircon enriched in REE and resulting from laser-induced fluorescence emissions of the luminophor impurities [58]. In some cases, Raman spectra of the Zrn-Bt rims are similar to Zrn-Znw spectra along the whole grain, which could indicate the same alteration processes. In the light CL-zones (principally corresponding with core zones) of the Zrn-Bt crystals, FWHM of the $\nu_3(SiO_4)$ band ranges from 8.7 to 18.4 cm^{-1}, which suggests only a partial amorphization of the crystal structure (so-called transit state). The peak reduction in the spectra of the zircon rim gives evidence of significant structural radiation damage (FWHM minimum value is 29.1 cm^{-1}), leading to an increase in the zircon reactive capacity and dramatic accumulation of trace elements [59].

The coincidence of Raman and ion microprobe analytical points make it possible to trace the relationship between the characteristics of the $\nu_3(SiO_4)$ peak with the composition of zircon, expressing them through appropriate correlations. As consequence, it was revealed that the peak broadening and the value of the Raman shift expectedly decrease with the increase in LREE, Li, F content and Th/U ratio. However, the increasing Ce-anomaly value is accompanied by an increase in Raman shift magnitude of $\nu_3(SiO_4)$ peak. Shown correlation is regular from the standpoint of Ce^{4+} incorporating into the zircon crystal structure according to the isovalent substitution model: due to the proximity of Ce^{4+} (0.097 nm) and Zr^{4+} (0.084 nm) ionic radius, isomorphic occurrence of Ce^{4+} in Zr^{4+} position does not lead to significant distortions of the zircon crystal lattice. The reduction of zircon crystallinity accompanied by the broadening and redshift of the $\nu_3(SiO_4)$ peak is also confirmed by the decrease in the degree of REE fractionation (LuN/LaN) specific to altered zircon.

5. Conclusions

The result of this study is a complex characteristic of zircon from the Verkhneurmiysky intrusive series with Li-F granites. A wide range of morphological and chemical features of zircon allowed to obtain new information on the formation and alteration of zircon from granitoids of different types and to determine a set of zircon characteristic features, which contribute to the correct identification of Li-F granites formed directly before the tin mineralization within the Verkhneurmiysky massif.

The division into varieties by the coefficient of crystal elongation contributed to an adequate interpretation of the morphotype distribution and facilitated the establishment of zircon stages of evolution even within a single sample.

We also found that the change in the trace elements content level on the way from zircon cores to its rims are quite different in zircon from biotite (Zrn-Bt) and zinwalditic (Zrn-Znw) granites with the maximum impurities' accumulation level in the rims of zircon from zinwalditic ones. Active fluid evaporation and their transporting role is the reason for the secondary alterations affected not only zircon from the rare-metal granites but also from the early biotite granites, which led to the accumulation of a wide range of trace elements (Ca, Ti, Ba, F, Cs, Eu, La, Li, Pr, P, Nb, Lu, Yb, Nd, and Hf) in the altered zircon rims. The main trend in the evolution of zircon composition is the progressive accumulation of volatile (H_2O, F), LIL (Cs, Sr), HFS (Hf, Nb), and rare-earth elements. The evolution of zircon morphology is not only the fact, that the low-temperature zircon morphotypes expectedly followed the high-temperature ones, but also the complication of the zircon structure in time—the formation of rare metal rims, the spread of secondary alteration zones with the crystal lattice disturbance, mineral inclusions, pores, and cracks. The Raman spectra of Zrn-Bt and Zrn-Znw show zircon crystallinity reduction through the formation of Verkhneurmiysky granitoid series. The reduction of zircon crystallinity is also verified by an increase in uranium accumulation gradient and resulting in an extension of the isomorphic capacity of zircon lattice with naturally determined accumulation of rare-earth (REE, Y) elements and abrupt increase in the concentration of volatiles (F) and rare alkalis (Li) at the stage of lithium–fluorine granites formation.

The composition and morphology of the studied zircon from zinnwaldite granites are similar to zircon from the Li-F granite of Severny Massif (Chukotka), the Mole Massif (Australia), the Erzgebirge (Germany, Czech Republic), which allows the possibility of using the identified complex of zircon features for the correct revealing of the Li-F granites.

Author Contributions: M.M.M.—methodology, investigation, writing—original draft preparation, supervision, visualization; A.V.A.—resources, data curation, conceptualization, software; J.B.—writing—review and editing, validation, formal analysis, project administration. All authors have read and agreed to the published version of the manuscript.

Funding: This research received no external funding.

Institutional Review Board Statement: Not applicable.

Informed Consent Statement: Not applicable.

Data Availability Statement: The data presented in this study are available in this article.

Acknowledgments: The studies were performed using the equipment of the Common Use Centre of the Saint Petersburg Mining University; we thank Yu. B. Marin, V.I. Alekseev, and S.G. Skublov. We express gratitude to the academic editor and three anonymous reviewers for critically reading the manuscript and suggesting substantial improvements.

Conflicts of Interest: The authors declare no conflict of interest. The funders had no role in the design of the study; in the collection, analyses, or interpretation of data; in the writing of the manuscript, or in the decision to publish the results.

References

1. Maiboroda, A.F.; Emelianenko, A.S.; Vtorushina, V.F. Scheme of magmatism of the Badzhal volcanic zone. *Rep. USSR Acad. Sci.* **1977**, *235*, 155–158.
2. Alekseev, V.I. Li-F Granites of the Far East. Ph.D. Thesis, Geological and Mineralogical Sciences, Mining University, Saint Petersburg, Russia, 2014.
3. Vdovina, I.A. *Evaluation of the Denudation Sheet of the Badzhal Ore Region by Morphostructural and Crystal Morphological Methods;* Candidate of Geography, Institute of Geography: Moscow, Russia, 2004.
4. Stepanov, V.A.; Melnikov, A.V. Deposits of gold-quartz formation of the Amur province. *J. Min. Inst.* **2017**, *223*, 20.
5. Brusnitsin, A.I.; Panova, E.G.; Smolensky, V.V. The discovery of Li-F granites within the Verkhneurmiysky ore cluster. *Izv. VUZov. Geol. Explor.* **1993**, *6*, 150–151.
6. Krivovichev, V.G.; Brusnitsin, A.I.; Zaitsev, A.N. Absolute age and geochemical characteristics of granites of the Verkhneurmiysky massif (Priamurye, Far East). *Geochemistry* **1996**, *2*, 106–111.
7. Rodionov, S.M. Tin Metallogeny of the Russian East. Ph.D. Thesis, Geological and Mineralogical Sciences, Lomonosov Moscow State University, Moscow, Russia, 2003.
8. Grigoriev, S.I. Features of the material composition of the Late Mesozoic granitoids of the Badzhal and Komsomolsk ore regions, their petrogenesis and connection with mineralization. *Reg. Geol. Metallog.* **1997**, *6*, 103–115.
9. Gonevchuk, V.G. *Tin-Bearing Magmatic Systems of the Far East: Magmatism and Ore Genesis;* Dalnauka: Vladivostok, Russia, 2002; pp. 140–167.
10. Alekseev, V.I. Rare metals in minerals of tin-bearing metasomatites of the Verkhneurmiysky ore cluster (Far East of Russia). *J. Min. Inst.* **2018**, *232*, 335–340.
11. Dobretsov, G.L.; Marin, Y.B.; Beskin, S.M.; Leskov, S.A. *Principles of Dissection and Mapping of Granitoid Intrusions and Identification of Petrological-Metallogenic Variants of Granitoid Series;* VSEGEI Publishing House: Saint Petersburg, Russia, 2007; pp. 39–56.
12. Marin, Y.B.; Beskin, S.M. Typification of rare-metal granites and forecast of rare-metal mineralization. In Proceedings of the International Petrographic Meeting "Petrography of the XXI Century", Petrology and Ore Content of the CIS Regions and the Baltic Shield, Apatity, Russia, 28–30 June 2005; pp. 47–48.
13. Beskin, S.M.; Marin, Y.B.; Matthias, V.V. So what exactly is "rare metal granite"? *Zapiski RMS* **1999**, *6*, 28–40.
14. Alekseev, V.I.; Sukhanova, K.G.; Marin, Y.B. Niobium Minerals as Indicators of a Genetic Link between Tin-Bearing Zwitter and Lithium-Fluorine Granite of the Verkhneurmiysky Massif in the Amur River Region. *Geol. Ore Depos.* **2019**, *8*, 698–707. [CrossRef]
15. Skublov, S.G.; Krasotkina, A.O.; Makeev, A.B.; Rizvanova, N.G.; Koiman, E. The first data on the age (U-Pb method, TIMS, LA-ICP-MS) of rutile from the polymineral ore occurrence Ichetyu, Middle Timan. *J. Min. Inst.* **2018**, *232*, 357–363.
16. Frost, B.R.; Barnes, C.G.; Collins, W.J.; Arculus, R.J.; Ellis, D.J.; Frost, C.D. A geochemical classification of granitic rocks. *J. Petrol.* **2001**, *42*, 2033–2048. [CrossRef]
17. Breiter, K.; Škoda, R. Vertical zonality of fractionated granite plutons reflected in zircon chemistry: The Cínovec A-type versus the Beauvoir S-type suite. *Geol. Carpathica* **2012**, *63*, 383–398. [CrossRef]

18. Nikolaeva, N.; Aleksandrova, T.; Romashev, A. Effect of grinding on the fractional composition of polymineral laminated bituminous shales. *Miner. Process. Extr. Metall. Rev.* **2018**, *39*, 231–234. [CrossRef]
19. Pashkevich, M.A.; Petrova, T.A. Recyclability of ore beneficiation wastes at the Lomonosov Deposit. *J. Ecol. Eng.* **2019**, *20*, 27–33. [CrossRef]
20. Sirotkin, A.N.; Talovina, I.V.; Duryagina, A.M. Mineralogy and geochemistry of alkaline lamprophyres of north-western Spitsbergen (Svalbard). *Chem. Erde* **2020**, *80*, 125508. [CrossRef]
21. Fedotova, A.A.; Bibikova, E.V.; Simakin, S.G. Zircon Geochemistry (Ionic Microprobe Data) as an Indicator of Mineral Genesis in Geochronological Research. *Geochemistry* **2008**, *9*, 980–997.
22. Popov, O.A.; Talovina, I.V.; Lieberwirth, H.; Duryagina, A.M. Quantitative microstructural analysis and X-ray computed tomography of ores and rocks—Comparison of results. *Minerals* **2020**, *2*, 129. [CrossRef]
23. McDonough, W.F.; Sun, S.S. The composition of the Earth. *Chem. Geol.* **1995**, *120*, 223–253. [CrossRef]
24. Dawson, P.; Hargreave, M.M.; Wilkinson, G.R. The vibrational spectrum of zircon ($ZrSiO_4$). *J. Phys. C* **1971**, *4*, 240–256. [CrossRef]
25. Gusik, A.; Zhang, M.; Koeberl, C.; Salje, E.; Redfern, S.; Pruenda, M. Infrared and Raman spectra of $ZrSiO_4$ experimentally shocked at high pressures. *Mineral. Mag.* **2004**, *65*, 801–811. [CrossRef]
26. Mikheeva, P.I.; Alekseev, V.I. Allanite in Granites of the Verkhneurmiysky massif (Amur region). In Proceedings of the XLIX International Scientific Conference "Student and Scientific and Technical Progress", Novosibirsk, Russia, 16–20 April 2011; p. 23.
27. Leontev, V.I.; Skublov, S.G.; Shatova, N.V.; Berezin, A.V. Zircon U-Pb Geochronology Recorded Late Cretaceous Fluid Activation in the Central Aldan Gold Ore District, Aldan Shield, Russia: First Data. *J. Earth Sci.* **2020**, *31*, 481–491. [CrossRef]
28. Kempe, U.; Wolf, D.; Sala, M. Members of the philipsbornite—Florencite and chernovite-xenotime solid solution series in metasomatic altered granites of the Zinnwald tin deposit (Erzgebirge, Germany). *Beih. Eur. J. Miner.* **1999**, *11*, 120.
29. Pupin, J.P. Zircon and granite petrology. *Contrib. Mineral. Petrol.* **1980**, *73*, 207–220. [CrossRef]
30. Brodskaya, R.L.; Marina, E.Y.; Shnay, G.K.; Saminin, I.A. Restoration of conditions and formations kinetics of rare metal granites according to crystallomorphology of accessory zircon. *Zapiski RMS* **1986**, *115*, 50–62.
31. Nasdala, L.; Wenzel, M.; Vavra, G.; Irmer, G.; Wenzel, T.; Kober, B. Metamictisation of natural zircon: Accumulation versus thermal annealing of radioactivity-induced damage. *Contrib. Mineral. Petrol.* **2001**, *141*, 125–144. [CrossRef]
32. Corfu, F.; Hanchar, J.M.; Hoskin, P.W.O.; Kinny, P. Atlas of zircon textures. *Rev. Mineral. Geochem.* **2003**, *53*, 469–500. [CrossRef]
33. Geisler, T.; Schaltegger, U.; Tomaschek, F. Re-equilibration of zircon in aqueous fluids and melts. *Elements* **2007**, *3*, 43–50. [CrossRef]
34. Hay, D.C.; Dempster, T.J. Zircon behaviour during low-temperature metamorphism. *J. Petrol.* **2009**, *50*, 571–589. [CrossRef]
35. Vasilev, E.A.; Zedgenizov, D.A.; Klepikov, I.V. The enigma of cuboid diamonds: The causes of inverse distribution of optical centers within the growth zones. *J. Geosci.* **2020**, *65*, 59–70. [CrossRef]
36. Förster, H.-J.; Ondrejka, M.; Uher, P. Mineralogical responses to subsolidus alteration of granitic rocks by oxidizing As-bearing fluids: REE arsenates and As-rich silicates from the Zinnwald granite, eastern Erzgebirge, Germany. *Can. Mineral.* **2011**, *49*, 913–930. [CrossRef]
37. Ondrejka, M.; Uher, P.; Pršek, J.; Ozdín, D. Arsenian monazite-(Ce) and xenotime-(Y), REE arsenates and carbonates from the Tisovec-Rejkovo rhyolite, Western Carpathians, Slovakia: Composition and substitutions in the (REE,Y)XO_4 system (X = P, As, Si, Nb, S). *Lithos* **2007**, *95*, 116–129. [CrossRef]
38. Pidgeon, R.T. Recrystallization of oscillatory-zoned zircon: Some geochronological and petrological implications. *Contrib. Mineral. Petrol.* **1992**, *110*, 463–472. [CrossRef]
39. Förster, H.-J. Composition and origin of intermediate solid solutions in the system thorite–xenotime–zircon–coffinite. *Lithos* **2006**, *88*, 35–55. [CrossRef]
40. Johan, Z.; Johan, V. Accessory minerals of the Cínovec (Zinnwald) granite cupola, Czech Republic: Indicators of petrogenetic evolution. *Mineral. Petrol.* **2005**, *83*, 113–150. [CrossRef]
41. Pettke, T.; Andreas, A.; Schaltegger, U.; Heinricha, C.A. Magmatic-to-hydrothermal crystallization in the W–Sn mineralized Mole Granite (NSW, Australia) Part II: Evolving zircon and thorite trace element chemistry. *Chem. Geol.* **2005**, *220*, 191–213. [CrossRef]
42. Hoskin, P.W.O.; Schaltegger, U. The Composition of Zircon and Igneous and Metamorphic Petrogenesis. *Rev. Mineral. Geochem.* **2003**, *53*, 27–62. [CrossRef]
43. Marin, Y.B. Accessory minerals of the granitoid series of tin and molybdenum provinces. *Zap. Ross. Mineral. O-va* **2004**, *6*, 1–7.
44. Tichomirowa, M.; Gerdes, A.; Lapp, M.; Leonhardt, D.; Whitehouse, M. The Chemical Evolution from Older (323–318 Ma) towards Younger Highly Evolved Tin Granites (315–314 Ma)—Sources and Metal Enrichment in Variscan Granites of the Western Erzgebirge (Central European Variscides, Germany). *Minerals* **2019**, *9*, 769. [CrossRef]
45. Hoskin, P.W.O. Trace-element composition of hydrothermal zircon and the alteration of Hadean zircon from the Jack Hills, Australia. *Geochim. Cosmochim. Acta* **2005**, *69*, 637–648. [CrossRef]
46. Kempe, U.; Gruner, T.; Renno, A.D.; Wolf, D.; René, M. Discussion on Wang. 'Chemistry of Hf-rich zircons from the Laoshan I- and A-type granites, Eastern China. *Mineral. Mag.* **2004**, *68*, 669–675. [CrossRef]
47. Poller, U.; Huth, J.; Hoppel, P.; Williams, I.S. REE, U, Th and Hf distribution in zircon from Western Carpathian Variscan granitoids: A combained cathodeluminescence and ion microprobe study. *Am. J. Sci.* **2001**, *301*, 585–876. [CrossRef]

48. Kudryashov, N.M.; Skublov, S.G.; Galankina, O.L.; Udoratina, O.V.; Voloshin, A.V. Abnormally high-hafnium zircon from rare-metal pegmatites of the Vasin-Mylk deposit (the northeastern part of the Kola Peninsula). *Chem. Erde* **2019**, *79*, 1–8. [CrossRef]
49. Watson, E.B.; Wark, D.A.; Thomas, J.B. Crystallization thermometers for zircon and rutile. *Contrib. Mineral. Petrol.* **2006**, *151*, 413–433. [CrossRef]
50. Ferry, J.M.; Watson, E.B. New thermodynamic models and revised calibrations for the Ti-in-zircon and Zr-in-rutile thermometers. *Contrib. Mineral. Petrol.* **2007**, *154*, 429–437. [CrossRef]
51. Fu, B.; Page, F.Z.; Cavosie, A.J.; Fournelle, J. Ti-in-zircon thermometry: Applications and limitations. *Contrib. Mineral. Petrol.* **2008**, *156*, 197–215. [CrossRef]
52. Watson, E.B.; Harrison, T.M. Zircon saturation revisited: Temperature and composition effects in a variety of crustal magma types. *Earth Planet Sci. Lett.* **1983**, *64*, 295–304. [CrossRef]
53. Hanchar, J.M.; Watson, E.B. Zircon Saturation Thermometry. *Rev. Mineral. Geochem.* **2003**, *53*, 89–112. [CrossRef]
54. Boehnke, P.; Watson, E.B.; Trail, D.; Harrison, T.M.; Schmitt, A.K. Zircon Saturation Re-Revisited. *Chem. Geol.* **2013**, *351*, 324–334. [CrossRef]
55. Cao, Y.; Li, S.R.; Zhang, H.F.; Liu, X.B.; Li, Z.Z.; Ao, C.; Yao, M.J. Significance of zircon trace element geochemistry, the Shihu gold deposit, western Hebei Province, North China. *J. Rare Earths* **2011**, *29*, 277–286. [CrossRef]
56. Pelleter, E.; Cheillets, A.; Gasquet, D. A tool for ion microprobe U-Pb dating of gold mineralization (Tamlalt-Menhouhou gold deposit—Morocco). *Chem. Geol.* **2007**, *245*, 135–161. [CrossRef]
57. Balashov, Y.A.; Skublov, S.G. Unique indicator capabilities of cerium in zircons of different genesis. In Proceedings of the Conference Dedicated to the 110th Anniversary of the Birth of Acad. D.S. Korzhinsky, IGEM RAS, Moscow, Russia, 7–9 October 2009; pp. 67–70.
58. Kozhevnikov, V.N.; Ivashevskaya, S.N.; Kevlich, V.I. Geochemistry and Raman spectra of zircons from ore (PGE*Au) amphibolites of the Travyanaya guba massif, North Karelia. *Proc. Karelian Res. Centre Russ. Acad. Sci.* **2015**, *7*, 36–53. [CrossRef]
59. Nasdala, L.; Irmer, G.; Wolf, D. The degree of metamictization in zircon: A Raman spectroscopic study. *Eur. J. Mineral.* **1995**, *7*, 471–487. [CrossRef]

minerals MDPI

Article

Relationships between Alluvial Facies/Depositional Environments, Detrital Zircon U-Pb Geochronology, and Bulk-Rock Geochemistry in the Cretaceous Neungju Basin (Southwest Korea)

Hyojong Lee [1], Min Gyu Kwon [1], Seungwon Shin [2], Hyeongseong Cho [3], Jong-Sun Kim [4], Yul Roh [5], Min Huh [5] and Taejin Choi [1,*]

1 Department of Energy and Resources Engineering, Chosun University, Gwangju 61452, Korea; barefox@gmail.com (H.L.); kmalsrb@naver.com (M.G.K.)
2 Geological Research Center, Geology Division, KIGAM, Daejeon 34132, Korea; ssw7304@kigam.re.kr
3 Department of Geology, Gyeongsang National University, Jinju 52828, Korea; choh@gnu.ac.kr
4 Ecological Survey Division, National Park Research Institute, Korea National Park Service, Wonju 26466, Korea; kjsun@pusan.ac.kr
5 Faculty of Earth Systems and Environmental Sciences, Chonnam University, Gwangju 61186, Korea; rohy@jnu.ac.kr (Y.R.); minhuh@jnu.ac.kr (M.H.)
* Correspondence: tchoi@chosun.ac.kr

Received: 3 September 2020; Accepted: 16 November 2020; Published: 17 November 2020

Abstract: Zircon U-Pb geochronology and bulk-rock geochemistry analyses were carried out to investigate their relationship with depositional environments of the non-marine Neungju Basin sediments in South Korea. The Neungju Basin was formed in an active continental margin setting during the Late Cretaceous with associated volcanism. Detrital zircon age distributions of the Neungju Basin reveal that the source rocks surrounding the basin supplied sediments into the basin from all directions, making different zircon age populations according to the depositional environments. Mudstone geochemistry with support of detrital zircon U-Pb age data reveals how the heterogeneity affects the geochemical characteristics of tectonic setting and weathering intensity. The sediments in the proximal (alluvial fan to sandflat) and distal (playa lake) environments differ compositionally because sediment mixing occurred exclusively in the distal environment. The proximal deposits show a passive margin signature, reflecting their derivation from the adjacent metamorphic and granitic basement rocks. The distal deposits properly indicate an active continental margin setting due to the additional supply of reworked volcaniclastic sediments. The proximal deposits indicate a minor degree of chemical weathering corresponding to fossil and sedimentological records of the basin, whereas the distal deposits show lower weathering intensity by reworking of unaltered volcaniclastic detritus from unstable volcanic and volcaniclastic terranes. Overall, this study highlights that compositional data obtained from a specific location and depositional environments may not describe the overall characteristic of the basin.

Keywords: provenance; compositional heterogeneity; major element; detrital zircon; nonmarine basin

1. Introduction

Geochemical analyses on sediments have been performed to reconstruct the depositional history of sedimentary basins (e.g., [1–5]). The composition of basin fills provides information about the type of source rocks, tectonic settings of the basin, and weathering conditions in the source areas [6–9]. However, sediments originated from different source rocks are progressively mixed during downstream transportation, generating a compositional contrast between the proximal and distal deposits [10,11]. In the proximal environments close to the basin margins, sediments tend to be deposited separately according to drainage systems, having a wide range of compositional variations. On the other hand, in the distal environments of a basin, sediments may be homogenized by the sediment mixing, having the average composition of surrounding source rocks.

Various tectonic setting discrimination methods and weathering indices have been proposed using the major element composition of sediments and sedimentary rocks. For example, the SiO_2–K_2O/Na_2O and K_2O/Na_2O–SiO_2/Al_2O_3 binary plots by Roser and Korsch [7] have been most widely applied to many provenance studies to discriminate tectonic settings (e.g., [12,13]). In the case of weathering indices, the hemical index of alteration (CIA), defined as molecular $[Al_2O_3/(Al_2O_3 + CaO + Na_2O + K_2O)] \times 100$ [8], has been successfully adopted for more than thirty years (e.g., [14,15]). Previous studies have indicated that these methods may provide inaccurate information due to the statistical limitations of the original dataset, the grain-size effect, and hydraulic sorting [16–18]. However, the influence of the compositional contrast between the proximal and distal deposits on the signatures of tectonic settings and weathering has yet to be tested in a nonmarine basin.

The Cretaceous Neungju Basin, an extensional basin located in the southwestern Korean Peninsula, is an excellent sample site for this study. The Neungju Basin is surrounded by source rocks of various lithology with different formation ages (Figure 1; [19–25]), possibly generating a compositional heterogeneity for the basin fills. Information about depositional environments and sediment transportation has been well constrained by previous studies [26]. Hence, in this study, we aim (1) to investigate variations of the source rocks of the Cretaceous Neungju Basin based on detrital zircon U-Pb ages and (2) to investigate how the compositional heterogeneity of sediments under different depositional environments influence geochemical signatures such as tectonic setting and weathering signatures.

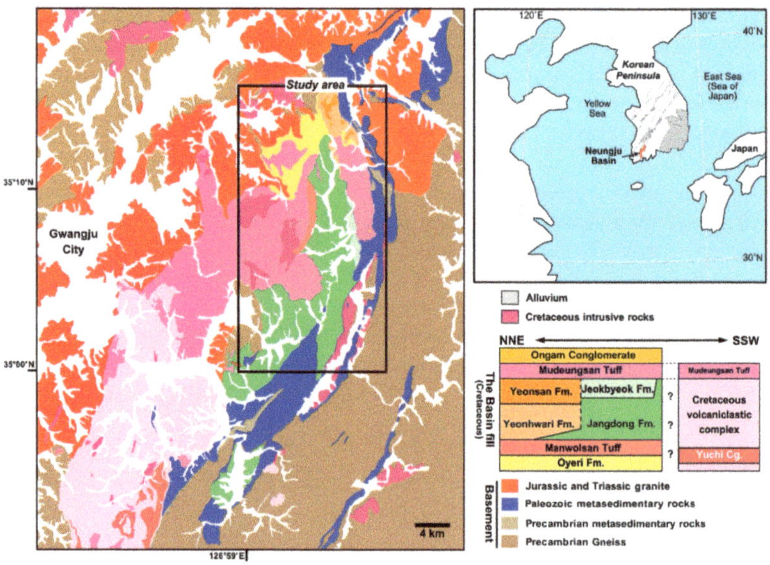

Figure 1. Geological map and stratigraphy of the Neungju Basin (modified after [19–25]). Precambrian granitic gneiss complexes broadly surround the basin. Paleozoic metasedimentary rocks are narrowly distributed along the eastern and southeastern basin margins. Triassic to Jurassic granites are mostly distributed on the northern and western side of the basin. In contrast to the northern part of the basin where epiclastic sediments were mainly deposited, the southern part of the basin was filled with volcaniclastic deposits, forming a volcaniclastic terrane.

2. Geological Settings

During the Cretaceous, NE-SW-trending sinistral strike-slip fault systems were developed along the East Asian continental margin due to oblique subduction of the Izanagi Plate beneath the East Asian continent [27]. Small nonmarine basins were formed under extensional or transtensional tectonic regimes along the NE-SW-trending fault systems in the southern part of the Korean Peninsula [28]. Subduction-related magmatic activity was initiated in central China and migrated eastward to Korea due to the oblique consumption of the ridge [29]. The Cretaceous nonmarine basins were filled with epiclastic and volcaniclastic deposits during extensional faulting and felsic to intermediate volcanic activity.

The Neungju Basin is one of the Cretaceous nonmarine strike-slip basins, located east of Gwangju City (Figure 1). The eastern and western basin margins are bounded by NNE-SSW-trending fault systems. However, the basin margins are unclear because a majority of the margins have been covered by volcaniclastic deposits and volcanic rocks supplied by eruptions during the late-stages of basin development (87–85 Ma; [30]). The basement rocks surrounding the Neungju Basin are mainly composed of three types: (1) Precambrian granitic gneiss complexes, (2) Paleozoic metasedimentary rocks, and (3) Jurassic and Triassic granites (Figure 1). A Precambrian granitic gneiss complex (1935 Ma; [31]) is broadly distributed on the eastern side of the basin, and is also found in areas over the western margin. Paleozoic metasedimentary rocks are narrowly distributed as an elongated belt along the eastern to southeastern basin margins. They are composed of quartzites, carbonates, schists, and coal-bearing shales. Although the depositional ages of the Paleozoic rocks are not clearly constrained, they are considered to be equivalent to the Upper Paleozoic sedimentary succession of Korea (Pyeongan Supergroup; [32]). The Precambrian and Paleozoic rocks have been intruded by Triassic to Jurassic granites (219–176 Ma; [31]).

The Neungju Basin is filled with an approximately 2000 m thick succession of epiclastic and volcaniclastic deposits. The northern part of the basin is mainly filled with epiclastic deposits, occasionally intercalated with volcaniclastic deposits (Figure 1; [20,22]). Whereas, the southern part of the basin is mainly filled with rhyolites, dacitic tuffs, and tuff breccias, subordinately with conglomerates, indicating development of volcaniclastic terrane on the southern part of the basin (Figure 1; [19,23]).

Sedimentological studies on the Neungju Basin have been focused on the northern part (e.g., [26,33,34]). In the northern part of the basin, alluvial fans were developed on the areas proximal to the basin margins, and transitioned to the alluvial plain to sandflat and to the central playa lake (Figure 2; [26]). Paleocurrent data indicate that the sediments derived from the surrounding highlands were transported toward the central playa lake [26]. Reworked volcaniclastic deposits were also supplied to the central playa lake through the alluvial plain and sandflat environments from the southern volcaniclastic terrane [26]. Volcanic activity occasionally intensified during basin development, resulting in the deposition of amalgamated tuff beds tens to hundreds of meters thick (the Manwolsan Tuff, middle Jangdong Tuff, and Mudeungsan Tuff) in the northern part of the basin (Figure 2; [26]).

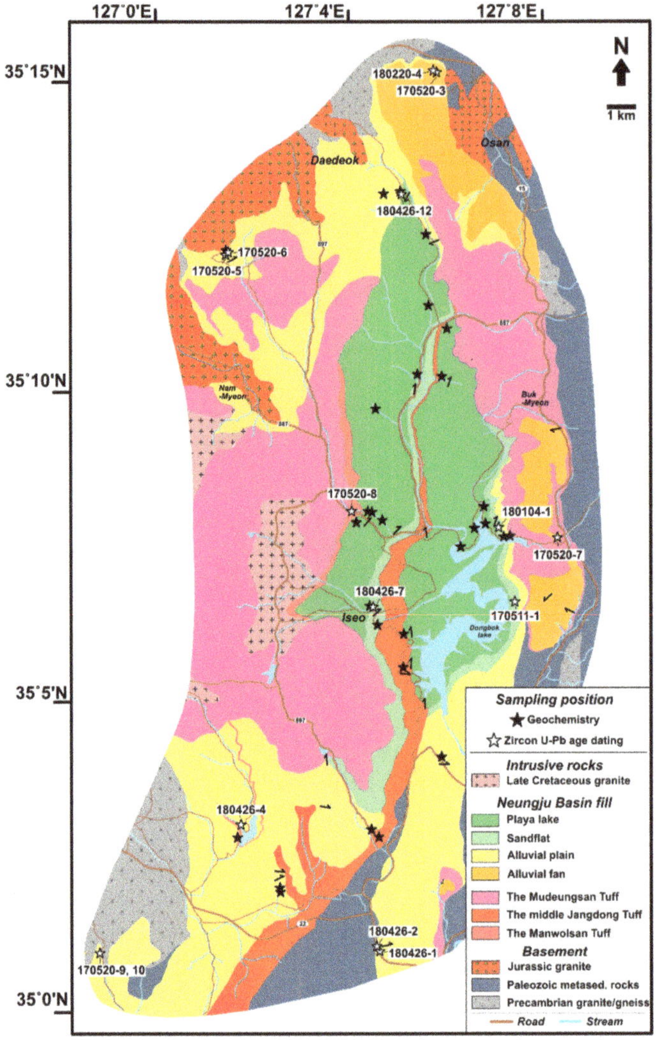

Figure 2. Spatial distribution of depositional environments in present outcrops of the Neungju Basin (modified after [26]). Sixteen mudstone samples were collected from the proximal depositional environments (alluvial plain and sandflat), and fourteen mudstone samples from the distal environment (playa lake). Samples were also collected from conglomerate, sandstone, and tuff and analyzed in terms of zircon U-Pb geochronology to constrain sediment provenance and the depositional age of the basin. Sampling positions for geochemical analysis and zircon U-Pb dating are indicated by closed and open stars, respectively. Black arrows indicate paleocurrent direction [26].

3. Sampling and Analytic Methods

Fifteen conglomerate, sandstone, and tuff samples were collected in the northern part of the Neungju Basin and were subsequently used for zircon U-Pb analysis (Figures 2 and 3). Zircon grains were separated from the collected samples by the conventional heavy mineral separation method. The zircon grains were mounted and polished to expose their cross-sections. Then, scanning-electron-microscope cathodoluminescence (SEM-CL) images of the zircon grains were obtained to identify their internal structures (Figure 4). Zircon U-Pb isotopic ratios were determined using a Nu Plasma II Multi Collector Inductively Coupled Plasma Mass Spectrometer (Nu Instruments, Wrexham, UK) equipped with a 193 nm ArF excimer Laser Ablation system (New Wave Research, Fremont, CA, USA) (LA-MC-ICPMS) at the Korea Basic Science Institute (KBSI). The Nu Plasma II mass spectrometer contains fixed collectors of sixteen Faraday detectors and five ion-counting electron multipliers. The collectors were arrayed to detect U-Pb isotopic ratios simultaneously as follows: ^{202}Hg (IC 4), ^{204}(Hg + Pb) (IC 3), ^{206}Pb (IC 2), ^{207}Pb (IC 1), ^{208}Pb (IC 0), ^{232}Th (high 7), and ^{238}U (high 9). ^{235}U was calculated from the signal at mass 238 using a natural ratio between ^{238}U/^{235}U = 137.88.

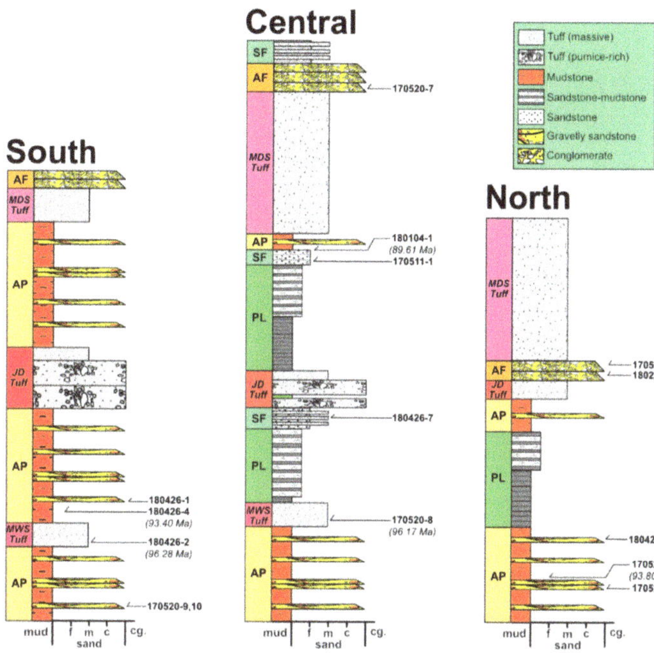

Figure 3. Schematic composite section of the study area (modified from [26]). AF = alluvial fan; AP = alluvial plain; SF = sandflat; PL = playa lake. The amalgamated tuff beds of tens to hundreds of meters thick is represented by MWS tuff = the Manwolsan Tuff; JD Tuff = the middle Jangdong Tuff; MDS Tuff = the Mudeungsan Tuff. Sample locations of the zircon U-Pb analyses are marked.

A mass number of 204 was used as a monitor for common ^{204}Pb after discarding the ^{204}Hg background. All analyses were carried out with a spot size of 15 μm in diameter, a 5 Hz repetition rate, and with an energy density of 3 J/cm^2 using He (650 mL/min.) as the carrier gas. Background intensities, dwell time, and wash out time were measured 30 s, 30 s, and 20 s, respectively. A Time-Resolved Analytical procedure was employed to monitor the measured isotope ratios. Signal intensities for each collector were collected every 0.2 s (integration time). Raw data were corrected for background noise, laser-induced elemental fractionation, mass discrimination, and drift in ion counter gains. U-Pb isotope ratios were calibrated against concordant reference zircons 91500 (1065 Ma; [35]). The reference zircon was measured at the beginning and end of each analytical session and at regular intervals during a session, using protocols adapted from Andersen et al. (2002) [36]. A correlation of signal vs. time was also assumed for the reference zircons. All ages were calculated with 2σ errors, and data reduction was conducted using Iolite 2.5 software package [37,38] and Isoplot 3.71 [39]. Because Mesozoic zircon grains occasionally show inaccurate ^{207}Pb/^{206}Pb ages, ^{206}Pb/^{238}U ages for <1000 Ma and ^{207}Pb/^{206}Pb ages for >1000 Ma were used for age probability curves, with a discordance of <±10%. The discordance

(%) was calculated in two ways: [%discordance = 100*(1 − ($^{206}Pb/^{238}U$ age/$^{207}Pb/^{235}U$ age))] for $^{206}Pb/^{238}U$ ages < 1000 Ma, [%discordance = 100*(1 − ($^{206}Pb/^{238}U$ age/$^{207}Pb/^{206}Pb$ age))] for $^{206}Pb/^{238}U$ ages > 1000 Ma.

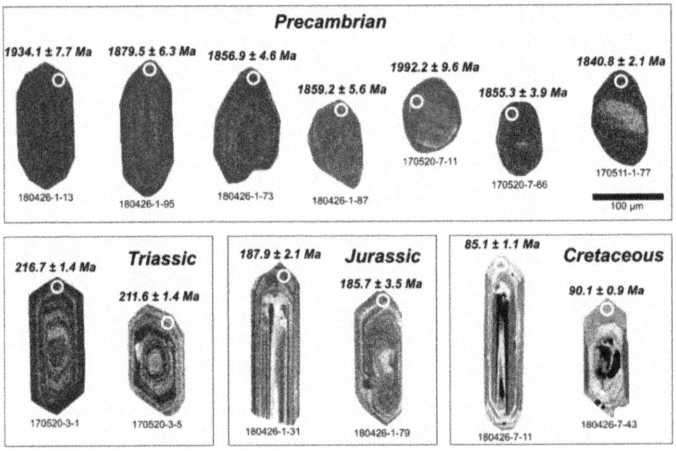

Figure 4. Scanning-electron-microscope (SEM) cathodoluminescence images of the selected zircon grains from the Neungju Basin. Mesozoic zircon grains are characterized by euhedral prismatic shapes, whereas Precambrian zircons tend to show more rounded shapes.

A total of 30 mudstone samples were collected from the proximal and distal deposits in the northern part of the Neungju Basin to investigate compositional contrast in tectonic setting and weathering signatures (Figure 2). Sixteen mudstone samples were collected from the proximal deposits representing alluvial plain (13) and sandflat (3) environments, and 14 mudstone samples were collected from the distal deposits representing the playa lake environment. The collected mudstone samples were powdered in agate mortar after they were washed with clean water and dried. The major element composition of the powdered samples was analyzed at KBSI. A PW2404 X-ray fluorescence spectrometer (XRF, Phillips, Andover, MA, USA) was used on fused glass beads to obtain major element contents, including SiO_2, Al_2O_3, Fe_2O_3 (total Fe), TiO_2, MnO, MgO, K_2O, Na_2O, and P_2O_4. Loss-on-ignition was estimated at 1000 °C for 60 min. The analytic uncertainty is less than 5% for the major element composition.

4. Results

4.1. Detrital Zircon U-Pb Ages

We obtained 1342 concordant (discordance < ± 10%) U-Pb ages from 15 sandstone, conglomerate, and tuff samples ranging from 85.1 ± 1.1 Ma to 3308 ± 3.5 Ma (Table S1). Most of the zircon ages in the conglomerates and sandstones fell into four dominant age groups: Cretaceous (12%; 100–85 Ma), Jurassic (33%; 200–162 Ma), Triassic (10%; 251–202 Ma), and Paleoproterozoic (43%; 2479–1742 Ma). 15 zircons exhibit Archean ages (1.6%; 3308–2510 Ma). A single zircon grain provided a Carboniferous age of 309 ± 17 Ma. The Cretaceous to Triassic zircons are characterized by euhedral prismatic shapes with oscillatory or patchy zonings in their CL images, suggesting a magmatic origin. In contrast, Precambrian zircons show dark and undulatory CL images and their shapes range from euhedral crystal to rounded grains reflecting abrasion during recycling processes (Figure 4). The age distribution of the sandstone samples is presented in the form of a probability density plots (Figure 5; [39]). U-Pb concordia diagrams of the tuffs in the Neungju Basin are shown in Figure 6, while those of the conglomerates and sandstones are shown as supplementary figures (Figures S1 and Figure S2).

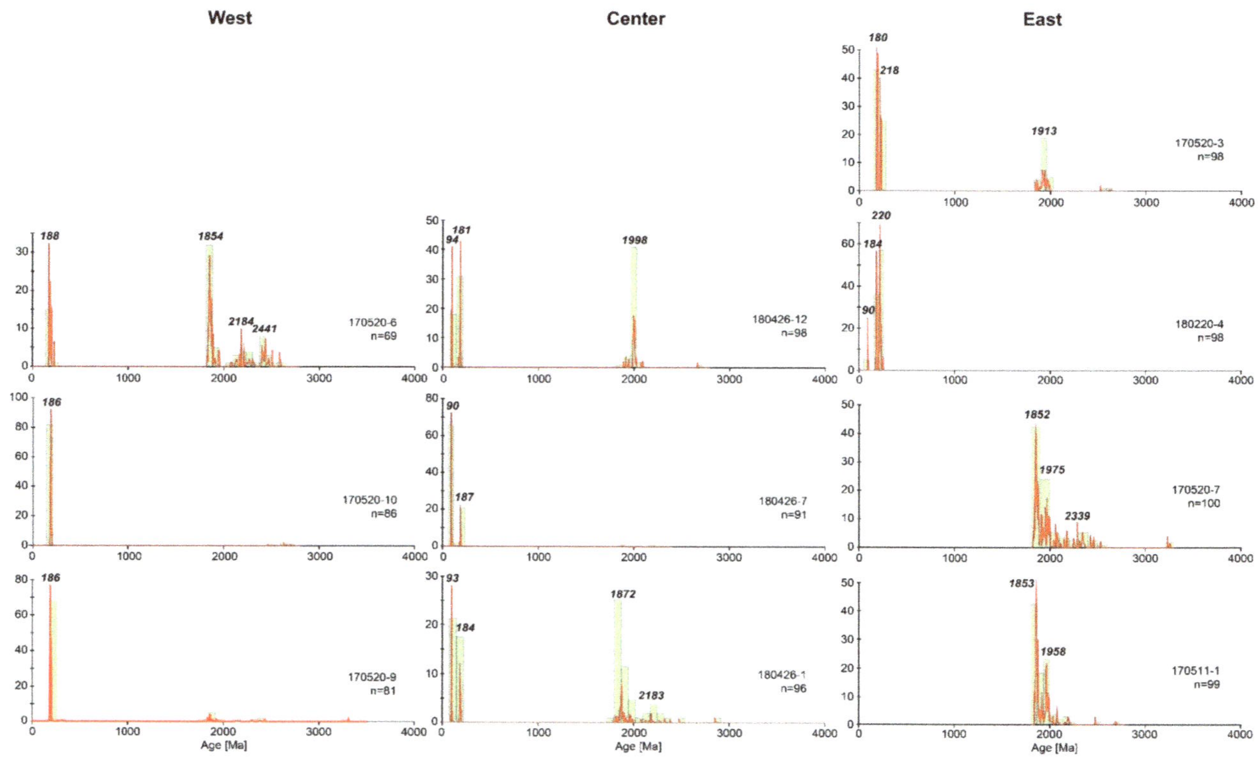

Figure 5. Probability density plots [39] of the detrital zircon grains in the conglomerate and sandstone samples in the Neungju Basin. $^{206}Pb/^{238}U$ ages for <1000 Ma and $^{207}Pb/^{206}Pb$ ages for >1000 Ma were used for age probability curves, with a discordance of <±10%. The plots show differences in their age population according to the sample location. Note that only the age spectra of the central samples and one eastern sample (180220-4) include Cretaceous ages.

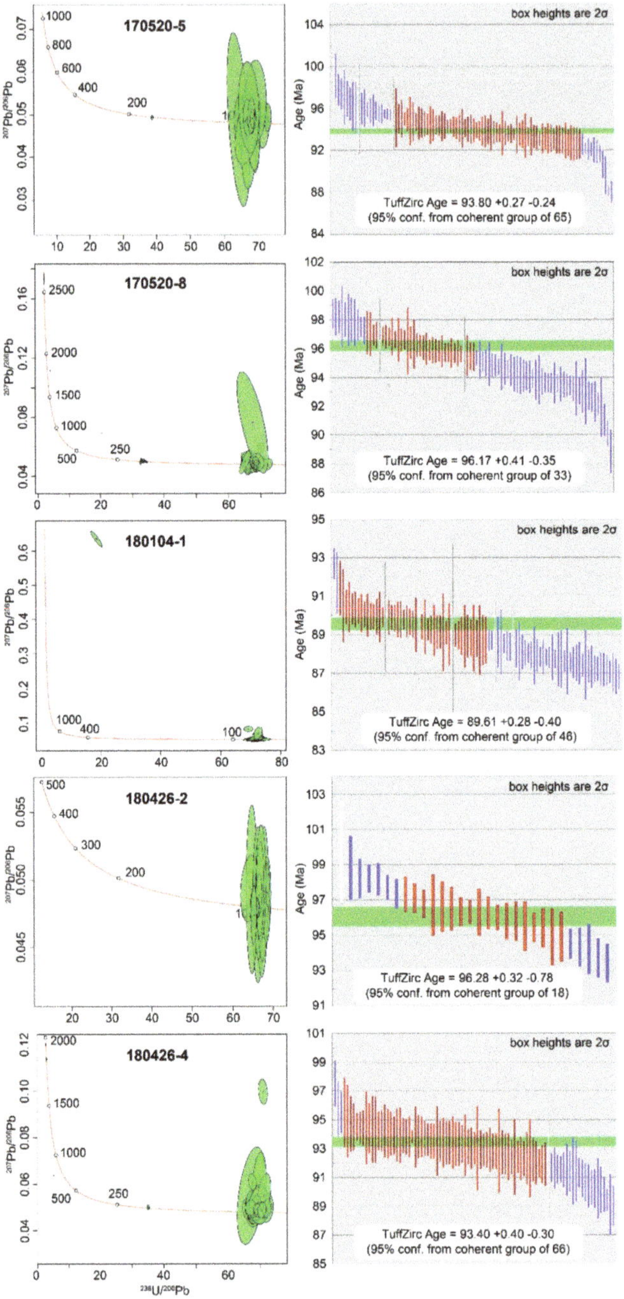

Figure 6. U-Pb Concordia diagrams and weighted mean ages [39] for zircons from tuff samples of the Neungju Basin. Errors are shown at 2-sigma level.

The detrital zircon age populations of the Neungju Basin sediments show variation according to the sample locations (Figures 2–4). The samples near the western and eastern margins have prominent zircon age peaks indicating their derivation from the adjacent basement rocks. In the northwestern part of the study area, sample 170520-6 showed Jurassic to Triassic ages between 173 and 228 Ma with an age peak at 188 Ma, and Paleoproterozoic to Archean ages between 1830 and 2596 Ma with major age peaks at 1854, 2184, and 2441 Ma. The southwestern samples (170520-9 and 170520-10) are exclusively comprised of Jurassic zircons (peaks at 186 Ma) with no Cretaceous but minor Paleoproterozoic grains. In the northeastern part of the basin, the zircon ages of sample 170520-3 mainly consist of Jurassic to Triassic ages between 178 and 226 Ma with an age peak at 180 and 218 Ma and Paleoproterozoic to Archean ages between 1939 and 2620 Ma with an age peak at 1913 Ma. Sample 180220-4 contains Cretaceous zircon grains between 89 and 99 Ma (5 grains), Jurassic to Triassic zircons between 172 and 251 Ma with prominent age peaks at 220 Ma. In the eastern part of the basin, the zircon age spectra of samples 170511-1 and 170520-7 comprise exclusively of Paleoproterozoic zircons with minor Archean. Sample 170511-1 has zircon grains mainly formed in 1905–2690 Ma with age peaks at 1853 and 1958 Ma. The zircon ages of sample 170520-7 consist of Paleoproterozoic zircons between 1846 and 3254 Ma with prominent age peaks at 1852, 1975, and 2339 Ma.

The basin-fill in the central part of the study area is characterized by significant amounts of zircons from the Late Cretaceous. The zircons of the northernmost sample (180426-12) include three main age groups: Late Cretaceous (95–88 Ma), Jurassic (186–162 Ma), and Paleoproterozoic to Archean (2663–1998 Ma) with age peaks at 94 Ma, 181 Ma and, 1998 Ma, respectively. Sample 180426-7 has zircon age populations of Cretaceous ages between 85 and 94 Ma with a prominent age peak at 90 Ma, Jurassic ages between 178 and 191 Ma with an age peak at 187 Ma, minor Paleoproterozoic ages between 1899 and 2215 Ma. The south sample (180426-1) shows zircon ages of Cretaceous (100–88 Ma) with an age peak at 93 Ma, Jurassic ages (193–179 Ma) with an age peak at 184 Ma, and Paleoproterozoic to Archean ages between 1742 and 2855 Ma with age peaks at 1872 and 2183 Ma.

The tuff samples (170520-5, 170520-8, 180426-2, 180426-4, and 180104-1) show their weighted mean ages as ca. 93.80 Ma, 96.17 Ma, 96.28 Ma, 93.40 Ma, and 89.61 Ma, respectively (Table S1 and Figure 6). Each sample contains several older zircon grains ranging in age from Late Archean (2593 Ma) to Jurassic (164 Ma). All the tuff samples have different eruption ages, indicative of frequent magmatic activity in the area adjacent to the Neungju Basin during its deposition.

4.2. Major Element Composition of Mudstones

The major element composition of the mudstone samples is presented in Table S2. In comparison to Post-Archean Average Shale (PAAS; [40]), the mudstone samples from the Neungju Basin are generally enriched in SiO_2, K_2O, Na_2O, and CaO, and depleted in MgO, Fe_2O_3, TiO_2, MnO and P_2O_5 (Figure 7). Although the samples show similar compositional characteristics for most of the major element contents regardless of their depositional environments, they are clearly distinguished in terms of the Na_2O content (Figure 8). The proximal deposits from the alluvial plain to sandflat environments show the Na_2O content ranging from 1.07% to 3.40%, with an average of 1.99%. In contrast, the distal deposits from the central playa lake show the Na_2O ranging from 1.35% to 4.95%, with an average of 3.50%. Each sample group show no distinctive compositional variation spatially. As presented in Figure 8, the proximal and distal deposits collected from different locations of the study area show similar ranges of major element composition, respectively.

New discriminant-function-based diagrams by Verma and Armstrong-Altrin (2013) [41] can discriminate tectonic settings of sedimentary rocks to three types: island or continental arc, continental rift, and collision. In this diagram, the proximal deposits are plotted near the boundary between the arc and collision fields (Figure 9c), possibly reflecting their derivation from source rocks, including volcanic rocks, metamorphic and metasedimentary rocks. The distal deposits are not overlapped with the proximal deposits in the diagram, indicating the compositional difference between the two sample groups. The distal deposits are plotted near the boundary between the collision and rift fields (Figure 9c). Source rocks for these samples are likely to be composed of metamorphic and metasedimentary rocks and felsic plutonic rocks.

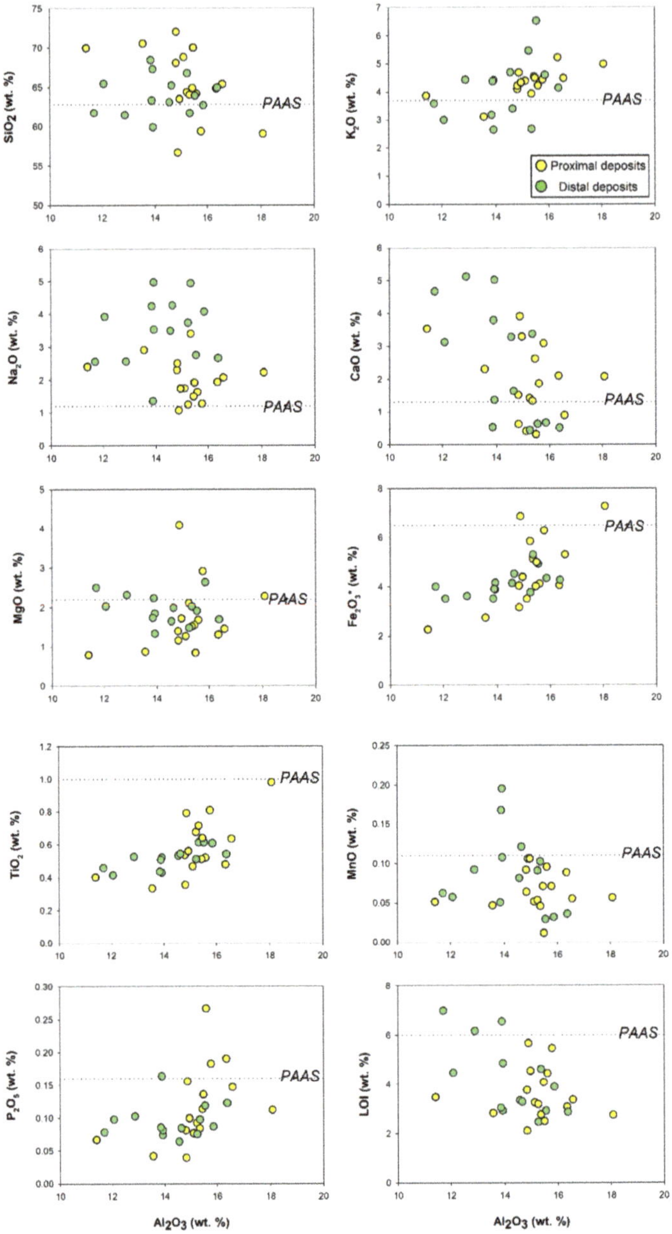

Figure 7. Binary plots of major elements versus Al_2O_3 for the mudstone samples. The proximal and distal deposits are clearly distinguished in terms of the Na_2O content.

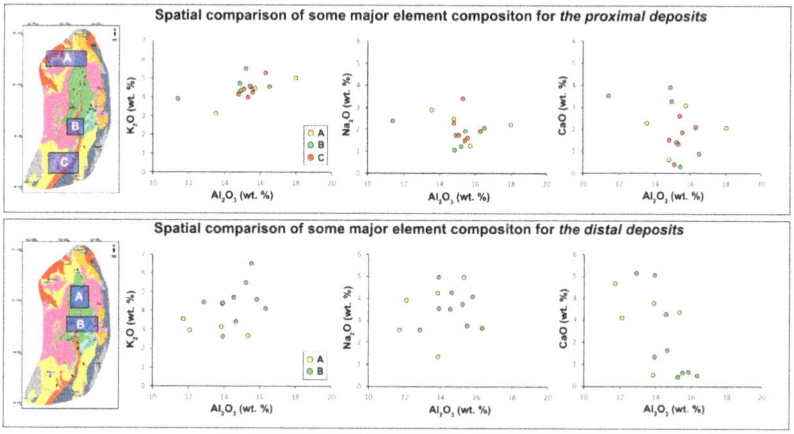

Figure 8. Spatial comparisons of some major element composition for the proximal and distal deposits. Each sample group shows similar range of major element composition, respectively.

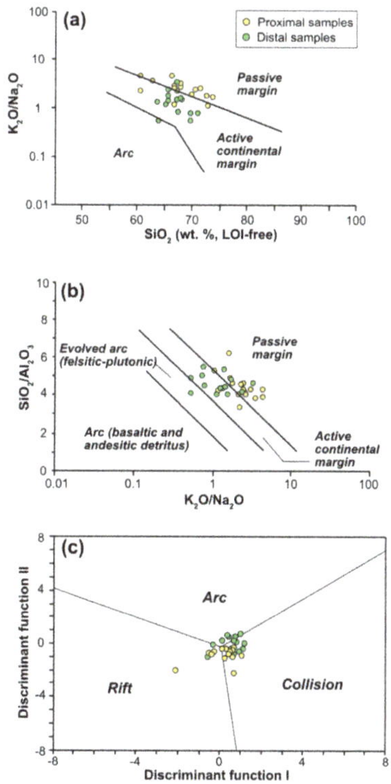

Figure 9. Tectonic setting discrimination diagrams for the mudstone samples. (**a**) K_2O/Na_2O–SiO_2 diagram [7]. (**b**) SiO_2/Al_2O_3–K_2O/Na_2O diagram [7] (**c**) Discriminant function diagram for high-silica [$(SiO_2)_{adj}$ = 63–95%] samples [41], where $(SiO_2)_{adj}$ refers to the modified SiO_2 value after volatile-free recalculation of the major elements to 100%. Two mudstone samples classified as low-silica [$(SiO_2)_{adj}$ = 35–63%] are plotted in the rift field (not presented). The proximal and distal deposits are plotted in these tectonic setting discrimination diagrams separately with minor overlap.

4.3. Weathering Intensities of Mudstones

Various weathering indices have been proposed to assess the degree of chemical weathering quantitatively for sediments and sedimentary rocks. The representative indices for the mudstones in the Neungju Basin are presented in Table 1 and Figure 10.

Table 1. Weathering indices for the mudstone samples from the Neungju Basin.

Depositional Environment	Sample Code	CIA	PIA	CPA	WIP	V	STI
Alluvial plain	180220-1	56	59	88	65	1.37	81.8
Alluvial plain	180426-4	57	61	80	64	1.92	85.1
Alluvial plain	180426-6	63	70	83	64	2.46	83.5
Alluvial plain	180426-12	60	66	78	64	2.37	86.8
Alluvial plain	180426-13	52	53	74	61	1.52	87.4
Alluvial plain	4-26-1050	64	74	84	58	2.92	85.5
Alluvial plain	4-26-1104	59	65	85	60	1.96	84.3
Alluvial plain	180426-11	59	63	83	74	1.78	80.0
Alluvial plain	180427-12-7	56	60	84	71	2.14	84.3
Alluvial plain	4-11-425	56	59	86	63	1.84	84.6
Alluvial plain	4-17-239-r	59	67	88	67	2.13	83.6
Alluvial plain	4-17-239-g	52	53	89	71	1.04	81.8
Alluvial plain	4-46-420	56	58	73	72	1.66	83.3
Sandflat	180426-7	64	74	83	59	3.49	84.5
Sandflat	4-11-1150	44	41	74	66	1.26	87.6
Sandflat	4-17-300	53	54	84	66	1.49	84.3
Playa lake	180426-8	47	47	65	82	0.94	83.5
Playa lake	4-10-239	44	42	65	75	0.89	86.8
Playa lake	4-10-445	41	38	74	73	0.82	86.1
Playa lake	0307-1-905	46	45	72	85	1.24	84.5
Playa lake	1129-1-1	52	52	68	78	1.22	84.8
Playa lake	1129-1-2	51	51	63	75	1.20	86.1
Playa lake	1129-3-2	41	38	71	88	0.96	84.4
Playa lake	180426-9	50	50	86	65	1.26	85.0
Playa lake	180427-14-37	55	60	77	88	2.16	83.8
Playa lake	180427-15-24	54	57	71	86	1.98	84.9
Playa lake	180427-15-39	55	58	70	86	1.43	83.5
Playa lake	4-10-408	55	57	67	72	1.40	86.3
Playa lake	4-11-110	63	69	79	66	2.18	84.0
Playa lake	4-17-340	41	37	75	81	0.91	85.1

The CIA is the most widely used weathering index, monitoring decomposition of feldspar grains and consequent formation of Al-rich clay minerals such as illite and kaolinite [8]. Fresh igneous rocks are found to have a CIA value between 45 and 55, whereas moderately weathered shales such as PAAS yield a value closer to 70 [8]. The CIA values of the mudstone samples range from 44 to 64, with an average of 56, indicating a minor degree of chemical weathering. Although the proximal and distal deposits show some overlap in their CIA values, they can be separated roughly at the CIA value of 54 (Figure 10). The CIA values of the proximal deposits range from 49 to 64, with an average of 59, indicating minor chemical weathering in the source areas. On the other hand, the value of the distal deposits ranges from 44 to 63, with an average of 52, indicating little alterations in the source areas. Weathering indices such as the Plagioclase Index of Alteration (PIA; [42]), the Chemical Proxy of Alteration (CPA; [43]), the Vogt's residual index (V; [44]), and the weathering Index of Parker (WIP; [45]) are also formulated to assess the leaching of mobile elements such as Na_2O, CaO, and MgO during weathering processes. In this study, they provide similar results with those of the CIA value (Figure 9). A low degree of chemical weathering is inferred for the distal deposits, and these samples are distinguished from the more weathered proximal deposits. The Silica–Titania Index (STI; [46]) is the only weathering index that showed a similar degree of chemical weathering for the proximal and distal deposits. The STI values of the proximal and distal deposits range from 80 to 88 (av. = 84) and 83 to 87 (av. = 85), respectively, indicating a slightly weathered condition [46].

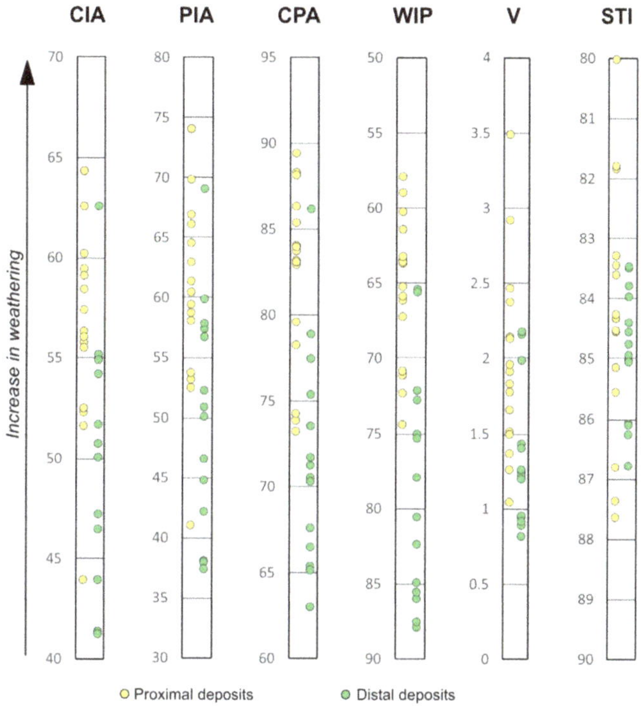

Figure 10. Conventional weathering indices for the mudstone samples. The weathering indices were formulated as molar calculation of some major elements: CIA = $[Al_2O_3/(Al_2O_3 + CaO + Na_2O + K_2O)] \times 100$; PIA = $(Al_2O_3 - K_2O)/(Al_2O_3 + CaO + Na_2O - K_2O) \times 100$; CPA = $Al_2O_3/(Al_2O_3 + Na_2O) \times 100$; WIP = $(2Na_2O/0.35 + MgO/0.9 + 2K_2O/0.25 + CaO/0.7) \times 100$; V = $(Al_2O_3 + K_2O)/(MgO + CaO + Na_2O)$; and STI = $(SiO_2/TiO_2)/[(SiO_2/TiO_2) + (SiO_2/Al_2O_3)] + (Al_2O_3/TiO_2)] \times 100$. In most of the weathering indices, except for the STI index, the distal deposits indicate lower degree of chemical weathering in compared to the proximal deposits.

5. Discussion

5.1. Depositional Age of the Neungju Basin

Our tuff samples show zircon age peaks ranging from ca. 96 to ca. 90 Ma, indicating the depositional age of the Neungju Basin. The sedimentation of the Neungju Basin began before ca. 96 Ma and continued after the Mudeungsan Tuff (Figures 2 and 3) erupted during 87–85 Ma [30]. After the eruption, the deposition occurred in the alluvial fan environment at the limited area in the basin's eastern part.

The frequent alternation of the Late Cretaceous volcaniclastic and epiclastic sediments in the Neungju Basin indicates the tectonic setting of the basin. Arc magmatism prevailed in the southwestern Korean Peninsula during the Late Cretaceous, forming a volcanic complex covering the region [47,48]. Kim et al. (2016) [47] subdivided the Cretaceous igneous rocks in the Korean Peninsula into four groups based on the formation timing: Group I (110–100 Ma), Group II (100–87 Ma), and Group III-IV (86–70 Ma). These groups are also distributed in a NW-SE direction; Group I in the northwest; Group II in the center; Group III-IV in the southeast. Group I-IV suggests the changes in the subduction modes of the paleo-Pacific Plate (Izanagi Plate) as slab rollback, slab steepening and melting, slab shallowing, and slab rollback and steepening, respectively [47]. The Cretaceous zircons in the Neungju Basin suggest that the basin was deposited during Group II magmatism, formed by inland-ward migrating arc magmatism in the Korean Peninsula.

5.2. Sediment Provenance of the Neungju Basin Based on Detrital Zircon Geochronology

Most detrital zircon grains in the Neungju Basin sandstones and conglomerates have age peaks at the Late Cretaceous (94–90 Ma), Jurassic (188–180 Ma), Triassic (218 Ma), and Paleoproterozoic (2441–1852 Ma) (Figure 5). Zircon ages of the tuff samples (96–90 Ma) in this study indicate that the Late Cretaceous detrital zircon grains in the Neungju Basin were derived from the synsedimentary volcanic or volcaniclastic rocks. Previous studies suggested that the tuff strata of the Neungju Basin derived from the southwest of the basin based on increasing southwestward thickness of the strata [20,22]. The Jurassic and Triassic zircon grains might be derived from the adjacent Mesozoic granites surrounding the basin (Figure 1). The Paleoproterozoic zircons could be supplied from Precambrian metamorphic rocks and Paleozoic metasedimentary rocks surrounding the basin. Most Precambrian metamorphic rocks in the Korean Peninsula formed in ca. 1.9–1.8 Ga [49–51], while the zircon grains older than 1.9 Ga are mainly found as detritus in metasedimentary rocks in Korea [52,53]. In addition, the detrital zircon age distribution of the Paleozoic metasedimentary basement rocks consists of Paleoproterozoic (a peak at ca. 1870 Ma) and Archean ages (a peak at ca. 2490 Ma) with no Paleozoic ages [54]. Hence, the source rocks of the Neungju Basin consists of Late Cretaceous volcanic/volcaniclastic rocks, Jurassic and Triassic granites, Paleozoic metasedimentary rocks, and Precambrian metamorphic rocks.

The detrital zircon U-Pb age data of the Neungju Basin show that sediment detritus in the proximal deposits close to the basin margin derived from nearby basement rocks with no or little Late Cretaceous volcanic material, in spite of their intercalation with Late Cretaceous tuff beds. The samples (170520-7 and 170511-1) near the eastern margin include zircon age clusters of Paleoproterozoic to Archean ages only, suggesting sediment derivation from the Precambrian metamorphic rocks distributed east of the basin. In contrast, the samples collected from the areas near the northeastern (170520-3 and 180220-4) and western margins (170520-6 and 170520-9) show zircon age clusters of Mesozoic ages between Jurassic and Triassic as well as Paleoproterozoic and Archean ages, although sample 180220-4 has five grains of Cretaceous aged zircons. Most of the sediment detritus were supplied from the Jurassic and Triassic granites and Precambrian metamorphic rocks broadly distributed to the north and west of the basin.

In contrast to the sandstone and conglomerate samples collected from the areas close to the basin margins, those collected from the southern and central parts of the study area (180426-1, 180526-7, and 180426-12) show distinct Late Cretaceous peak ages with variable amounts of Jurassic and Paleoproterozoic to Archean ages. A previous sedimentological study in the Neungju Basin reports that a playa lake had been developed in the central parts of the study area [26]. Sediment detritus derived from the respective source rocks were possibly mixed in the central areas of the playa lake and sandflat environments, resulting in more mixed zircon age peaks than those in the samples from proximal areas. The Cretaceous zircons were possibly reworked from the volcaniclastic terrane located at the south of the basin and transported northward to the central playa lake (e.g., [55,56]). Paleocurrent data support the derivation of the Cretaceous zircon grains from the south [26]. Direct derivation of the Cretaceous zircons by syndepositional volcanism is less reliable because tuff beds were not observed in the outcrop sections where the sandstone and conglomerate samples were collected, and their stratigraphic positions correspond to inter-eruption periods of the Neungju Basin [26].

5.3. Contrasting Geochemical Signatures of Tectonic Settings and Weathering Intensities in a Nonmarine Basin

The compositional contrast between the proximal and distal deposits can be explained based on the data collected from zircon age dating. The zircon age data indicate that Jurassic to Triassic granites and Precambrian metamorphic rocks as the primary sources in the marginal areas where alluvial drainage systems transported sediment detritus from the adjacent source areas. These source rocks have similar characteristics in major element composition [57]. Thus, the proximal deposits show similar major element composition regardless of variations in the relative contribution of Jurassic to Triassic granites and Precambrian metamorphic rocks. In contrast, the zircon age data indicate that considerable amounts of sediments were transported from the southern Cretaceous volcaniclastic terrane to the central playa lake. Cretaceous volcanic activity in the Neungju Basin areas was generated from calk-alkaline magma, resulting in the deposition of felsic pyroclastic, andesitic to dacitic rocks, and rhyolite [58–60]. Although various types of volcanic rocks were formed, they are characterized by abundant plagioclase content. Thus, it would be natural that the distal deposits show high Na_2O content than the proximal deposits.

Differences in weathering conditions in the source areas might have enhanced the compositional contrast between the proximal and distal deposits. Fossil and sedimentological evidence revealed in the Cretaceous Gyeongsang Basin, South Korea, indicated arid to semi-arid conditions in the Korean Peninsula during the Late Cretaceous (e.g., [34,61,62]). Besides, Lee and Lee (2003) [63] reported that the Hayang Group mudstones in the Gyeongsang Basin, correlate with the Neungju Basin fills, resulting in CIA values ranging from 54 to 67, with an average of 61. The CIA values in the Neungju Basin are consistent with the previous results, indicative of the weathering condition under the equilibrium state in the source areas. This results in a minor degree of compositional alteration for the proximal deposits due to loss of the mobile elements (e.g., Na_2O and CaO) during weathering processes. The weathering intensities of the distal deposits, however, are much lower than expected. To assess the weathering intensity properly, chemical loss of the mobile elements and the soil profile's physical erosion should occur at a stable rate in the source areas [64]. Active tectonism and storms may enhance the physical erosion of fresh bedrocks and less altered lower horizons of the soil profile, resulting in a lower estimate of weathering intensities (e.g., [65,66]). In the Neungju Basin, volcaniclastic terrane with high and unstable topography might have been formed in the southern part of the basin in a short time due to intense volcanic activities. After the rapid aggradation of the large volume of volcaniclastic sediments in the volcaniclastic terrane, the unconsolidated volcanic/volcaniclastic materials were easily eroded, reworked, and transported as debris-flows to the central playa lake far from the volcaniclastic terrane during the inter-eruption period (e.g., [67,68]). Thus, sediment detritus originated from the southern volcaniclastic terrane might be difficult to experience compositional alteration due to chemical weathering in the source areas, resulting in the distal deposits' low weathering intensities.

In this study, the differences in source rock composition cause the compositional contrast between the proximal and distal deposits, resulting in different tectonic setting signatures based on their major element composition. The central playa lake in the Neungju Basin acted as a mixing bowl for the sediments derived from various source rocks. In addition to sediments originated from Precambrian gneiss and Mesozoic granites, Na_2O-rich sediments from the southern Cretaceous volcaniclastic terrane were mixed in the playa lake, resulting in the active continental margin signatures for the distal deposits. On the contrary, in the basin's marginal areas, sediment mixing from different source rocks was inhibited. The proximal deposits reflect only the composition of adjacent source rocks, providing inadequate information about the tectonic setting of the basin. The compositional alteration due to chemical weathering might also influence the shift from active continental margin to the passive margin signature in the tectonic setting discrimination diagrams due to the loss of mobile elements. Overall, this study implies that geochemical indicators should be used with caution for nonmarine sediment because of the compositional heterogeneity of the basin-fill. The tectonic setting of a basin can be inferred by the composition of sediments deposited in the distal environments where sediments from various source rocks are mixed.

6. Conclusions

To investigate the compositional contrast of sediments in a nonmarine basin and its influence on the tectonic setting and weathering signatures, the nonmarine Neungju Basin was studied through the analysis of mudstone geochemistry supported by zircon U-Pb age dating. A total of 1342 zircon age data ranging from 3308 to 85 Ma mainly consists of Cretaceous (100-85 Ma), Jurassic (200–162 Ma), Triassic (251–202 Ma), Paleoproterozoic (2479–1742 Ma), and Archean (3308–2510 Ma). The results show that the evolution of the Neungju Basin was controlled by changes in subduction modes of the Paleo-Pacific Plate.

The Neungju Basin sediment was supplied from four different types of source rocks surrounding the basin; Precambrian metamorphic, Paleozoic metasedimentary, and Cretaceous volcanic rocks, as well as Jurassic and Triassic granites. These source rocks supplied sediments into the basin from all directions, making different zircon age distributions according to the depositional environments. The sediments in the proximal environments (alluvial fan to sandflat) close to the basin margins were derived from the adjacent basement rocks. On the contrary, in the distal environment (the central playa lake), sediments transported from the proximal deposits were mixed.

Mudstones deposited in the proximal deposits reflect the composition of nearby source rocks, whereas mudstones deposited in the distal deposits reflect the mixed characteristics of the source rocks. Due to the exclusive input of volcaniclastic sediment into the distal deposits, tectonic setting signatures differ significantly between the proximal and distal deposits. The proximal deposits show passive margin signatures, reflecting their derivation from metamorphic and granitic rocks, whereas the distal deposits show active continental margin signatures reflecting the overall source rock signatures. In the case of weathering, the proximal deposits indicate slightly higher weathering intensities than the distal deposits. Reworking of fresh volcanic detritus from the southern volcaniclastic terrane might have lowered the weathering intensities of the distal deposits.

Overall, this study implies that compositional analysis for reconstructing tectonic settings and weathering conditions should be carefully performed with support from detailed provenance and paleoenvironmental analyses. A nonmarine basin may have strong spatial contrast in sediment composition due to preferential mixing of source rocks, and a specific location and depositional environment may not reflect the overall characteristics of the basin.

Supplementary Materials: The following are available online at http://www.mdpi.com/2075-163X/10/11/1023/s1, Figure S1: U-Pb Concordia diagrams for zircons from conglomerate and sandstone samples of the Neungju Basin (continued to Figure S2). Errors are shown at 2-sigma level. Figure S2: U-Pb Concordia diagrams for zircons from conglomerate and sandstone samples of the Neungju Basin (continued from Figure S1). Errors are shown at 2-sigma level. Table S1: Zircon U-Pb isotopic ratios of the conglomerate, sandstone, and tuff samples from the Neungju Basin. Table S2: Major element composition of the mudstone samples from the Neungju Basin.

Author Contributions: Conceptualization, H.L. and T.C.; methodology, M.G.K., T.C., and H.L.; investigation, M.G.K., H.L., S.S., and H.C.; writing—original draft preparation, H.L.; writing—review and editing, T.C.; project administration, Y.R. and J.-S.K.; funding acquisition, M.H. Detrital zircon U-Pb age data of this work represents part of M.G.K.'s Masters Research performed at the Chosun University. All authors have read and agreed to the published version of the manuscript.

Funding: This research was funded by the National Research Foundation of Korea (NRF), grant number 2020R111A3072160, research grant from Damyang and Hwasun Counties, Republic of Korea to T.C. The APC was funded by the National Research Foundation of Korea (NRF), grant number 2020R111A3072160.

Acknowledgments: We thank the Editors Wilfried Winkler and Albrecht von Quadt and anonymous journal reviewers for their thoughtful comments.

Conflicts of Interest: The authors declare no conflict of interest. The funders had no role in the design of the study; in the collection, analyses, or interpretation of data; in the writing of the manuscript, or in the decision to publish the results.

References

1. Ohta, T. Geochemistry of Jurassic to earliest Cretaceous deposits in the Nagato Basin, SW Japan: Implication of factor analysis to sorting effects and provenance signatures. *Sediment. Geol.* **2004**, *171*, 159–180. [CrossRef]
2. Yang, J.; Du, Y.; Cawood, P.A.; Xu, Y. Modal and Geochemical Compositions of the Lower Silurian Clastic Rocks In North Qilian, NW China: Implications For Provenance, Chemical Weathering, and Tectonic Setting. *J. Sediment. Res.* **2012**, *82*, 92–103. [CrossRef]

3. Gu, X.X.; Liu, J.M.; Zheng, M.H.; Tang, J.X.; Qi, L. Provenance and tectonic setting of the Proterozoic turbidites in Hunan, South China: Geochemical evidence. *J. Sediment. Res.* **2002**, *72*, 393–407. [CrossRef]
4. Lee, J.I.; Lee, Y.I. Provenance of the lower Cretaceous Hayang Group, Gyeongsang Basin, southeastern Korea: Implications for continental-arc volcanism. *J. Sediment. Res.* **2000**, *70*, 151–158. [CrossRef]
5. Joo, Y.J.; Lee, Y.I.; Hisada, K.I. Provenance of Jurassic accretionary complex: Mino terrane, inner zone of south-west Japan—Implications for palaeogeography of eastern Asia. *Sedimentology* **2007**, *54*, 515–543. [CrossRef]
6. Roser, B.P.; Korsch, R.J. Provenance signatures of sandstone–mudstone suites determined using discriminant function analysis of major element data. *Chem. Geol.* **1988**, *67*, 119–139. [CrossRef]
7. Roser, B.P.; Korsch, R.J. Determination of tectonic settings of sandstone–mudstone suites using SiO_2 content and K_2O/Na_2O ratio. *J. Geol.* **1986**, *94*, 635–650. [CrossRef]
8. Nesbitt, H.W.; Young, G.M. Early Proterozoic climates and plate motions inferred from major element chemistry of lutites. *Nature* **1982**, *299*, 715–717. [CrossRef]
9. McLennan, S.M.; Hemming, S.; McDaniel, D.K.; Hanson, G.N. Geochemical approaches to sedimentation, provenance, and tectonics. *Geol. Soc. Am.* **1993**, *284*, 21–40.
10. Critelli, S.; Le Pera, E.; Ingersoll, R.V. The effect of source lithology, transport, deposition and sampling scale on the composition of southern California sand. *Sedimentology* **1997**, *44*, 653–671. [CrossRef]
11. Ingersoll, R.V. Actualistic sandstone petrofacies: Discriminating modern and ancient source rocks. *Geology* **1990**, *18*, 733–736. [CrossRef]
12. Huntsman-Mapila, P.; Tiercelin, J.J.; Benoit, M.; Ringrose, S.; Diskin, S.; Cotten, J.; Hémond, C. Sediment geochemistry and tectonic setting: Application of discrimination diagrams to early stages of intracontinental rift evolution, with examples from the Okavango and Southern Tanganyika rift basins. *J. Afr. Earth Sci.* **2009**, *53*, 33–44. [CrossRef]
13. Sarki Yandoka, B.M.; Abdullah, W.H.; Abubakar, M.B.; Hakimi, M.H.; Adegoke, A.K. Geochemical characterisation of Early Cretaceous lacustrine sediments of Bima Formation, Yola Sub-basin, Northern Benue Trough, NE Nigeria: Organic matter input, preservation, paleoenvironment and palaeoclimatic conditions. *Mar. Pet. Geol.* **2015**, *61*, 82–94. [CrossRef]
14. Kotlia, B.S.; Joshi, L.M. Late Holocene climatic changes in Garhwal Himalaya. *Curr. Sci.* **2013**, *104*, 911–919.
15. Rashid, S.A.; Ganai, J.A.; Masoodi, A.; Khan, F.A. Major and trace element geochemistry of lake sediments, India: Implications for weathering and climate control. *Arab. J. Geosci.* **2014**, *8*, 5677–5684. [CrossRef]
16. Ryan, K.M.; Williams, D.M. Testing the reliability of discrimination diagrams for determining the tectonic depositional environment of ancient sedimentary basins. *Chem. Geol.* **2007**, *242*, 103–125. [CrossRef]
17. Nesbitt, H.W.; Young, G.M.; McLennan, S.M.; Keays, R.R. Effects of chemical weathering and sorting on the petrogenesis of siliciclastic sediments, with implications for provenance studies. *J. Geol.* **1996**, *104*, 525–542. [CrossRef]
18. Ohta, T. Measuring and adjusting the weathering and hydraulic sorting effects for rigorous provenance analysis of sedimentary rocks: A case study from the Jurassic Ashikita Group, south-west Japan. *Sedimentology* **2008**, *55*, 1687–1701. [CrossRef]
19. Cheong, C.H.; Kim, G.S. *The Geological Map of Neungju Sheet (1:50,000) and Explanatory Text*; Geological Survey of Korea: Daejeon, Korea, 1966; p. 42. (In Korean and English)
20. Kim, B.K.; Park, B.K. *The Geological Map of Dongbok Sheet (1:50,000) and Explanatory Text*; Geological Survey of Korea: Daejeon, Korea, 1966; p. 33. (In Korean and English)
21. Kim, K.B.; Lee, B.J.; Hwang, S.K. *The Geological Map of Kwangju Sheet (1:50,000) and Geological Report*; Korea Institute of Energy and Resources: Daejeon, Korea, 1990; p. 20. (In Korean and English)
22. Son, C.M.; Kim, S.J. *The Geological Map of Changpyeong Sheet (1:50,000) and Explanatory Text*; Geological Survey of Korea: Daejeon, Korea, 1966; p. 30. (In Korean and English)
23. Choi, Y.K.; Yoon, H.D. *The Geological Map of Jangheung Sheet (1:50,000) and Explanatory Text*; Geological Survey of Korea: Daejeon, Korea, 1968; p. 18. (In Korean and English)
24. Hong, S.H.; Yun, U. *The Geological Map of Songjong Sheet (1:50,000) and Geological Report*; Korea Institute of Energy and Resources: Daejeon, Korea, 1986; p. 23. (In Korean and English)
25. Lee, S.M.; Kim, H.S. *The Geological Map of Bongnae Sheet (1:50,000) and Explanatory Text*; Geological Survey of Korea: Daejeon, Korea, 1966; p. 44. (in Korean and English)

26. Lee, H.; Sim, M.S.; Choi, T. Stratigraphic evolution of the northern part of the Cretaceous Neungju Basin, South Korea. *Geosci. J.* **2019**, *23*, 849–865. [CrossRef]
27. Chun, S.S.; Chough, S.K. Tectonic history of Cretaceous sedimentary basins in the southwestern Korean Peninsula and Yellow Sea. In *Sedimentary Basins in the Korean Peninsula and Adjacent Seas*; Chough, S.K., Ed.; Korean Sedimentology Research Group: Seoul, Korea, 1992; pp. 60–76, Special Publication 1.
28. Chough, S.K.; Kwon, S.T.; Ree, J.H.; Choi, D.K. Tectonic and sedimentary evolution of the Korea peninsula: A review and new view. *Earth Sci. Rev.* **2000**, *52*, 175–235. [CrossRef]
29. Wu, F.Y.; Han, R.H.; Yang, J.H.; Wilde, S.A.; Zhai, M.G.; Park, S.C. Initial constraints on the timing of granitic magmatism in North Korea using U–Pb zircon geochronology. *Chem. Geol.* **2007**, *238*, 232–248. [CrossRef]
30. Ahn, K.S.; Huh, M.; Son, J.M. Geological history and landscape of Mudeungsan National Park. *J. Geol. Soc. Korea* **2014**, *50*, 91–105. [CrossRef]
31. Turek, A.; Kim, C.B. U-Pb zircon ages of Mesozoic plutons in the Damyang–Geochang area, Ryongnam massif, Korea. *Geochem. J.* **1995**, *29*, 243–258. [CrossRef]
32. Seo, H.G.; Paik, S.H. *Researches on Coal Resources (II), North Part of the Iyang Area, Honam Coal Field, 83-Coal Resources-3-16*; Geological Survey of Korea: Daejeon, Korea, 1984. (In Korean)
33. You, H.S.; Moon, B.C.; Koh, Y.K.; Kim, H.G.; Lyu, S.O. A study on the paleodepositional environment of Cretaceous sedimentary rocks—Northeastern area of Chonnam. *J. Korean Earth Sci. Soc.* **1998**, *19*, 318–328.
34. Paik, I.S.; Huh, M.; So, Y.H.; Lee, J.E.; Kim, H.J. Traces of evaporites in Upper Cretaceous lacustrine deposits of Korea: Origin and paleoenvironmental implications. *J. Asian Earth Sci.* **2007**, *30*, 93–107. [CrossRef]
35. Wiedenbeck, M.; Alle, P.; Corfu, F.; Griffin, W.L.; Meier, M.; Oberli, F.; von Quadt, A.; Roddick, J.C.; Spiegel, W. Three natural zircon standards for U-Th-Pb, Lu-Hf, trace element and REE analyses. *Geostand. Geoanal. Res.* **1995**, *19*, 1–23. [CrossRef]
36. Andersen, T. Correction of common lead in U-Pb analyses that do not report ^{204}Pb. *Chem. Geol.* **2002**, *192*, 59–79. [CrossRef]
37. Paton, C.; Hellstrom, J.; Paul, B.; Woodhead, J.D.; Hergt, J. Iolite: Freeware for the visualization and processing of mass spectrometric data. *J. Anal. Atom. Spectrom.* **2011**, *26*, 2508–2518. [CrossRef]
38. Paton, C.; Woodhead, J.D.; Hellstrom, J.C.; Hergt, J.M.; Greig, A.; Maas, R. Improved laser ablation U-Pb zircon geochronology through robust downhole fractionation correction. *Geochem. Geophy. Geosyst.* **2010**, *11*, Q0AA06. [CrossRef]
39. Ludwig, K.R. *User's Manual for Isoplot 3.6: A Geochronological Toolkit for Microsoft Excel*; Berkeley Geochronology Center Special Publication: Berkeley, CA, USA, 2008.
40. Taylor, S.R.; McLennan, S.M. *The Continental Crust: Its Composition and Evolution*; Blackwell Scientific Publications: Oxford, UK, 1985; p. 315.
41. Verma, S.P.; Armstrong-Altrin, J.S. New multi-dimensional diagrams for tectonic discrimination of siliciclastic sediments and their application to Precambrian basins. *Chem. Geol.* **2013**, *355*, 117–133. [CrossRef]
42. Fedo, C.M.; Nesbitt, H.W.; Young, G.M. Unraveling the effects of potassium metasomatism in sedimentary rocks and paleosols, with implications for paleoweathering conditions and provenance. *Geology* **1995**, *23*, 921–924. [CrossRef]
43. Buggle, B.; Glaser, B.; Hambach, U.; Gerasimenko, N.; Marković, S. An evaluation of geochemical weathering indices in loess–paleosol studies. *Quat. Int.* **2011**, *240*, 12–21. [CrossRef]
44. Vogt, T. Sulitjelmafeltets geologi og petrografi. *Nor. Geol. Unders.* **1927**, *121*, 1–560. (In Norwegian with English abstract)
45. Parker, A. An index of weathering for silicate rocks. *Geol. Mag.* **1970**, *107*, 501–504. [CrossRef]
46. Jayawardena, U.D.S.; Izawa, E. A new chemical index of weathering for metamorphic silicate rocks in tropical regions: A study from Sri Lanka. *Eng. Geol.* **1994**, *36*, 303–310. [CrossRef]
47. Kim, S.W.; Kwon, S.; Park, S.I.; Lee, C.; Cho, D.L.; Lee, H.J.; Ko, K.; Kim, S.J. SHRIMP U–Pb dating and geochemistry of the Cretaceous plutonic rocks in the Korean Peninsula: A new tectonic model of the Cretaceous Korean Peninsula. *Lithos* **2016**, *262*, 88–106. [CrossRef]
48. Kwon, C.W.; Ko, K.; Gihm, Y.S.; Koh, H.J.; Kim, H. Late Cretaceous volcanic arc system in southwest Korea: Distribution, lithology, age, and tectonic implications. *Cretac. Res.* **2017**, *75*, 125–140. [CrossRef]
49. Gastil, R.G. The distribution of mineral dates in time and space. *Am. J. Sci.* **1960**, *258*, 1–35. [CrossRef]
50. Condie, K.C. Episodic continental growth and supercontinents: A mantle avalanche connection? *Earth Planet. Sci. Lett.* **1998**, *163*, 97–108. [CrossRef]

51. Kim, J.; Cho, M. Low-pressure metamorphism and leucogranite magmatism, northeastern Yeongnam Massif, Korea: Implication for Paleoproterozoic crustal evolution. *Precambrian Res.* **2003**, *122*, 235–251. [CrossRef]
52. Lee, Y.I.; Choi, T.; Orihashi, Y. LA-ICP-MS zircon U-Pb ages of the Precambrian Yuli Group. *J. Geol. Soc. Korea* **2011**, *47*, 81–87. (In Korean with English abstract)
53. Choi, T.; Lee, Y.I.; Orihashi, Y. Crustal growth history of the Korean Peninsula: Constraints from detrital zircon ages in modern river sediments. *Geosci. Front.* **2016**, *7*, 704–714.
54. Park, Y.J. SHRIMP U-Pb Age Distribution of the Detrital Zircons from the Sedimentary and Metasedimentary Rocks of the Southwestern Part of the Korean Peninsula. Master's Thesis, Pukyong National University, Busan, Korea, 2013. (In Korean with English abstract)
55. Nakayama, K.; Yoshikawa, S. Depositional processes of primary to reworked volcaniclastics on an alluvial plain; an example from the Lower Pliocene Ohta tephra bed of the Tokai Group, central Japan. *Sediment. Geol.* **1997**, *107*, 211–229.
56. Ashley, G.M.; Hay, R.L. Sedimentation patterns in a Plio-Pleistocene volcaniclastic rift-platform basin, Olduvai Gorge, Tanzania. In *Sedimentation in Continental Rifts*; Renaut, R.W., Ashley, G.M., Eds.; Society for Sedimentary Geology (SEPM): Broken Arrow, OK, USA, 2002. [CrossRef]
57. Lee, S.G.; Kim, D.Y. Geochemical composition of the continental crust in Korean Peninsula. *J. Pet. Soc. Korea* **2012**, *21*, 113–128. (In Korean with English abstract)
58. Koh, J.S.; Yun, S.H.; Kim, Y.L. Petrology of the Mt. Dungjuribong Volcanic Complex, Gurye-gun, southwest of Ryeongnam Massif. *J. Pet. Soc. Korea* **2009**, *18*, 349–370. (In Korean and English abstract)
59. Kim, Y.L.; Koh, J.S.; Lee, J.H.; Yun, S.H. Petrological study on the Cretaceous volcanic rocks in the southwest Ryeongnam Massif: (1) the Mt. Moonyu volcanic mass, Seungju-gun. *J. Pet. Soc. Korea* **2008**, *17*, 57–82. (In Korean with English abstract)
60. Jung, W.; Ki, Y.; Huh, M. A petrological study of the Mudeungsan Tuff focused on Cheonwangbong and Anyangsan. *J. Pet. Soc. Korea* **2014**, *23*, 325–336. (In Korean and English abstract)
61. Huh, M.; Paik, I.S.; Lockley, M.G.; Hwang, K.G.; Kim, B.S.; Kwak, S.K. Well-preserved theropod tracks from the Upper Cretaceous of Hwasun County, southwestern South Korea, and their paleobiological implications. *Cretac. Res.* **2006**, *27*, 123–138. [CrossRef]
62. Paik, I.S.; Lee, Y.I.; Kim, H.J.; Huh, M. Time, Space and Structure on the Korea Cretaceous Dinosaur Coast: Cretaceous Stratigraphy, Geochronology, and Paleoenvironments. *Ichnos* **2012**, *19*, 6–16. [CrossRef]
63. Lee, J.I.; Lee, Y.I. Geochemistry and provenance of Lower Cretaceous Sindong and Hayang mudrocks, Gyeongsang Basin, southeastern Korea. *Geosci. J.* **2003**, *7*, 107–122.
64. Nesbitt, H.W.; Fedo, C.M.; Young, G.M. Quartz and feldspar stability, steady and non-steady-state weathering, and petrogenesis of siliciclastic sands and muds. *J. Geol.* **1997**, *105*, 173–191. [CrossRef]
65. Lee, Y.I. Geochemistry of shales of the Upper Cretaceous Hayang Group, SE Korea: Implications for provenance and source weathering at an active continental margin. *Sediment. Geol.* **2009**, *215*, 1–12. [CrossRef]
66. Joo, Y.J.; Elwood Madden, M.E.; Soreghan, G.S. Anomalously low chemical weathering in fluvial sediment of a tropical watershed (Puerto Rico). *Geology* **2018**, *46*, 691–694. [CrossRef]
67. Smith, G.A. Facies sequences and geometries in continental volcaniclastic sediments. In *Sedimentation in Volcanic Settings*; Fisher, R.V., Smith, G.A., Eds.; Society for Sedimentary Geology (SEPM): Broken Arrow, OK, USA, 1991.
68. Smith, G.A. Sedimentology of proximal and distal volcaniclastics dispersed across an active foldbelt: Ellensburg Formation (late Miocene), central Washington. *Sedimentology* **1988**, *35*, 953–977. [CrossRef]

minerals

MDPI

Article

Zircon U–Pb Geochronology, Geochemistry and Geological Significance of the Anisian Alkaline Basalts in Gejiu District, Yunnan Province

Zhi Shang and Yongqing Chen *

School of Earth Sciences and Resources, China University of Geosciences, Beijing 100083, China; shangzhi6609@163.com

* Correspondence: yqchen@cugb.edu.cn

Received: 6 September 2020; Accepted: 16 November 2020; Published: 18 November 2020

Abstract: The Gejiu Anisian alkaline basalts (GAAB), distributed in the southern part of the Emeishan large igneous province (ELIP), are crucial to understand the tectonomagmatic activity during the Triassic. Geochronological, geochemical, and Sr-Nd-Pb isotopic analyses were systematically applied to explore the origin, petrogenesis, and tectonic setting of the GAAB, and how they relate to the ELIP. Zircon U-Pb dating set the eruption date at 244 Ma. Most of the samples belonged to alkaline basalts and had high TiO_2 (2.14–3.23 wt.%) and MgO (4.43–19.58 wt.%) contents. Large ion lithophile elements (LILEs) were enriched relative to high field strength elements (HFSEs). The rare earth elements (REEs) and trace element signatures in the normalized diagrams were similar to oceanic island basalts (OIB) and Emeishan high-Ti basalts. These samples had consistent Sr-Nd isotope compositions: the initial $^{87}Sr/^{86}Sr$ values ranged from 0.7044 to 0.7048 and $\varepsilon Nd(t)$ = 3.25–4.92. The Pb isotopes were more complex, the $(^{206}Pb/^{204}Pb)t$, $(^{207}Pb/^{204}Pb)t$, $(^{208}Pb/^{204}Pb)t$ ratios were 17.493–18.197, 15.530–15.722, and 37.713–38.853, respectively. Our results indicate that the GAAB originated from the deeper enriched mantle with 5% to 15% partial melting of garnet lherzolite and a segregation depth of 2 to 4 GPa (60–120 km). During the formation of the GAAB, clinopyroxene and Ti-Fe oxides were fractionally crystallized with insignificant crustal contamination. The GAAB were formed in a extensional regime that was related to the Gejiu-Napo rift event in the Triassic.

Keywords: gejiu basalts; zircon U–Pb dating; geochemistry; Sr–Nd–Pb isotope; petrogenesis; tectonic setting

---

1. Introduction

Basaltic magmatism is a fundamental geological process, and of the various kinds of basalts, alkaline basalts are thought to have unique geological implications [1,2]. Massive stretches of basaltic magma are distributed along the western margin of the South China Block in SW China, making up the widely known Emeishan large igneous province (ELIP), which covers an area of over 5×10^5 km^2 in SW China and northern Vietnam (Figure 1a) [3]. The basalts in the ELIP were mainly formed during the Late-Middle Permian (ca. 260 Ma) and have been identified as related to a mantle plume [4–12]. The Gejiu Anisian alkaline basalts (GAAB), formed later than the ELIP, are important in understanding the tectonic evolution of the western South China Block during post-ELIP tectonomagmatism.

However, the origin and mechanism behind the formation of the GAAB in the southern ELIP remain controversial and poorly understood, and the links between the GAAB and the ELIP have not been well established. Several formation mechanisms have been proposed for the GAAB, these have included an orogenic environment [13], a within-plate extensional environment [14,15], a back-arc rift basin [16–19], and a decompressive melting of the pre-existing Emeishan plume head induced by a rift event [20,21]. Nevertheless, no study has been able to provide sufficient evidence to confirm the origin or formation mechanism of the GAAB, likely due to unsystematic sampling and research methods.

In this paper, samples were collected from both outcrop and boreholes in Qilinshan and Laochang. Integrated geochronological, geochemical, and isotopic data were obtained to investigate the origin and formation mechanism of the GAAB, along with any links between the Middle Triassic igneous activity in Gejiu and the ELIP. This analytical data from the Gejiu area will contribute to a better understanding of these, as of yet, unsolved issues.

2. Geologic Background and Petrography

The Gejiu ore district is located in the South China block, East of the Ailaoshan belt [22] (Figure 1a). The collision of Indochina and South China Block took place during the Early Triassic, forming the widely known Ailaoshan suture zone, accompanied by complex tectonomagmatism [23]. It extends northwest to the Jinshajiang suture zone and southeastern to the Song Ma suture zone (Figure 1a) [24–26]. The Ailaoshan belt is an NW-SE striking fan-shaped area that narrows to the NW and widens to SE (500 km long, and 10 to 50 km wide) and this belt can be subdivided into Western, Central, and Eastern Ailaoshan zones from west to east [23].

The GAAB are situated in the eastern part of the Ailaoshan suture zone (Figure 1a) [3,20,21]. There are numerous faults in the Gejiu ore district (Figure 1b,c). The Gejiu fault as the southern extension of Xiaojiang fault, which has an SN-trending orientation, is the main fault dividing the Gejiu area into two sectors [22]. The secondary faults, which generally trend EW, show the signature of an equidistant distribution in the eastern sector and divide the Gejiu ore district into five main ore segments, namely the Malage, Songshujiao, Gaosong, Laochang, and Kafang from north to south (Figure 1c). The basalts of interest are primarily located in the eastern sector of the Gejiu ore district.

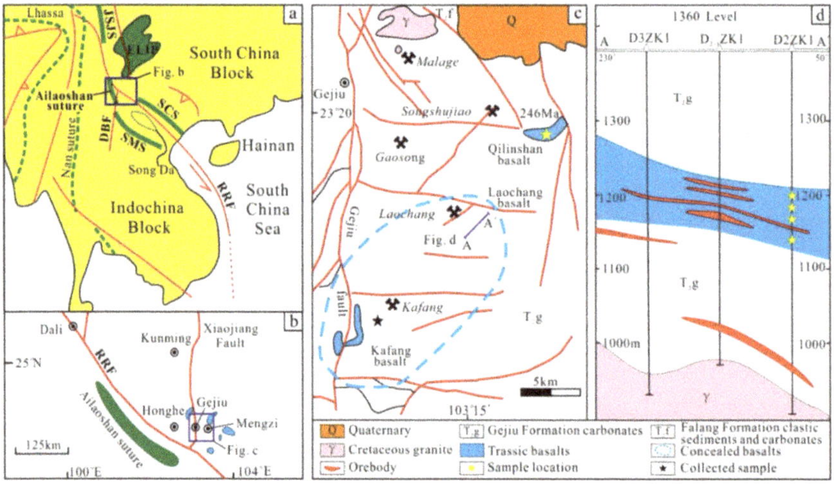

Figure 1. (**a**) Simplified tectonic map showing the study area in relation to major tectonic units in South China [23], JSJS: Jinshajiang suture, SCS: Song-Chay suture, SMS: Song Ma suture, RRF: Red River fault, DBF: Dien Bien Phu fault; (**b**) Simplified map showing the location of the study area [27]; (**c**) Geological map of the Gejiu ore district [21]; (**d**) Geological section of basalts in Laochang ore segment.

The Triassic GAAB cover an area of nearly 125 km^2 in the south of the ELIP, and are widely distributed in the Qilinshan, Laochang, and Kafang ore segments (Figure 1c) [18,20]. The basalts in Laochang and Kafang, which are considered as a unit, are concealed under the ground, while the Qilinshan basalts are exposed at the earth's surface. The Middle Triassic Gejiu Formation carbonates and the Falang Formation fine-grained clastic sediments and carbonates are the dominant strata (over

3000 m) in the Gejiu ore district. These basaltic lavas are found in the lower part of the Gejiu Formation where they were conformably interbedded within the Gejiu Formation. The Laochang and Kafang basalts exhibit a subhorizontal distribution with an overall dip to the northwest. The thickness of a single layer of the basalt ranges from 0.05 to 30 m, and the total thickness is about 100 m [13,21]. Due to the intrusion of the Mesozoic granitic magmatism, the basalts in Laochang and Kafang have undergone different degrees of metamorphism, forming olivine basalts with weak metamorphism and phlogopite metabasalts that have high degrees of metamorphism [17]. The Qilinshan basalts found in the eastern part of the Songshujiao ore segment underwent relatively low degrees of metamorphism and were less altered than the basalts in Laochang and Kafang.

In this research, outcrop samples were collected from Qilinshan and the borehole samples came from 1360 levels in the Laochang ore segment. Altogether, 59 samples were collected for zircon U-Pb dating, geochemical and Sr-Nd-Pb isotope analyses. The sample locations can be seen in Figure 1c,d. These basalts were dark green and had an abundance of amygdaloidal structures (Figure 2a). The oval-shaped amygdales were 0.2 to 2 cm in diameter and showed directional arrangements. These amygdales exhibited a nonuniform distribution in the basalt samples. Most of the amygdales were filled with silicic material and/or carbonate. Chloritization and actinolitization were common in the basalts. These basalts were mainly composed of pyroxene, plagioclase, olivine (Figure 2b,c), and accessory minerals such as ilmenite and magnetite. The alteration in Laochang basalts was more pervasive and severe than that in Qilinshan basalts, and the alteration minerals were mainly phlogopite, actinolite, chlorite, and tremolite (Figure 2d–f).

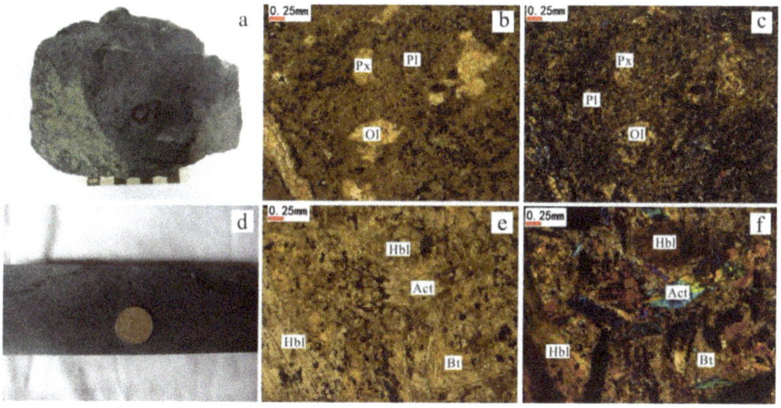

Figure 2. Typical hand specimen (**a**,**d**), single-polarized (**b**,**e**), and cross-polarized (**c**,**f**) light photomicrographs of basalts in Qilinshan (**a**–**c**) and Laochang (**d**–**f**). Px—pyroxene, Pl—plagioclase, Hbl—hornblende, Ol—olivine, Bt—biotite, Act—actinolite.

3. Methods

3.1. Zircon LA–ICP–MS Analyses

Thirty kilos of fresh basalts sample were collected to pick up zircons. The rock was dark green in color and had an amygdaloidal structure. The surface of the sample was first washed clean with water and dried naturally. The sample was then crushed to 80 meshes. After coarse washing with water, strong magnetic separation, electromagnetic separation, and fine washing with alcohol, the zircons were hand-picked under a double microscope.

Zircon U-Pb analyses were conducted at the Key Laboratory of Crust-Mantle Materials and Environments, University of Science and Technology of China. U-Pb abundance was measured using the latest Neptune Plus multiple collectors (ICP-MS) from Thermo Fisher Co. Ltd (Waltham, MA,

USA). The laser-ablation system used in the measurement was the GeoLasPro 193 nm developed by the Coherent Co. Ltd (Santa Clara, CA, USA). Helium was used as a carrier gas to enhance transport efficiency of ablated material. The spot size of the laser ablation beam was 32 μm. The LA-ICP-MS operating conditions were optimized based on measurements of the reference material zircon 91500. The accuracy of the data was verified by using GJ-1 as an auxiliary standard. The LA-ICP-MS measurement was carried out using time-resolved analysis operated in fast peak-hopping and DUAL detector mode with a short integration time. The Harvard standard zircon 91500 and GJ–1 were measured after every 5 to 10 sample spots, to ensure no drift was occurring. The data were analyzed in the ICPMSDataCal program [28] and the Isoplot program [29].

3.2. Major and Trace Element Analyses

Bulk-rock major and trace element compositions of the samples were measured at the Testing Center of Rock and Mineral in Henan province. The major elements were analyzed using a ZSX100e (Rigaku Co., Akishima, Tokyo, Japan) X-ray fluorescence spectrometer (XRF) on fused glass beads. The trace elements, including rare earth elements, were measured by inductively coupled plasma mass spectrometry (ICP-MS) with an XSERIES2 (Thermo Fisher Co. Ltd, Waltham, MA, USA). A detailed methodology of the analytical procedures can be found in Gao [30]. The analyses of the international standards returned values that agreed with their published values. Analyses of the international rock standards indicated that the precision and accuracy were better than 1.5% for all elements.

3.3. Whole Rock Sr–Nd–Pb Isotopic Analyses

The Sr-Nd isotope analyses were carried out at the Ministry of Education Key Laboratory of Orogenic Belts and Crustal Evolution, Peking University. Powdered bulk-rock samples were first spiked with mixed isotope tracers and then dissolved in a solution of HF and HNO3 in Teflon capsules before undergoing Rb-Sr and Sm-Nd isotope analyses. Rb, Sr, Sm, and Nd were separated using conventional ion exchange procedures, as described by Yan [31]. The Sr-Nd isotopic data were measured on a VG Axiom mass spectrometer. The Nd and Sr ratios were normalized to ^{146}Nd/^{144}Nd = 0.7219 and ^{86}Sr/^{88}Sr = 0.1194, respectively. The BCR-2 standards, which were ^{143}Nd/^{144}Nd = 0.512633 ± 0.000017 (2σ) and ^{87}Sr/^{86}Sr = 0.705013 ± 0.000019 (2σ), respectively, were used to assess analytical precision.

The Pb isotope analyses were tested at the Beijing Institute of Uranium Geology using a Isoprobe-T (GV, Manchester, UK) thermal ionization mass spectrometer (TIMS). These samples were first placed in crucibles and dissolved in HF and HClO4, then the purified Pb was extracted using basic anion exchange resin [32]. Analytical results for the NBS981 standard agreed with the reference value [33].

4. Results

4.1. Zircon U–Pb Dating

The LA–ICP–MS zircon U–Pb dating results are listed in Table 1. The sample ISKB–01 was collected from Qilinshan that was located at 22°19′40″ N, 103°16′43″ E. Twenty-five zircon grains were picked out from sample ISKB–01.

The majority of the zircon grains separated from the sample were light gray in color, euhedral in shape, and transparent in opacity under the microscope (Figure 3a). The grains were 50 to 200 μm long with elongation ratios ranging from 1:1 to 3:1. Almost all of the zircons exhibited oscillatory zoning in CL, implying a magmatic origin. The Th and U concentrations were 36.38 to 669.17 ppm and 71.73 to 721.1 ppm, respectively, with Th/U ratios ranging from 0.86 to 2.75. Twenty-five spots were tested, and all were from a single age group clustered in 256 to 237 Ma; the spots yielded a weighted mean ^{206}Pb/^{238}U age of 244 Ma (MSWD = 4.9) (Figure 3b). This age is considered to represent the crystallization age of the magmatic zircons.

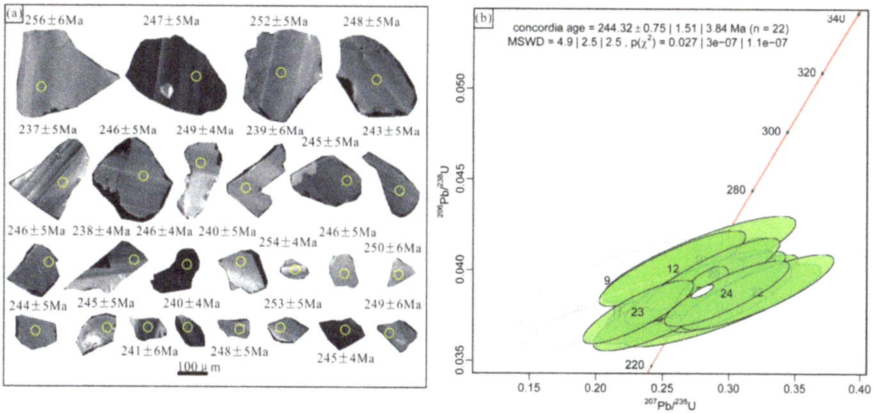

Figure 3. (**a**) Cathodoluminescence (CL) images of the selected zircons for LA–ICP–MS dating; (**b**) Concordia diagram of zircon U–Pb age for the Qilinshan basalts.

Table 1. U–Pb isotopic compositions for zircon from Qilinshan basalts sample ISKB–01.

Spot	Th (ppm)	U (ppm)	Th/U Ratio	$^{207}Pb/^{206}Pb$		$^{207}Pb/^{235}U$		$^{206}Pb/^{238}U$		$^{206}Pb/^{238}U$	
				Ratio	1σ	Ratio	1σ	Ratio	1σ	Age	1σ
KB1-1	198.92	184.46	0.9273075	0.05343	0.00316	0.27953	0.01568	0.03798	0.00069	240.3	4.3
KB1-2	41.16	97.41	2.3666181	0.04966	0.00452	0.25982	0.02294	0.03798	0.00084	240.3	5.19
KB1-3	55.46	143.07	2.5796971	0.05078	0.0038	0.27026	0.01945	0.03864	0.00077	244.4	4.77
KB1-4	175.69	170.82	0.9722807	0.05731	0.00345	0.30891	0.01764	0.03912	0.00073	247.4	4.52
KB1-5	318.8	289.82	0.9090966	0.05378	0.00258	0.29795	0.01329	0.04022	0.00067	254.2	4.14
KB1-6	48.42	79.03	1.6321768	0.05029	0.00514	0.27425	0.02729	0.03959	0.00094	250.3	5.82
KB1-7	40.24	80.26	1.9945328	0.05639	0.00597	0.29587	0.03048	0.03808	0.00095	240.9	5.91
KB1-8	182.34	280.02	1.5357025	0.05096	0.00251	0.27651	0.01273	0.03938	0.00065	249	4.04
KB1-9	50.75	97.65	1.9241379	0.04959	0.00457	0.26575	0.02376	0.03889	0.00086	246	5.35
KB1-10	37.2	77.65	2.0873656	0.03842	0.00561	0.20861	0.03004	0.03941	0.00095	249.1	5.88
KB1-11	36.38	71.73	1.9716877	0.04967	0.00543	0.27787	0.02968	0.04059	0.00099	256.5	6.12
KB1-12	50.71	139.24	2.7458095	0.05285	0.00389	0.28581	0.02024	0.03924	0.00078	248.1	4.86
KB1-13	41.08	109.27	2.6599318	0.04661	0.0042	0.25753	0.02254	0.04009	0.00086	253.4	5.36
KB1-14	391.5	337.49	0.8620434	0.05247	0.00231	0.28027	0.01134	0.03876	0.00062	245.1	3.83
KB1-15	53.54	134.48	2.5117669	0.05336	0.00384	0.28594	0.01977	0.03888	0.00076	245.9	4.74
KB1-16	57.97	143.68	2.4785234	0.05698	0.00373	0.30168	0.01882	0.03841	0.00075	243	4.68
KB1-17	50.91	130.35	2.5604007	0.05459	0.00377	0.29512	0.0195	0.03922	0.00076	248	4.72
KB1-18	41.88	81.05	1.9352913	0.04155	0.0055	0.21613	0.02815	0.03773	0.00089	238.8	5.56
KB1-19	112.93	137	1.2131409	0.04767	0.0039	0.24559	0.01947	0.03737	0.00075	236.5	4.68
KB1-20	57.63	148.17	2.5710567	0.04984	0.00356	0.26679	0.0183	0.03883	0.00075	245.6	4.64
KB1-21	47.7	123.4	2.5870021	0.05256	0.00389	0.28834	0.02049	0.03979	0.00082	251.5	5.08
KB1-22	669.17	721.1	1.0776036	0.05215	0.00177	0.27938	0.00836	0.03885	0.00057	245.7	3.52
KB1-23	48.31	92.96	1.9242393	0.06008	0.00467	0.32154	0.02398	0.03881	0.00086	245.5	5.32
KB1-24	109.27	141.25	1.2926695	0.04457	0.00337	0.23159	0.0169	0.03768	0.00072	238.5	4.47
KB1-25	45.9	120.01	2.614597	0.05587	0.00378	0.29842	0.01929	0.03873	0.00075	245	4.67

4.2. Whole Rock Geochemistry

Fifty samples were measured for major and trace element analyses (Supplementary Material, Table S1). These samples were collected from Qilinshan and Laochang, respectively. Almost all the samples showed different degrees of alteration with a large variation in loss on ignition (LOI) (1.35–9.08 wt.%), demonstrating relatively strong alteration.

For the samples from Qilinshan, the SiO_2 content ranged from 40 to 48.68 wt.%. Most rocks plotted fell in the alkaline basalts field, as seen in Figure 4a,b, and belonged to the potassic and sodic series (Figure 4c). These rocks had high TiO_2 contents (2.14–3.23 wt.%), with the TiO_2 contents of most samples being larger than 2.5, and belonging to the high-Ti basalts (Figure 4d), as defined by Xu [34]. The MgO contents were 7.76–13.92 wt.% and the Mg# = 53.27–67.77, which is characteristic of primary magma. The TFe_2O_3 content ranged from 10.18 to 16.06 wt.%. The K_2O contents (0.22–2.31 wt.%) in Qilinshan basalts were much lower than that in Laochang basalts. The LILEs were enriched relative to the HFSEs. The $(La/Yb)_N$ ratios varied from 8.20 to 23.68 (mean value = 13.34). As revealed by the primitive mantle normalized trace element diagram and the chondrite normalized rare earth element (REE) diagram (Figure 5), the patterns of the Qilinshan basalts were similar to the Emeishan basalts and strongly resembled those of OIB.

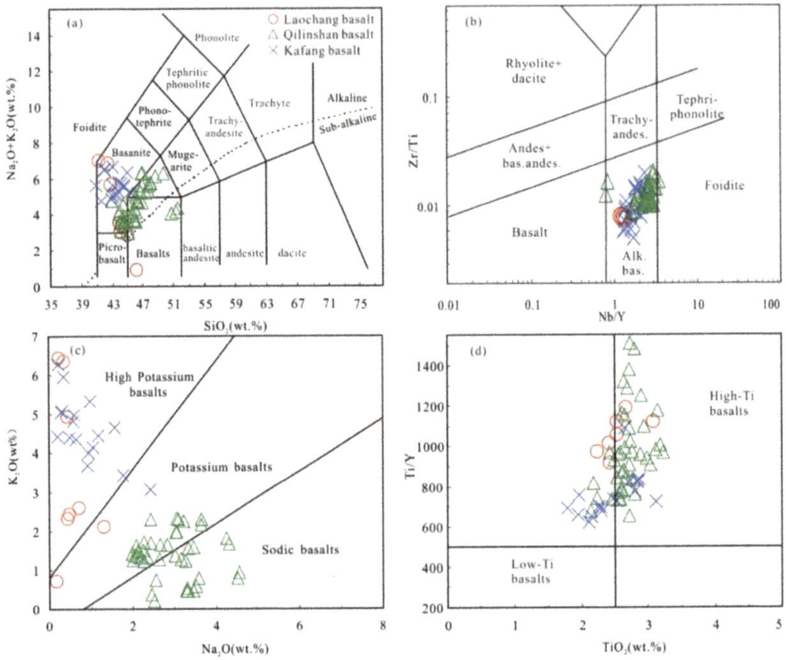

Figure 4. (**a**): SiO_2 vs. Na_2O+K_2O classification diagram; (**b**): Nb/Y vs. Zr/Ti diagram; (**c**): Na_2O vs. K_2O classification diagram; (**d**): TiO_2 vs. Ti/Y diagram. The data of Kafang from [16].

The Laochang basalt samples had extremely low SiO_2 contents (39.33–44.49 wt.%). Most samples fell in the basanite and basalt areas in Figure 4a, and belonged to high-Ti alkaline basalts in Figure 4b,d. In the diagram showing Na_2O vs. K_2O, the rocks had high-K characteristics (Figure 4c). The TiO_2 content ranged from 2.24 to 3.07 wt.%. The Laochang basalts had higher MgO content (12.16–19.58 wt.%) and Mg# values (64.39–76.95) relative to those of the Qilinshan basalts. The contents of TFe_2O_3 and K_2O were 11.05–14.02 wt.% and 2.12–6.45 wt.%, respectively, for the Laochang basalts. For the trace elements and REEs, the $(La/Yb)_N$ ratios varied from 7.43 to 18.96 (mean value = 12.24). These samples exhibited strongly positive Rb anomalies and negative Ba anomalies in the primitive mantle normalized trace element diagram (Figure 5). As can be seen in Figure 5, the patterns of immobile elements in Laochang and Kafang basalts were consistent with those of Emeishan basalts and OIB. Mobile elements such as Rb, Ba, Sr, K, and Pb have high activities and are sensitive to physical and chemical alterations. Therefore, the abnormities of these elements in Laochang and Kafang may have resulted from the process of metamorphism and/or hydrothermal alteration.

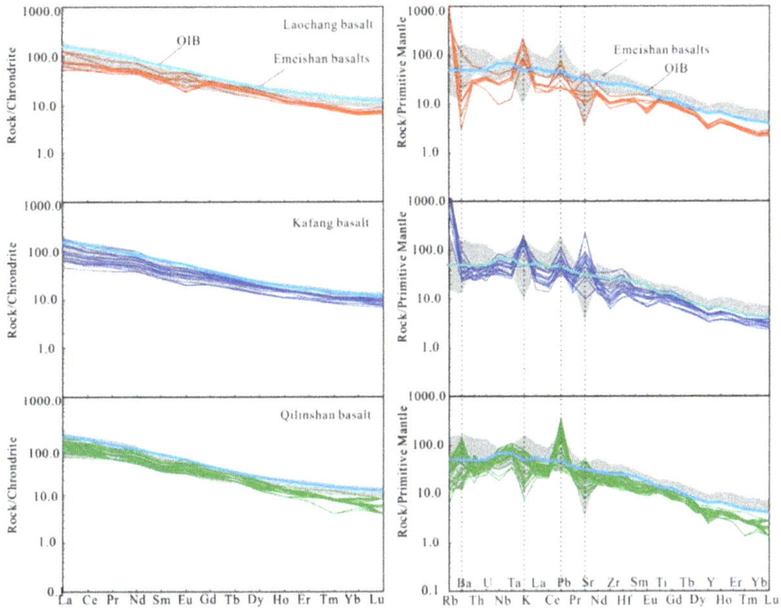

Figure 5. Chondrite normalized REE patterns and primitive mantle normalized trace element patterns of the GAAB. Normalization and OIB values followed [35]. The data of Emeishan basalts from [3]. The data of Kafang from [16].

4.3. Sr–Nd–Pb Isotopic Geochemistry

Whole-rock Sr-Nd-Pb isotopes of the samples from Qilinshan and Laochang are listed in Tables 2 and 3. The initial Sr-Nd-Pb isotopic ratios for all the samples were corrected to 244 Ma, according to our new zircon U-Pb dating age. The initial $^{87}Sr/^{86}Sr$ values of Qilinshan basalts ranged from 0.7044 to 0.7047 with εNd(t) values ranging from 4.69 to 4.92. The Laochang basalts had smaller initial $^{87}Sr/^{86}Sr$ values (0.7044–0.7048) and εNd(t) values (3.25–3.43). Lead isotopes are more complex because U, Th, and Pb are all mobile during alteration. The initial $^{206}Pb/^{204}Pb$, $^{207}Pb/^{204}Pb$, and $^{208}Pb/^{204}Pb$ ratios in Qilinshan basalts were 17.493–18.150, 15.530–15.613, and 37.713–38.853, respectively, and 17.976–18.197, 15.683–15.722, and 38.501–38.713 in Laochang basalts, respectively.

Table 2. Sr–Nd isotopic compositions of Qilinshan and Laochang basalts.

Sample	Age(Ma)	Rb(ppm)	Sr(ppm)	$^{87}Sr/^{86}Sr$	$(^{87}Sr/^{86}Sr)i$	Sm(ppm)	Nd(ppm)	$^{143}Nd/^{144}Nd$	$(^{143}Nd/^{144}Nd)i$	$\varepsilon Nd(t)$
Laochang-01	244	637	292.1	0.7263	0.7044	4.92	22.00	0.51272	0.51250	3.43
Laochang-02	244	710	83.0	0.7905	0.7048	5.46	24.61	0.51270	0.51249	3.25
Qilinshan-8	244	17	395.1	0.7052	0.7047	6.70	33.31	0.51276	0.51256	4.69
Qilinshan-9	244	13	549.0	0.7047	0.7044	6.56	31.07	0.51278	0.51257	4.92

Table 3. Pb isotopic compositions of Qilinshan and Laochang basalts.

Sample	Age (Ma)	$^{206}Pb/^{204}Pb$	$^{207}Pb/^{204}Pb$	$^{208}Pb/^{204}Pb$	$(^{206}Pb/^{204}Pb)t$	$(^{207}Pb/^{204}Pb)t$	$(^{208}Pb/^{204}Pb)t$	Φ	μ	Th/U
Laochang-01	244	18.554 ± 0.007	15.701 ± 0.005	39.106 ± 0.014	18.197	15.683	38.713	0.585	9.65	3.95
Laochang-02	244	18.658 ± 0.004	15.757 ± 0.004	39.252 ± 0.007	17.976	15.722	38.501	0.584	9.75	3.97
Qilinshan-08	244	18.780 ± 0.006	15.645 ± 0.006	39.282 ± 0.016	18.150	15.613	38.853	0.565	9.52	3.89
Qilinshan-09	244	18.859 ± 0.006	15.600 ± 0.005	39.145 ± 0.010	17.493	15.530	37.713	0.556	9.42	3.79

5. Discussion

5.1. Magma Source

In the past decade, the GAAB have attracted a lot of attention for its close relationship with the Cu deposits in the Gejiu ore district [22,36]. However, relatively little research has focused on how the GAAB were formed. For this reason, the magma source of the GAAB remain unresolved. While it is widely accepted that these basalts were the product of the melting of mantle peridotite, the identification of the mantle source has been proven difficult because of the complexity of lithospheric mantle [37]. The Sr-Nd-Pb isotopic characteristics of igneous rocks can often narrow the potential source regions of basalts because isotopic fractionation is not affected by magma crystallization and segregation in the magmatic process. The Sr and Nd isotopic characteristics of the GAAB were consistent, the initial $^{87}Sr/^{86}Sr$ = 0.7044–0.7048 and $^{143}Nd/^{144}Nd$ = 0.51249–0.51257, and the Pb isotopes exhibited an appropriate variation range. In the diagram of ($^{87}Sr/^{86}Sr$)i vs. εNd(t) (Figure 6a), all of the samples fell in the field of the Emeishan flood basalts and showed similarities with the OIB. Moreover, Pb isotopes also showed similar characteristics to Emeishan flood basalts (Figure 6b). The Sr-Nd-Pb isotopic data indicated that the GAAB were derived from the same OIB-like source as the Emeishan flood basalts.

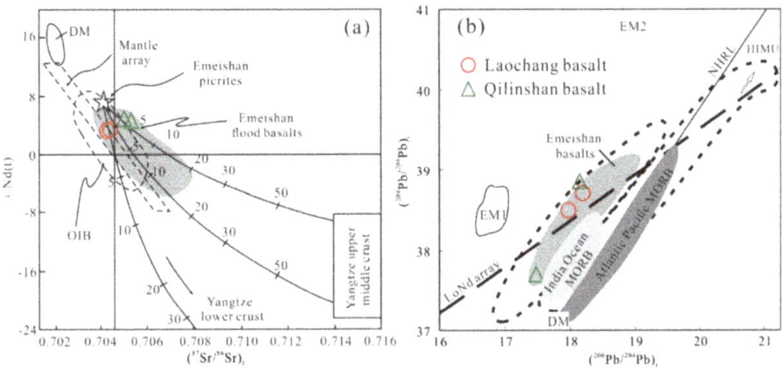

Figure 6. (**a**) ($^{87}Sr/^{86}Sr$)i vs. εNd(t) diagram (modified from [3]), OIB: ocean island basalt; (**b**) ($^{206}Pb/^{204}Pb$)t vs. ($^{208}Pb/^{204}Pb$)t diagram (t = 244 Ma, modified from [38]), EM1: enriched mantle 1, EM2: enriched mantle 2, DM: depleted mantle, HIMU: high $^{238}U/^{204}Pb$ mantle end member, NHRL: northern Hemisphere reference line; data of Emeishan basalts from [27].

Typically, the ratios of incompatible elements that share similar characteristics in the mantle (such as Zr/Nb and Nb/Th) are not strongly fractionated from each other during the process of partial melting or fractional crystallization. Hence, their ratios can be used to identify the source of magma [39,40]. The Nb/Th ratio is insensitive to various degrees of melting of garnet lherzolite, while the Zr/Nb ratio is moderately affected by different degrees of melting [40]. Therefore, the mantle source can be determined by comparing the Nb/Th and Zr/Nb ratios (Figure 7a). Most of the samples fell in the field near OIB, implying that the GAAB were derived from an enriched mantle sources similar to that of OIB.

Magma sources can also be identified by their Nb/Y and Zr/Y ratios, this is because of their predictability during the partial melting and fractional crystallization processes [41]. Thus, Fitton [42] proposed the formula: $\delta Nb = \log(Nb/Y) + 1.74 - 1.92 \times \log(Zr/Y)$, based on the Nb/Y and Zr/Y ratios. When $\delta Nb > 0$, the magma is related to an enriched mantle, but when $\delta Nb < 0$, the magma is related to a depleted mantle. Almost all of the samples had $\delta Nb > 0$, indicating an enriched mantle source. Likewise, enriched mantle sources can be identified by comparing Zr/Y and Nb/Y (Figure 7b).

These comparisons revealed that the majority of the samples fell in the OIB field, indicating the GAAB were derived from enriched mantle.

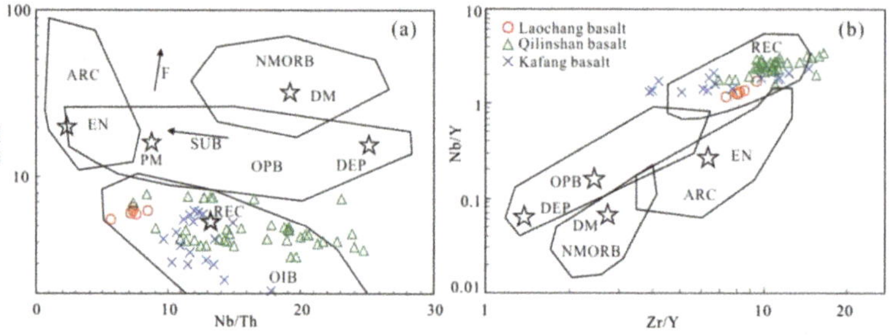

Figure 7. Diagram showing mantle components and fields for basalts from various tectonic settings [40,42]. (**a**) Nb/Th vs. Zr/Nb, SUB: subduction; (**b**) Zr/Y vs. Nb/Y. NMORB: normal mid ocean ridge basalts, DM: depleted mantle, REC: recycled components, EN: enriched, DEP: depleted components, ARC: arc basalts, OPB: oceanic plateau and island basalts. The data of Kafang from [16].

HFSEs and REEs are immobile during the alteration and weathering processes [43–45]. Therefore, the contents and ratios of light REEs and HFSEs are widely used to distinguish between magma sources and determine the degree of partial melting [46,47]. In the diagram showing La/Nb vs. La/Ba (Figure 8a), most samples can be seen to fall in the field of asthenosphere and OIB, with no indicators for crust contamination. Relative to heavy REEs (such as Yb), the enrichment of middle REEs (such as Sm) depends on the presence of garnet in the residual phase during the melting process. Sm is incompatible with garnet, whereas Yb is compatible, this means that the ratio of Sm/Yb can be intensely fractionated when there is a low degree of melting. Accordingly, using Sm vs. Sm/Yb, the degree of partial melting of the mantle source can be constrained using the non-modal batch melting equations [48,49]. Most of the samples had high MgO content (>8 wt.%), which could indicate that the historical source of the basaltic magma underwent partial melting [3]. As can be seen in Figure 8b, the GAAB originated from 5% to 15% partial melting of garnet lherzolite.

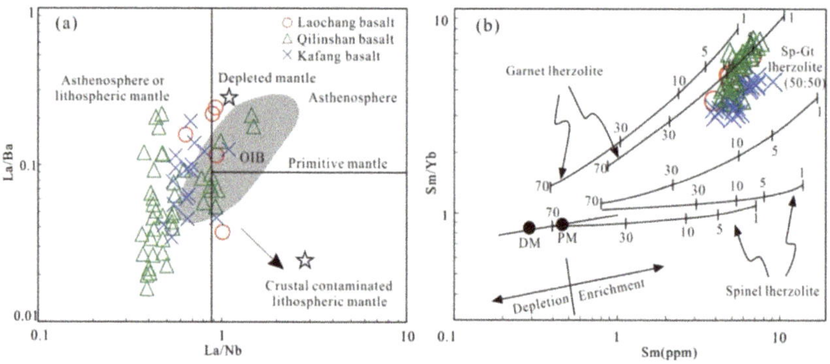

Figure 8. (**a**): La/Nb vs. La/Ba diagram (modified from [50]); (**b**): Sm vs. Sm/Yb diagram (modified from [3]). The data of Kafang from [16].

Generally, the TiO_2 content in magma from the asthenosphere is about 1.27 wt.%, while in magma from deeper mantle material it is generally more than 2 wt.% [37]. The GAAB had high TiO_2 content (TiO_2 > 2 wt.%), similar to the representative OIB (TiO_2 = 2.87 wt.%; [41]), and the Ti/Y ratios ranged from 620 to 1533, similar to the Emeishan high-Ti basalts [34], indicating they derived from the deeper enriched mantle. The OIB-like basalts studied here suggested that there was a period of within plate magmatism in the South China Block around 244 Ma. In the diagram of ($^{87}Sr/^{86}Sr$)i vs. ($^{143}Nd/^{144}Nd$)t (Figure 9f), the samples fell near the OIB field and displayed enriched mantle (EM1) evolutionary trends, which indicated they were related to an ancient subcontinental lithospheric mantle (SCLM) metasomatized by a small volume of volatile-enriched melt that originated from asthenosphere or deeper mantle [51–53]. As described above, based on the combination of geochemical and isotopic characteristics of the GAAB, it can be concluded that the GAAB were mainly derived from deeper enriched mantle material and experienced 5% to 15% partial melting of garnet lherzolite.

5.2. Petrogenesis

Parental magma compositions are determined by the nature of the mantle source, conditions during mantle melting, fractional crystallization, and the extent of crustal contamination [7]. REEs, HFSEs, and Sr-Nd-Pb isotope compositions are commonly used to estimate these variables for a given sample. Using basaltic magma, the conditions during mantle melting can be investigated to provide information about the processes by which they formed [54]. Although alteration is pervasive in the GAAB, the abundances of compatible and incompatible elements in basaltic melts are not strongly influenced by alteration or weathering [54], so they can still serve as accurate indicators of historical conditions.

A low degree of partial melting of pyrolite can lead to the enrichment of incompatible elements, but with an increasing degree of melting, the enrichment of incompatible elements is reduced. Light REEs and highly incompatible elements are mainly enriched in alkaline basalts with low SiO_2 content. Evidence has shown that alkaline basalts are always generated by 2% to 10% partial melting of the incompatible elements enriched mantle source [55–57]. The GAAB had trace element compositions and light REEs signatures similar to the OIB. The enrichment of the trace elements and light REEs indicated that the degree of partial melting was less than 10%. This conclusion can also be supported by the diagram of Sm vs. Sm/Yb (Figure 8b).

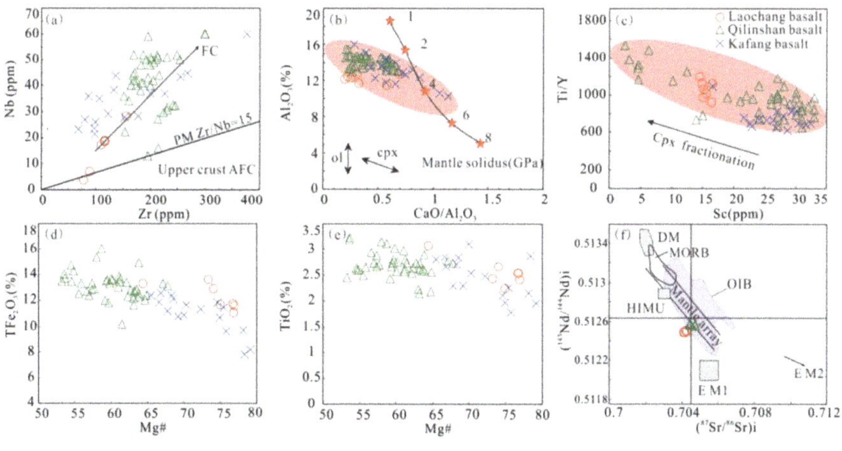

Figure 9. (**a**) Zr vs. Nb diagram (modified from [58]); (**b**) CaO/Al_2O_3 vs. Al_2O_3 diagram (modified from [40]), ol: olivine, cpx: clinopyroxene; (**c**) Sc vs. Ti/Y diagram (modified from [7]); (**d**) Mg# vs. TFe_2O_3 diagram; (**e**) Mg# vs. TiO_2 diagram; (**f**) ($^{87}Sr/^{86}Sr$)i vs. ($^{143}Nd/^{144}Nd$)i (modified from [58]). The data of Kafang from [16].

Fractional crystallization (FC) is integral to magma evolution. Hence, it is always used to interpret geochemical characteristics observed in evolved volcanic rocks. As can be seen in the Nb vs. Zr plot in Figure 9a, the FC process existed. Moreover, the negative correlations that can be observed in the diagrams of CaO/Al_2O_3 vs. Al_2O_3 (Figure 9b) and Sc vs. Ti/Y (Figure 9c) implied that the fractionation was of clinopyroxene, which was consistent with the petrographic features. Plagioclase fractionation appeared to be limited as there was no clear evidence for negative Eu anomalies, however most of the samples did have negative Sr anomalies (Figure 5). The near ubiquity of Sr anomalies might be attributed to the alteration or metamorphism during post-magmatic events. The negative correlations between Mg# and TFe_2O_3 (Figure 9d) and between Mg# and TiO_2 (Figure 9e) indicated the fractionation of Ti-Fe oxides. Based on the experimentally determined mantle solidus [40,59,60], magma segregation depth of basalts can be constrained by the CaO/Al_2O_3 vs. Al_2O_3 diagram. As seen in Figure 9b, the melts segregation depth of the GAAB was between 2 and 4 GPa (60–120 km).

Crustal contaminated mafic melts have characteristically high initial $^{87}Sr/^{86}Sr$ ratios, low εNd(t) values, and negative Nb and Ta anomalies [61]. However, the GAAB had relatively low initial $^{87}Sr/^{86}Sr$ ratios, high εNd(t) values, and no Nb and Ta anomalies (Figure 5), which suggested that the crustal material in the magma was limited. Weak correlations between SiO_2 and La/Nb (Figure 10a) and the non-linear relationship between SiO_2 and εNd(t) (Figure 10b) also indicated limited crustal contamination. The Nb/U ratio in primary mantle-derived magma is always greater than 47, and the La/Nb and La/Ta ratios are always smaller than 1.5 and 22, respectively. These ratios can be used to identify the existence of crustal material due to their sensitivity to crustal contamination [62–64].

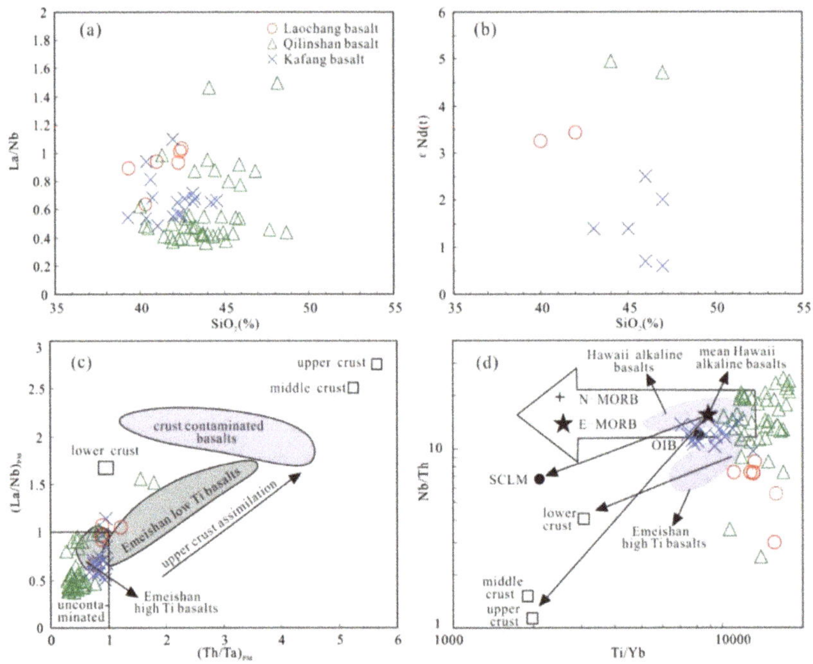

Figure 10. (**a**) SiO_2 vs. La/Nb diagram; (**b**) SiO_2 vs. εNd(t) diagram; (**c**) $(Th/Ta)_{PM}$ vs. $(La/Nb)_{PM}$ diagram (modified from [65]); (**d**) Ti/Yb vs. Nb/Th diagram (modified from [65]), N-MORB: Normal mid ocean ridge basalt, E-MORB: Enriched mid ocean ridge basalt. The data of Kafang from [16,21].

The Nb/U ratio in the GAAB was much higher than crustal material, but close to mantle-derived values. Likewise, the smaller La/Ta and La/Nb ratios (La/Ta ≤ 18; La/Nb ≤ 1.46) in most samples revealed similarities with primary magma and also indicated limited crustal contamination. We can determine the existence of crustal material in mantle-derived magma using the diagram of $(Th/Ta)_{PM}$ vs. $(La/Nb)_{PM}$. As can be seen in Figure 10c, most of the samples fell in the uncontaminated field of the plot, showing there was little or no crustal material in the magma. These signatures indicate that the assimilation of crust during the generation of the basalts was insignificant, and therefore the chemical compositions of these basaltic rocks accurately reflect the characteristics of mantle source regions. Nb/Th and Ti/Yb ratios can be used to identify the lithospheric mantle materials in basalts [66]. In the diagram of Ti/Yb vs. Nb/Th (Figure 10d), most of the samples fell in the field of the OIB and showing evolutionary trends consistent with SCLM. OIB are widely believed to be related to hotspots or mantle plumes, which are usually imaged as columns of heated material ascending from the core-mantle boundary or mantle transition zone [67–72]. The upwelling of asthenospheric mantle or asthenospheric melts generated by rift events in an extensional regime can provide enough heat for the melting of the SCLM. In this case, the OIB-like magma mixed with the SCLM magma, which then formed the primary magma of the GAAB and experienced limited crustal contamination during the ascent.

5.3. Tectonic Setting

The tectonic setting of the GAAB is controversial and various viewpoints still receive support. For example, Zhang [21] proposed that the GAAB were generated in an epeiric sea semi-restricted gulf environment, Wang [13] concluded that these basalts were formed in an orogenic environment. The proposed tectonic settings of the GAAB also include within-plate extensional environment [14,15]; back-arc rift basin [16–19]; and decompressive melting of Emeishan plume head induced by a rift event [20,21]. To clarify the tectonic setting, we provided new zircon dating, major, trace element, and Sr-Nd-Pb isotopic data from these basaltic rocks collected from several representative outcrops and drillholes in the Gejiu ore district.

In Gejiu district, the Gejiu-Napo sedimentary basin has mainly occurred in the Triassic. The lower Triassic Yongningzhen and Feixianguan Formations are composed of shallow water clastic rocks containing fuchsia greywacke and conglomerate. The middle Triassic Gejiu Formation consists mainly of thick-bedded dolomite, limestone, and mudstone. The deposition depth of this sequence deepens gradually upward. The GAAB mainly occurred within the lower section of the Gejiu Formation. The upper Triassic sequence comprises macker and coarse clastic rocks. This sedimentary sequence is characterized by shallow water coarse-grained sediments, complex gravel composition, and coarse gravel sediments from deep to surface. The molasse facies of the sea-land interaction environment marked the shrinkage and closure of the rift basin. The thickness of the Triassic strata (approximately 4000 m) increased dramatically in the Gejiu area [18]. The stratigraphic thickness thinned sharply on both the eastern and western sides of Gejiu (Figure 11). Besides, in Triassic, the bimodal magmatism is also developed in Gejiu district, including the alkaline basalt, dacite, and rhyolite [73]. The geological evidence indicates a rift-related volcanic-sedimentary event. Considering the zonal distribution of the sequence and the variation of stratigraphic thickness, the rift was likely controlled by a syngenetic fault.

Elements such as Ti, Zr, Y, Nb, Hf are useful in distinguishing basalts from different tectonic environments [74,75]. In the diagrams of Nb-Zr-Y (Figure 12a), Ti-Zr-Y (Figure 12b), and Hf-Th-Nb (Figure 12c), the majority of samples are distributed in the field of within-plate settings corresponding to the within-plate alkaline basalts. Moreover, the GAAB had characteristics of enrichment in light REEs, OIB-like geochemical signatures, and was similar to the high-Ti basalts of the ELIP. Young eruption age (244 Ma) excludes the possibility that the GAAB formed during the main event of ELIP (260–257 Ma) [76]. These basalts were coeval and genetically associated with the rift event in this area (Figure 11). Considering the facts described above, we propose that the GAAB erupted within the western margin of the South China Block in an extensional environment that was related to the Gejiu-Napo rift event in the Triassic.

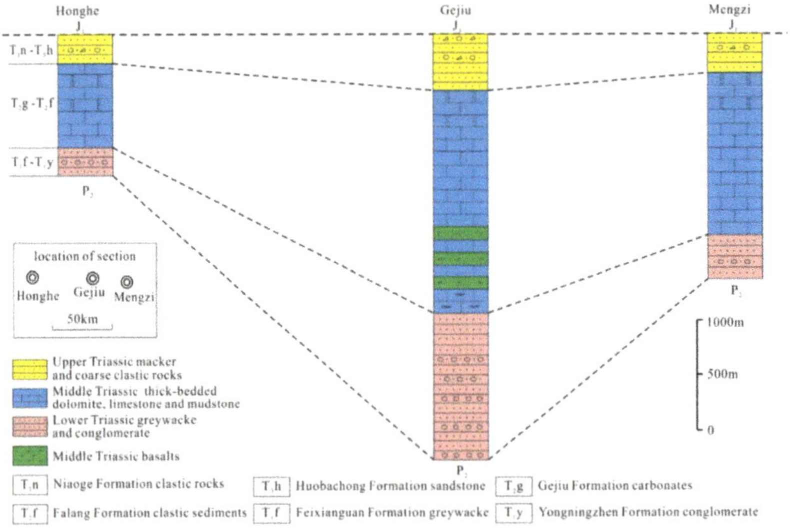

Figure 11. Stratigraphic correlation of the Triassic strata in the western South China block. The sedimentary data are from [3,18,73].

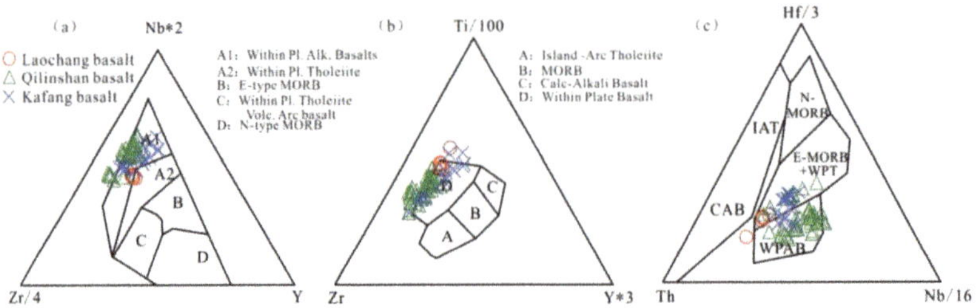

Figure 12. (**a**) Nb–Zr–Y diagram (modified from [74]); (**b**) Ti–Zr–Y diagram (modified from [75]); (**c**) Hf–Th–Nb diagram (modified from [77]), IAT: Island arc tholeiite, CAB: Calc-alkaline basalt, N-MORB: Normal mid ocean ridge basalt, E-MORB: Enriched mid ocean ridge basalt, WPT: Within plate tholeiite, WPAB: Within plate alkaline basalt. The data of Kafang from [16].

The Ailaoshan suture zone comprises several structurally juxtaposed tectonic units, including the Western Ailaoshan volcanic belts, Ailaoshan ophiolitic melange, Ailaoshan metamorphic complex from west to east [78–85]. Based on the comparison between the Permian and Triassic sedimentary rocks along the Ailaoshan suture zone, Xu [86] proposed they had different detrital provenance and formed in a depositional setting, which was likely caused by the closure of the Ailaoshan Ocean. Both the sedimentology and structural geology evidence show that the Ailaoshan suture zone has been a branch of the Paleo-Tethys Ocean and is the boundary between the South China Block and Indochina Block. The Paleo-Tethys Ailaoshan Ocean subducted beneath the Indochina Block in the Late Permian, and the onset of the collision between South China and Indochina blocks occurred in Early Triassic, around 250 Ma [23]. The Phia Bioc granites (248–245 Ma), Tongtiange leucogranites (~247 Ma), and rhyolites (247–246 Ma) in the Gaoshanzhai, Pantiange and formations are related to syn-collisional crust thickening [87–89]. The GAAB with the age of 244 Ma formed in the Gejiu-Napo rift basin were related to a post-collisional setting after the South China-Indochina collision (Figure 13).

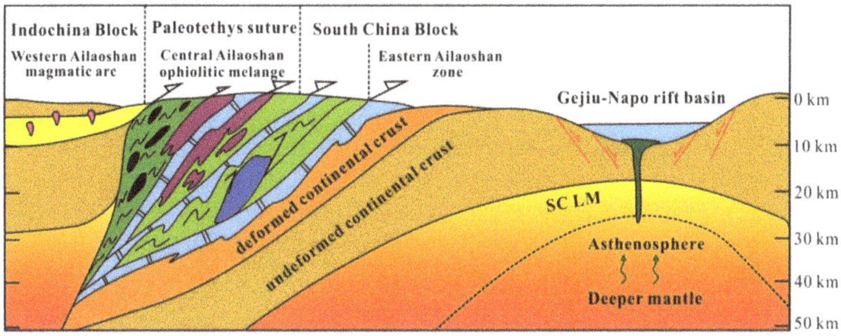

Figure 13. Schematic drawing of the tectonic evolution of the Gejiu-Napo rift event in Ailaoshan post-collisional setting (modified from [23]), SCLM: subcontinental lithospheric mantle.

5.4. Relationship with the ELIP

Traditionally, the GAAB were considered to be placed in the intermediate zone of the ELIP and most likely formed at the periphery of the ELIP with a low degree of melting of plume material [7,20,34]. Recent studies have suggested there was a post-ELIP magmatism in the period following 260 Ma [20,82]. The Triassic basalts in the southern ELIP have mainly been found in the Kaiyun, Qilinshan, and Laochang-Kafang areas. Based on the methods of geochronology, geochemistry, and Sr-Nd-Pb isotope analyses, Shellnutt [82] thought the magmatism that formed in 252 to 242 Ma in the spatial field of the ELIP was related to the post-ELIP magmatism caused by decompressive melting of the fossil Emeishan plume head beneath the South China crust that was associated with the collision between the South China and North China Blocks during the Middle Triassic. Zhang [20] concluded that, in Gejiu, these basalts were the product of the melting of the plume head after the main ELIP phase caused by the rift system along the southwestern margin of the South China Block during the Early Triassic. Zhang [16], on the other hand, thought there was no relationship between the GAAB and the ELIP because zircon U-Pb dating placed the age in Kafang at 214 Ma. It is noteworthy that the basalts in Kafang and Laochang both underwent intense alteration and metamorphism since their formation because of the extensive metallogenesis during the late Cretaceous tectonomagmatism, which may explain the inaccuracies in dating results.

Although the relationship between the GAAB and the ELIP has been discussed in recent publications [16,20], the GAAB, especially its eruption age, remain relatively understudied. Zhang [21] has reported the SIMS zircon U–Pb age of GAAB in Kafang was 244.4 Ma. The similar LA-ICP-MS U–Pb age of Kaiyuan basalts in adjacent area was 247 Ma [20]. In this study, the zircon U–Pb age reveals that the non-mineralized GAAB in Qilinshan were formed in 244 Ma. Generally, the zircon grains in basalts are small and dark in color. In addition, the existence of oscillatory zoning in zircon grains indicates their magmatic origin. The characteristics of trace elements show that the assimilation of crust in generation of the basalts was insignificant. Besides, the GAAB are interbedded within the Triassic Gejiu Formation, which was considered to be formed in 245 to 237 Ma [21]. Combining with the regional geology, the GAAB and the Gejiu Formation are the important constituents of the volcanic-sedimentary sequence in Middle Triassic in Gejiu area. Thus, we thought the zircon U–Pb age can reflect the magmatic history of the GAAB.

After the formation of the ELIP in ~260 Ma, minor magmatic events occurred around the center of ELIP between 252 to 242 Ma. However, these post-260 Ma magmatic events were not directly related to Emeishan magmatism. This is because most large igneous provinces worldwide had relatively short durations (1–5 Ma). ID-TIMS zircon U–Pb ages show that the Karoo large igneous province in South Africa lasting about 1 Ma or less for the major pulse of magmatism [90,91]. The emplacement of large volumes of volcanic rocks during the main pulse of the Siberian Traps in Russia occurred within <1 Ma [92]. New CA-ID-TIMS U-Pb zircon ages from the Malwa Plateau indicate that the Deccan large igneous province in India erupted from 65.9 to 66.5 Ma [93].

As discussed above, the geochemical characteristics of the GAAB are similar to the Emeishan high-Ti basalts. However, the U-Pb dating age of 244 Ma obtained from Qilinshan basalts in Gejiu reveals a later date than the ELIP (ca. 260 Ma), as constrained by various dating methods in different kinds of rocks [4,27,94]. It is generally accepted that the magmatism in ELIP began at, and ended, shortly after 260 Ma [4,8,82,95–97]. Considering the age discrepancies and difference in tectonic setting, there is no geodynamic or genetic link between GAAB and ELIP. The similar geochemical characteristics between the GAAB and the Emeishan high-Ti basalts are most likely due to the similar magma source.

6. Conclusions

Combined with geochronological, petrological, and geochemical studies of the alkaline basalts in the Gejiu area, our conclusions are as follows:

1. Zircon U-Pb dating results showed that the GAAB were formed in 244 Ma (MSWD = 4.9, n = 22).
2. The GAAB, similar to OIB and Emeishan high-Ti basalts, displayed enriched LREE patterns, with LILEs enriched relative to HFSEs. The GAAB had consistent Sr-Nd isotope compositions: the $(^{87}Sr/^{86}Sr)$ = 0.7044–0.7048, and εNd(t) = 3.25–4.92. The Pb isotopes were more complex, the $(^{206}Pb/^{204}Pb)t$ = 17.493–18.197, $(^{207}Pb/^{204}Pb)t$ = 15.530–15.722, $(^{208}Pb/^{204}Pb)t$ = 37.713–38.853, respectively.
3. The GAAB originated from deeper enriched mantle material with 5% to 15% partial melting of garnet lherzolite and a segregation depth between 2 and 4 GPa (60–120 km). During the formation of the GAAB, clinopyroxene and Ti-Fe oxides were fractionally crystallized with insignificant crustal contamination.
4. The GAAB were the product of the decompressive melting of the deeper enriched mantle in a extensional environment that was related to the Gejiu-Napo rift event formed in a post-collisional setting after the South China-Indochina coillision.

Supplementary Materials: The following is available online at http://www.mdpi.com/2075-163X/10/11/1030/s1, Table S1: Data of major and trace elements in this paper.

Author Contributions: Data curation, Z.S.; formal analysis, Z.S.; funding acquisition, Y.C.; investigation, Z.S.; project administration, Y.C.; supervision, Y.C.; writing—original draft, Z.S.; writing—review and editing, Y.C. All authors have read and agreed to the published version of the manuscript.

Funding: This research was funded by the National Natural Science Foundation of China (Grant No. 41972312, 41672329), the National Key Research and Development Project of China (Grant No. 2016YFC0600509), and the Project of China Geological Survey (Grant No. 1212011120341).

Acknowledgments: We are thankful to Guopei Mo, Zhenfei Ma, Lina Zhang, and Luxue Qin for their assistance with samples collection, Shibo Liu for his help with zircon U–Pb dating. We also thank Xiangguo Guo for his help with thin sections observation. We are also indebted to the academic editor and three anonymous reviewers for their critical and constructive comments which substantially improved an early version of the manuscript.

Conflicts of Interest: The authors declare no conflict of interest.

References

1. Condie, K.C. Secular variation in the composition of basalts, an index to mantle evolution. *J. Petrol.* **1985**, *26*, 545–563. [CrossRef]
2. Condie, K.C. Geochemical changes in basalts and andesites across the Archean-Proterozoic boundary: Identification and significance. *Lithos* **1989**, *23*, 1–18. [CrossRef]
3. Zhang, J.W.; Huang, Z.L.; Luo, T.Y.; Qian, Z.K.; Zhang, Y. Origin of early Triassic rift-related alkaline basalts from Southwest China: Age, isotope, and trace-element constraints. *Int. Geol. Rev.* **2013**, *55*, 1162–1178. [CrossRef]
4. Zhou, M.F.; Malpas, J.; Song, X.Y.; Robinson, P.T.; Sun, M.; Kennedy, A.K.; Lesher, C.M.; Keays, R.R. A temporal link between the Emeishan large igneous province (SW China) and the end-Guadalupian mass extinction. *Earth Planet. Sci. Lett.* **2002**, *196*, 113–122. [CrossRef]

5. Zhou, M.F.; Robinson, P.T.; Lesher, C.M.; Keays, R.R.; Zhang, C.J.; Malpas, J. Geochemistry, petrogenesis, and metallogenesis of the Panzhihua gabbroic layered intrusion and associated Fe–Ti–V-oxide deposits, Sichuan Province, SW China. *J. Petrol.* **2005**, *46*, 2253–2280. [CrossRef]
6. Zhou, M.F.; Zhao, J.H.; Qi, L.; Su, W.; Hu, R.Z. Zircon U–Pb geochronology and elemental and Sr–Nd isotopic geochemistry of Permian mafic rocks in the Funing area, SW China. *Contrib. Mineral. Petrol.* **2006**, *151*, 1–19. [CrossRef]
7. Zhou, M.F.; Arndt, N.T.; Malpas, J.; Wang, C.Y.; Kennedy, A.K. Two magma series and associated ore deposit types in the Permian Emeishan large igneous province, SW China. *Lithos* **2008**, *103*, 352–368. [CrossRef]
8. Guo, F.; Fan, W.; Wang, Y.; Li, C. When did the Emeishan mantle plume activity start? Geochronological and geochemical evidence from ultramafic-mafic dikes in southwestern China. *Int. Geol. Rev.* **2004**, *46*, 226–234. [CrossRef]
9. Zhong, H.; Zhu, W.G. Geochronology of layered mafic intrusions from the Pan-Xi area in the Emeishan large igneous province, SW China. *Miner. Depos.* **2006**, *41*, 599–606. [CrossRef]
10. Wang, C.Y.; Zhou, M.F.; Keays, R.R. Geochemical constraints on the origin of the Permian Baimazhai mafic–ultramafic intrusion, SW China. *Contrib. Mineral. Petrol.* **2006**, *152*, 309–321. [CrossRef]
11. He, B.; Xu, Y.G.; Huang, X.; Luo, Z.; Shi, Y.; Yang, Q.; Yu, S. Age and duration of the Emeishan flood volcanism, SW China: Geochemistry and SHRIMP zircon U–Pb dating of silicic ignimbrites, post-volcanic Xuanwei Formation and clay tuff at the Chaotian section. *Earth Planet. Sci. Lett.* **2007**, *255*, 306–323. [CrossRef]
12. Shellnutt, J.G.; Zhou, M.F. Permian peralkaline, peraluminous and metaluminous A-type granites in the Panxi district, SW China: Their relationship to the Emeishan mantle plume. *Chem. Geol.* **2007**, *243*, 286–316. [CrossRef]
13. Wang, X.K. The geological and geochemical characteristics of the basic volcanic rocks of Kafang deposit in Gejiu district. *J. Kunming Univ. Sci. Technol.* **1993**, *18*, 1–9. (In Chinese)
14. Zheng, S.H. An Analysis of the genesis of Cu deposit of Meta–Basic rock type Cu deposit of east Gejiu. *Yunnan Geol.* **2012**, *31*, 315–330. (In Chinese)
15. Zhang, X.L. Geochemical characteristics of basic volcanic rock in Indo-Chinese epoch in Gejiu, Yunnan, and its geotectonic background. *Miner. Resour. Geol.* **2011**, *25*, 429–435. (In Chinese)
16. Zhang, G.S.; Fang, W.X.; Peng, R.; Zheng, H.Y. Zircon U-Pb Chronology, Origin and Tectonic Significance of the Triassic High Potassic Volcanic Rock from Gejiu, Yunnan, Southwestern China. *Geotecton. Metallog.* **2019**, *43*, 1219–1235. (In Chinese)
17. Fang, W.X.; Jia, R.X. Characteristics of the Alkaline Picritic Volcanic Rocks in the Gejiu Superlarge Tin-Copper Deposit and their Continental Dynamic Implications. *Geotecton. Metallog.* **2011**, *35*, 137–148, (In Chinese with English Abstract).
18. Li, Y.S.; Qin, D.X.; Guo, N.N. Geotectonic setting and mineralization significance of indochina epoch basalt in eastern Gejiu of Yunnan. *Nonferrous Met.* **2009**, *61*, 104–109. (In Chinese)
19. Zhang, H.; Fang, W.X.; Du, Y.L. Geochemical Characteristics and Geological Significance of the Kafang Alkaline Volcanic Rocks in Gejiu Area, Yunnan Province. *Geotecton. Metallog.* **2014**, *38*, 885–897. (In Chinese)
20. Zhang, J.W.; Huang, Z.L.; Luo, T.Y.; Yan, Z.F. LA-ICP-MS zircon geochronology and platinum-group elements characteristics of the Triassic basalts, SW China: Implications for post-Emeishan large igneous province magmatism. *J. Asian Earth Sci.* **2014**, *87*, 69–78. [CrossRef]
21. Zhang, J.W.; Dai, C.G.; Huang, Z.L.; Luo, T.Y.; Qian, Z.K.; Zhang, Y. Age and petrogenesis of Anisian magnesian alkali basalts and their genetic association with the Kafang stratiform Cu deposit in the Gejiu supergiant tin-polymetallic district, SW China. *Ore Geol. Rev.* **2015**, *69*, 403–416. [CrossRef]
22. Cheng, Y.B.; Carl, S.; Mao, J.W. Granite, gabbro and mafic microgranular enclaves in the Gejiu area, Yunnan Province, China: A case of two-stage mixing of crust- and mantle-derived magmas. *Contrib. Mineral. Petrol.* **2012**, *164*, 659–676. [CrossRef]
23. Faure, M.; Lin, W.; Chu, Y.; Lepvrier, C. Triassic tectonics of the Ailaoshan Belt (SW China): Early Triassic collision between the South China and Indochina Blocks, and Middle Triassic intracontinental shearing. *Tectonophysics* **2016**, *683*, 27–42. [CrossRef]
24. Wang, X.; Metcalfe, I.; Jian, P.; He, L.; Wang, C. The Jinshajiang–Ailaoshan Suture Zone, China: Tectonostratigraphy, age and evolution. *J. Asian Earth Sci.* **2000**, *18*, 675–690. [CrossRef]

25. Wang, Q.; Deng, J.; Li, C.; Li, G.; Yu, L.; Qiao, L. The boundary between the Simao and Yangtze blocks and their locations in Gondwana and Rodinia: Constraints from detrital and inherited zircons. *Gondw. Res.* **2014**, *26*, 438–448. [CrossRef]
26. Faure, M.; Nguyen, V.V.; Hoai, L.T.T.; Lepvrier, C. Early Paleozoic or Early-Middle Triassic collision between the South China and Indochina blocks: The controversy resolved? Structural insights from the Kon Tum massif (Central Vietnam). *J. Asian Earth Sci.* **2018**, *166*, 162–180. [CrossRef]
27. Xu, Y.G.; He, B.; Chung, S.L.; Menzies, M.A.; Frey, F.A. Geologic, geochemical, and geophysical consequences of plume involvement in the Emeishan flood-basalt province. *Geology* **2004**, *32*, 917–920. [CrossRef]
28. Liu, Y.S.; Gao, S.; Hu, Z.C.; Gao, C.G.; Zong, K.Q.; Wang, D.B. Continental and oceanic crust recycling-induced melt-peridotite interactions in the Trans-North China Orogen: U-Pb dating, Hf isotopes and trace elements in zircons from mantle xenoliths. *J. Petrol.* **2010**, *51*, 537–571. [CrossRef]
29. Ludwig, K.R. *ISOPLOT 3.0: A Geochronological Toolkit for Microsoft Excel*; Special Publication No. 4; Berkeley Geochronology Center: Berkeley, CA, USA, 2003; Volume 4, pp. 1–71.
30. Gao, S.; Liu, X.M.; Yuan, H.L.; Hattendorf, B.; Günther, D.; Chen, L.; Hu, S. Determination of Forty Two Major and Trace Elements in USGS and NIST SRM Glasses by Laser Ablation-Inductively Coupled Plasma-Mass Spectrometry. *Geostand. Newsl.* **2002**, *26*, 181–196. [CrossRef]
31. Yan, Q.R.; Wang, Z.Q.; Liu, S.W. Opening of the Tethys in southwest China and its significance to the breakup of East Gondwanaland in late Paleozoic: Evidence from SHRIMP U-Pb zircon analyses for the Garzê ophiolite block. *Chin. Sci. Bull.* **2005**, *50*, 256–264. [CrossRef]
32. Zhang, H.F.; Sun, M.; Zhou, X.H.; Fan, W.M.; Zhai, M.G.; Yin, J.F. Mesozoic lithosphere destruction beneath the North China craton: Evidence from major-, trace-element and Sr–Nd–Pb isotope studies of Fangcheng basalts. *Contrib. Mineral. Petrol.* **2002**, *144*, 241–253. [CrossRef]
33. Belshawet, N.S.; Freedman, P.A.; O'Nions, R.K.; Frank, M.; Guo, Y. A new variable dispersion double-focusing plasma mass spectrometer with performance illustrated for Pb isotopes. *Int. J. Mass Spectrom.* **1998**, *181*, 51–58. [CrossRef]
34. Xu, Y.G.; Chung, S.L.; Jahn, B.M.; Wu, G. Petrologic and geochemical constraints on the petrogenesis of Permian-Triassic Emeishan flood basalts in southwestern China. *Lithos* **2001**, *58*, 145–168. [CrossRef]
35. Sun, S.S.; McDonough, W.F. Chemical and isotopic systematics of oceanic basalts: Implications for mantle composition and processes. In *Magmatism in the Ocean Basins*; Geological Society London Special Publications: London, UK, 1989; pp. 313–345.
36. Yang, Z.X.; Mao, J.W.; Chen, M.H.; Cheng, Y.B.; Chang, Y. Geology, geochemistry and genesis of Kafang copper deposit in Gejiu, Yunnan Province. *Acta Petrol. Sin.* **2010**, *26*, 830–844, (In Chinese with English Abstract).
37. Zhang, Y.T.; Liu, J.Q.; Guo, Z.F. Permian basaltic rocks in the Tarim basin, NW China: Implications for plume–lithosphere interaction. *Gondwana Res.* **2010**, *18*, 596–610. [CrossRef]
38. Wang, Y.F.; Zhang, Z.C.; Wang, L.J.; Lv, L.S.; Li, H.B. Geochemical characteristics of Permian basalt from Hutiaoxia and Jin'an area of the Emeishan Large Igneous Province and constraints on their source region. *Acta Petrol. Sin.* **2013**, *29*, 4387–4403, (In Chinese with English Abstract).
39. Jung, S.; Hoernes, S. The major and trace element and isotope (Sr, Nd, O) geochemistry of Cenozoic alkaline rift-type volcanic rocks from the Rhon area (central Germany): Petrology, mantle source characteristics and implications for asthenosphere-lithosphere interactions. *J. Volcanol. Geotherm. Res.* **2000**, *99*, 27–53. [CrossRef]
40. Condie, K.C. Incompatible element ratios in oceanic basalts and komatiites, tracking deep mantle sources and continental growth rates with time. *Geochem. Geophys. Geosyst.* **2003**, *4*, 1005. [CrossRef]
41. Zhu, D.C.; Mo, X.X.; Wang, L.Q.; Zhao, Z.D.; Liao, Z.L. Hotspot-ridge interaction for the evolution of Neo-Tethys: Insights from the Late Jurassic-Early Cretaceous magmatism in southern Tibet. *Acta Petrol. Sin.* **2008**, *24*, 225–237. (In Chinese)
42. Fitton, J.G.; Saunders, A.D.; Norry, M.J.; Hardarson, B.S.; Talor, R.N. Thermal and chemical structure of the Iceland plume. *Earth Planet. Sci. Lett.* **1997**, *153*, 197–208. [CrossRef]
43. Staudigel, H.; Hart, S.R. Alteration of basaltic glass: Mechanisms and significance for the oceanic crust–seawater budget. *Geochim. Cosmochim. Acta.* **1983**, *47*, 337–350. [CrossRef]

44. Wang, C.Y.; Zhou, M.F.; Qi, L. Permian flood basalts and magmatic intrusions in the Jinping (SW China)–Song Da (northern Vietnam) district: Mantle sources, crustal contamination and sulfide segregation. *Chem. Geol.* **2007**, *243*, 317–343. [CrossRef]
45. Zhang, B.L.; Lv, G.X.; Su, J.; Shen, X.L.; Liu, R.L.; Liu, J.G.; Hai, L.F.; Zhang, G.L. A study of the tectono-lithofacies mineralization regularities of the Gejiu tin polymetallic orefield, Yunnan, and prospecting in its western part. *Earth Sci. Front.* **2015**, *22*, 078–087. (In Chinese)
46. Zhao, J.H.; Zhou, M.F. Geochemistry of Neoproterozoic mafic intrusions in the Panzhihua district (Sichuan Province, SW China): Implications for subduction-related metasomatism in the upper mantle. *Precambrian Res.* **2007**, *152*, 27–47. [CrossRef]
47. Yang, G.X.; Li, Y.J.; Tong, L.L.; Li, G.Y.; Shen, R.; Li, Z.; Wu, L. Geochronology, geochemistry and petrogenesis of pillow basalts from Mayile region in West Junggar. *Actor Petrol. Sin.* **2016**, *32*, 522–536. (In Chinese)
48. Qi, L.; Zhou, M.F. Platinum-group elemental and Sr–Nd–Os isotopic geochemistry of Permian Emeishan flood basalts in Guizhou Province, SW China. *Chem. Geol.* **2008**, *248*, 83–103. [CrossRef]
49. Shaw, D.M. Trace element fractionation during anatexis. *Geochim. Cosmochim. Acta* **1970**, *34*, 237–243. [CrossRef]
50. Saunders, A.D.; Storey, M.; Kent, R.W.; Norry, M.J. Consequences of plume-lithosphere interactions. In *Magmatism and the Causes of Continental Break-up*; Storey, B.C., Alabaster, T., Pankhurst, R.J., Eds.; Geological Society: London, UK, 1992; Volume 68, pp. 41–60.
51. Song, Y.; Frey, F.A.; Zhi, X. Isotopic characteristics of Hannuoba basalts, eastern China: Implications for their petrogenesis and the composition of subcontinental mantle. *Chem. Geol.* **1990**, *85*, 35–52. [CrossRef]
52. Basu, A.R.; Wang, J.W.; Huang, W.K.; Xie, G.H.; Tatsumoto, M. Major element, REE and Pb, Nd and Sr isotopic geochemistry of Cenozoic volcanic rocks of eastern China: Implication for their origin from suboceanic type mantle reservoirs. *Earth Planet. Sci. Lett.* **1991**, *105*, 149–169. [CrossRef]
53. Tatsumoto, M.; Nakamura, Y. DUPAL anomaly in the Sea of Japan: Lead, neodymium, and strontium isotopic variations at the eastern Eurasian continental margin. *Geochim. Cosmochim. Acta* **1991**, *55*, 3697–3708. [CrossRef]
54. Brenhin, K.; Blair, S. Plate tectonics and continental basaltic geochemistry throughout Earth history. *Earth Planet. Sci. Lett.* **2018**, *481*, 290–304.
55. Zhou, P.B.; Mukasa, S.B. Nd-Sr-Pb isotopic, and major and trace-element geochemistry of Cenozoic lavas from the Khorat Plateau, Thailand: Sources and petrogenesis. *Chem. Geol.* **1997**, *137*, 175–193. [CrossRef]
56. Jaques, A.L.; Green, D.H. Anhydrous melting of peridotite at 0—15 kb pressure and the genesis of tholeiitic basalts. *Contrib. Mineral. Petrol.* **1980**, *73*, 287–310. [CrossRef]
57. Takahashi, E.; Kushiro, I. Melting of a dry peridotite at high pressures and basalt magma genesis. *Am. Miner.* **1983**, *68*, 859–879.
58. Derakhshi, M.; Ghasemi, H.; Miao, L.C. Geochemistry and petrogenesis of Soltan Maidan basalts (E Alborz, Iran): Implications for asthenosphere-lithosphere interaction and rifting along the N margin of Gondwana. *Geochemistry* **2017**, *77*, 131–145. [CrossRef]
59. Herzberg, C. Depth and degree of melting of komatiites. *J. Geophys. Res.* **1992**, *97*, 4521–4540. [CrossRef]
60. Herzberg, C. Generation of plume magmas through time: An experimental perspective. *Chem. Geol.* **1995**, *126*, 1–16. [CrossRef]
61. DePaolo, D.J. Trace element and isotopic effects of combined wallrock assimilation and fractional crystallization. *Earth Planet. Sci. Lett.* **1981**, *53*, 189–202. [CrossRef]
62. Lemdjou, Y.B.; Zhang, D.; Jean, P.T.; Hu, J.X.; Nicaise, B.T.N.; Landry, S.T.; Yuan, Y. Elemental and Sr–Nd–Pb isotopic compositions, and K–Ar ages of transitional and alkaline plateau basalts from the eastern edge of the West Cameroon Highlands (Cameroon Volcanic Line). *Lithos* **2020**, 105–414. [CrossRef]
63. Hofmann, A.W.; Jochum, K.; Seufert, M.; White, W.M. Nb and Pb in oceanic basalts: New constraints on mantle evolution. *Earth Planet. Sci. Lett.* **1986**, *79*, 33–45. [CrossRef]
64. Hart, W.K.; WoldeGabriel, G.; Walter, R.C.; Mertzman, S.A. Basaltic volcanism in Ethiopia: Constraints on continental rifting and mantle interactions. *J. Geophys. Res. Solid Earth* **1989**, *94*, 7731–7748. [CrossRef]
65. Zhu, D.C.; Pan, G.T.; Mo, X.X.; Wang, L.Q.; Zhao, Z.D.; Liao, Z.L.; Geng, Q.R.; Dong, G.C. Identification for the Mesozoic OIB-type Basalts in Central Qinghai-Tibetan Plateau: Geochronology, Geochemistry and Their Tectonic Setting. *Acta Geol. Sin.* **2006**, *80*, 1312–1328, (In Chinese with English Abstract).

66. Li, X.H.; Li, Z.X.; Zhou, H.W.; Liu, Y.; Kinny, P.D. U-Pb zircon geochronology, geochemistry and Nd isotopic study of Neoproterozoic bimodal volcanic rocks in the Kangdian Rift of South China: Implications for the initial rifting of Rodinia. *Precambrian Res.* **2002**, *113*, 135–154. [CrossRef]
67. Niu, Y.L. Some basic concepts and problems on the petrogenesis of intra-plate ocean island basalts. *Chin. Sci. Bull.* **2009**, *54*, 4148–4160. [CrossRef]
68. Hofmann, A.W. Mantle geochemistry: The message from oceanic volcanism. *Nature* **1997**, *385*, 219–229. [CrossRef]
69. Maruyama, S.; Santosh, M.; Zhao, D. Superplume, supercontinent, and post-perovskite: Mantle dynamics and anti-plate tectonics on the core-mantle boundary. *Gondwana Res.* **2007**, *11*, 7–37. [CrossRef]
70. Davies, G.F. *Mantle Convection for Geologists*; Cambridge University Press: Cambridge, UK, 2011; p. 240.
71. Safonova, I.; Santosh, M. Accretionary complexes in the Asia-Pacific region: Tracing archives of ocean plate stratigraphy and tracking mantle plumes. *Gondwana Res.* **2014**, *25*, 126–158. [CrossRef]
72. Safonova, I.; Kojima, S.; Nakae, S.; Romer, R.L.; Seltmann, R.; Sano, H.; Onoue, T. Oceanic island basalts in accretionary complexes of SW Japan: Tectonic and petrogenetic implications. *J. Asian Earth Sci.* **2015**, *113*, 508–523. [CrossRef]
73. Fang, W.X.; Zhang, H.; Jia, R.X. Dynamics of Triassic Back-Arc Rift Basin and its Metallogenic Sequence in Gejiu of Yunnan Province to Napo of Guangxi Zhuang Autonomous Region, China. *Geotecton. Metallog.* **2011**, *35*, 552–566, (In Chinese with English Abstract).
74. Meschede, M. A method of discrimination between different types of mid–ocean ridge basalts and continental tholeiites with the Nb–Zr–Y diagram. *Chem. Geol.* **1986**, *56*, 207–218. [CrossRef]
75. Pearce, J.A.; Cann, J.R. Tectonic setting of basic volcanic rocks determined using trace element analyses. *Earth Planet. Sci. Lett.* **1973**, *19*, 290–300. [CrossRef]
76. Shellnutt, J.G.; Denyszyn, S.W.; Mundil, R. Precise age determination of mafic and felsic intrusive rocks from the Permian Emeishan large igneous province (SW China). *Gondwana Res.* **2012**, *22*, 118–126. [CrossRef]
77. Wood, D.A. The application of a Th–Hf–Ta diagram to problems of tectonomagmatic classification and to establishing the nature of crustal contamination of basaltic lavas of the British tertiary volcanic province. *Earth Planet. Sci. Lett.* **1980**, *50*, 11–30. [CrossRef]
78. Lai, C.K.; Meffre, S.; Crawford, A.J.; Zaw, K.; Xue, C.D.; Halpin, J.A. The Western Ailaoshan Volcanic Belts and their SE Asia connection: A new tectonic model for the Eastern Indochina Block. *Gondwana Res.* **2014**, *26*, 52–74. [CrossRef]
79. Liu, J.; Tran, M.D.; Tang, Y.; Nguyen, Q.L.; Tran, T.H.; Wu, W.; Chen, J.; Zhang, Z.; Zhao, Z. Permo-Triassic granitoids in the northern part of the Truong Son belt, NW Vietnam: Geochronology, geochemistry and tectonic implications. *Gondwana Res.* **2012**, *22*, 628–644. [CrossRef]
80. Lepvrier, C.; Maluski, H.; Tich, V.V.; Leyreloup, A.; Thi, P.T.; Nguyen, V.V. The Early Triassic Indosinian orogeny in Vietnam (Truong Son Belt and Kontum Massif); implications for the geodynamic evolution of Indochina. *Tectonophysics* **2004**, *393*, 87–118. [CrossRef]
81. Yan, D.P.; Zhou, M.F.; Wang, C.Y.; Xia, B. Structural and geochronological constraints on the tectonic evolution of the Dulong-Song Chay tectonic dome in Yunnan province, SW China. *J. Asian Earth Sci.* **2006**, *28*, 332–353. [CrossRef]
82. Shellnutt, J.G.; Zhou, M.F.; Yan, D.P.; Wang, Y. Longevity of the Permian Emeishan mantle plume (SW China): 1 Ma, 8 Ma or 18 Ma? *Geol. Mag.* **2008**, *145*, 373–388. [CrossRef]
83. Wang, Y.J.; Qian, X.; Cawood, P.A.; Liu, H.C.; Feng, Q.L.; Zhao, G.C.; Zhang, Y.H.; He, H.Y.; Zhang, P.Z. Closure of the East Paleotethyan Ocean and amalgamation of the Eastern Cimmerian and Southeast Asia continental fragment. *Earth Sci. Rev.* **2018**, *186*, 195–230. [CrossRef]
84. Lai, C.K.; Meffre, S.; Crawford, A.J.; Zaw, K.; Halpin, J.A.; Xue, C.D.; Salam, A. The Central Ailaoshan ophiolite and modern analogs. *Gondwana Res.* **2014**, *26*, 75–88. [CrossRef]
85. Zaw, K.; Meffre, S.; Lai, C.K.; Burrett, C.; Santosh, M.; Graham, I.; Ml, T.; Salam, A.; Kamvong, T.; Cromie, P. Tectonics and metallogeny of mainland Southeast Asia—A review and contribution. *Gondwana Res.* **2014**, *26*, 5–30.
86. Xu, J.; Xia, X.; Huang, C.; Cai, K.; Yin, C.; Lai, C.-K. Changes of provenance of Permian and Triassic sedimentary rocks from the Ailaoshan suture zone (SW China) with implications for the closure of the eastern Paleotethys. *J. Asian Earth Sci.* **2019**, *170*, 234–248. [CrossRef]

87. Liu, C.; Deng, J.F.; Liu, J.L.; Shi, Y.L. Characteristics of volcanic rocks from Late Permian to Early Traissic in Ailaoshan tectonomagmatic belt and implications for tectonic settings. *Acta Petrol. Sin.* **2011**, *27*, 3590–3602, (in Chinese with English abstract).
88. Zi, J.W.; Cawood, P.A.; Fan, W.M.; Wang, Y.J.; Tohver, E.; McCuaig, T.C.; Peng, T.P. Triassic collision in the Paleo-Tethys Ocean constrained by volcanic activity in SW China. *Lithos* **2012**, *144*, 145–160. [CrossRef]
89. Roger, F.; Maluski, H.; Lepvrier, C.; Van, T.V.; Paquette, J.L. LA-ICPMS zircons U/Pb dating of Permo-Triassic and Cretaceous magmatisms in Northern Vietnam-geodynamical implications. *J. Asian Earth Sci.* **2012**, *48*, 72–82. [CrossRef]
90. Ivanov, A.V.; Meffre, S.; Thompson, J.; Corfu, F.; Vadim, S.; Kamenetsky, M.B.; Demonterova, E.I. Timing and genesis of the Karoo-Ferrar large igneous province: New high precision U-Pb data for Tasmania confirm short duration of the major magmatic pulse. *Chem. Geol.* **2017**, *455*, 32–43. [CrossRef]
91. Svensen, H.; Corfu, F.; Polteau, S.; Hammer, O.; Planke, S. Rapid magma emplacement in the Karoo Large Igneous Province. *Earth Planet. Sci. Lett.* **2012**, *325–326*, 1–9. [CrossRef]
92. Augland, L.E.; Ryabov, V.V.; Vernikovsky, V.A.; Planke, S.; Polozov, A.G.; Callegaro, S.; Jerram, D.A.; Svensen, H.H. The main pulse of the Siberian Traps expanded in size and composition. *Sci. Rep.* **2019**, 18723. [CrossRef]
93. Michael, P.E.; Blair, S.; Kyle, M.S.; Gerta, K.; Thierry, A.; Syed, F.R.K. U-Pb zircon age constraints on the earliest eruptions of the Deccan Large Igneous Province, Malwa Plateau, India. *Earth Planet. Sci. Lett.* **2020**, *540*, 116249.
94. Shellnutt, J.G.; Jahn, B.M. Origin of late Permian Emeishan basaltic rocks from the Panxi region (SW China): Implications for the Ti-classification and spatial–compositional distribution of the Emeishan flood basalts. *J. Volcanol. Geotherm. Res.* **2011**, *199*, 85–95. [CrossRef]
95. Lo, C.H.; Chung, S.L.; Lee, T.Y.; Wu, G. Age of the Emeishan flood magmatism and relations to Permian–Triassic boundary events. *Earth Planet. Sci. Lett.* **2002**, *198*, 449–458. [CrossRef]
96. Boven, A.; Pasteels, P.; Punzalan, L.E.; Liu, J.; Luo, X.; Zhang, W.; Guo, Z.; Hertogen, J. $^{40}Ar/^{39}Ar$ geochronological constraints on the age and evolution of the Permo-Triassic Emeishan volcanic province, southwest China. *J. Asian Earth Sci.* **2002**, *20*, 157–175. [CrossRef]
97. Fan, W.; Wang, V.; Peng, T.; Miao, L.; Guo, F. Ar–Ar and U–Pb geochronology of late Paleozoic basalts in western Guangxi and its constraints on the eruption age of Emeishan basalt magmatism. *Chin. Sci. Bull.* **2004**, *49*, 2318–2327. (In Chinese) [CrossRef]

minerals

MDPI

Article

Tectonic Evolution of the West Bogeda: Evidences from Zircon U-Pb Geochronology and Geochemistry Proxies, NW China

Yalong Li [1], Wei Yue [2], Xun Yu [1], Xiangtong Huang [1], Zongquan Yao [3], Jiaze Song [1], Xin Shan [4], Xinghe Yu [5] and Shouye Yang [1,*]

[1] State Key Laboratory of Marine Geology, Tongji University, Shanghai 200092, China; 1710583@tongji.edu.cn (Y.L.)
[2] School of Geography, Geomatics and Planning, Jiangsu Normal University, Xuzhou 221116, China
[3] School of geological and mining engineering, Xinjiang University, Urumqi 830047, China
[4] The First Institute of Oceanography, Soa, Qingdao 266000, China
[5] School of Energy Resource, China University of Geosciences (Beijing), Beijing 100083, China
* Correspondence: syyang@tongji.edu.cn

Received: 2 March 2020; Accepted: 8 April 2020; Published: 10 April 2020

Abstract: The Bogeda Shan (Mountain) is in southern part of the Central Asian Orogenic Belt (CAOB) and well preserved Paleozoic stratigraphy, making it an ideal region to study the tectonic evolution of the CAOB. However, there is a long-standing debate on the tectonic setting and onset uplift of the Bogeda Shan. In this study, we report detrital zircon U-Pb geochronology and whole-rock geochemistry of the Permian sandstone samples, to decipher the provenance and tectonic evolution of the West Bogeda Shan. The Lower-Middle Permian sandstone is characterized by a dominant zircon peak age at 300–400 Ma, similar to the Carboniferous samples, suggesting their provenance inheritance and from North Tian Shan (NTS) and Yili-Central Tian Shan (YCTS). While the zircon record of the Upper Permian sandstone is characterized by two major age peaks at ca. 335 Ma and ca. 455 Ma, indicating the change of provenance after the Middle Permian and indicating the uplift of Bogeda Shan. The initial uplift of Bogeda Shan was also demonstrated by structural deformations and unconformity occurring at the end of Middle Permian. The bulk elemental geochemistry of sedimentary rocks in the West Bogeda Shan suggests the Lower-Middle Permian is mostly greywacke with mafic source dominance, and tectonic setting changed from the continental rift in the Early Permian to post rift in the Middle Permian. The Upper Permian mainly consists of litharenite and sublitharenite with mafic-intermediate provenances formed in continental island arcs. The combined evidences suggest the initial uplift of the Bogeda Shan occurred in the Late Permian, and three stages of mountain building include the continental rift, post-rift extensional depression, and continental arc from the Early, Middle, to Late Permian, respectively.

Keywords: detrital zircon U-Pb geochronology; bulk geochemistry; provenance; tectonic setting; West Bogeda Shan

1. Introduction

The Central Asian Orogenic Belt (CAOB) is located between the Siberia Craton to the north and the Tarim and North China Craton to the south. It is regarded as the largest (extending 7000 km from west to east) accretionary orogenic belt on Earth. It was formed by a series of amalgamation events of several micro-continents and island arcs during the Late Carboniferous to Permian periods (Figure 1a) [1–3]. The Tian Shan forms the southern part of the CAOB, with an average elevation of ca. 2000 m and summits >7000 m. It is the key orogenic belt to study the tectonic evolution of the

CAOB due to its well-preserved stratigraphic units of ophiolites, volcanic rocks, granitoids, high-grade metamorphic rocks, and sedimentary sequences [4–7]. The Chinese Tian Shan can be further divided into three different tectonic units, including the North Tian Shan (NTS), Yili-Central Tian Shan (YCTS), and South Tian Shan (STS) from north to south [8,9]. Among them, the NTS, especially the Bogeda Shan in the northeast, are considered as the essential part and an ideal region to examine the evolution of the CAOB due to its well-preserved and exposed Late Paleozoic to Mesozoic strata.

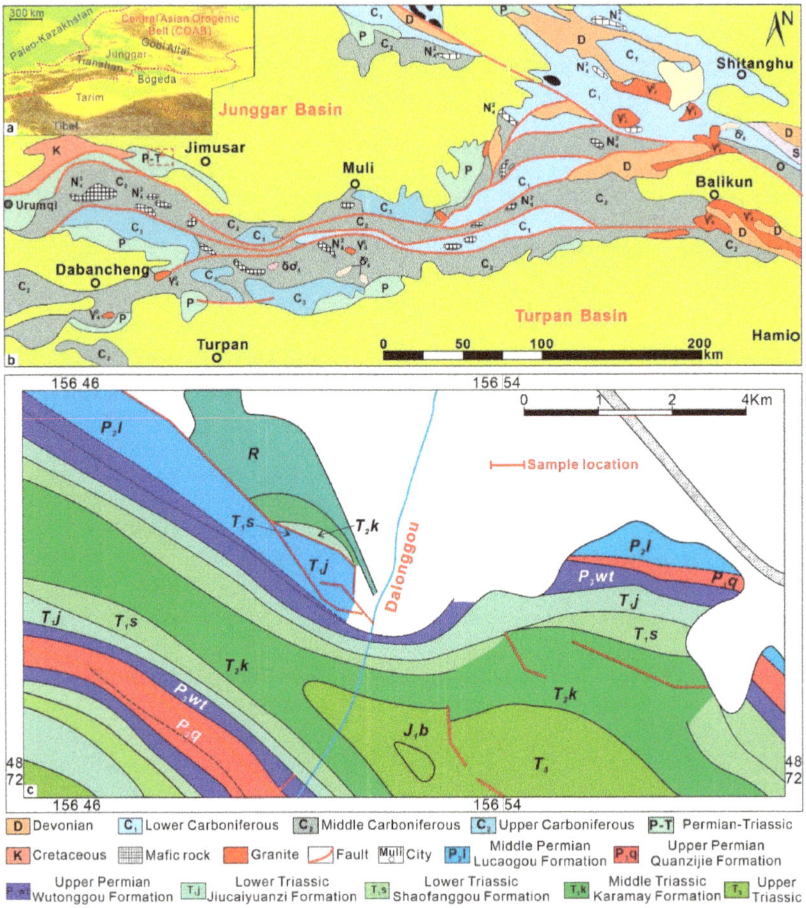

Figure 1. The sketch map of Junggar basin and the strata of Dalongkou section. (**a**) Maps show the study region, the Central Asian Orogenic Belt, Tianshan Mountains, and Bogeda Shan; (**b**) a map of Bogeda Shan; (**c**) strata distribution of the Dalongkou section.

The Late Paleozoic witnessed significant climate and tectonic changes in Northwest China, including the Bogeda region. Despite previous research on the tectonic evolution of Bogeda Shan, the timing of initial uplift and its tectonic setting are still enigmatic. Some authors first suggested that a small-scale orogeny occurred in the Bogeda area during the Carboniferous [10], based on zircon U-Pb geochronology, while the initial uplift of the West Bogeda Shan occurred in the Late Permian [11,12]. However, other researchers suggested that the first significant uplift of Bogeda Shan occurred during the Early to Middle Jurassic, according to the configuration of sedimentary units [13–15]. Besides, the Late Jurassic has been considered as the most critical stage for the Bogeda Shan uplift [16–18]. As to the tectonic setting, two contrasting viewpoints exist, i.e., a continental rift or an island arc, and are still highly debated [19].

It is widely accepted that the Bogeda Shan was uplifted during the Mesozoic, but the onset time remains largely controversial. Therefore, this study aims to illustrate the provenances, tectonic setting, and evolution of the West Bogeda Shan during the Permian based on the integrated analyses of petrology, detrital zircon U-Pb geochronology, and bulk geochemistry of sedimentary rocks from the Lower to Upper Permian. The timing of the initial uplift of the Bogeda Shan will be revealed with all these analyses.

2. Geological Setting and Stratigraphy

The study area of Bogeda Shan belongs to the NTS Belt which is composed of volcanic arcs, intra-arc basin, and accretionary complex in relation to the final closure of the Northern Tianshan Ocean [20]. The NTS Belt can be further divided into the Bogeda-Harlik Belt in the north and Juelotage Belt in the south. The Bogeda Shan is a part of the Bogeda-Harlik Belt and extends >250 km in the east–west direction (Figure 1b). It is the geographical boundary of the two largest basins in northwest China, with the Junggar Basin in the north and Turpan-Hami Basin in the south. The Paleozoic orogenic movements formed the high topography of Bogeda Shan due to India–Asia collision during the early Cenozoic [21–23]. The central part of the Bogeda Shan consists of Carboniferous sedimentary rocks (Lower Carboniferous composed by marine volcanic ignimbrite and bimodal volcanic lava, while Upper Carboniferous is dominated by felsic ignimbrite and marine basaltic lava), while the north and south parts are composed mainly by Mesozoic and Cenozoic sediments. The lithology of the East Bogeda Shan is more variable than in the western part, and the latter was formed during the Early Carboniferous (Mississippian) [24–27]. The Dalongkou section investigated in this study is distributed in the Fukang Depression in the West Bogeda Shan, with the clear exposure of the Permian and Triassic strata for field work (Figure 1c).

The Permian strata at the anterior of West Bogeda Shan are made up of the following eight central units: the Shirenzigou Formation (P_1s), Tashkula Formation (P_1t) in the Lower Permian, Ulupo Formation (P_2w), Jingjingzigou Formation (P_2j), Lucaogou Formation (P_2l) and Hongyanchi Formation (P_2h) in the Middle Permian, as well as Quanzijie Formation (P_3q) and Wutonggou Formation (P_3wt) in the Upper Permian (Figure 2). The paleontologic record suggest the depositional environments in the anterior region of the West Bogeda Shan were normal marine and shelf depositions during the Carboniferous to Early Permian (Figure 2), with sandstone and siltstone representing the primary lithologies. In the Middle Permian, the depositional environment changed to terrestrial and lacustrine facies, resulting in a variable lithology of sandstone, siltstone, mudstone, and oil-bearing mudstone. The Upper Permian strata is composed of conglomerate and sandstone deposited in alluvial fan and braided river environment [28–33].

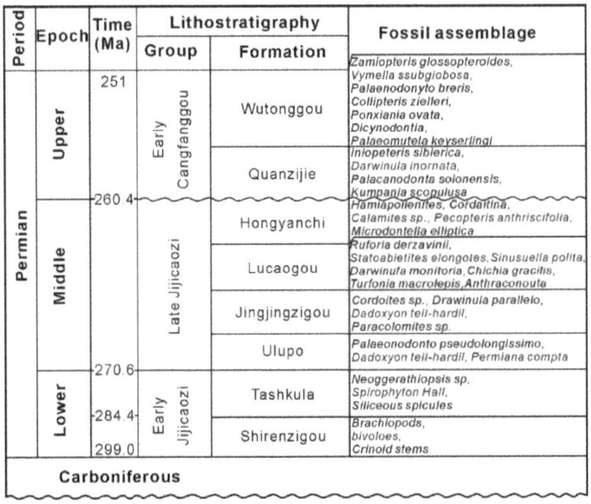

Figure 2. Chronology and lithostratigraphy of the Permian strata around the Bogeda Shan, southern Junggar Basin, NW China [33].

3. Sampling and Analytical Methods

This study collected new data on the samples from the Upper Permian, and a total of eleven sandstone samples were collected from the Dalongkou section (43°57′21″ N, 88°51′44″ E) with the thickness of 117.53 m (Figure 3a,b). Among them, seven sandstone samples were chosen for zircon U-Pb dating, and eleven samples for whole-rock geochemistry analyses. The samples are mainly from quartz-rich medium to fine grained sandstone (Figure 3h,i), with poor to moderate sorting and roundness (Figure 3c,d).

3.1. Zircon U-Pb Geochronology

The samples for detrital zircon U-Pb dating were prepared following the previous procedures [34,35]. The sandstone samples were first crushed with the agate mortar. Then, the grain size fraction of 63–125 μm was separated by the wet-sieving method. After wet sieving, tribromomethane liquid ($CHBr_3$) was used to separate heavy minerals, followed by magnetic separation. Later, detrital zircon grains were then identified and picked out from non-magnetic or weak magnetic minerals under a binocular microscope. About 200–300 grains of zircon were randomly selected, pasted on adhesive tapes, and enclosed in epoxy resin followed by polishing to yield a smooth flat surface. Before being ablated by a laser, cathodoluminescence (CL) images were used to check the internal structures of zircons by the electron microprobe of JEOL JXA-8230 (JEOL, Tokyo, Japan).

The measurements of zircon U-Pb ratios were performed at Tong University using a 193 nm excimer laser (Resonetics M50L) (Resonetics, Nashua, NH, USA) coupled with a quadrupole inductively coupled plasma mass spectrometry (ICP-MS, Agilent 7900, Agilent, Santa Clara, CA, USA). The zircon grains were ablated with a laser spot size of 26 μm at the repetition of 6 Hz and the fluence of 4 J cm^{-2}. Masses 206, 207, 208, 232, 235, and 232 were acquired by the ICP-MS. Reference zircon materials 91500 and Plešovice were measured periodically to carry out U-Pb age external calibration and monitor the measurements. The U-Pb isotope ratios and the corresponding ages were calibrated using UranOS software [36]. The brief calibration procedure included blank subtraction, calculation of ratio of means, instrumental drift correction, and normalization by primary reference material (91500). The uncertainties of U-Pb isotopes ratios and ages were propagated in the calibration and results were reported with 2σ uncertainties [36]. As we did not acquired mass 204, the common Pb correction was performed using the Stacey-Kramers method on the basis of the measured $^{206}Pb/^{238}U$ ages [37]. To minimize the uncertainty due to some poor-quality ages, U-Pb ages with discordance larger than 10% were excluded from the following discussion. The discordance of $^{206}Pb/^{238}U$ age less than 1.4 Ga is defined as 100*(1-$^{206}Pb/^{238}U/^{207}Pb/^{235}U$) and the discordance of $^{206}Pb/U^{238}$ age greater than 1.4 Ga is defined as 100*(1-$^{206}Pb/^{238}U/^{207}Pb/^{206}Pb$) [38]. The weighted mean ages of reference zircons 91500 and Plešovice are 1062.8 ± 9.9 Ma and 336.1 ± 3.1 Ma, respectively (Figure 6a), which are consistent with the reference ages within the uncertainties [39–41].

Figure 3. Panorama of the Upper Permian Wutonggou Formation and sampling location for zircon chronology analysis. (**a**) Panorama of Dalong kou section. (**b**) Lithology column of P_3wt. (**c**) Photomicrographs of the sandstone samples of P_3wt. (**d**) Plant fragements were found in thin section. (**e**) Serious strata deformation in P_2l (the end of Middle Permian). (**f**) and (**g**) Conglomerates of P_3q. (**h**) and (i) Sandstone of P_3wt. (**j**) and (**k**) Bivalve and plant fragements were distributed in sandstone.

3.2. Major and Trace Elemental Analysis

For the measurements of major and trace elemental compositions in the bulk samples, eleven samples were first ground by an agate mortar, and then organic matter was removed in a muffle furnace at the temperature of 600 °C. A mixture of 1:1 HF and HNO_3 acids was added to the samples and kept in dissolution bombs in an oven of 190 °C for 48 h for digestion. After the digestion, 30% HNO_3 was added to the samples before putting into the oven for at least 12 h at 190 °C. The completely digested samples were measured for major and trace elements by ICP-OES (IRIS Advantage) and ICP-MS (Agilent 7900), respectively. Four kinds of geo-standards (BCR-2, BHVO-2, AGV-2, and GSP-2) [42] were used for the analytic quality control, which yields the analytical uncertainties less than 5%. The Si concentration was calculated by assuming the total content of major oxides and trace elements is 100% according to previous research [43]. All of the above sample preparations and measurements were conducted at the State Key Laboratory of Marine Geology, Tongji University, Shanghai, China.

4. Results

4.1. Detrital Zircon U-Pb Geochronology

To examine the sediment provenances of the West Bogeda Shan from the Carboniferous to Triassic periods, all available detrital zircon U-Pb ages from previous studies are compiled and presented in Table S1 [24,32,44–47]. The U-Pb ages of the Upper Permian detrital zircons from seven sandstone samples were measured in the present study. Most of the zircon grains have oscillatory zoning and are generally euhedral to subhedral on the CL images (Figure 4), suggesting that these zircons from the Upper Permian originated mostly from acidic magmatic rocks. The Th/U ratios of these zircons are mostly >0.1, further diagnostic of their magmatic origin (Figure 5) [48]. Results of all samples within 90–110% concordance (or less than 10% discordance) are plotted in the U-Pb concordia diagrams (Appendix A), and the data with concordance <90% and >110% are excluded during subsequent analysis.

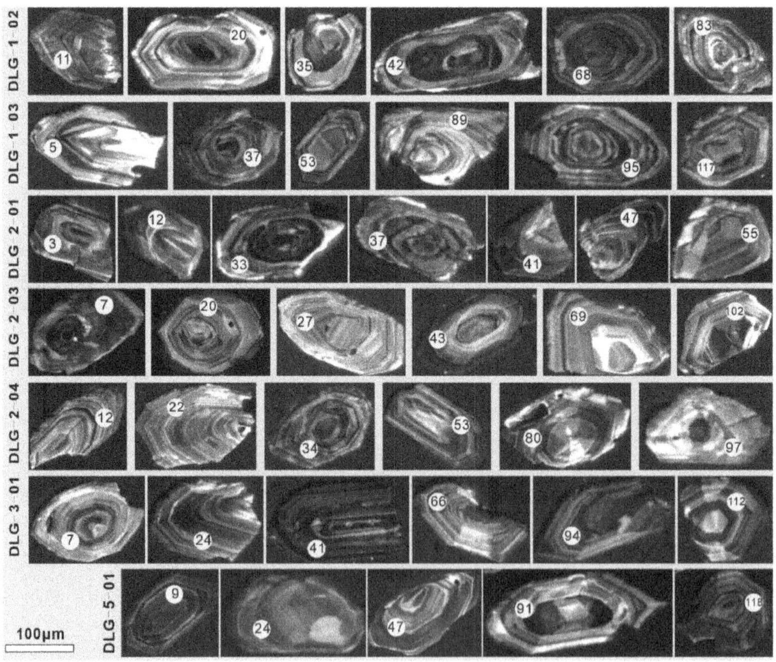

Figure 4. Representative cathodoluminescence (CL) images of detrital zircons from the Upper Permian rocks, west of Bogeda Shan, NW China.

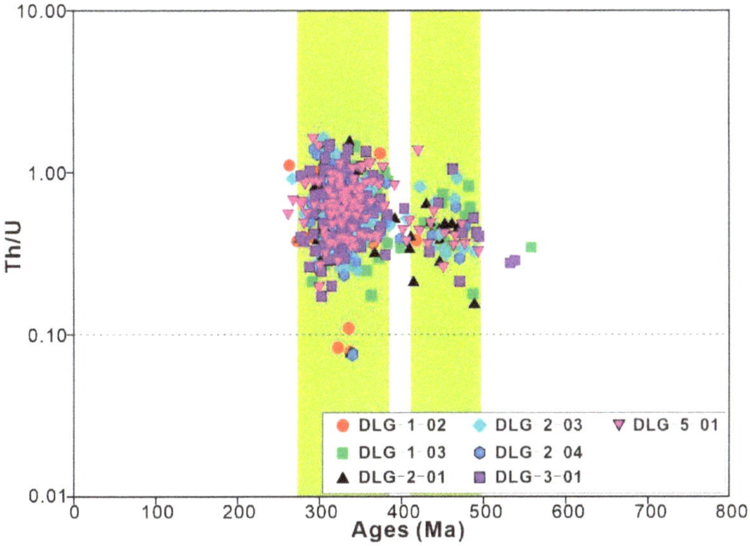

Figure 5. Zircon U-Pb ages versus Th/U ratios of the Upper Permian Wutonggou Formation. The dashed line is Th/U = 0.1.

The U-Pb age populations of detrital zircon grains of Carboniferous and Lower-Middle Permian are similar, mostly in the range of 300–541 Ma [8,47,49]. While, the zircons from the Upper Permian in the west Bogeda yield the U-Pb ages varying from ca. 262 Ma to ca. 558 Ma, with the oldest zircon from the Cambrian, and the youngest from the Permian. Most zircon grains are older than Carboniferous (>295 Ma), which account for 94.3%. To all of the zircon U-Pb ages, the Carboniferous zircons with U-Pb ages of 295–354 Ma have a mean proportion of 69.4%, bracketing the dominant period of zircon generation. Devonian zircons with U-Pb ages of 354–410 Ma are the second highest population, accounting for 13.7%. In comparison, zircon ages younger than the Carboniferous (<295 Ma) only account for 5.7%. Overall, the peak ages of zircon from the Upper Permian sedimentary rocks are ca. 335 Ma and ca. 455 Ma (Figure 6b).

4.2. Major and Trace Elements

The compositions of major and trace elements of eleven sandstone samples from the West Bogeda Shan are given in Table 1. Geochemical data of samples from the Lower-Middle Permian were collected from previous studies [8,50,51] and presented in Table S2.

The contents of SiO_2 exhibit a wide range from 42.9% to 89.2% with a notable increase from the Lower to Upper Permian. Al_2O_3 contents, however, yield lower values in the Upper Permian samples and high in the Middle Permian. No apparent linear relationships exist between Na_2O, K_2O, Fe_2O_3, and SiO_2, while both MgO and CaO are negatively correlated with SiO_2. TiO_2 contents increase with SiO_2 in the Lower-Middle Permian samples, but decrease in the Upper Permian. MnO contents have no linear relationship with SiO_2 in the Lower-Middle Permian, and a negative correlation in the Upper Permian. Trace element compositions of rock samples in West Bogeda Shan are shown in Table 1. Elements including Th, Zr Hf, Sr, and Ba have positive correlations with SiO_2, showing increasing trends towards the Upper Permian, while Rb and Sr have lower concentrations in the Upper Permian samples. The Harker Variation Diagram of major and trace elements are shown in Appendix A.

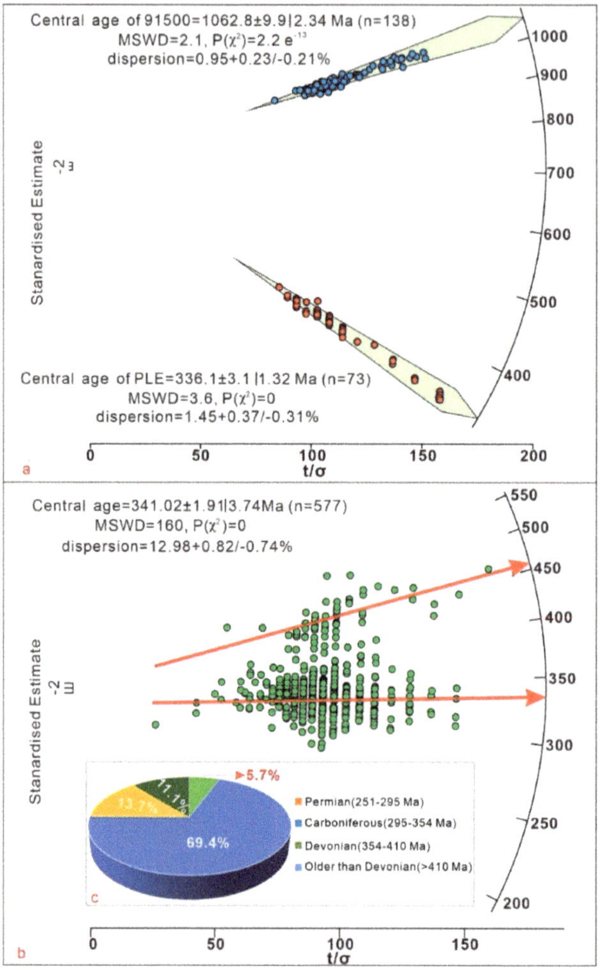

Figure 6. Age distribution of all the samples. (**a**) Age of standard samples 91500 and Plešovice. The age of 91500 and PLE were 1062.8 ± 9.9 Ma and 336.1 ± 3.1 Ma respectively; which were very close to reference ages. (**b**) Distribution characteristics of samples. Most of the zircon grains formed during the Carboniferous period in the West Bogeda Shan. The t/σ on X-axis indicates the precision.

Table 1. Compositions of major (unit: wt.%) and trace elements (unit: ppm) in the sedimentary rocks of the Upper Permian in Dalongkou section, West Bogeda Shan. LOI is loss on ignition. Eu/Eu* indicates the anomaly of europium. See text for the details.

Elements	DLG-1-01	DLG-1-02	DLG-1-03	DLG-2-01	DLG-2-02	DLG-2-03	DLG-2-04	DLG-2-05	DLG-3-01	DLG-3-03	DLG-5-01
Al_2O_3	6.4	6.8	6.9	5.5	7.3	6.0	2.6	4.4	4.7	4.5	5.9
CaO	0.5	21.7	1.9	0.8	0.6	2.4	16.4	0.8	0.7	19.9	1.0
Fe_2O_3	3.9	3.0	2.9	3.1	3.2	2.8	3.1	1.7	3.0	2.3	3.3
K_2O	1.5	1.6	1.4	1.7	1.9	1.4	1.8	1.0	1.8	1.6	2.0
MgO	0.8	0.7	0.5	0.7	0.7	0.4	0.7	0.4	0.6	0.5	0.7
MnO	0.1	0.3	0.2	0.1	0.1	0.2	0.8	0.1	0.1	0.6	0.1
Na_2O	2.4	2.1	3.2	1.9	2.1	2.7	2.4	1.9	3.3	2.1	2.9
P_2O_5	0.1	0.1	0.1	0.1	0.1	0.1	0.1	0.2	0.1	0.1	0.1
Ti_2O	0.6	0.4	0.5	0.4	0.5	0.5	0.6	0.4	0.5	0.3	0.5
SiO_2	81.5	58.6	78.0	84.3	81.6	82.4	66.2	86.3	83.3	64.7	81.6
LOI	1.5	3.9	3.7	1.0	1.5	0.5	4.0	2.0	1.4	2.5	1.5
Total	99.4	99.2	99.4	99.5	99.5	99.2	98.5	99.1	99.5	99.1	99.5

Table 1. *Cont.*

Elements	DLG-1-01	DLG-1-02	DLG-1-03	DLG-2-01	DLG-2-02	DLG-2-03	DLG-2-04	DLG-2-05	DLG-3-01	DLG-3-03	DLG-5-01
Sc	6.6	8.1	6.5	4.7	4.9	7.3	4.4	8.3	6.9	4.9	5.6
Ti	3937	2708	3559	2923	3197	3899	4808	5694	3342	2325	3531
Cr	68.9	46.1	41.5	35.0	57.6	78.7	84.2	134.4	31.2	20.7	37.5
Cu	19.7	14.0	20.3	13.7	14.8	22.7	28.4	25.8	20.6	15.3	19.4
Zn	63.2	49.1	60.6	99.2	56.3	60.7	72.6	93.5	56.3	60.1	72.6
Sr	152.7	650.5	145.1	179.6	135.4	470.9	705.0	243.3	187.7	526.3	147.8
Y	14.6	18.7	13.9	16.9	14.9	17.3	17.3	19.9	11.4	10.0	15.1
Zr	167.2	114.3	167.7	139.0	149.0	198.1	187.0	163.7	151.3	109.9	169.9
Nb	7.1	5.6	7.7	5.4	5.8	7.8	9.0	9.2	7.4	5.4	7.3
La	20.0	16.1	15.4	19.1	19.4	15.6	22.8	24.3	17.8	18.7	22.0
Ce	48.0	34.3	33.9	49.5	45.6	35.5	46.5	56.9	36.5	38.8	52.3
Pr	5.0	3.7	3.9	4.8	4.8	4.1	5.4	6.4	4.1	4.1	5.4
Nd	19.4	14.6	15.5	18.7	18.9	16.2	21.0	26.0	15.9	15.5	20.6
Sm	3.8	3.0	3.3	3.9	3.7	3.7	4.3	5.6	3.1	3.0	4.0
Eu	1.0	1.0	1.1	1.1	1.1	1.1	1.4	1.6	1.0	1.0	1.1
Gd	3.7	3.1	3.2	4.0	3.7	3.6	4.4	5.5	3.0	3.0	3.8
Tb	0.5	0.4	0.5	0.6	0.5	0.6	0.6	0.8	0.4	0.4	0.5
Dy	3.1	2.5	2.7	3.2	3.0	3.3	3.6	4.2	2.4	2.3	3.0
Ho	0.6	0.5	0.5	0.6	0.6	0.7	0.7	0.8	0.5	0.5	0.6
Er	1.9	1.7	1.8	1.9	1.8	2.1	2.2	2.2	1.4	1.4	1.9
Tm	0.3	0.2	0.2	0.3	0.3	0.3	0.3	0.3	0.2	0.2	0.3
Yb	1.9	1.8	1.7	1.8	1.7	2.3	2.1	1.9	1.4	1.2	1.8
Lu	0.3	0.3	0.3	0.3	0.3	0.3	0.3	0.3	0.2	0.2	0.3
Th	5.9	4.9	4.5	6.1	6.3	4.7	5.8	5.6	5.0	4.1	6.4
Eu/Eu*	0.8	1.0	1.0	0.8	0.9	1.0	1.0	0.9	1.0	1.1	0.9
La/Th	3.4	3.3	3.4	3.1	3.1	3.3	4.0	4.4	3.6	4.6	3.5
Zr/Sc	25.3	14.1	26.0	29.9	30.1	27.1	42.8	19.8	22.0	22.6	30.4
Th/Sc	0.9	0.6	0.7	1.3	1.3	0.6	1.3	0.7	0.7	0.8	1.1

5. Discussion

5.1. *Provenance Variation of Sedimentary Rocks in the West Bogeda Shan*

Detrital zircon geochronology has been used as a proxy for sedimentary provenance analysis due to zircon's stability during weathering and transporting [52,53]. Detrital zircon U-Pb ages of the Upper Permian samples have two notable age peaks at ca. 335 Ma and ca. 455 Ma (Figure 7a), whereas only one main age population was observed for the Carboniferous to Middle Permian samples (Figure 7b). Therefore, we infer that the sediment provenances in the West Bogeda Shan were derived from relatively homogeneous sources with a narrow zircon age population in the Carboniferous to Middle Permian periods, but obviously changed to multiple sources during the Late Permian and Triassic. Detrital zircons collected from the Upper Carboniferous to Middle Permian with major peaks at ca. 342.0 Ma, ca. 310.2 Ma and ca. 311.7 Ma, respectively (Figure 7b). Detrital zircons with ages of 360–320 Ma and 320–300 Ma may source from the magmatic belts of the YCTS and NTS, respectively [8]. The paleocurrents in the Permian and Triassic were mainly north directed, implying that the source of the detrital zircon grains was suited to the south in the Tian Shan area (Figure 7c) [8,15,51,54]. Therefore, the Carboniferous volcanic rocks in the NTS and magmatic belt of YCTS are considered as the sources of deposits in the Upper Carboniferous to Middle Permian [55]. Two major age peaks at ca. 330.2 Ma and ca. 448.0 Ma could be identified for Upper Permian deposits and three major age peaks for Lower Triassic deposits at ca. 245.9 Ma, ca. 314.7 Ma and ca. 458.1 Ma, which may indicate the initial uplift of Bogeda Shan and could be the source of the Junggar Basin [55]. This could also be demonstrated by poorly-sorted and rounded conglomerate and pebbly sandstone of Late Permian age, suggesting that the deposits were near the source area without long distance of transporting (Figure 3f).

Moreover, as shown in the Dickinson diagram (Figure 8), the Permian sedimentary rocks in the West Bogeda Shan are mostly composed of lithic arkose, feldspathic litharenite, and litharenite. Besides, the tectonic setting in which the sedimentary rocks formed is significantly different for the Lower-Middle to Upper Permian. The Lower-Middle Permian sedimentary rocks were developed on magmatic arcs, while the data for the Upper Permian rocks indicate a recycled orogenic setting. These observations further indicate the provenance and tectonic setting variation from Lower-Middle Permian to Upper Permian.

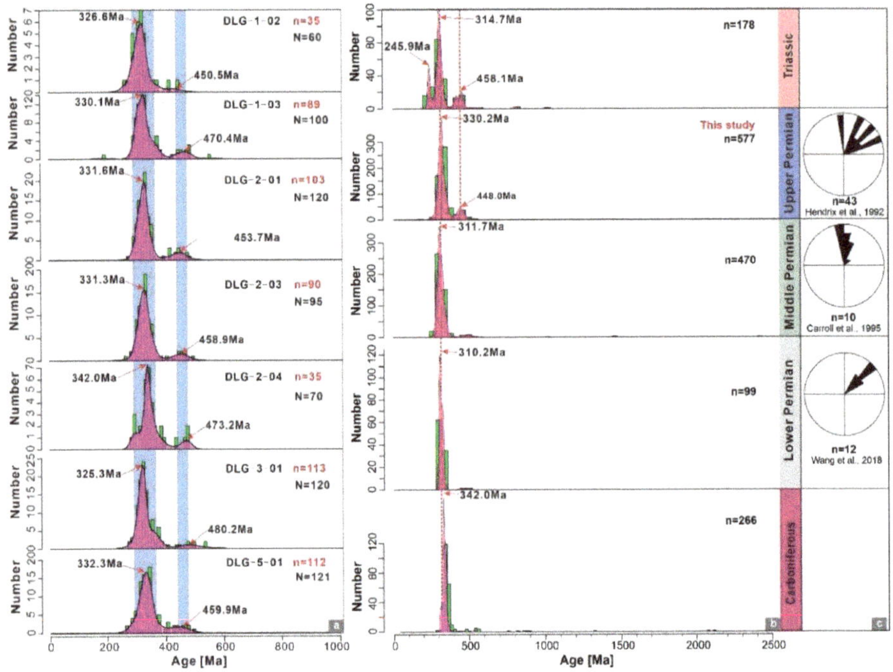

Figure 7. Probability-density-frequency plots and number histograms of U-Pb ages of detrital zircons. (**a**) Upper Permian, (**b**) Upper Carboniferous to Lower Triassic. Data of Carboniferous are from [24,45–47]; Data of Lower Permian and Middle Permian are from [8,44]; Data for Triassic are from [56]. (**c**) Paleo-current of Lower Permian [57], Middle Permian [58] and Upper Permian to Lower Triassic [15]. N is total measured grains, and n is grains with 90–110% concordance.

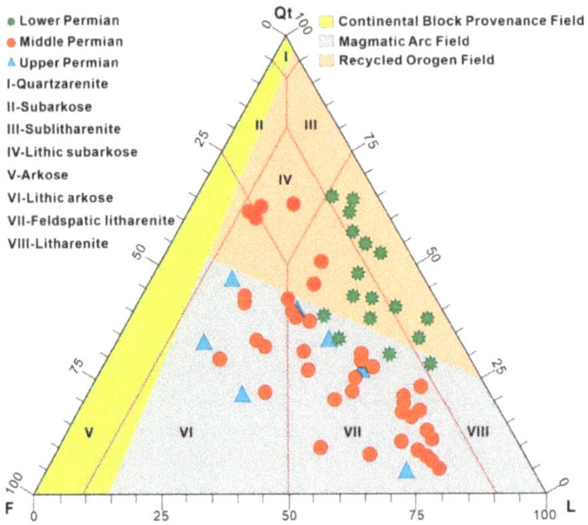

Figure 8. Compositions of sedimentary rock samples in the Dickinson ternary diagrams of Lower, Middle, and Upper Permian, West Bogeda Shan, NW China, modified after [57,59]. Qt: Quartz; F: Feldspar; L: Lithic.

Rare earth elements (REEs) and high-field-strength elements (HFSEs) such as Nb, Ta, Zr, and Hf are relatively conservative during sediment weathering, transport, and post-depositional processes, and thus are treated as reliable tracers for sediment provenances [60,61]. Even though the REE could be affected by grain size and chemical weathering, provenance composition plays a key role on the REE geochemistry of sediments [62]. Chondrite-normalized REE patterns of Permian samples (Figure 9), show the enrichment of light REE (LREE) relative to the heavy REE (HREE) [63]. The LREE/HREE ratios of the Lower, Middle, and Upper Permian samples have ranges of 5.0–10.0, 4.5–5.7, and 5.8–8.9, respectively. Relatively negative Eu anomalies (Eu/Eu* generally <0.7) are observed in the samples from the Lower-Middle Permian (Figure 9a,b), while the Upper Permian samples have weak or no Eu anomalies (Eu/Eu* around 0.9) (Figure 9c). This could further indicate the provenance variation occurred from Lower-Middle Permian to Upper Permian [60].

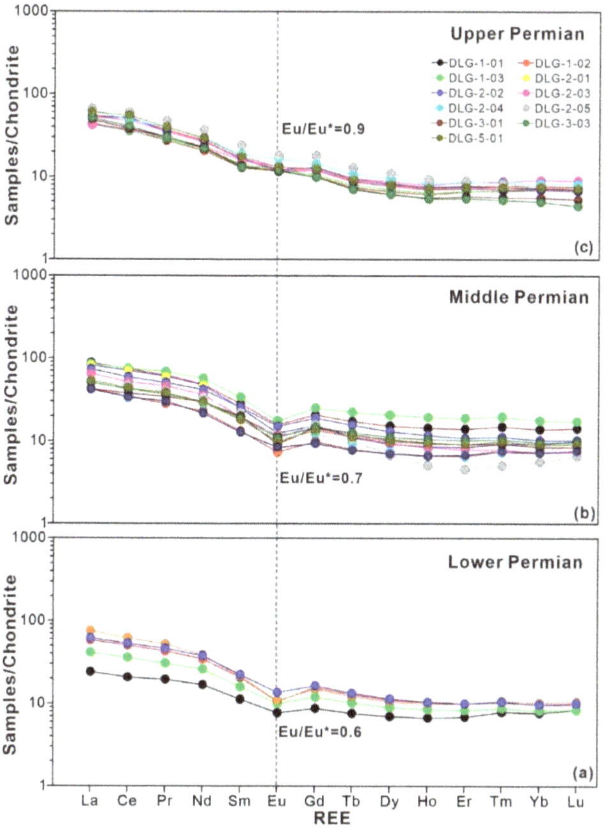

Figure 9. Chondrite-normalized rare earth element (REE) diagrams for the sedimentary samples [63]. (**a**) Chondrite-normalized REE diagrams of Lower Permian. (**b**) Chondrite-normalized REE diagrams of Middle Permian. (**c**) Chondrite-normalized REE diagrams of Upper Permian. Data of Lower-Middle Permian from previously published papers, while the data of Upper Permian is from this study.

All these observations of bulk geochemistry suggest a provenance change from the Early-Middle to Late Permian. We therefore argue that the Middle Permian could be the crucial period for the variations of sediment provenance and tectonic setting in the West Bogeda Shan, which is synchronous with the tectonic evolutions of Bogeda Shan during the Late Carboniferous (Pennsylvanian) and Early Permian periods [64], corresponding to the tectonic setting of the Harlik-Dananhu arc [65]. Sediments deposited in the basins where the West Bogeda Shan is currently located during the Late Carboniferous and Early Permian witnessed the tectonic evolution at that time.

5.2. Source Rock Composition and Paleoclimate

In sedimentary rocks, accessory minerals such as zircon, monazite, and apatite are rich in REEs. Generally, felsic rocks have higher LREE/HREE ratios and strong Eu depletions, whereas mafic rocks display relatively low LREE/HREE ratios and moderate Eu anomalies [63]. The Lower-Middle Permian sandstone samples have higher LREE/HREE ratios and weak negative Eu anomalies, while the Upper Permian samples have weak or no Eu anomalies (Figure 9). This observation apparently suggests that the Permian sedimentary rocks in the West Bogeda Shan might have been were derived from the multiple sources, albeit with the dominance of mafic components.

Ratios of Zr/Sc and Th/Sc are useful proxies for identifying the effects of sedimentary recycling and source compositions of sedimentary rocks [60,66]. The large variations but overall positive correlations of Zr/Sc and Th/Sc ratios suggest the variable provenance rock compositions of the West of Bogeda Shan, rather than sedimentary recycling (Figure 10a). The plot of REE versus La/Yb suggests that the Permian samples in the West Bogeda Shan are dominated by sedimentary rocks (Figure 10b). Meanwhile, the lithology of sedimentary rock in the Lower-Middle Permian is different from that of the Upper Permian. The Lower-Middle Permian was mostly composed of greywacke, while shale was subordinate. The Upper Permian was dominated by sublitharenite (Figure 10c). Based on the plot of Hf versus La/Th, the sedimentary rocks in the West Bogeda originated from mixed mafic sources during the Early-Middle Permian but changed to a mixture of mafic and acidic arc sources in the Upper Permian (Figure 10d) [67]. Similarly, the discrimination plot of Co/Th versus La/Sc [19,45,63] suggests that most of the Permian sandstones are classified into mafic volcanic and andesitic sources, although some of the Lower-Middle Permian samples are of felsic volcanic origins (Figure 10e). The combined analyses all demonstrate that most of the Permian sedimentary rocks inherit mafic detritus in the West Bogeda Shan. Besides, the sedimentary rocks in the Early-Middle Permian formed in an arid climate, and then were transferred to a humid climate zones along with the increasing chemical maturity in the Upper Permian (Figure 10f). This could be demonstrated by widely distributed bivalve and plant fragments in the Upper Permian sandstones (Figure 3d,j,k).

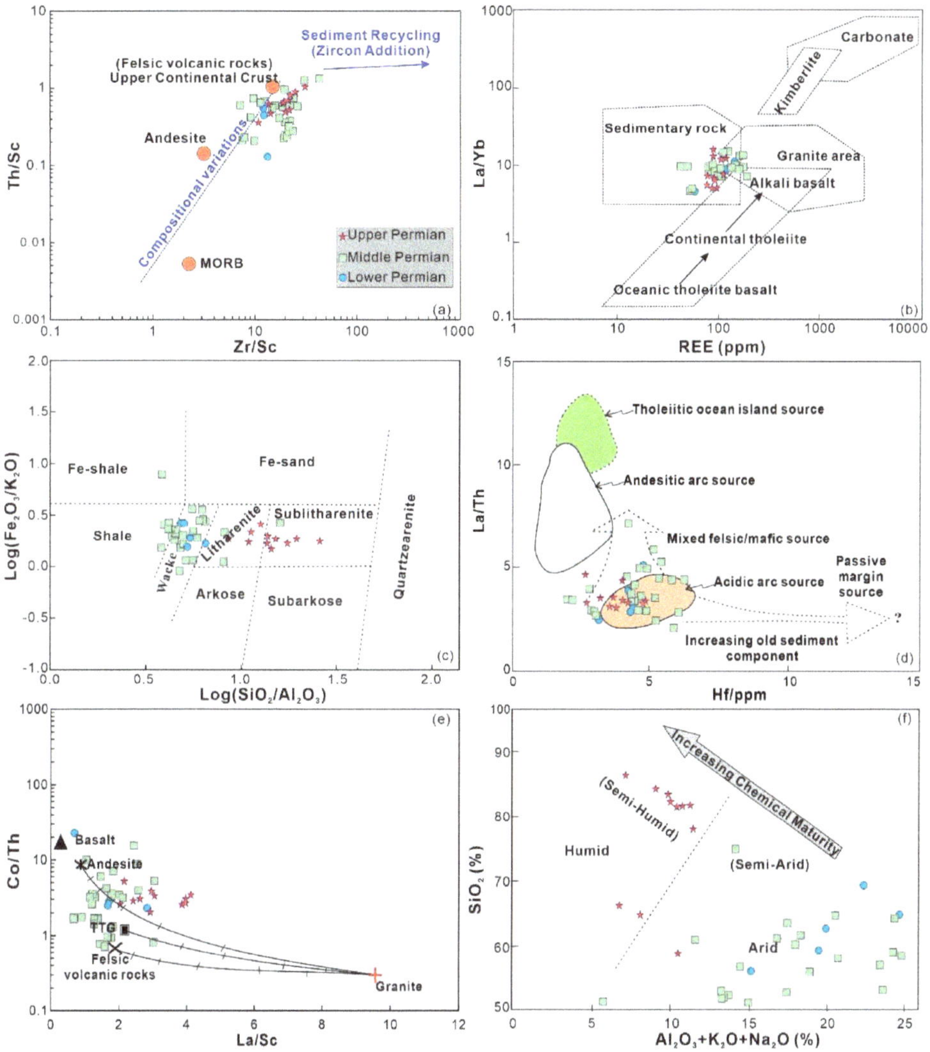

Figure 10. Discrimination provenance diagrams for sedimentary rocks from Carboniferous to Upper Permian of West Bogeda Shan. (**a**) Diagram of Th/Sc vs Zr/Sc [66]; (**b**) REE vs La/Yb [68]; (**c**) geochemical classification of clastic rocks [69,70]; (**d**) diagram of La/Th vs Hf [67]. (**e**) Diagram of Co/Th vs. La/Sc. Average compositions of volcanic rocks from [71]; (**f**) binary diagram SiO_2 versus Al_2O_3 + K_2O + Na_2O to discriminate the climatic conditions during the period of Carboniferous to Late Permian [72].

5.3. Rapid Change in Lithology and Depositional Environment

The lithology and depositional environment of West Bogeda area experienced a multiphase evolution during the Permian. During the Late Carboniferous and Early Permian, the depositional environment of West Bogeda area was dominated by semi-deep to deep marine environment, and gravity flow deposits was the main lithology [33,57]. At the end of the Early Permian, the shallow water deposited sandstone directly overlying on the mudstone. The depositional environment transitioned from deep water environment to shallow water environment [73,74]. Further, the depositional environment was transferred from marine to nonmarine environment in the Middle Permian with the main lithology of fine sandstone, mudstone and oil-bearing mudstone. Meanwhile, deformation during the Middle Permian and unconformity contact of Middle and Upper Permian could be found in the area (Figure 3e), which may relate to the uplift of West Bogeda Shan. The deposits of Late Permian age are mainly composed of purplish-reddish conglomerate in P_3q (Figure 3f,g) and pebbly sandstone in P_3wt (Figure 3h,i) with poorly-sorted and rounded pebbles. They are typical molasse formations and close to the source area. Combined with our field works and previous studies [31,57], the depositional environment of Upper Permian was alluvial fan and braided river, which was significantly different from Lower-Middle Permian. The rapid change of lithology and environment also suggest the initial uplift of the West Bogeda Shan during the Late Permian.

5.4. Tectonic Setting and Basin Evolution in the West Bogeda Shan

Discrimination diagrams for tectonic setting of siliciclastic sediments and sedimentary rocks are mostly based on geochemical compositions such as the contents of major and trace elements and their ratios [75–77]. To analyze the tectonic settings of the West Bogeda Shan, proxies of DF1 and DF2 are defined based on major elements components according to previous researches [78,79]. The discrimination diagram of DF1 versus DF2 suggests that the tectonic setting of the West Bogeda Shan in the Early Permian was dominated by a continental rift, and partly changed to island arc in the Middle Permian (Figure 11a) towards an active continental margin and continental island arc in the Late Permian (Figure 11b).

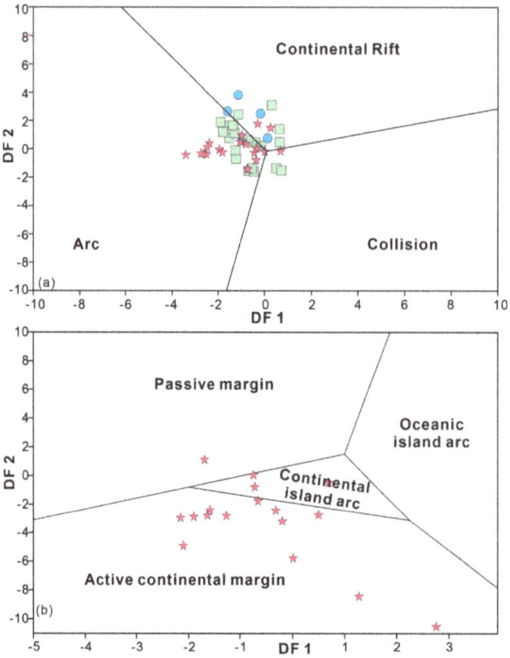

Figure 11. Tectonic discrimination diagrams with major elements of clastic rocks from Carboniferous to Upper Permian. (**a**) Discriminant-function multi-dimensional diagram for high and low silica clastic sediments from three tectonic settings (equation for DF1 and DF2 based on Surendra P. Verma et al. [78]. (**b**) Discriminant-function multi-dimensional diagram for high and low silica clastic sediments from three plots of discriminant scores along Function 1 versus Function 2, to discriminate rocks suites of West Bogeda Shan (equation for DF1 and DF2 based on Mukul R. Bhatia [79]). Symbols are the same as those in Figure 10.

Abundant evidences of abrupt changes of depositional environments, provenance area and source rock composition and development of bimodal volcanic-sedimentary rock series corroborate to the hypothesized rift setting during the Early Permian [27,30]. The discrimination results based on the trace element compositions, such as La–Th–Sc and Th–Sc–Zr/10 ternary diagrams also suggest that almost all sedimentary rocks in the West Bogeda Shan were derived from mafic sources in the Permian. Meanwhile, a continental island arc is the preferred tectonic setting at the epoch of the Late Permian (Figure 12; Table 2).

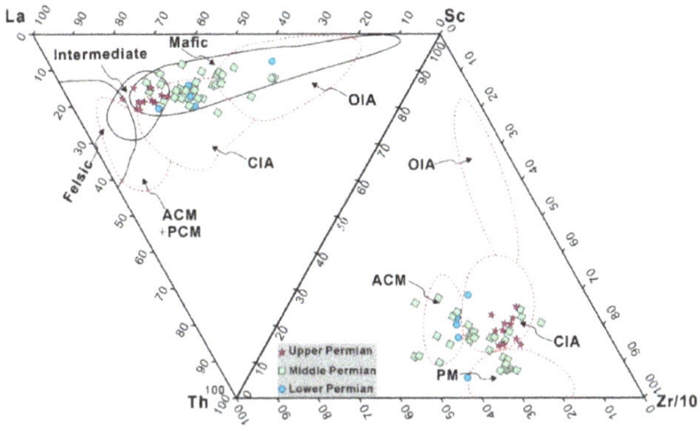

Figure 12. Tectonic discrimination diagrams with trace elements of clastic rocks from Carboniferous to Upper Permian, modified after [75]; OIA = oceanic island arc; CIA = continental island arc; ACM = active continental margin; PM = passive continental margin. Symbols are the same as in Figure 10.

Table 2. Comparison of representative REE characteristics of clastic rocks from Carboniferous to Upper Permian clastic rocks with greywacke from various tectonic settings [75]. The REE was normalized with chondrite [63].

Tectonic Setting	Provenance	REE Parameters						
		La	Ce	REE	La/Yb	$(La/Yb)_N$	LREE/HREE	Eu/Eu*
Ocean Island Arc	Undissected magmatic arc	8 ± 1.7	19 ± 3.7	58 ± 10	4.2 ± 1.3	2.8 ± 0.9	3.8 ± 0.9	1.04 ± 0.1
Continental Island Arc	Dissected magmatic arc	27 ± 4.5	59 ± 8.2	146 ± 20	11 ± 3.6	7.5 ± 2.5	7.7 ± 1.7	0.79 ± 0.1
Active Continental Margin	Uplifted basement	37	78	186	12.5	8.5	9.1	0.6
Passive Margin	Craton-interior tectonic highland	49	85	210	15.9	10.8	8.5	0.6
Lower Permian (Average)		16.6	36.8	88.9	8.9	6.0	7.1	0.6
Middle Permian (Average)		16.3	35.1	96.9	6.3	4.2	5.1	0.7
Upper Permian (Average)		19.2	43.4	102.6	11.0	7.5	7.6	0.9

In summary, the combination of detrital zircon geochronology, whole-rock geochemical and sedimentary characters suggest that the initial uplift of Bogeda Shan occurred in the Late Permian. Combined with previous studies, three tectonic phases characterized the basin evolution from a continental rift, post-rift extensional depression to continental arc (initial uplift).

5.4.1. Inheritance from Upper Carboniferous (Lower Permian)

The tectonic setting of Bogeda Shan during the Carboniferous has been long debated [80]. Geochemical data suggests that it was not an island arc as proposed by Sébastien Laurent-Charvet et al. [81] but could have been a continental rift during the Carboniferous and Early Permian. Geochemical investigations of volcanic rocks and turbiditic deposits as well as gravimetric and magnetic data suggest that Turpan Block and Junggar Block were separated in the end of the Early

Permian due to extension and rift of the Bogeda area [56,58,82]. The rift setting was also demonstrated by a series of coarse clastic rocks with intercalations of pillow lava-vesicular basalt [57]. Then, the extension of the belt started to rollback, which formed the Paleo-Bogeda Shan. The combined evidences shown above indicate that the tectonic setting of continental rift in the Early Permian is similar to the Late Carboniferous setting (Figure 13a) [27,30]. The tectonic setting transformed from continental rift to inland arc was a result of the collision of the Junggar and Tarim Blocks at the end of the Early Permian. From this collision, the tectonic setting of West Bogeda Shan changed to intracontinental tectonic evolution stage, and the original terrain of West Bogeda formed (Figure 13b). At the end of the Early Permian, most parts of the terrain were still a submarine environment, while only small parts were lifted up [32,74,83–85].

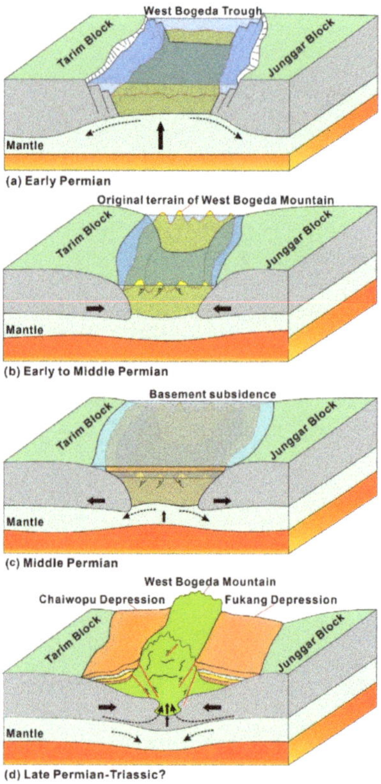

Figure 13. Tectonic evolution modal of West Bogeda Shan during the Permian. The initial uplift started at the end of Middle Permian. (**a**) Early Permian; (**b**) Early to Middle Permian; (**c**) Middle Permian; (**d**) Late Permian to Triassic?

5.4.2. Transitional Period (Middle Permian)

During early Middle Permian, depositional environment and tectonic setting were relatively stable, without intensive deformation and only some small scale of tectonic activities in the West Bogeda Shan [32,86]. Due to the relaxation of the compression and rebound of crust deformation, the island arc in the West Bogeda area received a large volume of sediments [54]. With a large sediment supply and incessant basement subsidence, the West Bogeda Shan basin closed during the Middle Permian (Figure 13c) [15,87]. Hence, the tectonic setting was post-rift extensional depression in the Middle Permian. This observation is consistent with some previous studies that reported the wide distribution of bimodal volcanic rocks in the Bogeda Shan [73] and some submarine olistostrome in the West Bogeda Shan [30]. Deformation and unconformity at the end of the Middle Permian also implies the onset tectonic evolution of the West Bogeda Shan.

5.4.3. Initial Uplift of West Bogeda (Upper Permian)

The integrated data from lithological observation and sedimentary geochemical analyses indicates that the provenance characteristics, geochemical composition, and tectonic setting of rocks in the Permian obviously changed from the Early-Middle to Late Permian. Meanwhile, according to previous studies, the powerful intracontinental collision occurred between the Junggar and Tarim Blocks occurred and initiated West Bogeda Shan uplift in the Late Permian (Figure 13d) [32,88]. Thus, this study confirms the previous recognition that the initial uplift of the Bogeda Shan happened in the Late Permian. Meanwhile, the depositional environment, sediment provenance and depositional center greatly changed as a response to the uplift of the Bogeda Shan.

6. Conclusions

This study presents the data of detailed zircon U-Pb geochronology and whole-rock geochemical compositions of Permian sandstones from the West Bogeda Shan, and discusses the provenances, source rock compositions, tectonic settings and basin evolution history. Several conclusions are summarized here.

(1) Detrital zircon U-Pb chronology suggests the sediments in the Lower-Middle Permian were inherited from the Carboniferous, showing one dominant age population with NTS and YCTS as the main source. However, two or three age populations are notable in the West Bogeda Shan during the periods of Middle Permian to Triassic, suggesting changing sediment provenances. REE series, especially Eu anomalies, also indicate the changes of sediment provenance in the Upper Permian.

(2) The sedimentary rocks in the West Bogeda originated from the mafic-dominant sources during the Early-Middle Permian but changed to lithologies are mixture of mafic and acidic arc sources in the Upper Permian. Besides, the Lower-Middle Permian dominated by wacke and the Upper Permian by litharenite and sublitharenite. The different source rock compositions between the Lower-Middle Permian and the Upper Permian resulted from the complex tectonic evolution of the Bogeda Shan in the Upper Permian.

(3) The provenance, lithology, and depositional environment were significantly changed from the Late Carboniferous to Late Permian. Strata deformation and unconformity also occurred at the end of Middle Permian, which was closely related to the uplift of West Bogeda Shan. Three stages characterized the tectonic evolution of the West Bogeda Shan, showing the continental rift in the Early Permian (inherited from the Upper Carboniferous), post-rift extensional depression in the Middle Permian, and continental arc in the Late Permian. With the initial uplift of Bogeda Shan in the Upper Permian, the depositional environment and sediment provenance changed significantly.

Supplementary Materials: The following are available online at http://www.mdpi.com/2075-163X/10/4/341/s1, Table S1: Carboniferous, Permian, Jurassic, and Standard sample, Table S2: major and trace elements.

Author Contributions: Writing—Original Draft Preparation by Y.L., Writing—Review and Editing by S.Y., X.H., W.Y., X.Y. and X.S., Data analysis and Figures by Y.L. and J.S., Sampling and Experiments by Y.L., X.Y. and Z.Y. All authors have read and agreed to the published version of the manuscript.

Funding: This work was partly supported by the National Natural Science Foundation of China (Grant No. 41730531, 41991324) and the National Programme on Global Change and Air-Sea Interaction (GASI-GEOGE-03).

Acknowledgments: The authors would like to thank S.L., Z.Y. and L.J. for their help during the field work.

Conflicts of Interest: The authors declare no conflict of interest.

Appendix A

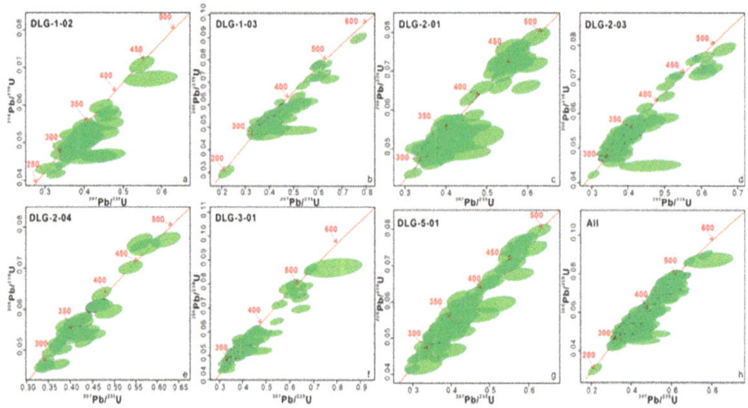

Figure A1. U-Pb Concordia diagrams for zircon grains of the 7 sandstone samples. All of these data are excluded out of discordance >10%.

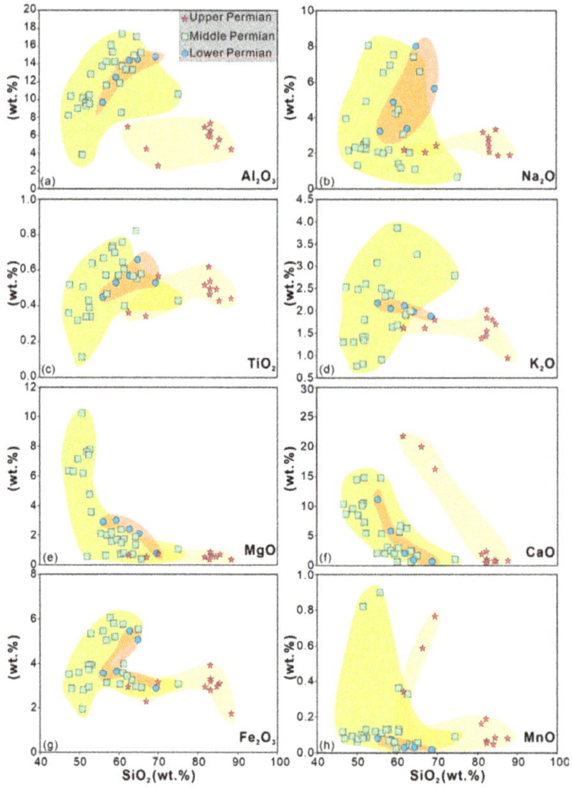

Figure A2. Harker Variation Diagram of major elements for rock samples in the West of Bogeda Mountain, South Junggar. Data of Early and Middle Permian from Liu et al. [51]; Liu et al. [50].

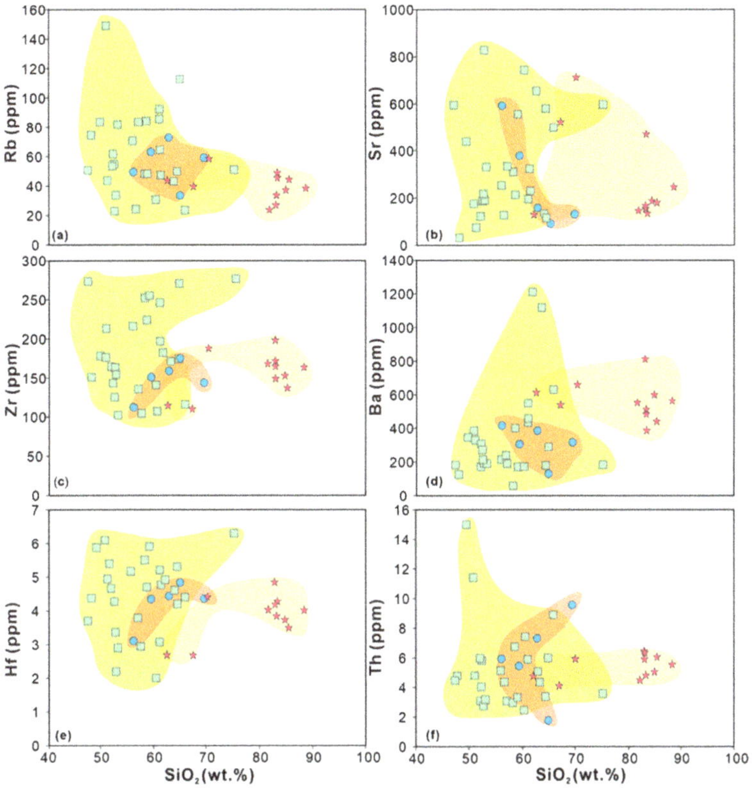

Figure A3. Selected Harker Variation Diagram of trace elements for rock samples in the West of Bogeda Mountain, South Junggar. Symbols are the same with Figure A2.

References

1. Charvet, J.; Shu, L.S.; Laurent-Charvet, S.; Wang, B.; Faure, M.; Cluzel, D.; Chen, Y.; De Jong, K. Palaeozoic tectonic evolution of the Tianshan belt, NW China. *Sci. China Earth Sci.* **2011**, *54*, 166–184. [CrossRef]
2. Kröner, A.; Kovach, V.; Belousova, E.; Hegner, E.; Armstrong, R.; Dolgopolova, A.; Seltmann, R.; Alexeiev, D.V.; Hoffmann, J.E.; Wong, J.; et al. Reassessment of continental growth during the accretionary history of the Central Asian Orogenic Belt. *Gondwana Res.* **2014**, *25*, 103–125. [CrossRef]
3. Xiao, W.J.; Windley, B.F.; Allen, M.B.; Han, C.M. Paleozoic multiple accretionary and collisional tectonics of the Chinese Tianshan orogenic collage. *Gondwana Res.* **2013**, *23*, 1316–1341. [CrossRef]
4. Gao, J.; Long, L.L.; Klemd, R.; Qian, Q.; Liu, D.Y.; Xiong, X.M.; Wei, W.S.; Liu, Y.T.; Yang, W.F. Tectonic evolution of the South Tianshan orogen and adjacent regions, NW China: Geochemical and age constraints of granitoid rocks. *Int. J. Earth Sci.* **2009**, *98*, 1221–1238. [CrossRef]
5. Hu, A.Q.; Jahn, B.; Zhang, G.X.; Chen, Y.B.; Zhang, Q.F. Crustal evolution and Phanerozoic crustal growth in northern Xinjiang: Nd isotopic evidence. Part I. Isotopic characterization of basement rocks. *Tectonophysics* **2000**, *328*, 15–51. [CrossRef]
6. Jahn, B.; Wu, F.Y.; Chen, B. Granitoids of the Central Asian Orogenic Belt and continental growth in the Phanerozoic. *Earth Environ. Sci. Trans. R. So.* **2000**, *91*, 181–193.
7. Wong, K.; Sun, M.; Zhao, G.C.; Yuan, C.; Xiao, W.J. Geochemical and geochronological studies of the Alegedayi Ophiolitic Complex and its implication for the evolution of the Chinese Altai. *Gondwana Res.* **2010**, *18*, 438–454. [CrossRef]

8. Zhang, Y.; Li, Z.S.; Nie, F.; Tian, X.L.; Shi, Y.H. Age, provenance and tectonic evolution of Late Paleozoic strata in Bogda Mountain, Xinjiang: Evidence from detrital zircon U-Pb geochronology. *China J. Geol.* **2015**, *50*, 155–181. (In Chinese with English Abstract).
9. Jolivet, M.; Gloria, H.; Robin, C.; Barrier, L.; Bourquin, S.; Guo, Z.; Jia, Y.; Guerit, L.; Yang, W.; Fu, B. Reconstructing the Late Palaeozoic-Mesozoic topographic evolution of the Chinese Tian Shan: Available data and remaining uncertainties. *Adv. Geosci.* **2013**, *37*, 7–18. [CrossRef]
10. Wang, Z.X.; Li, T.; Zhou, G.Z.; Lu, M.A.; Liu, Y.Q.; Li, Y. Geological record of the late-carboniferous orogeny in Bogedashan, Northern Tianshan Mountains, Northwest China. *Earth Sci. Front.* **2003**, *10*, 63–70. (In Chinese with English Abstract).
11. Choulet, F.; Chen, Y.; Wang, B.; Faure, M.; Cluzel, D.; Charvet, J.; Lin, W.; Xu, B. Late Paleozoic paleogeographic reconstruction of Western Central Asia based upon paleomagnetic data and its geodynamic implications. *J. Asian Earth Sci.* **2011**, *42*, 867–884. [CrossRef]
12. Gao, J.; Li, M.S.; Xiao, X.C.; Tang, Y.Q.; He, G.Q. Paleozoic tectonic evolution of the Tianshan Orogen, northwestern China. *Tectonophysics* **1998**, *287*, 213–231. [CrossRef]
13. Yu, C.H.; Jiang, Y.Q.; Liu, S.H. Jurassic sedimentary boundary between the Junggar and Turpan-Hami basins in Xinjiang. *Sediment. Facies Palaeogeogr* **1996**, *16*, 48–54. (In Chinese with English Abstract).
14. Greene, T.J.; Carroll, A.R.; Hendrix, M.S.; Graham, S.A.; Wartes, M.A.; Abbink, O.A. *Sedimentary Record of Mesozoic Deformation and Inception of the Turpan-Hami Basin, Northwest China*; Geological Society of America: Boulder, CO, USA, 2001; pp. 317–340.
15. Hendrix, M.S.; Graham, S.A.; Carroll, A.R.; Sobel, E.R.; Mcknight, C.L.; Schulein, B.J.; Wang, Z.X. Sedimentary record and climatic implications of recurrent deformation in the Tian Shan: Evidence from Mesozoic strata of the north Tarim, south Junggar, and Turpan basins, northwest China. *Geol. Soc. Am. Bull.* **1992**, *104*, 53–79. [CrossRef]
16. Chen, K.; Wei, L.; Wang, Q. The Bogeda Shan uplifting: Evidence from multiple phases of deformation. *J. Asian Earth Sci.* **2015**, *99*, 1–12. [CrossRef]
17. Fang, S.H.; Song, Y.; Jia, C.Z.; Wang, X.L.; Yuan, Q.D. The Mesozoic–Cenozoic clastic composition of Bogda area, Xinjiang: Implications on the evolution of basin–range pattern. *Acta Geol. Sin.* **2007**, *81*, 1229–1237. (In Chinese with English Abstract).
18. Zhang, C.H.; Liu, D.B.; Zhang, C.L.; Wang, Z.Q.; Wang, Q.X. Stratigraphic constraints on the initial uplift age of Bogda Shan, Xinjiang, northwest China. *Earth Sci. Front.* **2005**, *12*, 294–302. (In Chinese with English Abstract).
19. Guo, W.; Zhou, D.W.; Ouyang, Z.J.; Zhou, X.H. Structure features of olistostrome under Permian extension environment in Bogda Mountain Region. *J. Northwest Univ. (Natural Sci.)* **2011**, *41*, 658–675. (In Chinese with English Abstract).
20. Xiao, W.J.; Zhang, L.C.; Qin, K.Z.; Sun, S.; Li, J.L. Paleozoic accretionary and collisional tectonics of the Eastern Tianshan (China): Implications for the continental growth of central Asia. *Am. J. Sci.* **2004**, *304*, 370–395. [CrossRef]
21. Burtman, V.S. Structural geology of variscan Tien Shan, USSR. *Am. J. Sci.* **1975**, *275*, 157–186.
22. Tapponnier, P.; Peltzer, G.; Armijo, R. On the mechanics of the collision between India and Asia. *Geol. Soc. Lond. Spec. Publ.* **1986**, *19*, 113–157. [CrossRef]
23. Windley, B.F.; Allen, M.B.; Zhang, C.; Zhao, Z.Y.; Wang, G.R. Paleozoic accretion and Cenozoic redeformation of the Chinese Tien Shan range, central Asia. *Geology* **1990**, *18*, 128–131. [CrossRef]
24. Chen, D.C.; Zhao, X.M.; Deng, J. U-Pb dating of Carboniferous sandstone detrital zircon from the north of the Bogda Mountains, eastern Xingjiang, and its geological significances. *Acta Geol. Sin.* **2010**, *84*, 1770–1780.
25. XBGMR. *Regional Geology of Xinjiang Uygur Autonomy Region*; Geological Publishing House: Beijing, China, 1993.
26. Wang, J.L.; Wu, C.D.; Li, Z.; Zhou, T.Q.; Wu, J.; Wang, J. The tectonic evolution of the Bogda region from Late Carboniferous to Triassic time: Evidence from detrital zircon U–Pb geochronology and sandstone petrography. *Geol. Mag.* **2018**, *155*, 1063–1088. [CrossRef]
27. Yang, X.F.; He, D.F.; Wang, Q.C.; Tang, Y. Tectonostratigraphic evolution of the Carboniferous arc-related basin in the East Junggar Basin, northwest China: Insights into its link with the subduction process. *Gondwana Res.* **2012**, *22*, 1030–1046. [CrossRef]

28. Lei, W.S.; Guo, J.F.; Chen, D.L.; Chu, J.P.; Li, Y. Ductile shearing and LA-ICP MS zircon U–Pb dating of S a'erqiaoke ductile shear belt in the B ogda orogen: Constraints on the process of mountain building of eastern Tianshan. *Geol. J.* **2018**, *53*, 206–222. [CrossRef]
29. Obrist-Farner, J.; Yang, W.; Hu, X. Nonmarine time-stratigraphy in a rift setting: An example from the Mid-Permian lower Quanzijie low-order cycle Bogda Mountains, NW China. *J. Palaeogeogr.* **2015**, *4*, 27–51. [CrossRef]
30. Shu, L.S.; Wang, B.; Zhu, W.B.; Guo, Z.J.; Charvet, J.; Zhang, Y. Timing of initiation of extension in the Tianshan, based on structural, geochemical and geochronological analyses of bimodal volcanism and olistostrome in the Bogda Shan (NW China). *Int. J. Earth Sci.* **2011**, *100*, 1647–1663. [CrossRef]
31. Thomas, S.G.; Tabor, N.J.; Yang, W.; Myers, T.S.; Yang, Y.; Wang, D. Palaeosol stratigraphy across the Permian–Triassic boundary, Bogda Mountains, NW China: Implications for palaeoenvironmental transition through earth's largest mass extinction. *Palaeogeogr. Palaeoclimatol. Palaeoecol.* **2011**, *308*, 41–64. [CrossRef]
32. Wang, J.L.; Wu, C.D.; Li, Z.; Zhou, T.Q.; Wu, J.; Wang, J. Whole-rock geochemistry and zircon Hf isotope of L ate C arboniferous–T riassic sediments in the Bogda region, NW C hina: Clues for provenance and tectonic setting. *Geol. J.* **2018**, *54*, 1853–1877. [CrossRef]
33. Wartes, M.A.; Carroll, A.R.; Greene, T.J. Permian sedimentary record of the Turpan-Hami basin and adjacent regions, northwest China: Constraints on postamalgamation tectonic evolution. *Geol. Soc. Am. Bull.* **2002**, *114*, 131–152. [CrossRef]
34. Yue, W.; Jin, B.F.; Zhao, B.C. Transparent heavy minerals and magnetite geochemical composition of the Yangtze River sediments: Implication for provenance evolution of the Yangtze Delta. *Sediment. Geol.* **2018**, *364*, 42–52. [CrossRef]
35. Guo, L.; Zhang, H.F.; Harris, N.; Xu, W.C.; Pan, F.B. Detrital zircon U–Pb geochronology, trace-element and Hf isotope geochemistry of the metasedimentary rocks in the Eastern Himalayan syntaxis: Tectonic and paleogeographic implications. *Gondwana Res.* **2017**, *41*, 207–221. [CrossRef]
36. Dunkl, I.; Mikes, D.; Frei, D.; Gerdes, A.; Tolosana-Delgado, R.; Von Eynatten, H. *UranOS: Data Reduction Program for Time-resolved U/Pb Analyses*. Available online: http://www.sediment.uni-goettingen.de/staff/dunkl/software/uranos.html (accessed on 10 April 2020).
37. Stacey, J.S.; Kramers, J.D. Approximation of terrestrial lead isotope evolution by a two-stage model. *Earth Planet. Sci. Lett.* **1975**, *26*, 207–221. [CrossRef]
38. Gehrels, G. Detrital zircon U-Pb geochronology applied to tectonics. *Annu. Rev. Earth Planet. Sci.* **2014**, *42*, 127–149. [CrossRef]
39. Page, F.Z.; Fu, B.; Kita, N.T.; Fournelle, J.; Spicuzza, M.J.; Schulze, D.J.; Viljoen, F.; Basei, M.A.S.; Valley, J.W. Zircons from kimberlite: New insights from oxygen isotopes, trace elements, and Ti in zircon thermometry. *Geochim. Cosmochim. Acta* **2007**, *71*, 3887–3903. [CrossRef]
40. Wiedenbeck, M.P.; Corfu, A.F.; Griffin, W.L.; Meier, M.; Oberli, F.; Von, Q.A.; Roddick, J.C.; Spiegel, W. Three natural zircon standards for U-Th-Pb, Lu-Hf, trace element and REE analyses. *Geostand. Newsl.* **1995**, *19*, 1–23. [CrossRef]
41. Sláma, J.; Košler, J.; Condon, D.J.; Crowley, J.L.; Gerdes, A.; Hanchar, J.M.; Horstwood, M.S.A.; Morris, G.A.; Nasdala, L.; Norberg, N.; et al. Plešovice zircon—A new natural reference material for U–Pb and Hf isotopic microanalysis. *Chem. Geol.* **2008**, *249*, 1–35. [CrossRef]
42. Raczek, I.; Stoll, B.; Hofmann, A.W.; Peter Jochum, K. High-Precision Trace Element Data for the USGS Reference Materials BCR-1, BCR-2, BHVO-1, BHVO-2, AGV-1, AGV-2, DTS-1, DTS-2, GSP-1 and GSP-2 by ID-TIMS and MIC-SSMS. *Geostand. Newsl.* **2001**, *25*, 77–86. [CrossRef]
43. Guo, Y.L.; Yang, S.Y.; Su, N.; Li, C.; Yin, P.; Wang, Z.B. Revisiting the effects of hydrodynamic sorting and sedimentary recycling on chemical weathering indices. *Geochim. Cosmochim. Acta* **2018**, *227*, 48–63. [CrossRef]
44. Liu, D.D.; Marc, J.; Yang, W.; Zhang, Z.Y.; Cheng, F.; Zhu, B.; Guo, Z.J. Latest Paleozoic–Early Mesozoic basin–range interactions in South Tian Shan (northwest China) and their tectonic significance: Constraints from detrital zircon U–Pb ages. *Tectonophysics* **2013**, *599*, 197–213. [CrossRef]
45. Hou, T.; Zhang, Z.C.; Santosh, M.; Encarnacion, J.; Zhu, J.; Luo, W.J. Geochronology and geochemistry of submarine volcanic rocks in the Yamansu iron deposit, Eastern Tianshan Mountains, NW China: Constraints on the metallogenesis. *Ore Geol. Rev.* **2014**, *56*, 487–502. [CrossRef]

46. Xie, W.; Luo, Z.Y.; Chen, Y.B.; Hong, L.B.; Ma, L.; Ma, Q. Petrogenesis and geochemistry of the Late Carboniferous rear-arc (or back-arc) pillow basaltic lava in the Bogda Mountains, Chinese North Tianshan. *Lithos* **2016**, *244*, 30–42. [CrossRef]
47. Xie, W.; Xu, Y.G.; Chen, Y.B.; Luo, Z.Y.; Hong, L.B.; Ma, L.; Liu, H.Q. High-alumina basalts from the Bogda Mountains suggest an arc setting for Chinese Northern Tianshan during the Late Carboniferous. *Lithos* **2016**, *256*, 165–181. [CrossRef]
48. Kirkland, C.L.; Smithies, R.H.; Taylor, R.; Evans, N.; McDonald, B. Zircon Th/U ratios in magmatic environs. *Lithos* **2015**, *212*, 397–414. [CrossRef]
49. Su, B.X.; Qin, K.Z.; Sakyi, P.A.; Li, X.H.; Yang, Y.H.; Sun, H.; Tang, D.M.; Liu, P.P.; Xiao, Q.H.; Malaviarachchi, P.K. U–Pb ages and Hf–O isotopes of zircons from Late Paleozoic mafic–ultramafic units in the southern Central Asian Orogenic Belt: Tectonic implications and evidence for an Early-Permian mantle plume. *Gondwana Res.* **2011**, *20*, 516–531. [CrossRef]
50. Liu, D.D.; Zhang, C.; Yao, E.D.; Song, Y.; Jiang, Z.X.; Luo, Q. What generated the Upper Permian to Triassic unconformities in the southern Junggar Basin and western Turpan Basin; tectonic uplift, or increasing aridity? *Palaeogeogr. Palaeoclimatol. Palaeoecol.* **2017**, *468*, 1–17. [CrossRef]
51. Liu, D.D.; Kong, X.Y.; Zhang, C.; Wang, J.B.; Yang, D.X.; Liu, X.Y.; Wang, X.P.; Song, Y. Provenance and geochemistry of Lower to Middle Permian strata in the southern Junggar and Turpan basins: A terrestrial record from mid-latitude NE Pangea. *Palaeogeogr. Palaeoclimatol. Palaeoecol.* **2018**, *495*, 259–277. [CrossRef]
52. Cawood, P.A.; Nemchin, A.A.; Freeman, M.; Sircombe, K. Linking source and sedimentary basin: Detrital zircon record of sediment flux along a modern river system and implications for provenance studies. *Earth Planet. Sci. Lett.* **2003**, *210*, 259–268. [CrossRef]
53. Deng, K.; Yang, S.Y.; Li, C.; Su, N.; Bi, L.; Chang, Y.P.; Chang, S.C. Detrital zircon geochronology of river sands from Taiwan: Implications for sedimentary provenance of Taiwan and its source link with the east China mainland. *Earth-Sci. Rev.* **2017**, *164*, 31–47. [CrossRef]
54. Wang, Z.X.; Li, T.; Zhang, J.; Liu, Y.Q.; Ma, Z.J. The uplifting process of the Bogda Mountain during the Cenozoic and its tectonic implication. *Sci. China Ser. D.* **2008**, *51*, 579–593. [CrossRef]
55. Greene, T.J.; Carroll, A.R.; Wartes, M.; Graham, S.A.; Wooden, J.L. Integrated provenance analysis of a complex orogenic terrane: Mesozoic uplift of the Bogda Shan and inception of the Turpan-Hami Basin, NW China. *J. Sediment. Res.* **2005**, *75*, 251–267. [CrossRef]
56. Ji, H.J.; Tao, H.F.; Wang, Q.; Ma, D.X.; Hao, L.W. Petrography, geochemistry, and geochronology of Lower Jurassic sedimentary rocks from the Northern Tianshan (West Bogda area), Northwest China: Implications for provenance and tectonic evolution. *Geol. J.* **2018**, *54*, 1688–1714. [CrossRef]
57. Wang, J.; Cao, Y.C.; Wang, X.T.; Liu, K.Y.; Wang, Z.K.; Xu, Q.S. Sedimentological constraints on the initial uplift of the West Bogda Mountains in Mid-Permian. *Sci. Rep.* **2018**, *8*, 1–14. [CrossRef] [PubMed]
58. Carroll, A.R.; Graham, S.A.; Hendrix, M.S.; Ying, D.; Zhou, D. Late Paleozoic tectonic amalgamation of northwestern China: Sedimentary record of the northern Tarim, northwestern Turpan, and southern Junggar basins. *Geol. Soc. Am. Bull.* **1995**, *107*, 571–594. [CrossRef]
59. Lewin, A.; Meinhold, G.; Hinderer, M.; Dawit, E.L.; Bussert, R. Provenance of sandstones in Ethiopia during Late Ordovician and Carboniferous–Permian Gondwana glaciations: Petrography and geochemistry of the Enticho Sandstone and the Edaga Arbi Glacials. *Sediment. Geol.* **2018**, *375*, 188–202. [CrossRef]
60. McLennan, S.M.; McCulloch, M.T.; Taylor, S.R.; Maynard, J.B. Effects of sedimentary sorting on neodymium isotopes in deep-sea turbidites. *Nature* **1989**, *337*, 547–549. [CrossRef]
61. Nelson, B.K.; DePaolo, D.J. Comparison of isotopic and petrographic provenance indicators in sediments from Tertiary continental basins of New Mexico. *J. Sediment. Res.* **1988**, *58*, 348–357.
62. Yang, S.Y.; Jung, H.S.; Choi, M.S.; Li, C.X. The rare earth element compositions of the Changjiang (Yangtze) and Huanghe (Yellow) river sediments. *Earth Planet. Sci. Lett.* **2002**, *201*, 407–419. [CrossRef]
63. Taylor, S.R.; McLennan, S.M. *The Continental Crust: Its Composition and Evolution*; Blackwell: Oxford, UK, 1985; pp. 1–312.
64. Zhang, X.R.; Zhao, G.C.; Paul, R.E.; Sun, M.; Han, Y.G.; Hou, W.Z.; Liu, D.X.; Wang, B.; Liu, Q.; Xu, B. Paleozoic magmatism and metamorphism in the Central Tianshan block revealed by U–Pb and Lu–Hf isotope studies of detrital zircons from the South Tianshan belt, NW China. *Lithos* **2015**, *233*, 193–208. [CrossRef]

65. Huang, B.; Fu, D.; Kusky, T.; Ruan, K.; Zhou, W.X.; Zhang, X.H. Sedimentary provenance in response to Carboniferous arc-basin evolution of East Junggar and North Tianshan belts in the southwestern Central Asian Orogenic Belt. *Tectonophysics* **2018**, *722*, 324–341. [CrossRef]
66. McLennan, S.M.; Hemming, S.; McDaniel, D.K.; Hanson, G.N. *Geochemical Approaches to Sedimentation, Provenance, and Tectonics*; Geological Society of America: Boulder, CO, USA, 1993; p. 21.
67. Floyd, P.A.; Leveridge, B.E. Tectonic environment of the Devonian Gramscatho basin, south Cornwall: Framework mode and geochemical evidence from turbiditic sandstones. *J. Geol. Soc.* **1987**, *144*, 531–542. [CrossRef]
68. Allegre, C.J.; Minster, J.F. Quantitative models of trace element behavior in magmatic processes. *Earth Planet. Sci. Lett.* **1978**, *38*, 1–25. [CrossRef]
69. Herron, M.M. Geochemical classification of terrigenous sands and shales from core or log data. *J. Sediment. Res.* **1988**, *58*, 820–829.
70. Pettijohn, F.J.; Potter, P.E.; Siever, R. *Production and Provenance of Sand and Sandstone*; Springer: Berlin/Heidelberg, Germany, 1972; pp. 294–326.
71. Condie, K.C. Chemical composition and evolution of the upper continental crust: Contrasting results from surface samples and shales. *Chem. Geol.* **1993**, *104*, 1–37. [CrossRef]
72. Suttner, L.J.; Dutta, P.K. Alluvial sandstone composition and paleoclimate; I, Framework mineralogy. *J. Sediment. Res.* **1986**, *56*, 329–345.
73. Chen, X.J.; Shu, L.S.; Santosh, M. Late Paleozoic post-collisional magmatism in the Eastern Tianshan Belt, Northwest China: New insights from geochemistry, geochronology and petrology of bimodal volcanic rocks. *Lithos* **2011**, *127*, 581–598. [CrossRef]
74. Xie, W.; Xu, Y.G.; Luo, Z.Y.; Liu, H.Q.; Hong, L.B.; Ma, L. Petrogenesis and geodynamic implications of the Late Carboniferous felsic volcanics in the Bogda belt, Chinese Northern Tianshan. *Gondwana Res.* **2016**, *39*, 165–179. [CrossRef]
75. Bhatia, M.R.; Crook, K.A. Trace element characteristics of graywackes and tectonic setting discrimination of sedimentary basins. *Contrib. Mineral. Petr.* **1986**, *92*, 181–193. [CrossRef]
76. Roser, B.P.; Korsch, R.J. Determination of tectonic setting of sandstone-mudstone suites using SiO_2 content and K_2O/Na_2O ratio. *J. Geol.* **1986**, *94*, 635–650. [CrossRef]
77. Gabo, J.A.S.; Dimalanta, C.B.; Asio, M.G.S.; Queaño, K.L.; YumulJr, G.P.; Imaie, A. Geology and geochemistry of the clastic sequences from Northwestern Panay (Philippines): Implications for provenance and geotectonic setting. *Tectonophysics* **2009**, *479*, 111–119. [CrossRef]
78. Verma, S.P.; Armstrong-Altrin, J.S. New multi-dimensional diagrams for tectonic discrimination of siliciclastic sediments and their application to Precambrian basins. *Chem. Geol.* **2013**, *355*, 117–133. [CrossRef]
79. Bhatia, M.R. Plate tectonics and geochemical composition of sandstones. *J. Geol.* **1983**, *91*, 611–627. [CrossRef]
80. Wang, X.W.; Wang, X.W.; Ma, Y.S. The tectonic evolution of Bogda Mountain, Xinjiang, Northwest China and its relationship to oil and gas accumulation. *Geoscience* **2007**, *21*, 116–124. (In Chinese with English Abstract).
81. Laurent, C.S.; Charvet, J.; Monié, P.; Shu, L. Late Paleozoic strike-slip shear zones in eastern Central Asia (NW China): New structural and geochronological data. *Tectonics* **2003**, *22*, 1–24.
82. Yang, W.; Marc, J.; Guillaume, D.N.; Guo, Z.J.; Zhang, Z.C.; Wu, C.D. Source to sink relations between the Tian Shan and Junggar Basin (northwest China) from Late Palaeozoic to Quaternary: Evidence from detrital U-Pb zircon geochronology. *Basin Res.* **2013**, *25*, 219–240. [CrossRef]
83. Simonov, V.A.; Mikolaichuk, A.V.; Safonova, I.Y.; Kotlyarov, A.V.; Kovyazin, S.V. Late Paleozoic–Cenozoic intra-plate continental basaltic magmatism of the Tienshan–Junggar region in the SW Central Asian Orogenic Belt. *Gondwana Res.* **2015**, *27*, 1646–1666. [CrossRef]
84. Carroll, A.R.; Liang, Y.H.; Graham, S.A.; Xiao, X.C.; Hendrix, M.S.; Chu, J.C.; McKnight, C.L. Junggar basin, northwest China: Trapped Late Paleozoic Ocean. *Tectonophysics* **1990**, *181*, 1–14. [CrossRef]
85. Xiao, W.J.; Han, C.M.; Yuan, C.; Sun, M.; Lin, S.F.; Chen, H.L.; Li, Z.L.; Li, J.L.; Sun, S. Middle Cambrian to Permian subduction-related accretionary orogenesis of Northern Xinjiang, NW China: Implications for the tectonic evolution of central Asia. *J. Asian Earth Sci.* **2008**, *32*, 102–117. [CrossRef]
86. Gao, H.H.; He, D.F.; Tong, X.G.; Wen, Z.X.; Wang, Z.M. Tectonic-depositional environment and proto-type basins evolution of the Late Ordovician in the Tarim Basin. In Proceedings of the 19th EGU General Assembly, Vienna, Austria, 23–28 April 2017; Volume 19, p. 11218.

87. Yang, W.; Feng, Q.; Liu, Y.Q.; Neil, T.; Dan, M.; James, L.C.; Lin, J.Y.; Thomas, S. Depositional environments and cyclo-and chronostratigraphy of uppermost Carboniferous–Lower Triassic fluvial–lacustrine deposits, southern Bogda Mountains, NW China—A terrestrial paleoclimatic record of mid-latitude NE Pangea. *Glob. Planet. Chang.* **2010**, *73*, 15–113. [CrossRef]
88. Obrist-Farner, J.; Yang, W. Provenance and depositional conditions of fluvial conglomerates and sandstones and their controlling processes in a rift setting, mid-Permian lower and upper Quanzijie low order cycles, Bogda Mountains, NW China. *J. Asian Earth Sci.* **2017**, *138*, 317–340. [CrossRef]

minerals
MDPI

Article

U–Pb Zircon Geochronological and Petrologic Constraints on the Post-Collisional Variscan Volcanism of the Tiddas-Souk Es-Sebt des Aït Ikko Basin (Western Meseta, Morocco)

Ismail Hadimi [1,*], Nasrrddine Youbi [1,2], Abdelhak Ait Lahna [1], Mohamed Khalil Bensalah [1,3], Oussama Moutbir [1], João Mata [3], Miguel Doblas [4], Colombo Celso Gaeta Tassinari [5], Laura Gaggero [6], Miguel Angelo Stipp Basei [7], Kei Sato [5], Warda El Moume [1] and Moulay Ahmed Boumehdi [1,3]

[1] Department of Geology, Faculty of Sciences-Semlalia, Cadi Ayyad University, Prince Moulay Abdellah Boulevard, Marrakech P.O. Box 2390, Morocco; youbi@uca.ac.ma (N.Y.); aitlahna.abdelhak@gmail.com (A.A.L.); bensalah@uca.ac.ma (M.K.B.); oussama.moutbir2@gmail.com (O.M.); warda.elmoume@gmail.com (W.E.M.); boumehdi@uca.ac.ma (M.A.B.)
[2] Faculty of Geology and Geography, Tomsk State University, 634050 Tomsk, Russia
[3] Instituto DomLuiz, Faculdade de Ciências, Universidade de Lisbona, 1749-016 Lisboa, Portugal; jmata@fc.ul.pt
[4] Instituto de Geociencias (CSIC-UCM), Ciudad Universitaria, c/Doctor Severo Ochoa 7, 28040 Madrid, Spain; m.doblas@igeo.ucm-csic.es
[5] Centro de Pesquisas Geocronológicas (CPGeo), Instituto de Geociências (IG), Universidade de São Paulo—USP, São Paulo 05422-970, Brazil; ccgtassi@usp.br (C.C.G.T.); keisato@usp.br (K.S.)
[6] Department of Earth, Environment and Life Sciences, University of Genoa, Corso Europa 26, 16132 Genoa, Italy; laura.gaggero@unige.it
[7] Instituto de Geociências (IG), Departamento de Mineralogia e Geotectônica (GMG), Universidade de São Paulo—USP, Rua do Lago, 562 Cidade Universitária, São Paulo 05508-080, Brazil; baseimas@usp.br
* Correspondence: ismail.hadimi@gmail.com

Citation: Hadimi, I.; Youbi, N.; Ait Lahna, A.; Bensalah, M.K.; Moutbir, O.; Mata, J.; Doblas, M.; Tassinari, C.C.G.; Gaggero, L.; Basei, M.A.S.; et al. U–Pb Zircon Geochronological and Petrologic Constraints on the Post-Collisional Variscan Volcanism of the Tiddas-Souk Es-Sebt des Aït Ikko Basin (Western Meseta, Morocco). *Minerals* **2021**, *11*, 1099. https://doi.org/10.3390/min11101099

Academic Editors: Wilfried Winkler and Albrecht von Quadt

Received: 29 August 2021
Accepted: 30 September 2021
Published: 7 October 2021

Abstract: The NE–SW trending Tiddas Souk Es-Sebt des Ait Ikko (TSESDAI) basin, located at 110 km southeast of Rabat, in the region of Khmesset between the village of Tiddas and Souk Es-Sebt des Ait Ikko, is the third largest late Palaeozoic continental trough in the northern Central Moroccan Meseta. It is a ~20 km long and ~2–3 km wide basin, comprising mainly mixed volcano-sedimentary reddish-purple continental Permian rocks laying with an angular unconformity on Visean deep marine siliciclastic sediments and unconformably overlain by the Triassic and Cenozoic formations. In this study we aim to better determine the age of Permian volcanics and their chemical and mineralogical characteristics, as well as assess the provenance of inherited zircons, thus contributing to the understanding of the late stages of the Variscan orogeny in Morocco. The standard volcanic succession includes the following terms: (i) andesites, lapilli tuffs and andesitic ash deposits; (ii) accumulations of rhyolitic lavas; (iii) lapilli tuffs and rhyolitic ash (formation F1); (iv) flows and breccias of dacites; (v) andesite flows; and (vi) basaltic flows. The various volcanic and subvolcanic studied rocks display calc-alkaline-series characteristics with high contents of SiO_2, Al_2O_3, CaO, MgO, and relatively abundant alkalis, and low contents of MnO. In the classification diagram, the studied facies occupy the fields of andesites, trachy-basalts, dacites, trachydacites, and rhyolites and display a sub-alkaline behavior. These lavas would be derived from a parental mafic magma (basalts) produced by partial fusion of the upper mantle. Specific chemical analyses that were carried out on the mineralogical phases (biotite and pyroxene) revealed that the examined biotites can be classified as magnesian and share similarities with the calc-alkaline association-field, while the clinopyroxenes are mainly augites and plot on the calc-alkaline orogenic basalt field. Andesites and dacites of TSESDAI show similarities with the rocks of the calc-alkaline series not linked to active subduction and which involve a continental crust in their genesis. The existence of enclaves in the lavas of the TSESDAI massif; the abnormally high contents of Rb, Ba, Th, and La; and the systematic anomalies in TiO_2 and P_2O_5 indicate also a crustal contamination mechanism. Three magmatic episodes are distinguished with two episodes that correspond to an eruptive cycle of calc-alkaline andesites and rhyolites followed by a basaltic episode. The SHRIMP U–Pb geochronologic data of zircons recovered from the rhyolite dome of Ari El Mahsar in TSESDAI basin show a Concordia age of 286.4 ± 4.7 Ma interpreted to date the magmatic crystallization of this dome. Thus, the rhyolite likely belongs to the third magmatic episodes of TSESDAI.

Keywords: Western Hercynian Meseta; Permian basin; Tiddas Souk Es-Sebt des ait ikko volcanic basin; U–Pb geochronology; mineralogy; petrology

1. Introduction

The European Northwest African Magmatic Province (EUNWA or EUNWAMP; e.g., [1–3] was emplaced during Late Carboniferous–Permian times and has been linked to the gravitational collapse of the previously overthickened and weakened Variscan orogenic belt. The gravitational collapse of the Variscan edifice occurred through simple pure shear low-angle extensional detachments during the late Carboniferous–Permian, giving rise to a Basin and Range type extensional province in Europe, and northwestern Africa involving the unroofing of large metamorphic core complexes and synextensional plutonic bodies, dike and sill swarms, and volcanic successions. Coevally with this extensional scenario, Europe and northwestern Africa were affected by a complex system of conjugate strike-slips faults (NE–SW sinistral and NW–SE dextral), which partially disrupted the Variscan edifice, resulting in new Permo-Carboniferous stress patterns with the principal compressional axis oriented N–S [4,5]. This episode was accompanied by sediment deposition and volcanism in transtensional and pull-apart basins [1,6,7]. This episode resulted from a major dextral transcurrent movement along an intracontinental shear zone located between Gondwana and Laurussia [4,5,8,9]. To date, a range of chronometers has been applied to determine crystallization ages from across the region of the EUNWA Large Igneous Province, including whole-rock Rb–Sr and K–Ar dating; $^{40}Ar/^{39}Ar$ dating of mineral separates; and U–Pb dating of zircon, titanite, and perovskite (e.g., [10]). The duration of activity is currently estimated to span a period of ca. 100 million years, from the Early Carboniferous to the Late Permian–Early Triassic (350–250 Ma), with several hiatuses [11]. Three main pulses can be distinguished at ca. 300 Ma, 290–275 Ma, and 250 Ma, and each of these pulses can be considered a separate LIP within the overall EUNWA Large Igneous Province. These eruptive cycles are well represented in Morocco in northwestern Africa and also in southern Scandinavia and northern Germany. The huge volume of extruded and intruded magmatic products of the EUNWA province (in the Oslo Graben, the estimated volume is 35,000 km^3, while in the North German Basin, the total volume of felsic volcanic rocks, mainly rhyolites and rhyodacites, was of the order of 48,000 km^3) has led to suggestions of a thermally anomalous mantle plume to explain this magmatic pulse [12,13]. A significant problem of the plume hypothesis is the duration of activity and the helium isotope signature of lithospheric mantle xenoliths from the Scottish Permo-Carboniferous dikes, sills, and vents [14]. The EUNWA magmatic province may have contributed to the great Gondwanan glaciation that occurred from the late Devonian to the late Permian [15–17]. Glaciers achieved their maximum paleolatitudinal range between the middle Stephanian (ca. 305 Ma ago) and the end of the Sakmarian (ca. 284 Ma ago) [17]. This hypothesis is termed the icehouse-silicic large igneous province (SLIP) hypothesis [18].

In Morocco, the Permian sedimentation was continental and occurred in small isolated basins (Khenifra, Boudoufoud, Bou-Achouch, Tiddas Souk Es-Sebt des Aït Ikko (TSESDAI), Chougrane, Mechraa Ben Abbou, Senhaja, Ouelad Maachou, Ourika, and Argana; see Figure 1), mostly originated as NE–SW to NNE–SSW half-graben-like structures, during the gravitational collapse of the Variscan orogenic belt [19–22]. They are the remnants of strongly subsided intramountainous troughs predominantly filled with red-colored detritus, often associated with andesitic and rhyolitic lava flows, domes, pyroclastic rocks, and shallow-level dikes and sills. These volcanic basins and their associated plumbing systems belong to the European Northwest African Magmatic Province (EUNWAMP); e.g., [1]).

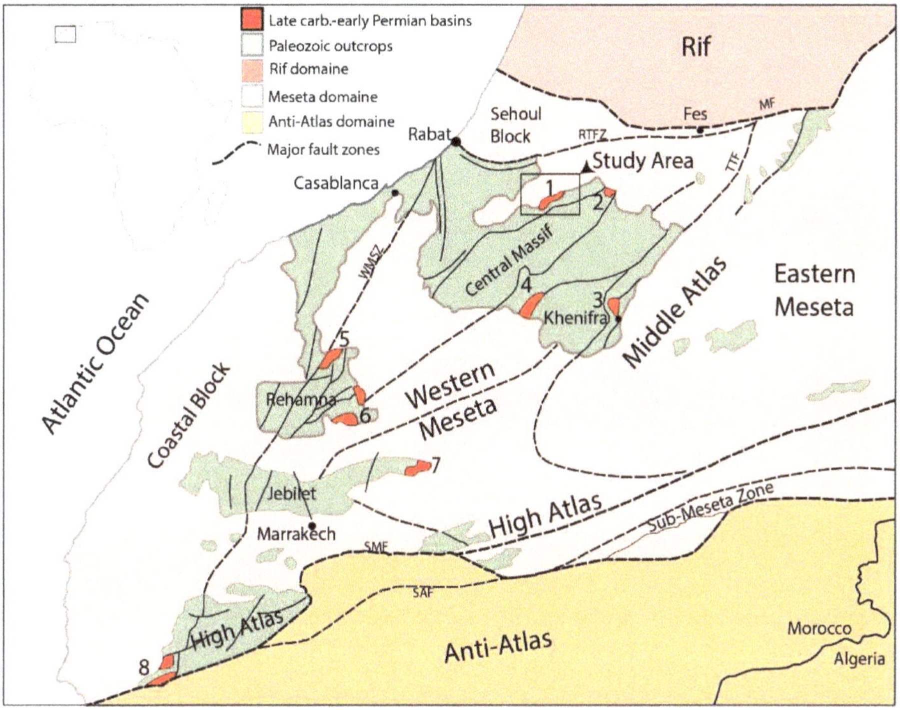

Figure 1. Principal tectonic domains of central Morocco (modified after [23–26]). Numbers 1–8 correspond to Carboniferous–Early Permian basins (1: the studied outcrops of Tiddas; 2: Bou Achouch, 3: Khenifra, 4: Chougrane, 5: Mechra ben Abbou, 6: Nzalet el Hararcha, 7: Senhaja, 8: Souss). Abbreviations: MF, Maghrebide Front; RTFZ, Rabat Tiflet Fault Zone; SAF, South Atlas Fault; SMF, South Meseta Fault; TTF, Tizi n' Tretten Fault Zone; WMSZ, Western Meseta Shear Zone.

While several Late Carboniferous–Permian basins have been extensively studied for their stratigraphic, paleontological, and sedimentological characteristics, very little is known on the petrology, geochemistry, and geochronology, with the exception of the recent work of [23]. In this study, we focus on TSESDAI Permian volcanism affecting the northwestern Gondwanaland margin in order to (1) determine, using a sensitive high-resolution ion microprobe (SHRIMP) the zircon U–Pb age of volcanic crystallization; (2) assess the provenance of inherited zircons; (3) describe the mineral chemistry of the main rock-forming minerals; (4) characterize geochemically the magma affinities; and (5) discuss the petrogenesis of volcanic rocks.

2. Geological Background

The NE–SW TSESDAI basin is located 110 km to the SE of Rabat (Figure 1), in the region of Khemisset (between the villages of Tiddas and Souk Es Sebt des Ait Ikko) and is the third-largest late Palaeozoic continental trough in the northern Central Moroccan Meseta. It is a ~20 km long and ~2–3 km wide basin, comprising mainly mixed volcanic–sedimentary reddish-purple Permian rocks laying with an angular unconformity on Visean deep marine siliciclastic sediments and being unconformably overlain by the Triassic and Cenozoic formations (Figure 2). The volcanic/subvolcanic formations of this basin consist of rhyolitic domes with associated rhyolitic dike swarms and andesitic to dacitic flows with a relatively reduced extension. These magmatic formations occupy an area of about 48 km^2 and display massive textures and tabular surfaces. They are aligned along a NE–SW direction corresponding to the orientation of the major Hercynian to late Hercynian structures. These volcanics are intruded by a dike swarm with a NE–SW dominant direction and covered by Triassic and Tertiary deposits [19,27–29]. The red sedimentary package (40 to 300 m thick) of this basin is generally incomplete due to faulting. Three main formations have been recognized on the TSESDAI basin by [30–33]: (i) A 100 m-thick lower formation (F1) constituted by purple–red silty argillites, alternating with conglomerates including andesites, quartz, quartzite, sandstone, and rare limestones pebbles; (ii) an intermediate formation (F2) of about 120 m thick, consisting of silty argillites gullied and interspersed with channel deposits (e.g., puddingstones and coarse sandstone) with lenticular stratification and oblique tabular/horizontal bedding (the paleocurrents are directed towards the SW and S); and (iii) an upper formation (F3) of 50 to 80 m thick that is marked by thick red silty argillites interspersed with sandstone in centimetric to decametric benches. Determining the Permian age of these formations was constrained by using plant remains and vertebrate ichnofossils (e.g., [33,34]). The age has been recently confirmed by [35]. The occurrence of volcanics in this basin was first reported and termed by [36], being considered part, with those of the Khenifra Basin, of the so-called "Central Moroccan volcanic axis", extending from Khenifra to TSESDAI basins and displaying felsic and mafic volcanics overlain by Neogene/Quaternary formations. Until the 1980s, this "axial-model" has been accepted by several authors (e.g., [37] and references therein). Later on, many contributions largely improved the geological knowledge of this basin (e.g., [6,27,29]). The main conclusions of these authors are as follows: (i) The age of the TSESDAI basin volcanic complex is Permian s.l. (late-Carboniferous and pre-Triassic, probably Stephanian/Permian). (ii) There is a distinction of three major volcanic pulses, the first one (mainly andesitic) predating the Permian series. Witnesses of this pulse have been observed in the Tiddas region; however, most of this material is reworked within the conglomerates. The second pulse is synsedimentary and displays various interesting petrographic characteristics and remarkable andesitic lava flows interbedded within the detrital sediments. The third pulse consists of intrusive rhyolitic dikes that precede the Ari el Mahsar rhyolitic dome emplacement and the dacitic flows. (iii) There has been identification of three intra-Permian tectonic events. The close relationship between volcanism and fracturation is obvious; dikes and effusive centers are aligned along N40–N70, mostly strike-slip faults.

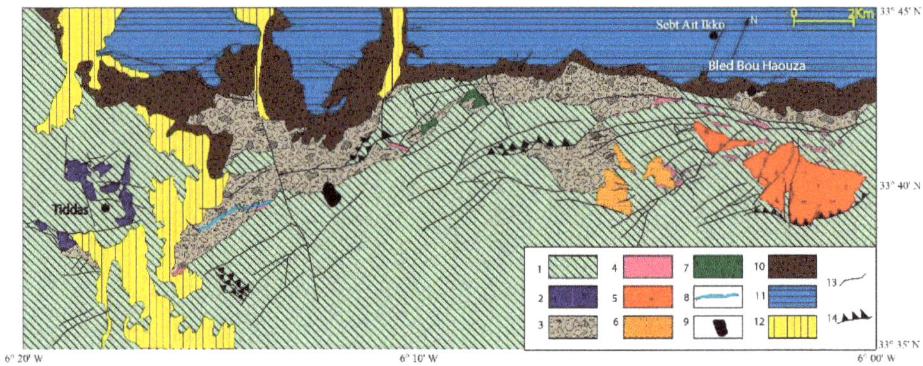

Figure 2. Geological map of the TSESDAI volcanic complex and its environs (based on [28], completed by [29,38]): 1. Upper Visean Paleozoic terrains (schists and quartzites); 2. pyroxene andesties (Tiddas andesites); 3. Permian conglomerates, sandstones, and argillites; 4. biotite rhyolite dikes; 5. biotite rhyolitic dome (Ari Almahsar and Agrued El khebiza); 6. pyroxene dacites (Bled Bou Haouza); 7. olivine and pyroxene andesites (Sebt Ait ikko); 8. pyroxene dolerite vein; 9. pyroxene basalt flows of El Gitoune; 10. Triassic detrital and volcanic formations; 11. Messinian and Pliocene conglomeratic and marl formations; 12. Quaternary; 13. undefined faults; 14. inverse faults.

The assignment of the TSESDAI basin to the Permian was based on (a) the intrusion of rhyolites into the Visean deep marine siliciclastics (e.g., shark-thin rhyolitic extrusive dome of Ari el Mahsar and its satellite dikes); (b) the angular unconformity between the Visean siliciclastics folded during the late Westphalian A and red detrital sediments; (c) the presence of reworked rhyolitic/andesitic pebbles in the red sediments of this basin and their absence in the detrital sediments of the Westphalian C/D preserved nearby in the Sidi Kacem graben; (d) the angular unconformity between the Triassic and the red detrital sediments; and (e) a strike-slip fault that postdates the basin sedimentary/volcanic formations and predates the Triassic/Cenozoic.

3. Sampling and Analytical Procedures

3.1. Sampling

In order to investigate the nature, the tectonic setting, and the petrogenetic processes of the post-collisional Variscan volcanism of the TSESDAI basin, we collected and analyzed a total of eighteen samples from the least weathered outcrops (major and trace elements; Table 1). Nine chemicals analyses of [19,28] were used for comparison. In addition, a sample from the shark-fin-like rhyolitic extrusive dome of Ari el Mahsar (sample ARM15; N 30°48′45.7″, W 7°32′29.1″) was dated using SHRIMP U–Pb zircon geochronology (Figure 14, Table 2). Specific chemical analyzes were carried out on the mineralogical phases (pyroxene; Tables S1 and S2, biotite; Table S3) to characterize them and to determine the physicochemical conditions of the crystallization of the studied rocks. The pyroxenes selected for this study were exclusively from dacites given their better preservation in comparison with those of the andesites, which are mostly transformed into secondary minerals (chlorite, calcite, and silica).

3.2. Analytical Procedures

3.2.1. U–Pb Geochronology: Sensitive High-Resolution Ion Microprobe (SHRIMP IIe)

Igneous zircon grains from the shark-fin-like rhyolitic extrusive dome of the Ari el Mahsar sample ARM15 was separated for U–Pb age determinations at the CPGeo-USP, Brazil. The crystals were further mounted, together with the TEMORA-2 standard [39], in epoxy and polished to expose the interior of the grains. After coating with Au, the polished mounts were comprehensively examined with a FEI Quanta 250 scanning electron microscope equipped with secondary electrons and cathodoluminescence (CL) detectors at CPGeo-USP; the most common conditions used in CL analysis were as follows: 60 μA emission current, 15.0 kV accelerating voltage, 7 μm beam diameter, 200 μs acquisition time, and a resolution of 1024 × 884. The same mounts were analyzed afterwards by the U–Pb isotopic technique using a SHRIMP II following the analytical procedures of [40]. Correction for common Pb was made based on the ^{204}Pb measured, and the typical error for the $^{206}Pb/^{238}U$ ratio was less than 2%; uranium abundance and U/Pb ratios were calibrated against the TEMORA standard, and the ages were calculated using the Isoplot® (version$^{3.0}$, Berkeley Geochronology Center, Berkeley, CA, USA) application of [41]. Errors were reported as 1σ deviations, and ages were calculated at the 95% confidence level.

Table 1. Chemical compositions in major and traces elements of Tiddas-Souk Es Sebt des Ait Ikko volcanic complex magmatic rocks added to nine whole rock chemical analyses available in the literature (identified by an asterisk (*); [19,28]). Mg # = (MgO/40.32)/(Fe2O3t*0.9/71.8) + (MgO/40.32).

No Fac.	1	1	2	2	2	2	2	2	2	2	3	3	3	3
No Samp.	**YK48**	**YK49**	**YK37**	**YK44**	**YK43**	**YK32**	**YK27**	**YK1**	**W184 ***	**W186 ***	**YK6**	**W183A ***	**W181 ***	**W183B ***
SiO_2	54.12	56.21	63.90	64.83	64.85	65.39	65.85	65.86	66.47	67.85	67.18	68.35	68.82	68.82
TiO_2	1.04	1.03	0.73	0.69	0.69	0.59	0.64	0.59	0.70	0.66	0.24	0.29	0.32	0.35
Al_2O_3	15.99	15.88	15.79	15.33	15.45	15.22	15.10	14.93	14.20	14.05	14.73	13.99	14.55	14.12
Fe_2O_3t	5.62	5.41	3.26	4.04	3.90	3.63	3.83	3.53	2.26	3.72	1.42	2.39	2.41	1.72
Fe_2O_3	-	-	-	-	-	-	-	-	2.09	2.72	-	1.44	2.04	0.96
FeO	-	-	-	-	-	-	-	-	0.15	0.90	-	0.85	0.33	0.68
MnO	0.06	0.07	0.01	0.05	0.04	0.02	0.04	0.02	0.02	0.03	0.02	0.03	0.06	0.01
MgO	2.55	2.09	0.37	1.51	1.79	1.02	0.61	0.88	0.18	0.4	0.43	0.41	0.25	0.57
CaO	6.32	5.93	2.45	2.62	2.74	3.02	2.59	2.26	1.97	2.22	3.86	2.18	2.02	1.98
Na_2O	3.39	3.36	2.06	3.15	3.18	3.31	2.53	2.93	1.95	2.95	4.01	5.11	4.49	5.1
K_2O	3.6	3.28	7.33	3.92	3.86	4.02	5.11	5.17	8.37	5.61	1.89	2.17	2.8	2.48
P_2O_5	0.31	0.32	0.18	0.15	0.16	0.12	0.15	0.13	0.16	0.12	0.07	0.07	0.12	0.11
LOI	7.1	6.13	3.07	3	3.1	3.09	3.3	2.97	3.51	2.61	5.52	4.36	3.68	3.97
Total	100.1	99.71	99.15	99.29	99.76	99.43	99.75	99.27	99.77	100.12	99.37	99.25	99.48	99.15
Rb	134	125	163	135	134	146	157	150	-	-	44	-	-	-
Ba	655	650	653	494	520	630	521	600	-	-	442	-	-	-
Sr	578	502	342	265	275	289	239	250	-	-	455	-	-	-
Th	4.5	4.1	4.1	3.8	3.8	4.3	4	3.9	-	-	2.5	-	-	-
U	1	2	2	2	1	1	2	2	-	-	2	-	-	-
Pb	2	2	2	2	2	2	2	2	-	-	2	-	-	-
Zr	298	288	202	214	219	214	191	214	-	-	151	-	-	-
Hf	4	4	2	3	3	3	1	3	-	-	2	-	-	-
Nb	5	5	5	5	5	5	5	5	-	-	5	-	-	-
La	45	41	41	38	38	43	40	39	-	-	25	-	-	-
Ce	80	80	79	69	75	77	75	77	-	-	52	-	-	-
Nd	33	35	30	25	27	27	27	24	-	-	9	-	-	-
Y	20	20	13	16	16	17	16	16	-	-	-	-	-	-
Cr	110	111	37	54	56	52	54	49	-	-	4	-	-	-
Co	18	19	9	13	13	9	11	10	-	-	-	-	-	-
Ni	41	54	20	25	25	23	32	23	-	-	-	-	-	-
V	110	97	69	65	69	58	53	55	-	-	7	-	-	-
Cu	36	-	-	-	-	-	-	-	-	-	-	-	-	-
Zn	58	37	28	52	57	45	49	49	-	-	35	-	-	-
Mg #	0.961	0.816	0.234	0.777	0.953	0.581	0.330	0.515	0.162	0.223	0.610	0.350	0.212	0.671

Table 1. *Cont.*

No Fac.	3	3	3	3	3	3	3	3	3	3	3	3	4	4
No Samp.	**W147 ***	**YK4**	**W144 ***	**YK13**	**YK19**	**YK7**	**YK8**	**YK5**	**W182 ***	**W145 ***	**YK12**	**ARM15**	**YK14F**	**YK36F**
SiO_2	69.1	69.55	69.76	69.8	69.95	70.3	70.3	70.42	70.45	70.7	70.81	71.69	70.06	70.29
TiO_2	0.24	0.23	0.29	0.23	0.23	0.25	0.25	0.23	0.31	0.54	0.23	0.26	0.24	0.23
Al_2O_3	15.35	15.07	15.5	15.17	15.29	15.41	15.27	15.25	14.05	13.72	15.33	15.58	14.72	14.73
Fe_2O_3t	1.73	1.5	1.63	1.47	1.53	1.68	1.39	1.5	2.40	1.44	1.48	-	1.62	1.61
Fe_2O_3	1.36	-	1.07	-	-	-	-	-	2.08	0.18	-	1.63	-	-
FeO	0.33	-	0.5	-	-	-	-	-	0.29	1.13	-	-	-	-
MnO	0.03	0.01	0.03	0.02	0.01	0.02	0.02	0.01	0.04	0.02	0.01	0.03	0.03	0.03
MgO	0.7	0.43	0.45	0.43	0.42	0.11	0.18	0.43	0.26	0.61	0.44	0.21	0.7	0.7
CaO	2.55	2	1.12	1.93	1.83	1.85	1.92	1.94	0.49	1.93	1.76	0.19	1.88	1.87
Na_2O	4.11	4.11	4.38	4.12	4.21	4.72	4.33	4.11	4.85	4.75	4.14	3.75	4.11	4.11
K_2O	0.27	2.6	2.49	2.57	2.5	2.74	2.79	2.61	2.48	2.24	2.54	3.76	1.83	1.82
P_2O_5	0.07	0.07	0.08	0.07	0.07	0.07	0.07	0.07	0.11	0.08	0.07	0.12	0.05	0.05
LOI	5.51	3.45	3.6	3.31	3.2	2.69	3.1	3.34	3.33	2.91	3.22	1.98	4.15	4.07
Total	99.62	99.02	99.27	99.12	99.24	99.84	99.62	99.91	98.74	98.81	100.03	99.20	99.39	99.51
Rb	-	55	-	55	54	59	63	55	-	-	55	56.54	56	56
Ba	-	668	-	636	618	538	751	698	-	-	595	627.54	65	630
Sr	-	373	-	376	373	382	428	380	-	-	370	343.00	492	494
Th	-	2.7	-	2.6	2.7	2.7	2.6	2.9	-	-	2.7	3.69	2.9	2.7
U	-	2	-	2	2	2	2	2	-	-	2	0.63	2	2
Pb	-	2	-	2	2	2	2	2	-	-	2	30.93	2	2
Zr	-	153	-	156	156	153	155	157	-	-	158	183.00	153	155
Hf	-	2	-	2	2	2	2	2	-	-	2	3.60	2	2
Nb	-	5	-	5	5	5	5	5	-	-	5	2.21	5	5
La	-	27	-	26	27	27	26	29	-	-	27	20.45	29	27
Ce	-	49	-	50	54	48	48	43	-	-	49	37.85	52	51
Nd	-	8	-	7	13	13	10	9	-	-	9	15.58	14	12
Y	-	-	-	-	-	-	-	-	-	-	-	4.64	-	-
Cr	-	2	-	1	1	3	2	10	-	-	3	-	1	-
Co	-	-	-	-	-	-	-	-	-	-	-	-	-	-
Ni	-	-	-	-	-	-	-	-	-	-	-	23.00	-	-
V	-	9	-	11	9	14	12	12	-	-	11	-	9	11
Cu	-	-	-	-	-	-	-	-	-	-	-	-	-	-
Zn	-	15	-	13	13	38	98	14	-	-	16	-	76	76
Mg #	0.819	0.578	0.559	0.589	0.554	0.132	0.261	0.578	0.221	0.855	0.599	-	0.872	0.878

Table 2. SHRIMPII zircon U–Th–Pb results of the rhyolitic dome of TSESDAI.

Unit/ Sample	Spot Name	Concentrations U (ppm)	Th (ppm)	$^{232}Th/^{238}U$	± %	$^{206}Pb^*$ (ppm)	^{206}Pbc (%)	Ratios $^{238}U/^{206}Pb^*$	1 σ %	$^{207}Pb^*/^{206}Pb^*$	1 σ %	$^{207}Pb^*/^{235}U$	1 σ %	$^{206}Pb^*/^{238}U$	1 σ %	Rho	Ages (Ma) $^{206}Pb/^{238}U$	1 σ (Ma)	$^{207}Pb/^{206}Pb$	1 σ (Ma)	Conc %
Tiddas Souk Es-Sebt Des Ait Ikko/Ari el Mahsar/ARM15	1.1	283	171	0.62	2.31	20.3	0.30	11.945	3.468	0.0594	0.991	0.69	3.607	0.084	3.468	0.962	518	17	582	22	111
	2.1	615	365	0.61	1.59	24.2	−0.02	21.851	3.083	0.0519	0.721	0.33	3.166	0.046	3.083	0.974	288	9	282	16	98
	3.1	632	468	0.76	0.78	23.4	0.56	23.206	3.121	0.0538	1.527	0.32	3.474	0.043	3.121	0.898	272	8	362	34	125
	4.1	675	383	0.59	1.34	25.3	0.15	22.856	3.055	0.0522	0.904	0.32	3.186	0.044	3.055	0.959	276	8	295	21	107
	5.1	619	260	0.43	4.79	24.0	0.14	22.116	3.053	0.0529	0.807	0.33	3.158	0.045	3.053	0.967	285	9	323	18	112
	6.1	497	219	0.46	0.69	47.1	4.08	8.711	3.447	0.0941	0.508	1.49	3.484	0.115	3.447	0.989	701	23	1510	10	157
	7.1	399	169	0.44	0.22	16.1	0.16	21.211	3.237	0.0525	1.400	0.34	3.527	0.047	3.237	0.918	297	9	309	32	104
	8.1	165	97	0.60	0.66	64.5	1.21	2.174	3.087	0.1669	0.285	10.59	3.101	0.460	3.087	0.996	2439	63	2527	5	104
	8.2	669	380	0.59	0.22	25.5	0.23	22.522	3.063	0.0520	1.496	0.32	3.409	0.044	3.063	0.899	280	8	284	34	101
	9.1	564	744	1.36	0.49	41.8	0.80	11.541	3.057	0.0606	1.227	0.72	3.294	0.087	3.057	0.928	536	16	625	26	115
	10.1	229	50	0.23	1.47	49.8	1.37	3.902	3.165	0.1030	0.784	3.64	3.260	0.256	3.165	0.971	1471	42	1679	14	114
	10.2	484	160	0.34	2.51	47.3	2.26	8.583	3.371	0.0809	0.722	1.30	3.448	0.117	3.371	0.978	710	23	1219	14	144
	11.1	557	323	0.60	0.71	22.0	0.56		3.223	0.0527	1.871	0.33	3.727	0.046	3.223	0.865	290	9	318	43	109
	11.2	176	86	0.50	9.59	10.1	0.31	14.943	4.013	0.0588	2.546	0.54	4.753	0.067	4.013	0.844	418	16	561	55	126
	12.1	582	387	0.69	0.83	23.2	0.35	21.585	3.531	0.0516	1.963	0.33	4.039	0.046	3.531	0.874	292	10	267	45	90
	13.1	195	81	0.43	0.65	39.9	2.04	4.114	3.089	0.1055	0.495	3.53	3.129	0.243	3.089	0.987	1403	39	1722	9	121
	13.2	652	332	0.53	0.73	26.2	0.18	21.409	3.143	0.0520	1.543	0.33	3.501	0.047	3.143	0.898	294	9	284	35	96
	14.1	497	171	0.36	0.41	28.9	0.14	14.786	3.937	0.0556	2.050	0.52	4.439	0.068	3.937	0.887	422	16	438	46	104
	14.2	105	56	0.55	0.39	8.3	0.55	10.896	3.138	0.0589	2.984	0.74	4.330	0.092	3.138	0.725	566	17	562	65	99
	15.1	243	116	0.49	0.29	11.7	0.54	17.881	3.087	0.0548	2.342	0.42	3.875	0.056	3.087	0.797	351	11	402	52	113
	16.1	300	139	0.48	0.27	24.2	0.07	10.638	3.133	0.0603	0.876	0.78	3.254	0.094	3.133	0.963	579	17	616	19	106
	17.1	471	186	0.41	0.81	19.5	0.58	20.707	3.190	0.0541	2.456	0.36	4.026	0.048	3.190	0.792	304	9	376	55	120
	18.1	336	104	0.32	0.55	29.2	0.09	9.864	3.072	0.0611	0.754	0.85	3.163	0.101	3.072	0.971	622	18	644	16	104
	19.1	49	52	1.10	0.46	4.1	0.39	10.235	3.248	0.0609	3.219	0.82	4.573	0.098	3.248	0.710	601	19	637	69	106
	19.2	440	186	0.44	2.01	17.0	0.87	22.216	3.150	0.0521	4.986	0.32	5.897	0.045	3.150	0.534	284	9	290	114	102
	20.1	361	4	0.01	1.34	101.3	1.44	3.023	3.071	0.1235	0.307	5.63	3.086	0.331	3.071	0.995	1842	49	2007	5	125
	20.2	1129	410	0.37	0.35	71.3	1.01	13.605	3.706	0.0566	4.970	0.57	6.199	0.074	3.706	0.598	457	16	475	110	104
	21.1	774	478	0.64	0.19	31.6	0.14	21.065	3.058	0.0522	1.213	0.34	3.290	0.047	3.058	0.930	299	9	294	28	98
	22.1	461	45	0.10	2.34	112.5	2.73	3.424	3.063	0.1227	0.274	4.94	3.075	0.292	3.063	0.996	1652	45	1996	5	120
	22.2	1082	920	0.88	0.19	40.7	0.14	22.807	3.162	0.0517	1.283	0.31	3.413	0.044	3.162	0.927	277	9	273	29	99
	23.1	649	399	0.63	0.46	25.3	0.30	22.068	3.223	0.0520	1.643	0.33	3.618	0.045	3.223	0.891	286	9	287	38	101

Errors are 1-sigma; Pbc and Pb* indicate the common and radiogenic portions, respectively. Common Pb corrected using measured ^{204}Pb.

3.2.2. Whole-Rock Geochemistry

Chemical compositions of the analyzed whole rocks are given in (Table 1), which also presented nine whole rock chemical analyses available on the literature (identified by an asterisk; [19,28]). Whole rocks of 18 samples were analyzed for major oxides and trace elements by X-ray fluorescence (XRF) spectroscopy using a Phillips PW 1400 at the Centre d'Analyses Minérale, University of Lausanne, Lausanne, Switzerland. After reducing the sample to centimeter sized chips in a hydraulic press, the freshest pieces were selected and crushed using a jaw crusher and then powdered in an agate swing mill. Routine analyses of major elements and traces were performed on fused disks prepared from 1.2 g of calcined sample powder mixed with lithium-tetraborate. Calibrations were based on certified international standards. The limits of detection depend on the element concerned and the matrix but typically range between 1 to 7 ppm. The dated sample (ARM 15) was analyzed for bulk rock major and trace element analyses using the analytical facilities of the Instituto Andaluz De Ciencias De La Tierra (IACT), CSIC–UGR, Armilla, Granada, Spain. The analytical techniques were described in detail by [42].

3.2.3. Mineral Chemistry: Electron Probe MicroAnalyser (EPMA)

Major and trace element analyses of minerals (pyroxenes and biotites) were performed at the CAMP-Paris of the University of Pierre et Marie Curie (Laboratoire de Géochimie Comparée et Systématique, Paris, France) using a Cameca SX100 electron microprobe calibrated with a combination of natural and synthetic standards including sanidine (Al), titanite (Ti, Si, and Ca), halite (Cl), chromite (Cr), topaz (F), andradite (Fe), orthoclase (K), olivine (Mg), spessartine (Mn), and albite (Na). Analyses were carried out using a beam size of 1 μm, an accelerating voltage of 15 kV, and a probe current of 20 nA. Counting time varied depending on elements and minerals (see [38] for more details).

4. Results

4.1. Field Observations and Petrographic Features

The major types of volcanic rocks outcropping in the TSESDAI basin and their petrographic features are summarized below.

4.1.1. Dolerites

These rocks crop out in the northeastern sector of the Tiddas region as a NE–SW kilometric length dolerite dike (50 m thick). Their primary mineralogical composition consists of pyroxene and plagioclase (60–70% An), both presenting variable degrees of alteration. Pyroxene are often epigenized into secondary minerals such as chlorite, calcite, and silica, while plagioclases are partially or completely sericitized.

4.1.2. Sebt Ait Ikko Andesites

The andesites crop out southwest of Sebt Ait Ikko, on both sides of the Oued BouAzza valley. They consist of a 15 m thick flow pile uncomfortably resting on the F1 formation of Permian detrital deposits [38]. These andesites are mainly composed of plagioclase, pyroxene, and olivine, presenting variable degrees of alteration.

4.1.3. Tiddas Andesites and Associated Pyroclastites

In the vicinity of Tiddas, the andesites and associated pyroclastites crop out at the Kikene and Glimis Mountains (Figure 3A). They consist of 15 to 50 m thick flows and pyroclastites lying with an angular unconformity directly on the Visean basement [38]. The summit is systematically underlined by metric-scale horizons of breccias, sometimes covered by tuffs of ash lapillis of the same nature and Permian detritic deposits of the F1 formation. The mineralogical assemblage essentially consists of pyroxenes and plagioclases.

Figure 3. Field photographs: (**A**) Tiddas andesites located to the SW of this village; (**B**) dacites of Bled Bou Haouza located to the SE of Sebt Ait Ikko; (**C**) Ari el Mahsar dome laying on the Visean basement; (**D**) a close view of the rhyolites in the Ari el Mahsar dome. The location of sample ARM15 is indicated.

4.1.4. Pyroxenes Dacites

These rocks outcrop exclusively in the area of Bled Bou Haouza (Figure 3B). They are either laying directly or with an angular discordance over the folded or laminated upper Visean basement. Occasionally they are separated from the latter by Autunian molasses formed conglomeratic and sandstone deposits with mainly F1 volcanic elements. Sometimes, they are overlain by a level of breccias of the same nature. The mineralogical assemblage is composed of pyroxene, plagioclase, and sanidine.

4.1.5. Biotite Rhyolites

The biotite rhyolites outcrop in the eastern part of TSESDAI as extrusive domes associated with a network of rhyolitic dikes with a NE–SW general trend. The dome and the dikes are both systematically intrusive in the pre-Permian basement. The rhyolitic domes (Figure 3C,D) have generally prismatic joints, developing vertical prisms of about 15 to 20 m in height. Sometimes the prisms are imperfect as a result of the development of horizontal joints. The prismatic feature locally disappears due to the latter development of spheroidal disjunction. The rhyolitic dikes have a general trend of N50–60 and a metric to kilometer extension, with frequent columnar jointing with contiguous sharp edge columns characterized by centimetric polygonal sections. Massive, rounded blocks with "onion peel" structures are common.

4.1.6. Lapilli Tuffs and Rhyolitic Ash Deposits

These 5 m thick lenticular pyroclastic deposits are interbedded within the detrital deposits of the formation F1. They are characterized by a decimetric stratification with internal horizontal laminations, normally or inversely graded, with occasional accretionary lapilli levels.

4.2. Lithostratigraphic Organization

Figure 4 illustrates the spatiotemporal distribution of the described facies. Depending on the areas, 50 to 350 m thicknesses of sedimentary, lava flows, and pyroclastic deposits have been accumulated, representing about 24 km^3 of the volcanic and sedimentary material. The buildup of these piles is the result of a volcanological and sedimentary evolution that differs very little from one sector to another. The standard volcanic succession includes (i) a 15 to 50 m thick stack of andesites breccias (Tiddas andesites), castings and lapilli tuff, and andesitic ash deposits; (ii) 100 m of rhyolitic lavas structured in a series of domes and preceded by the intrusion of rhyolitic dikes, a phase that forms stratigraphically and that is the lateral equivalent of the "Tiddas andesites"; (iii) 2 m deposits of lapilli tuffs and rhyolitic ashes that are contemporaneous with the filling of the basin (deposits of the F1 formation); (iv) 60 m of flows and breccias of dacite; (v) 15 m of andesite flows; and (vi) 50 m of basaltic flows.

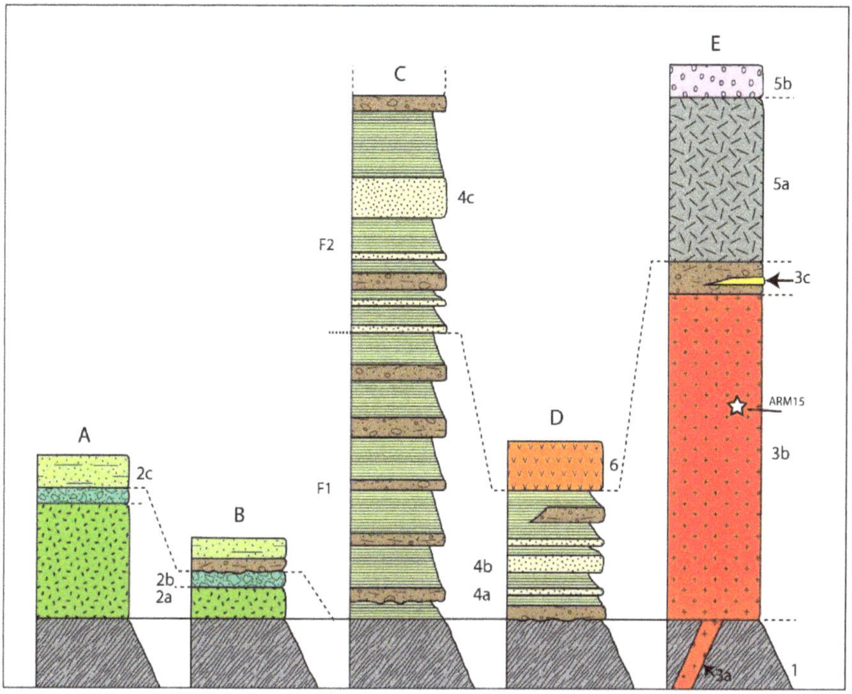

Figure 4. Lithostratigraphic columns of the Permian volcanic complex of TSESDAI: 1. Upper Visean Paleozoic terrains; 2. Tiddas pyroxene andesites in flows (2a), flow breccias (2b), and lapilli tuff and ash deposits (2c); 3. biotite rhyolitic dikes (3a), domes (3b), and lapilli tuffs and ash deposits (3c); 4. detrital formations of Permian conglomerates (4a), argillites (4b), and sandstones (4c); 5. Bled Bou Haouza dacites with pyroxenes (in flows (5a) and flow breccias (5b); 6A Sebt Ait Ikko andesites with olivine and pyroxenes. F1, lower formation; F2, intermediate formation. Location of sections; (**A**), west of Tiddas (Glimis); (**B**), west–southwest of Tiddas (KiKene); (**C**), northeast of Tiddas; (**D**), southwest of Sebt Ait Ikko; (**E**), south of Sebt Ait Ikko.

4.3. Mineral Chemistry

4.3.1. Clinopyroxenes of Dacites

The clinopyroxenes (Cpx) are colorless with polysynthetic twins and/or sectorial zoning structures constituting 5% of the total volume of the dacitic lava flows. They appear as phenocrysts (2%) and microlites (3%) (Figure 5A) and are euhedral or subhedral, sometimes corroded. Cationic proportions based on six oxygen atoms per formula unit of the studied clinopyroxenes allowed them to be classified mainly as augites, using [43] systematics, with compositions clustered around an average of En 42–Wo 42–Fs 16 (Figure 6, Tables S1 and S2). They are characterized by (i) a high SiO_2 content (49.66 wt.% up to 52.66 wt.% and an average of 51.22 wt.%); (ii) a low TiO_2 content (0.31 up to 1.70 wt.% with an average of 0.60 wt.%; (iii) a low or high Al_2O_3 content (1.04 to 5.5 wt.% with an average of 2.09 wt.%); (iv) and a CrO_3 content ranging from 0.025 to 0.31 wt.% with an average of 0.10 wt.%. Al^{IV} is usually low (0.034 to 0.141) indicating the lack of significant non-quadrilateral substitutions. These are accomplished by the incorporation of small amounts of aluminum (Al^{VI} = 0 to 0.101) and titanium (Ti = 0.009 to 0.048) in the octahedral site.

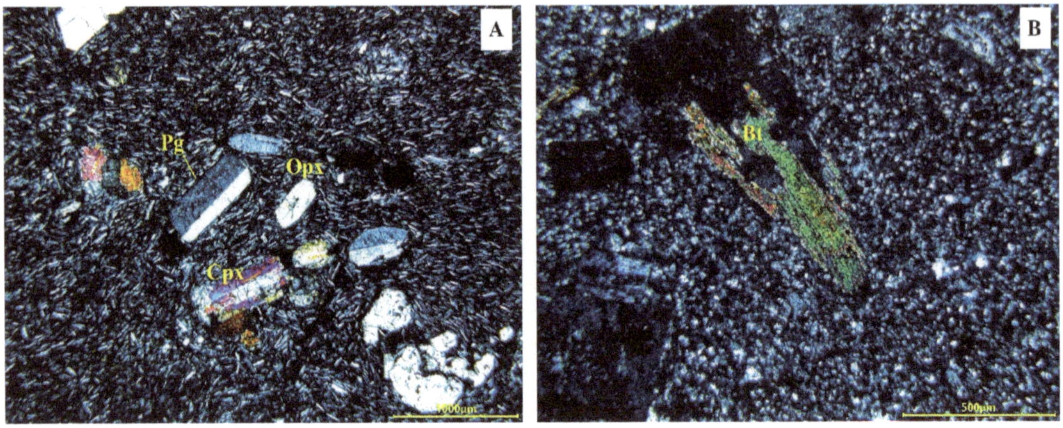

Figure 5. Thin-section photomicrographs of andesites from Tiddas and the ARM 15 sample. (**A**) Tiddas andesite with plagioclase (Pl), clinopyroxene (Cpx), and orthopyroxene (Opx); (**B**) sample ARM15 of a rhyolitic dome with biotites (Bt).

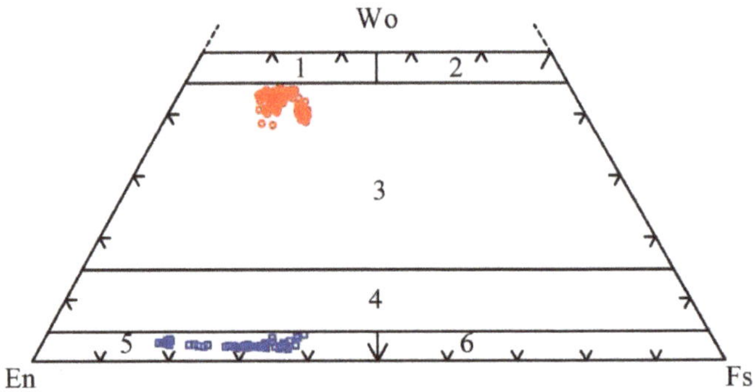

Figure 6. Pyroxene composition of the dacite flows of the TSESDAI eruptive complex in the enstatite (En), Wollastonite (Wo), and ferrosilite (Fs) triangle of [43]. Blue circles indicate the analyzed orthopyroxenes. Red circles indicate the analyzed clinopyroxenes. The numbers (1 to 6) indicate respectively; Diopside, Hedenbergite, Augite, Pigeonite, Clinoenstatite, Clinoferrosilite.

4.3.2. Biotite of Rhyolites

Biotite constitutes about 6% of the total rock volume and appears isolated or grouped in glomeruli with feldspar phenocrysts (2%) being euhedral to subhedral with rarely corroded contours (Figure 5B), and as microlites (4%). Biotite crystals retain generally their primary optical characteristics, but occasionally they are partially or totally transformed into chlorite and invaded by oxide granules (Figure 5B). The chemical analyzes of the biotites and their structural formulas calculated on the basis of 22 oxygen atoms (Table S3) show that they are characterized by (i) low iron oxide content (mean average = 17.45 wt.%) and calcium (CaO = 0.09 wt.%); (ii) relatively high levels of titanium (TiO_2 = 3.13 wt.%), which is compatible with a medium magmatic formation, alumina (Al_2O_3 = 15.28 wt.%), and magnesium (MgO = 11.72 wt.%) (e.g., [44–47]; and (iii) Fe/Mg + Fe ratios ranging between 0.44 and 0.49 allowing their classification as magnesian biotites (see Figure 7; [48]).

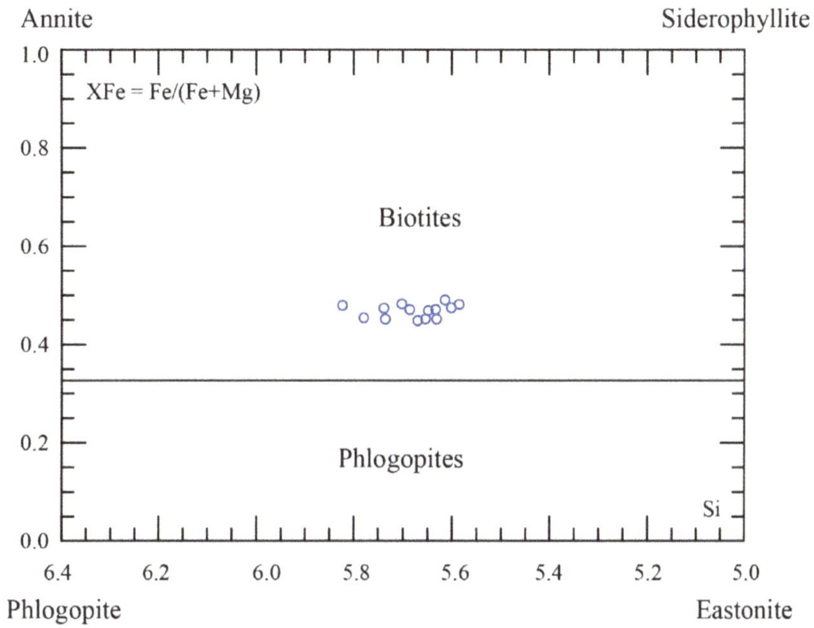

Figure 7. Biotite's compositional domain of the TSESDAI rhyolitic dome in the Fe/Fe + Mg diagram of [48]. Blue circles indicate the analyzed Biotites.

4.4. *Whole-Rock Geochemistry*

4.4.1. Major Elements

The studied volcanic and subvolcanic rocks display variable contents of SiO_2 (54.11 to 70.80 wt.%), Al_2O_3 (13.42 to 15.99 wt.%), CaO (0.49 to 6.32 wt.%), MgO (0.11 to 2.55 wt.%), and alkalis ($Na_2O + K_2O$ = 4.38 to 10.32 wt.%). Loss on ignition (LOI) contents are moderately high and variable (LOI = 1.98 to 6.13 wt.%) suggesting the operation of post-magmatic processes, which requires caution when interpreting the concentrations of the more mobile elements (e.g., alkalis). The classification diagram of [49] shows that the studied rocks occupy the fields of basaltic trachyandesite, trachydacites, and rhyolites. The rhyolites (in domes and dikes) from TSESDAI display a sub-alkaline behavior, while the andesites and dacites display an alkaline one [50] (Figure 8). The andesites and a large part of the dacites have a very high K_2O content due to the alteration of feldspar to sericite. This explains the migration of their representative points in domains not identical to those indicated by their petrographic and mineralogical features, and this is mainly due to the alteration of feldspar to sericite and clays and devitrification glasses made up of chlorite and other clay minerals. The evolution diagram of [51] (Figure 9), where the content of SiO_2 is taken as an index of differentiation, allows for the following remarks:

- The TiO_2, Al_2O_3, FeO*, MgO, CaO, and P_2O_5 oxides have a negative correlation with SiO_2, while K_2O and Na_2O have an opposite or a random behavior.
- A negative correlation of Al_2O_3 and SiO_2 in the dacitic terms could be a result of the fractionation of alumino-silicate minerals such as plagioclase. The same correlation is observed in the transition from dacites to rhyolites.
- The occurrence of a hiatus between the mafic terms (andesites), the intermediate and the felsic terms (dacites and rhyolites). The intermediate terms of the felsic andesite type are absent.

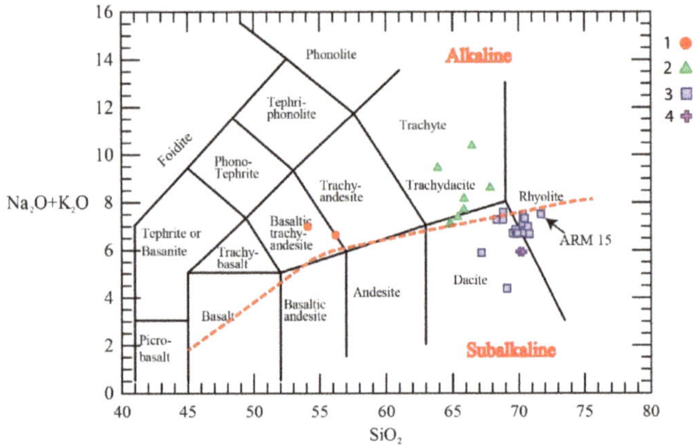

Figure 8. Chemical nomenclature of the volcanic rocks of the TSESDAI eruptive complex. Legend: (1) andesites; (2) dacites (Bled Bou Haouza); (3) rhyolitic domes; (4) rhyolitic dike.

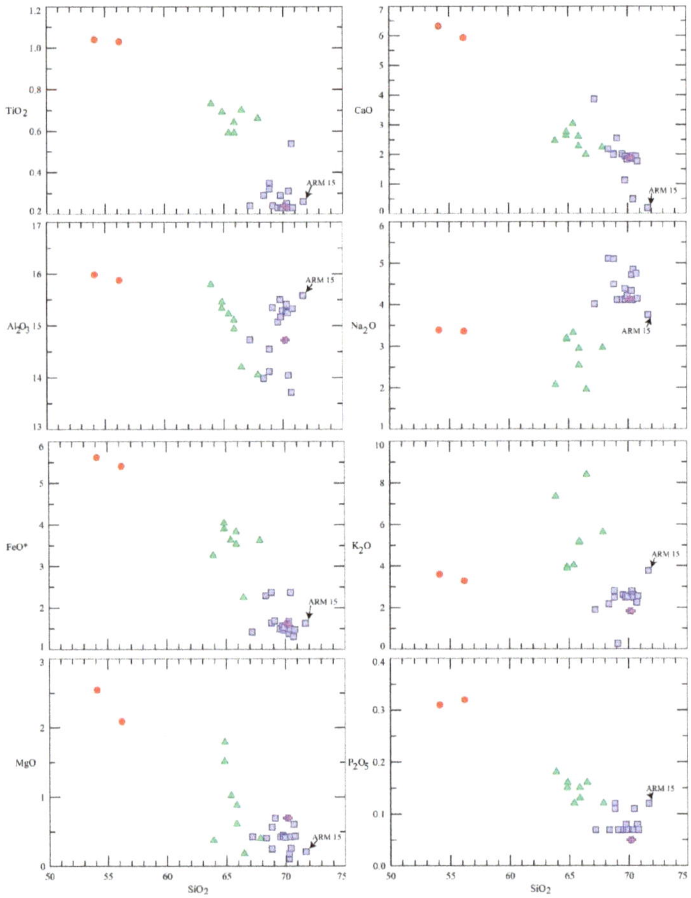

Figure 9. Diagrams of the variation of some oxides versus SiO_2. See symbol legend in Figure 8.

4.4.2. Trace Elements

The contents of large-ion lithophile element (LILE) including low field strength (LFS) (Rb, K, Ba, Sr, Th) and light rare earth elements (La, Ce) occasionally display large variations. However, the same order of magnitude for the various facies and for a given element is preserved. The behavior of Sr seems compatible with plagioclase fractionation in the andesite–dacite series (Figure 10) (correlation between CaO and SiO_2 in these rocks). Ba shows a slightly positive correlation with SiO_2 in andesites. The decrease of Ba in dacites and rhyolites is related to the fractionation of Ba-rich minerals such as Sanidine. The Th and rare earths (e.g., La, Ce, Nd) have a negative correlation with SiO_2 (Figure 10).

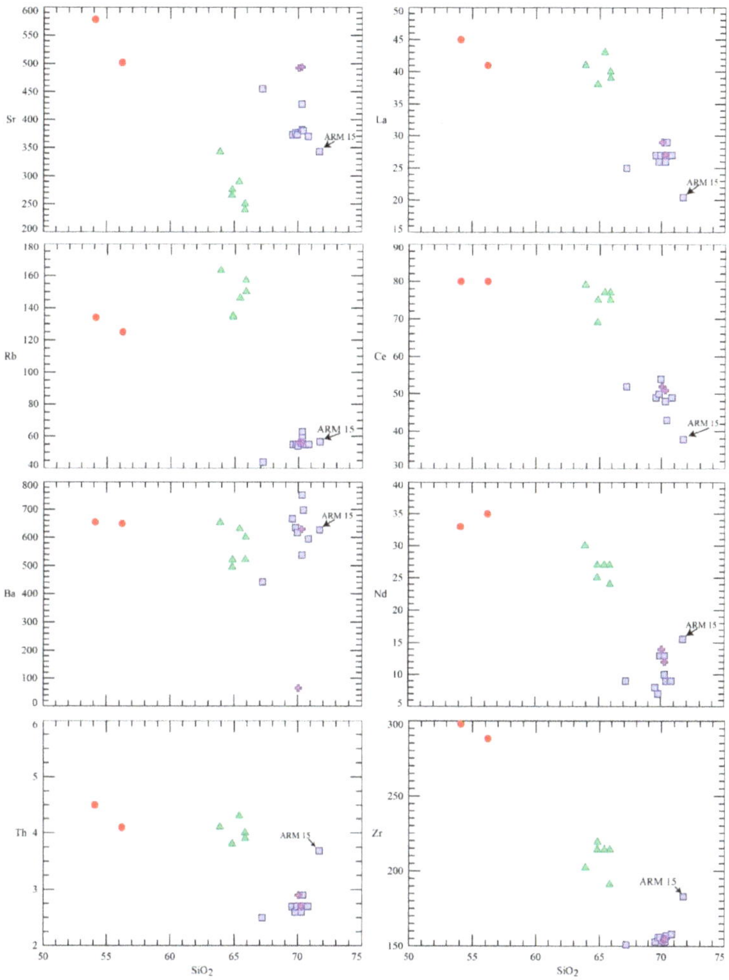

Figure 10. Diagram of the variation of some trace elements versus SiO_2. See symbol legend in Figure 8.

The high field strength elements (HFSE), such as Zr, Hf, Ti, P, Nb, and Ta are considered immobile during the alteration [52–54]. In the studied volcanic and subvolcanic facies, Zr and Hf contents are also high. The Zr contents range between 151 to 298 and the Hf contents vary from 1 to 4. The latter shows a negative correlation with SiO_2. The spectra obtained for andesites and dacites (Figure 11A), although corresponding to rocks of different mineralogical compositions and degrees of differentiation, have very similar traces, when compared to MORB N, all these profiles presenting important enrichments in strongly incompatible elements (LILE), an important decrease between Th and Y (Th/Y = 0.56 to 0.84, average = 0.66), with a well-marked negative anomaly in Nb and a depletion in Ti and transition elements of the 3d series; in addition, the values of the La/Nb ratios are between 2 and 7 (La/Nb = 2.93 to 4.3, average 3.73). Compared to the andesites, the dacites show more pronounced negative TiO_2 and P_2O_5 anomalies, higher La/Th ratios, and a depletion of less incompatible elements, particularly the transition elements of the 3d series. Similarly, the LILE (Sr, Ba, Rb, U, Th) appear low in comparison to the Nb, La, and Ce contents. All the spectra of rhyolitic domes show, when compared to the ORG (Figure 11B), an enrichment in LILE (Sr, K, Rb, Ba, Th, La, and Ce), an increased deficit in weakly incompatible elements or HFSE (Zr, Hf, and Nb), with a significant decrease between Th and Nb that triggers a very marked negative anomaly in Nb (Th/Nb varies from 0.5 to 0.58 with an average of 0.54). An important fractionation of light rare earths is underlined in the Trace elements spidergram by the slope of the line between La and Nd (La/Nd varies from 2.07 to 3.71, average of 2.71) and the enrichment in LILE. The spectra of rhyolites when compared to the ORG shows Rb/Zr ratios very similar to those of [55], with values that go from 0.29 to 0.41, averaging 0.36.

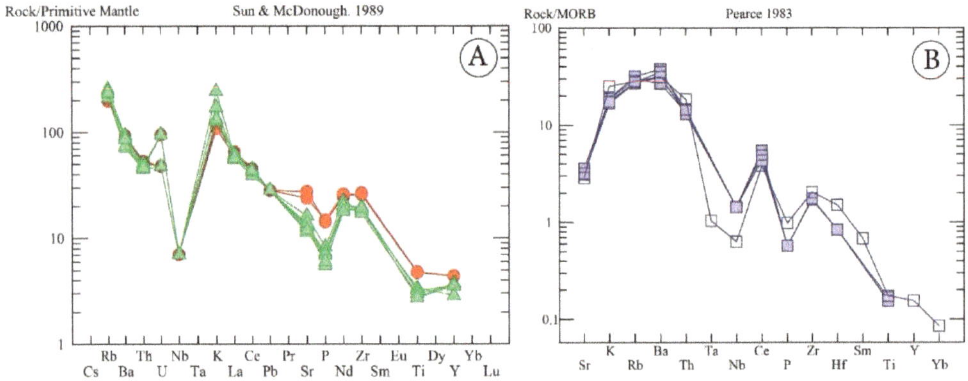

Figure 11. Trace elements spidergram of TSESDAI andesites and dacites normalized to N-type MORB (**A**), and of a TSESDAI rhyolitic dome normalized to ORG (**B**). See symbol legend in Figure 8.

4.5. Geochronology: The Shark Fin Rhyolitic Extrusive Dome of Ari el Mahsar, Sample ARM15

Zircons from sample ARM15 exhibit mean lengths and widths of 240 µm and 98 µm, respectively; they are euhedral to subhedral. Cathodo-luminescence images show oscillatory zoning patterns in most zircons. Some grains show an overgrowth by continuous non-luminescent rims with cores that are mostly zoned in an oscillatory manner, which can be interpreted as inherited (Figure 12). In addition, the morphological typology of the zircons shows the omnipresence of the S type around an S13 nucleus, the average points IA and IT [56], respectively, reaching values of 385 and 420, with formation temperatures between 750 °C and 800 °C (Figure 13). The Th/U ratio ranges from 0.01 to 1.36, indicating a mostly igneous population [57,58]. The Concordia diagram shows that the shark-fin-like rhyolitic extrusive dome of Ari el Mahsar (sample ARM15) yielded two predominant Concordia ages (603 $\pm$ 25 Ma and 286.4 $\pm$ 4.7 Ma) (Figure 14, Table 2). The young age is more concordant, being interpreted as the crystallization age of rhyolites, while the 603 Ma age indicates inherited zircon from the Pan-African crust under the Rehamna.

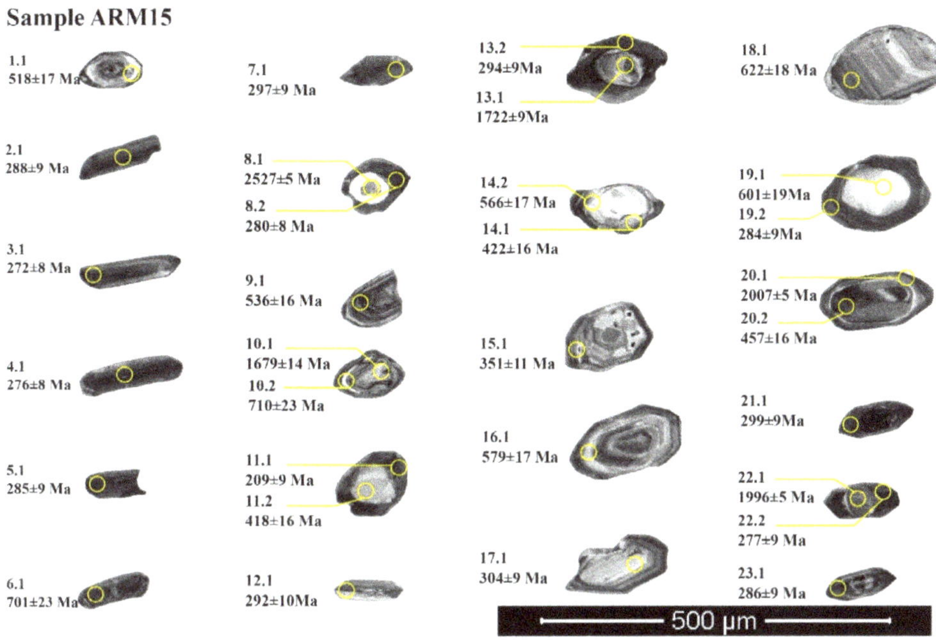

Figure 12. Cathodoluminescence (CL) images of zircon grains obtained from the rhyolitic sample (ARM15) showing their internal texture, spots (yellow circles), and apparent $^{206}Pb/^{238}U$ ages in Ma (1σ error).

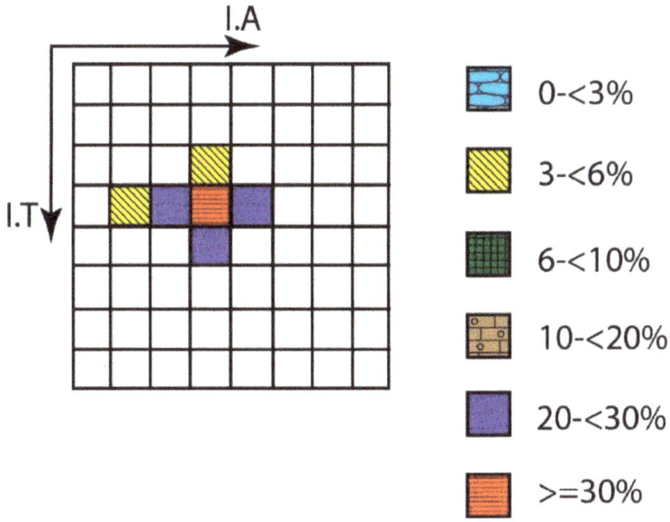

Figure 13. Typological distribution diagram of zircons from biotites of Ari el Mahsar rhyolitic dome.

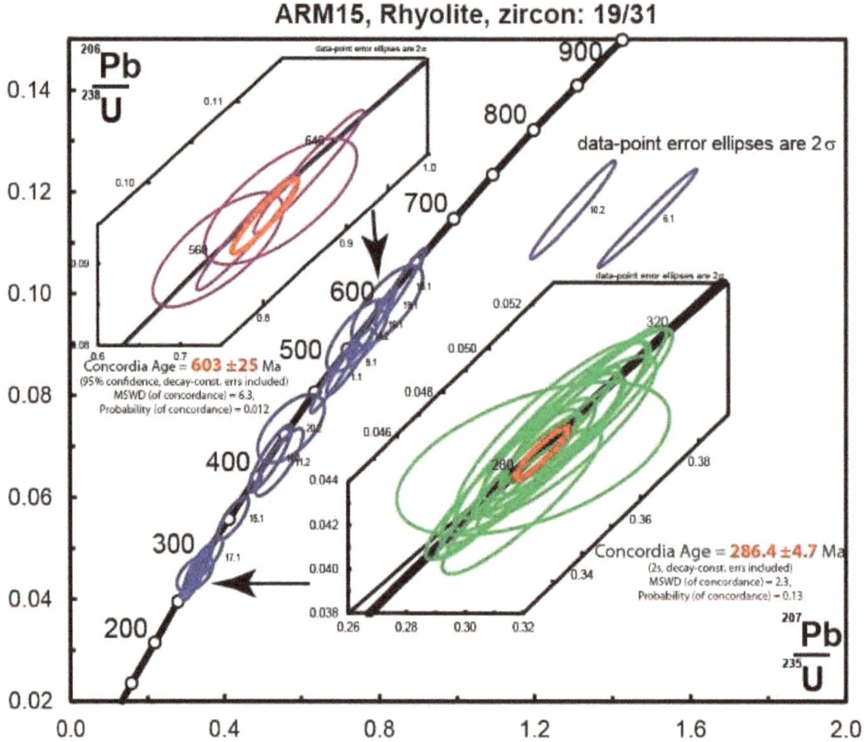

Figure 14. Zircon U–Pb SHRIMP geochronology. Wetherill Concordia plots of U–Pb zircon data for the rhyolitic sample (ARM15).

5. Discussion

5.1. The Calc-Alkaline Affinity of TSESDAI Rocks

Given the evidence for alteration of the studied rocks (LOI up to 6.13 wt.%), the use of alkali concentrations to decipher the magmatic affinities of the rocks was discarded. Instead, we used the ratio (Nb/Y) between two elements, which being characterized by high field strength are as such considered immobile during meteoric alteration and low-grade metamorphism and are also considered proxies of the degree of alkalinity of magmas (e.g., [59,60]). The studied rocks are characterized by Nb/Y ratios clearly below 0.6, the threshold usually considered as a divider between sub-alkaline and alkaline affinities, clearly pointing to the sub-alkaline characteristics of these rocks. Additionally, all the volcanic rocks of Tiddas show an absence of enrichment in FeO and TiO_2 with the increase of the magmatic evolution, as represented by the index FeO*/MgO, clearly pointing, according the criteria defined by [61,62], to the calc-alkaline characteristics of these rocks. This is confirmed by the low contents in TiO_2 (0.23 to 1.04%) and by the negative anomalies in Nb and Ti in the primordial mantle normalized diagrams (Figure 15) and by Zr/Y ratios clearly above 7 (see [59]). The chemical data collected on clinopyroxenes (cf. supra) confirm this calc-alkaline affinity with clinopyroxenes plotting on the calc-alkaline orogenic basalt field in the discrimination diagrams of [63] (Figure 15) and [64] (Figure 16). Additionally, the composition of biotites is typical from those usually found on calc-alkaline rocks (Figure 17; [46]).

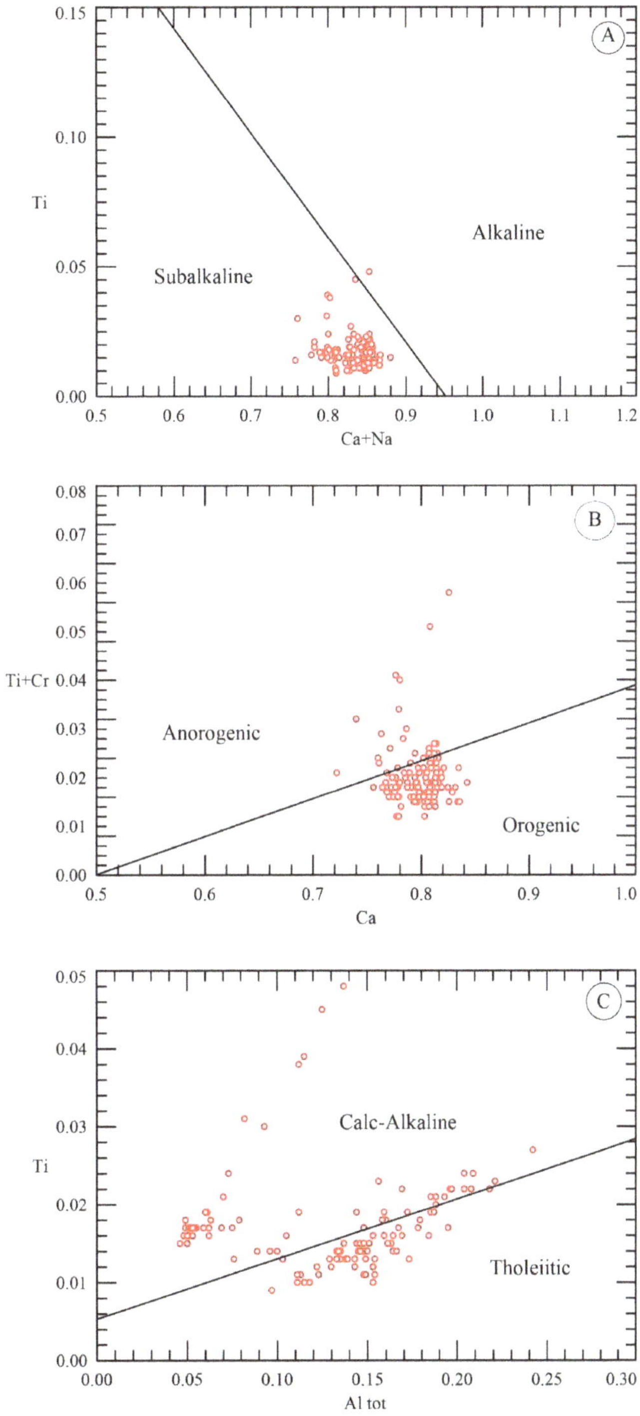

Figure 15. Distribution of the studied clinopyroxenes in three different diagrams: (**A**) Ca + Na–Ti, (**B**) Ca–Ti + Cr, (**C**) Altot–Ti of [63]. Red circles indicate the analyzed clinopyroxenes.

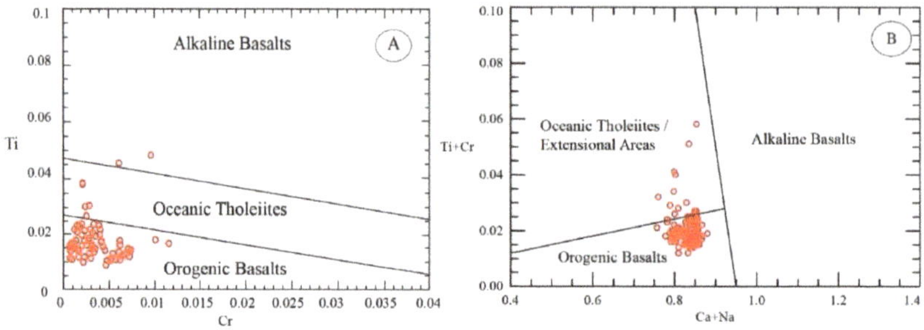

Figure 16. Position of the analyzed clinopyroxenes in the diagrams of [64]: (**A**) Ti–Cr, (**B**) Ti + Cr–Ca + Na. Red circles indicate the analyzed clinopyroxenes.

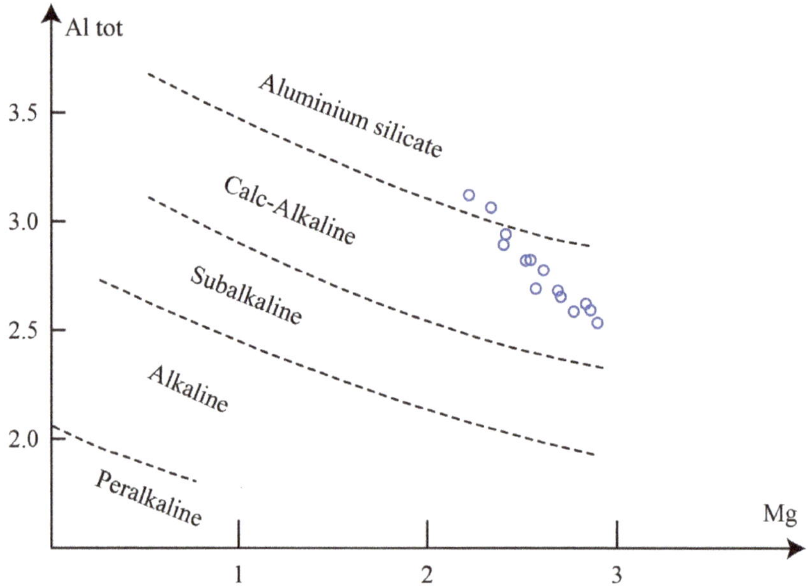

Figure 17. Position of the analyzed biotites in the Altot–Mg diagram of [46]. Blue circles indicate the analyzed Biotites.

5.2. Petrogenesis of TSESDAI Rocks

Calc-alkaline rocks are abundant along destructive plate margins, but calc-alkaline geochemistry is not an indisputable indicator of sin-subduction processes [65]. Indeed calc-alkaline rocks are also known from regions undergoing extension without a time correlated subduction, such as the Basin and Range province [66], the Mexican Volcanic Belt [65], or the northern Da Hinggan Mountains in China [67].

The Paleozoic geology of Morocco was shaped by the Variscan orogeny, with continental collision, between 360 and 290 Ma, having occurred at circa 330–300 Ma, i.e., some 44–14 or even 4 Ma before the emplacement of the studied rocks. This indicates that the generation of the magmas at the origin of the studied rocks cannot be associated with an operating subduction process [25].

When compared to some reference calc-alkaline volcanic series (e.g., calc-alkaline basalts of island arcs [68]; calc-alkaline andesites and dacites of active margins [69]; calc-alkaline andesites of the Tertiary volcanic province of the Basin and Range in the western USA [70]), the chemical composition means and element ratios of andesites and dacites of the TSESDAI volcanics demonstrate that they display greater similarities with the calc-alkaline series not linked to active subduction, being more similar to those generated in intracontinental extensional settings, as is the case of the Tertiary volcanic province of the Basin and Range (e.g., [71]). However, the Permian andesites of Tiddas is enriched in TiO_2, P_2O_5, Rb, and Cr and depleted in SiO_2, MnO, Sr, Ba, Nb, Hf, and Y. On the other hand, the elemental ratios of the Moroccan Permian andesites and those of the Basin and Range are usually very similar and thus confirm the intraplate calc-alkaline character of the volcanic rocks studied in the present paper. On a first approach, these comparative results suggest that the studied rocks have the same genetic origin as the lavas of the Basin and Range, i.e., petrogenesis involving both fractional crystallization and crustal contamination processes [70].

5.2.1. Fractional Crystallization

Based on their petrographic and geochemical data/criteria, the least differentiated terms of the volcanic materials (the andesites) do not represent primary liquids which have been in equilibrium with the residual paragenesis resulting from partial melting of the common mantle sources. Indeed, they are characterized by low Mg# (0.816 to 0.961) and concentrations in the incompatible element Ni (41 to 58). The POAM (plagioclase–olivine/orthopyroxene–augite–magnetite) crystal fractionation model is a widely accepted process to explain the genesis of andesitic magmas from primary mantle liquids of basaltic composition, and by extension, the mode of evolution of calc-alkaline magmas [72]. The sub-parallel normalized trace elements distribution, as well as the behavior of major and trace elements as a function of silica and Th used as differentiation indices and the decrease of transition elements during the magma evolution clearly show the role played by fractional crystallization in the petrogenesis of the andesites–dacites–rhyolites series of the TSESDAI basin

5.2.2. Crustal Contamination

The existence of enclaves in the lavas of the TSESDAI massif led to the assumption of a magmatic evolution partially controlled by crustal contamination mechanisms. The participation of such processes is also suggested by two observations: (i) the abnormally high contents of Rb, Ba, Th, and La; and (ii) the systematic anomalies in TiO_2 and P_2O_5. The intervention of the continental crust in the petrogenesis could also explain the following facts:

(i) The absence of basalts and the low volume of andesites compared to dacites and rhyolites. According to [72,73], a thick continental crust would have the effect of prolonging the ascent of basaltic and andesitic flows, and thus the fractional crystallization of minerals at low pressure, therefore favoring differentiation and assimilation until reaching the ultimate rhyodacitic to rhyolitic terms;
(ii) The specific chemical characteristics of the andesites such as the enrichment in highly incompatible elements (Rb, Ba, Th, Sr, K_2O, and La) and high ratios (La/Th = 2, 93 to 3; 2.97 on average).

It is interesting to note that Ti and Nb negative anomalies increase with the increasing degree of magmatic evolution, which argues in favor of the conjunction of assimilation and fractional crystallization processes (the ACF process of [74]), a consequence of the exothermic behavior of crystallization processes. The fact that even the least fractionated rocks (the andesites) present such Nb and Ti anomalies suggests that the source of magmas could have some supra-subduction fingerprints inherited from previous subduction events, most probably the Variscan subduction that occurred some 420–330 Ma before [25].

5.3. Volcanological History and Formation Model of the TSESDAI Rocks

The geological and paleogeographical context of the TSESDAI region during the Permian period (deposits within intramontanous continental half-grabens) and the low volume of the outcropping materials (24 km^3) are indicative of an intraplate continental environment. In addition, the lava flows show features suggesting aerial-type eruption for the TSESDAI volcanism such as the absence of pillow lavas and the presence of flows breccias. Indeed, the pyroclastic formations have juvenile pyroclastic fragments with extensive vesiculation and accreted lapilli. These characteristics are generally ascribed to aerial to sub-aerial environments. Intra-volcanic epiclastites have features of continental fluvial sedimentary rocks. The volcanism of the TSESDAI basin is bimodal, and it is characterized by a predominance of rhyolitic and andesitic facies over dacitic ones. The emissions of andesitic and dacitic magmas take place mainly in the form of lava flows of limited extent, whereas the rhyolitic facies occur in the form of dikes, domes, and pyroclastic fall deposits (deposits of aerial fallout). The volcanological history of the TSESDAI complex is relatively simple when compared to other Moroccan complexes (e.g., Khenifra). It includes three well constrained episodes. The first episode is found in the southwestern boundary (columns A and B, Figure 4) and in the northeastern zone (column E, Figure 4). The volcanic activity begins with the emplacement of pyroxene-rich andesitic flows ("Tiddas andesites"), followed by weak deposits of lapilli, lapilli tuff, and andesitic ashes with volcanological characteristics similar to typical volcanic aerial fallout deposits. Far from the emission centers (characterized by relatively powerful accumulations (50m) and facies diversities; Glimis sector) (F1), a fluvial sedimentation is established including conglomerates with basal pebbles and andesites (alluvial fan deposits) and fallout of aerial projections of andesitic characteristics (andesites to the SW and rhyolites to the NE). The thickness of the andesitic units is between 15 and 40 m, and the total pile reaches nearly 50 m. Contemporaneously with the emplacement of the Tiddas andesitic flows, a dome-like structure is established at the northwestern limit of the volcanic complex (column E, Figure 4), followed by the emplacement of rhyolitic dikes. These domes might be classified as "dome-pistachios", "dome-pistons", or "plug domes" according to the classification of [75]. The second episode, recorded in the central part of the basin (columns C and D, Figure 4) consists of dacite and andesite flows with olivine and pyroxene. The only witnesses of the latter episode are the Bled Bou Haouza and Tabahart outcrops. The last episode was recorded only in the southeastern part of the complex, NE of Tiddas (Figure 2). It is characterized by the emplacement of NE–SW-oriented pyroxene/dolerite dikes. The basaltic flows of El Gitoune (50 m) probably constitute the effusive witnesses of this episode. This episode would be equivalent to the pyroxene dolerites of Khenifra. The vent system is distributed over the entire volcanic domain and is generally oriented parallel to the NE–SW direction of the structures and fractures. Even if this volcanism is basically decentralized, a restricted area (Ari Al Mahsar Sector, southeast of Sebt Ait ikko) has a relatively high density of dikes and domes with high petrographic diversities, i.e., a volcanic apparatus. Based on all the characteristics previously described, the volcanism of the TSESDAI basin can be ascribed to a continental strato-volcano model. However, as in Khenifra, the absence of associated pyroclastic flows (ignimbrites) and pyroclastite deposits in the TSESDAI region is a peculiarity that differentiates the Permian volcanism of western Morocco from the Viseo–Namurian volcanism of eastern Morocco, where caldera-type structures associated with the emission of important ignimbritic layers are notorious [76–79].

5.4. Links between Volcanism and Tectonics in the TSESDAI Basin

The Tiddas-Souk Es Sebt region was subjected during the Permian to an essentially compressive regime [29,80,81], with four deformation episodes, the first three mainly compressive and the fourth corresponding to an extensional event. Among the compressional episodes, the first one is transpressional, with a N50 to N60 compressional direction. This episode is responsible for the opening of the TSESDAI Permian basin following left-lateral N60–N90 strike-slip accidents and the extrusion of the first andesitic and rhyolitic volcanism. A N10–N40 to N70 syn-sedimentary normal-fault system can be clearly associated with this episode. This fault system guides the collapse of the blocks and thus lead to the individualization of Permian grabens and half-grabens. The NW–SE extension, developed by this transtensional regime and guided by the rework of normal faults inherited from the Hercynian major phases, gave rise to a series of collapsed blocks. Thus, the Permian graben and half-graben structures clearly resulted from this phase. The opening of deep extensional fractures in the Paleozoic substratum allowed the extrusion of the first volcanic series (the rhyolitic domes and associated rhyolitic dikes to the northeast and the Tiddas andesitic flows) and conditioned the Permian sedimentation. The second episode is a WNW–ESE-oriented compressional regime. It triggered conjugate strike-slip fault systems affecting the Permian deposits: N80 to N90 right-lateral faults and N140–150 left-lateral faults. The later fault system transected the normal faults triggered during the first tectonic event (NW–SE extension). The third tectonic episode involved NNW–SSE to N–S regional compression, reactivating some N40–N70 normal faults into inverse ones. This episode is also responsible for the N80-oriented-axis of the large-radius folds deforming the Permian detrital deposits. The fourth episode is a NW–SE extensional event accompanied by N40 to N60 normal faults that gave rise to a series of collapsed blocks. The age of this extensional episode remains unknown. However, the fact that the faults generated during this episode affect only the basal conglomerates of the Triassic series and that its extension direction is compatible with the dolerite dikes of the El Gnoune sector suggest a Late Permian to Early/Middle Triassic age. On the map of Figure 18, we represented different characteristics of the emission centers (cumulo-domes, sills, maximum thicknesses of flows, and the diversity of eruptive mechanisms) and the major accidents that acted as rifts during the Permian. We suggest that the dikes and the emission points are mostly aligned along N40–N70 accidents whose complex and polyphase play triggered the chaotic peeling off characteristics of the deformation. The magma emission would occur by injection, either in relays of major accidents, or as en echelon tension cracks or normal faults. The direction of shortening Z or compression $\sigma 1$ can be deduced from the direction of the dikes: N50 to N90 with a clear predominance of the N60 direction, which coincides with the alignment of the emission centers. This direction is similar to the N50–60 compressional vector that controlled the opening of the basin: relay-basins along left-lateral accidents. It should also be noted that the inclination of the veins is compatible with the strike-slip movement along these accidents.

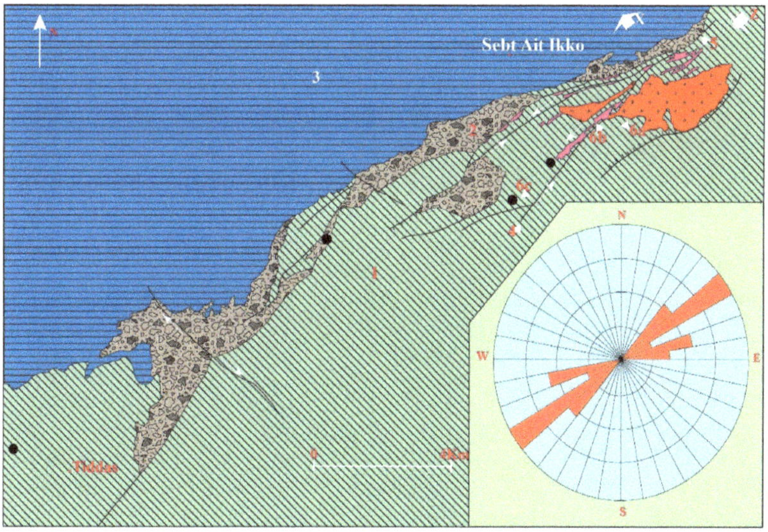

Figure 18. Structural control of emission centers in the Permian volcanics of the TSESDAI volcanic complex: (1) Pre-Permian Paleozoic basement; (2) Autunian conglomerates, sandstones, and mudstones; (3) Mesozoic and Tertiary formations; (4) left-lateral fault; (5) normal faults; (6) emission centers, domes (6a), veins (6b), and maximum thickness of flows (6c). Z, direction of shortening; X, direction of extension. The inlet displays a rosette of the directions and cumulative lengths of the rhyolitic veins.

5.5. Geochronological Constraints

The SHRIMP U–Pb geochronologic data of zircons recovered from the rhyolite dome of Ari El Mahsar in the TSESDAI basin shows a Concordia age of 286.4 ± 4.7 Ma (MSWD of concordance = 0.47, probability of concordance = 0.49) interpreted to date the magmatic crystallization of this dome (Figure 14, Table 2). According to the stratigraphic framework of [6,27], this age should be older than the andesites of the first pulse and the dacites of the second one [23]. The 286.4 ± 4.7 Ma dating agrees with an Artinskian–Kungurian age attributed to the clastic succession in relation with their flora and macrofauna (e.g, ichnofauna) content of [23,33,34,82]. At the Bou Achouch basin, the C–P sedimentary series is considered by [34] as Kungurian in age based on macroflora assemblages found in their fine-grained clastic units and reworked ash deposits. In addition, similarities between the flora of Bou Achouch and the Lower Rotliegend of the Saar-Nahe Basin (Germany; [83]) extend their age back into the latest Carboniferous. Recently, authors in [23] obtained a weighted mean $^{206}Pb/^{238}U$ age of 301.50 ± 1.24 Ma for an andesite (Table S4). This age is considered the best estimate of the eruption moment of the andesite. According to [23], the flora studied by [34] are indeed older than Kungurian (~283–273 Ma), similar to what occurs with the flora of the Lower Rotliegened (latest Carboniferous to early Permian) sediments of the Saar-Nahe Basin in Germany [83], considering that this andesite represents indeed the top of the stratigraphic sequence at Bou Achouch as suggested by the stratigraphic analyses of [6]. However, according to [23], the latter stratigraphic framework is composite, as the basin stratigraphy is disturbed by thrusts. Therefore, the possibility that some andesites stratigraphically lie below the fossil-bearing units cannot be excluded. Previous ages of the volcanic rocks of the Khenifra basin comprise a K–Ar whole rock age of 264 ± 10 Ma obtained from a rhyolitic breccia at Jbel Bou Hayati [84]. At the same locality, the volcanic pile that was assigned to the early Permian contains fossil wood [34,85,86]. Recently LA–ICP–MS U–Pb (on zircon) ages from several volcanic localities in Khenifra were published by [87], including a dacite from the lower lava flows at Jbel Taghat (295.1 ± 2.9 Ma) and interpreted as the oldest volcanic pulse of Khenifra. The dome of Sidi Tir delivered ages ranging from 290.3 ± 2.1 Ma to 287.9 ± 3.8 Ma, thus suggesting that the dome was emplaced during ~2 Ma. Rhyolitic pyroclastic fall deposits from Sidi Tir yielded two predominant age clusters at 307.3 ± 2.6 Ma and 290.6 ± 2.6 Ma. The latter age was thought to be more concordant with the age of the eruption. At Talat Mechtal, inferred to post-date Cisuralian sedimentation in Khenifra, authors in [87] obtained an age of 280.3 ± 2.1 Ma for the upper dacitic lavas, supposedly marking the last volcanic pulse in agreement with the lithostratigraphic sequence [7]. From the same locality, authors in [23] reported a $^{206}Pb/^{238}U$ weighted mean-age of 305.59 ± 2.68 Ma for an andesite lava laying at the base of the Talat Mechtal volcanic succession, overlain by 290.6 ± 2.6 Ma pyroclastic fall deposits [87]; this implies that volcanism at Talat Mechtal began ~15 Ma earlier than previously thought [87]. In the Chougrane and Souk el Had Bouhsoussène basins, the age of the sedimentary package is not well constrained; on the basis of its similarity with the facies of the Khenifra and TSESDAI basins it is assumed that the date might be early Permian. The sedimentary package is overlain by the Bir el Gassaa andesites dated to 270 ± 17 Ma by whole rock K–Ar techniques [88]. Two samples, an andesite from Bir el Gassaa that lies on the top of the C–P stratigraphic sequence of Chougrane and an andesite from Souk el Had Bouhsoussène, were recently dated by [23]. The first sample only provided three concordant ages that range from 307.8 ± 4 to 295.5 ± 3.4 Ma, implying that the underlying C–P sediments are earliest Permian or late Carboniferous [23]. The andesitic sample from Souk el Had Bouhsoussène sub-basin has $^{206}Pb/^{238}U$ ages ranging between 295.6 ± 2.9 and 267.9 ± 3.9 Ma. According to existing stratigraphic models of the Chougrane CPB, the andesitic sample from Bir el Gassaa is considered to lie stratigraphically above the andesitic sample from Souk el Had Bouhsoussène sub-basin and is thus expected to be younger. According to [23], the existing stratigraphic model would require consideration as the crystallization age of the youngest value of 295.5 ± 3.4 Ma and 295.6 ± 2.9 Ma from the two dated samples. Even if these geochronological results are clearly inadequate to obtain solid conclusions, they provisionally suggest that the Souk el Had Bouhsoussène sub-basin is younger than the Chougrane one, and therefore the two basins evolved independently, in contrast to the above-mentioned stratigraphic models. The youngest age of the andesitic sample from Souk el Had Bouhsoussène sub-basin could offer a valid constraint on the final phase of the Variscan tectonism, since this basin was tilted during this stage [23]. In the Mechra Ben Abbou basin of "Bled Mekrach", authors in [89] obtained a 611 ± 20 Ma U–Pb SHRIMP age on zircon from a rhyolitic dome, supposedly inherited from zircons of the Pan-African basement of the Rehamna massif and a younger more concordant age of 285.3 ± 4.9 Ma interpreted as the age of crystallization of the rhyolitic dome [89]. South of the basin, a microgranitic/microdioritic dike swarm intruding the pre-Permian basement has been dated to 285.4 ± 6.1 Ma (U–Pb on zircon) [89–92]. Two samples, from an andesite flow and an andesite plug (Douar Ouled Said Ben Ali lavas), provided concordant $^{206}Pb/^{238}U$ (on zircon) ages of 284.2 ± 4.6 and 294.63 ± 0.67 Ma, respectively. The younger age of 284.2 ± 4.6 Ma would imply that volcanism could have begun in this basin at least 10 Ma earlier [23]. In the Nzalet el Hararcha basin, sediments are sealed by Permian lavas dominated by andesites and calc-alkaline rhyolitic domes [38,89,93–95], and the entire basinal sequence is unconformably covered by Cretaceous and Cenozoic rocks [95]. A rhyolitic sample (Sidi Bou Yahia dome) provided a robust $^{206}Pb/^{238}U$ weighted mean age of 277.07 ± 0.61 Ma, which represents the first absolute age constraint from this basin [23].

6. Conclusions

The main conclusions of this study are summarized below:

(1) The volcanic characteristics of the TSESDAI volcanic complex and the distribution of the emission centers and their structural control suggest a multi-vent continental stratovolcano-type facies model.

(2) The Ari Al Mahssar area has a relatively high density of dikes and domes and a remarkable petrographic diversity indicative of a possible volcanic apparatus.

(3) The mineralogy of the primary crystals (biotite and clinopyroxenes) and the morphological typology of the zircons of the TSESDAI volcanic rocks suggest a calc-alkaline orogenic magmatic setting.

(4) The geochemical study of major and trace elements confirms the intracontinental calc-alkaline signature of the studied rocks, similar to the andesites of the American Basin and Range extensional realm.

(5) The geochemical study also indicates an evolution controlled by fractional crystallization and crustal contamination (assimilation).

(6) The petrographic and volcanological characteristics of the TSESDAI volcanic series (in particular the first and the second episodes) suggest a similarity with the well-defined first Permian cycle of the Khenifra region, with the exception of the dolerite dikes and the basaltic flows of El Gitoune (third episode) that belong to the second Permian cycle, despite their similar NE–SW alignment to the first and the second episodes studied here.

(7) The TSESDAI basin recorded four Permian deformation episodes. The first one is responsible for the opening of the Permian basin along N60–N90 left-lateral strike-slip accidents and the establishment of the first andesitic and rhyolitic volcanic episode. The second one is defined by a conjugated system of faults affecting the Permian deposits. The third compressive episode (NNW–SSE to N–S orientation of $\sigma1$) reactivates some N40–N70 normal faults into inverse faults. The fourth episode is a NW–SE trending extensional event triggering N40 to N60 normal faults and consequent collapsed blocks.

(8) The SHRIMP U–Pb geochronologic data of zircons recovered from the rhyolitic dome of Ari El Mahsar in the TSESDAI basin show a Concordia age of 286.4 ± 4.7 Ma interpreted to date the magmatic crystallization of this dome. Thus, this rhyolite belongs likely to the third magmatic episode of the region.

Supplementary Materials: The following are available online at https://www.mdpi.com/article/10.3390/min11101099/s1, Table S1: Chemical analysis of dacite clinopyroxenes from Tiddas-Souk Es Sebt des Ait Ikko volcanic complex, and their structural formula calculated on the basis of 6 oxygens, Table S2: Chemical analysis of dacites orthopyroxenes from Tiddas-Souk Es Sebt des Ait Ikko volcanic complex, and their structural formula calculated on the basis of 6 oxygens, Table S3: Chemical analysis of rhyolite biotites in dome of Tiddas-Souk Es Sebt des Ait Ikko volcanic complex and their structural formulas calculated on the basis of 22 oxygens, Table S4: Overview of new and published geochronological data from Carboniferous–Early Permian basins of central Morocco (after [23]).

Author Contributions: Conceptualization, I.H., N.Y., and A.A.L.; methodology, N.Y.; software, I.H., A.A.L., O.M., and W.E.M.; validation, N.Y., J.M., and M.D.; formal analysis, C.C.G.T., M.A.S.B., and K.S; investigation, I.H., N.Y., and A.A.L.; resources, N.Y.; data curation, N.Y., A.A.L., and W.E.M.; writing—original draft preparation, I.H., N.Y., and A.A.L.; writing—review and editing, O.M., J.M., M.D., L.G., M.A.S.B., K.S., W.E.M., and M.A.B.; visualization, I.H.; supervision, N.Y., M.K.B., and M.A.B.; project administration, N.Y., M.K.B., and M.A.B.; funding acquisition, C.C.G.T. All authors have read and agreed to the published version of the manuscript.

Funding: This work was funded by the Universidade de São Paulo: USP-UCAM/2016, the Russian Science Foundation: Mega-Grant 14.Y26.31.0012, and by the FCT—Fundação para a Ciência e a Tecnologia: UIDB/50019/2020.

Data Availability Statement: Not Applicable.

Acknowledgments: Most of this work was carried out at the Department of Geology of the Faculty of Sciences-Semlalia, Cadi Ayyad University of Marrakech and the "Centro de Pesquisas Geocronológicas (CPGeo), Instituto de Geociências (IG), Universidade de São Paulo-USP, São Paulo (SP), Brazil" within the framework of the scientific and technical agreement between the São Paulo University and the Cadi Ayyad University (responsibles Colombo Celso Gaeta Tassinari and Nasrrddine Youbi). João Mata acknowledges the financial FCT support through project UIDB/50019/2020-IDL. Nasrrddine Youbi is supported by Russian Mega-Grant 14.Y26.31.0012.

Conflicts of Interest: The authors declare no conflict of interest.

References

1. Doblas, M.; Oyarzun, R.; López-Ruiz, J.; Cebriá, J.M.; Youbi, N.; Mahecha, V.; Lago, M.; Pocoví, A.; Cabanis, B. Permo-Carboniferous volcanism in Europe and northwest Africa: A superplume exhaust valve in the centre of Pangaea? *J. Afr. Earth Sci.* **1998**, *26*, 89–99. [CrossRef]
2. Wilson, M.; Neumann, E.R.; Davies, G.R.; Timmerman, M.J.; Heeremans, M.; Larsen, B.T. Permo-carboniferous magmatism and rifting in Europe: Introduction. *Geol. Soc. Spec. Publ.* **2004**, *223*, 1–10. [CrossRef]
3. Youbi, N.; Ernst, R.E.; Mitchell, R.N.; Boumehdi, M.A.; El Moume, W.; Lahna, A.A.; Bensalah, M.K.; Söderlund, U.; Doblas, M.; Tassinari, C.C.G. Preliminary Appraisal of a Correlation Between Glaciations and Large Igneous Provinces Over the Past 720 Million Years. In *Large Igneous Provinces: A Driver of Global Environmental and Biotic Changes*; John Wiley & Sons: Hoboken, NJ, USA, 2021; pp. 169–190. [CrossRef]
4. Arthaud, F.; Matte, P. Les decrochements tardi-hercyniens du sud-ouest de l'europe. Geometrie et essai de reconstitution des conditions de la deformation. *Tectonophysics* **1975**, *25*, 139–171. [CrossRef]
5. Arthaud, F.; Matte, P. Late Paleozoic strike-slip faulting in southern Europe and northern Africa: Result of a right-lateral shear zone between the Appalachians and the Urals. *Bull. Geol. Soc. Am.* **1977**, *88*, 1305–1320. [CrossRef]
6. Cailleux, Y.; Gonord, H.; Le guern, M.; Sauvage, M. Taphrogenèse et magmatisme permien dans le Maroc Central. *Bull. Fac. Sci. Marrakech.* **1983**, *1*, 24–39.
7. Youbi, N.; Cabanis, B.; Chalot-Prat, F.; Cailleux, Y. Histoire volcano-tectonique du massif permien de Khénifra (Sud-Est du Maroc central). *Geodin. Acta* **1995**, *8*, 158–172. [CrossRef]
8. Van Hilten, D. Evaluation of some geotectonic hypotheses by paleomagnetism. *Tectonophysics* **1964**, *1*, 3–71. [CrossRef]
9. Elter, F.M.; Gaggero, L.; Mantovani, F.; Pandeli, E.; Costamagna, L.G. The Atlas-East Variscan -Elbe shear system and its role in the formation of the pull-apart Late Palaeozoic basins. *Int. J. Earth Sci.* **2020**, *109*, 739–760. [CrossRef]
10. Timmerman, M.J.; Heeremans, M.; Kirstein, L.A.; Larsen, B.T.; Spencer-Dunworth, E.A.; Sundvoll, B. Linking changes in tectonic style with magmatism in northern Europe during the late Carboniferous to latest Permian. *Tectonophysics* **2009**, *473*, 375–390. [CrossRef]
11. Upton, B.G.J.; Stephenson, D.; Smedley, P.M.; Wallis, S.M.; Fitton, J.G. Carboniferous and Permian magmatism in Scotland. *Geol. Soc. Spec. Publ.* **2004**, *223*, 195–218. [CrossRef]

12. Ernst, R.E.; Buchan, K.L. Layered mafic intrusions: A model for their feeder systems and relationship with giant dyke swarms and mantle plume centres. *S. Afr. J. Geol.* **1997**, *100*, 319–334.
13. Torsvik, T.H.; Steinberger, B.; Cocks, L.R.M.; Burke, K. Longitude: Linking Earth's ancient surface to its deep interior. *Earth Planet. Sci. Lett.* **2008**, *276*, 273–282. [CrossRef]
14. Kirstein, L.A.; Dunai, T.J.; Davies, G.R.; Upton, B.G.J.; Nikogosian, I.K. Helium isotope signature of lithospheric mantle xenoliths from the Permo-Carboniferous magmatic province in Scotland—No evidence for a lower-mantle plume. *Geol. Soc. Spec. Publ.* **2004**, *223*, 243–258. [CrossRef]
15. Veevers, J.J.; Powell, C.M. Late Paleozoic glacial episodes in Gondwanaland reflected in transgressive- regressive depositional sequences in Euramerica. *Geol. Soc. Am. Bull.* **1987**, *98*, 475–487. [CrossRef]
16. Crowell, J.C. *Pre-Mesozoic Ice Ages: Their Bearing on Understanding the Climate System*; Geological Society of America, Inc.: Boulder, CO, USA, 1999; Volume 192, ISBN 0813711924.
17. Isbell, J.L.; Miller, M.F.; Wolfe, K.L.; Lenaker, P.A. *Timing of Late Paleozoic Glaciation in Gondwana: Was Glaciation Responsible for the Development of Northern Hemisphere Cyclothems?* Special Paper of The Geological Society of America; Geological Society of America, Inc.: Boulder, CO, USA, 2003; Volume 370. [CrossRef]
18. Cather, S.M.; Dunbar, N.W.; McDowell, F.W.; McIntosh, W.C.; Scholle, P.A. Climate forcing by iron fertilization from repeated ignimbrite eruptions: The icehouse-silicic large igneous province (SLIP) hypothesis. *Geosphere* **2009**, *5*, 315–324. [CrossRef]
19. El Wartiti, M.; Broutin, J.; Freytet, P.; Larhrib, M.; Toutin-Morin, N. Continental deposits in Permian basins of the Mesetian Morocco, geodynamic history. *J. Afr. Earth Sci.* **1990**, *10*, 361–368. [CrossRef]
20. Saber, H.; El-Wartiti, M.; Broutin, J.; Toutin Morin, N. L'intervalle stephano-permien (fin du cycle varisque au Maroc). *Gaia* **1995**, *11*, 57–71.
21. Saber, H.; El Wartiti, M. Histoire sédimentaire et tectonique tardi-hercynienne des bassins de l'Oued Zat et Ida Ou Zal (Haut Atlas occidental, Maroc): Bassins en transtension sur décrochements. *J. Afr. Earth Sci.* **1996**, *22*, 301–309. [CrossRef]
22. Abbou, M.B.; Soula, J.C.; Brusset, S.; Roddaz, M.; N'Tarmouchant, A.; Driouch, Y.; Christophoul, F.; Bouadbelli, M.; Majesté-Menjoulas, C.; Béziat, D.; et al. Contrôle tectonique de la sédimentation dans le système de bassins d'avant-pays de la Meseta marocaine. *C. R. Acad. Sci. Ser. IIA Earth Planet. Sci.* **2001**, *5*, 703–709. [CrossRef]
23. Domeier, M.; Font, E.; Youbi, N.; Davies, J.; Nemkin, S.; Van der Voo, R.; Perrot, M.; Benabbou, M.; Boumehdi, M.A.; Torsvik, T.H. On the Early Permian shape of Pangea from paleomagnetism at its core. *Gondwana Res.* **2021**, *90*, 171–198. [CrossRef]
24. Hoepffner, C.; Soulaimani, A.; Piqué, A. The Moroccan Hercynides. *J. Afr. Earth Sci.* **2005**, *43*, 144–165. [CrossRef]
25. Michard, A.; Soulaimani, A.; Hoepffner, C.; Ouanaimi, H.; Baidder, L.; Rjimati, E.C.; Saddiqi, O. Tectonophysics The South-Western Branch of the Variscan Belt: Evidence from Morocco. *Tectonophysics* **2010**, *492*, 1–24. [CrossRef]
26. Michard, A.; Ouanaimi, H.; Hoepffner, C.; Soulaimani, A.; Baidder, L. Comment on Tectonic relationships of Southwest Iberia with the allochthons of Northwest Iberia and the Moroccan Variscides by J.F. Simancas et al. [C. R. Geoscience 341 (2009) 103–113]. *C. R. Geosci.* **2010**, *342*, 170–174. [CrossRef]
27. Gonord, H.; Le Guern, M.; Turiot, D.; Rebourd, H. Mise en évidence d'un volcanisme rhyolitique stéphano-permien sur la bordure nord du massif hercynien central du Maroc extension et importance du volcanisme tardy-hercynien. *C. R. Acad. Sci.* **1980**, *291*, 51–54.
28. El Wartiti, M. Les Terrains Permo-Carbonifères et Leur Couverture dans la Zone de Tiddas-Souk es Sebt (Bordure Nord-Ouest de la Méséta Marocaine. Nord du Maroc Central). Ph.D. Thesis, Mohamed V, Rabat, Morocco, 1981.
29. Zouine, E.M. Evolution structurale tardi-hercynienne de la bordure septentrionale du Maroc Central entre Tiddas et Jbel Triona. Ph.D. Thesis, E.N.S. Souissi, Rabat, Morocco, 1986.
30. Larhrib, M. Les Formations Permiennes et Triasico-Mio-Quaternaires des Régions de Tiddas, Maaziz et Sebt Aït Ikko (NW du Maroc Central). Stratigraphie, Sédimentologie et Reconstitution des Paléoenvironnements Sédimentaires. Ph.D. Thesis, Ecole Normale Supérieure, Rabat, Morocco, 1988.
31. Larhrib, M. Les formations fluviatiles rouges violacées du bassin permien (Autunien inférieur) de Tiddas-Sebt Aït Ikkou (NW du Maroc Central). In Proceedings of the 8ème Colloque Bassins Sédimentaires, El Jadida, Morocco, 23–25 May 1990; pp. 3–4.
32. Larhrib, M. Flore fossile et séquences des formations rouges fluviatiles du bassin autunien de Tiddas-Sebt Aït Ikkou (nord-ouest du Maroc central). In *Le Permien et le Trias du Maroc: État des Connaissances*; Medina, F., Ed.; Presses Universitaires du Maghreb: Marrakech, Morocco, 1996; pp. 19–29.
33. Voigt, S.; Lagnaoui, A.; Hminna, A.; Saber, H.; Schneider, J.W. Revisional notes on the Permian tetrapod ichnofauna from the Tiddas Basin, central Morocco. *Palaeogeogr. Palaeoclimatol. Palaeoecol.* **2011**, *302*, 474–483. [CrossRef]
34. Broutin, J.; Aassoumi, H.; El Wartiti, M.; Freytet, P.; Kerp, H.; Quesada, C.; Toutin-morin, N. The Permian basins of Tiddas, Bou Achouch and Khenifra (central Morocco). Biostratigraphic and palaeophytogeographic implications. *Mémoires Muséum Natl. d'Histoire Nat.* **1998**, *179*, 257–278.
35. Marchetti, L.; Ronchi, A.; Santi, G.; Schirolli, P.; Conti, M.A. Revision of a classic site for Permian tetrapod ichnology (Collio Formation, Trompia and Caffaro valleys, N. Italy), new evidences for the radiation of captorhinomorph footprints. *Palaeogeogr. Palaeoclimatol. Palaeoecol.* **2015**, *433*, 140–155. [CrossRef]
36. Termier, H. Etudes géologiques sur le Maroc Central et le Moyen Atlas septentrional, Rabat. 1936. Available online: https://www.persee.fr/doc/rga_0035-1121_1937_num_25_3_3982_t1_0525_0000_1 (accessed on 20 May 2021).

37. Michard, A. *Elements de Geologie Marocaine;* Notes et Memoires du Service Geologique No. 252; Editions du Service Geologique du Maroc: Rabat, Morroco, 1976; p. 408.
38. Youbi, N. Le Volcanisme "Post-Collisionnel" un Magmatisme Intraplaque Relié à des Panaches Mantelliques. Etude Volcanologique et Géochimique. Exemple d'application dans le Néoprotérozoique Terminal (PIII) de l'Anti-Atlas et le Permien du Maroc. Ph.D. Thesis, Faculty SEMLALIA, CAdi AYYAD, Marrakesh, Morocco, 1998.
39. Black, L.P.; Kamo, S.L.; Allen, C.M.; Davis, D.W.; Aleinikoff, J.N.; Valley, J.W.; Mundil, R.; Campbell, I.H.; Korsch, R.J.; Williams, I.S.; et al. Improved 206Pb/238U microprobe geochronology by the monitoring of a trace-element-related matrix effect; SHRIMP, ID-TIMS, ELA-ICP-MS and oxygen isotope documentation for a series of zircon standards. *Chem. Geol.* **2004**, *205*, 115–140. [CrossRef]
40. Williams, I.S. U-Th-Pb geochronology by ion microprobe. In *Applications of Microanalytical Techniques to Understanding Mineralising Processes;* McKibben, M., Shanks, W., Riley, W., Eds.; Reviews in Economic Geology; Society of Economic Geologists: Littleton, CO, USA, 1998; pp. 1–35.
41. Ludwig, K.R. Isoplot 3: A Geochronological Toolkit for Microsoft Excel. *Berkeley Geochronol. Cent.* **2003**, *4*, 70.
42. Ouabid, M.; Garrido, C.J.; Ouali, H.; Harvey, J.; Hidas, K.; Marchesi, C.; Acosta-Vigil, A.; Dautria, J.M.; El Messbahi, H.; Román-Alpiste, M.J. Late Cadomian rifting of the NW Gondwana margin and the reworking of Precambrian crust–evidence from bimodal magmatism in the early Paleozoic Moroccan Meseta. *Int. Geol. Rev.* **2020**, 1–24. [CrossRef]
43. Morimoto, N. Nomenclature of Pyroxenes. *Mineral. Petrol.* **1988**, *39*, 55–76. [CrossRef]
44. Tiba, T. Titaniferous Biotites and Associated Phenocrysts in Dike Rocks From Dozen, Oki Islands. *J. Jpn. Assoc. Mineral. Petrol. Econ. Geol.* **1972**, *67*, 357–369. [CrossRef]
45. Le Bel, L. Micas magmatiques et hydrothermaux dans l'environnement du porphyre cuprifère de Cerro Verde-Santa Rosa, Pérou. *Bull. Minéral.* **1979**, *102*, 35–41. [CrossRef]
46. Nachit, H.; Razafimahefa, N.; Stussi, J.; Carron, J. Composition chimique des biotites et typologie magmatique des granitoïdes. *C. R. Acad. Sci.* **1985**, *301*, 813–818.
47. Abdel-Rahman, A.F.M. Nature of biotites from alkaline, calc-alkaline, and peraluminous magmas. *J. Petrol.* **1994**, *35*, 525–541. [CrossRef]
48. Deer, W.; Howie, R.A.; Zussman, J. *An Introduction to Rock-Forming Minerals;* Mineralogical Society of Great Britain & Ireland: London, UK, 1966.
49. Bas, M.J.L.; Maitre, R.W.L.; Streckeisen, A.; Zanettin, B. A chemical classification of volcanic rocks based on the total alkali-silica diagram. *J. Petrol.* **1986**, *27*, 745–750. [CrossRef]
50. Miyashiro, A. Nature of alkalic volcanic rock series. *Contrib. Mineral. Petrol.* **1978**, *66*, 91–104. [CrossRef]
51. Harker, A. Igneous rock-series and mixed igneous rocks. *J. Geol.* **1900**, *8*, 389–399. [CrossRef]
52. Saunders, A.; Tarney, J.; Marsh, N.; Wood, D. Ophiolites as oceanic crust or marginal basin crust: A geochemical approach. In Proceedings of the Ophiolites International Ophiolite Symposium, Nicosia, Chyprus, 1–8 April 1979; pp. 193–204.
53. Cabanis, B. Identification des Séries Magmatiques dans les Socles Métamorphiques sur la Base de Critères Géologiques, Pétrographiques et Géochimiques. Ph.D. Thesis, P. et M.-Curie, Paris, France, 1986.
54. Thorpe, R.S.; Leat, P.T.; Mann, A.C.; Howells, M.F.; Reedman, A.J.; Campbell, S.D.G. Magmatic evolution of the ordovician snowdon volcanic centre, North Wales (UK). *J. Petrol.* **1993**, *34*, 711–741. [CrossRef]
55. Harris, N.B.W.; Pearce, J.A.; Tindle, A.G. *Geochemical Characteristics of Collision-Zone Magmatism;* Special Publications; Geological Society: London, UK, 1986; Volume 19, pp. 67–81. [CrossRef]
56. Pupin, J.P. Zircon and granite petrology. *Contrib. Mineral. Petrol.* **1980**, *73*, 207–220. [CrossRef]
57. Hoskin, P.W.O.; Schaltegger, U. The composition of zircon and igneous and metamorphic petrogenesis. *Rev. Mineral. Geochem.* **2003**, *53*, 27–62. [CrossRef]
58. Linnemann, U.; Gerdes, A.; Drost, K.; Buschmann, B. The continuum between Cadomian orogenesis and opening of the Rheic Ocean: Constraints from LA-ICP-MS U-Pb zircon dating and analysis of plate-tectonic setting (Saxo-Thuringian zone, northeastern Bohemian Massif, Germany). *Spec. Pap. Geol. Soc. Am.* **2007**, *423*, 61–96. [CrossRef]
59. Winchester, J.A.; Floyd, P.A. Geochemical discrimination of different magma series and their differentiation products using immobile elements. *Chem. Geol.* **1977**, *20*, 325–343. [CrossRef]
60. Pearce, J.A. Geochemical fingerprinting of oceanic basalts with applications to ophiolite classification and the search for Archean oceanic crust. *Lithos* **2008**, *100*, 14–48. [CrossRef]
61. Miyashiro, A. The troodos ophiolitic complex was probably formed in an island arc. *Earth Planet. Sci. Lett* **1973**, *19*, 218–224. [CrossRef]
62. Miyashiro, A. Volcanic rock series in island arcs and active continental margins. *Am. J. Sci.* **1974**, *274*, 321–355. [CrossRef]
63. Leterrier, J.; Maury, R.C.; Thonon, P.; Girard, D.; Marchal, M. Clinopyroxene composition as a method of identification of the magmatic affinities of paleo-volcanic series. *Earth Planet. Sci. Lett.* **1982**, *59*, 139–154. [CrossRef]
64. Mollard, J.; Maury, R.C.; Leterrier, J.; Bourgois, J. Teneurs en chrome et titane des clinopyroxenes calciques des basaltes; Application à l'identification des affinités magmatiques de roches paléovolcaniques. *C. R. Acad. Sci.* **1983**, *296*, 903–908.
65. Sheth, H.C.; Torres-Alvarado, I.S.; Verma, S.P. What is the "calc-alkaline rock series"? *Int. Geol. Rev.* **2002**, *44*, 686–701. [CrossRef]
66. Hawkesworth, C.; Turner, S.; Gallagher, K.; Hunter, A.; Bradshaw, T.; Rogers, N. Calc-alkaline magmatism, lithospheric thinning and extension in the Basin and Range. *J. Geophys. Res.* **1995**, *100*, 10271–10286. [CrossRef]

67. Fan, W.M.; Guo, F.; Wang, Y.J.; Lin, G. Late Mesozoic calc-alkaline volcanism of post-orogenic extension in the northern Da Hinggan Mountains, northeastern China. *J. Volcanol. Geotherm. Res.* **2003**, *121*, 115–135. [CrossRef]
68. Dupuy, C.; Dostal, J.; Marcelot, G.; Bougault, H.; Joron, J.L.; Treuil, M. Geochemistry of basalts from central and southern New Hebrides arc: Implication for their source rock composition. *Earth Planet. Sci. Lett.* **1982**, *60*, 207–225. [CrossRef]
69. Deruelle, B. Petrology of the plio-quaternary volcanism of the South-Central and Meridional Andes. *J. Volcanol. Geotherm. Res.* **1982**, *14*, 77–124. [CrossRef]
70. Gans, P.B.; Mahood, G.A.; Schermer, E. Synextensional magmatism in the Basin and Range Province; A case study from the eastern Great Basin. *Spec. Pap. Geol. Soc. Am.* **1989**, *233*, 1–53. [CrossRef]
71. Putirka, K.; Platt, B. Basin and Range volcanism as a passive response to extensional tectonics. *Geosphere* **2012**, *8*, 1274–1285. [CrossRef]
72. Gill, J. *Orogenic Andesites and Plate Tectonics*; Springer: Berlin/Heidelberg, Germany, 1981.
73. Coulon, C. Le volcanisme calco-alcalin cénozoïque de Sardaigne (Italie). Pétrographie, géochimie et génèse des laves andésitiques et des ignimbrites. Signification géodynamique. Ph.D. Thesis, AIX-MARSEILLE III, Marseille, France.
74. DePaolo, D.J. Trace element and isotopic effects of combined wallrock assimilation and fractional crystallization. *Earth Planet. Sci. Lett.* **1981**, *53*, 189–202. [CrossRef]
75. Emami, M.H.; Michel, R. Les Volcans Dôméens du Néogène de la Région de Qom (Iran Central). Essai de Classification de l'Activité Volcanique Dôméenne. *Bull. Volcanol.* **1982**, *45*, 317–332. [CrossRef]
76. Kharbouch, F. Pétrographie et Géochimie des Laves Dinantiennes de la Méséta Nord-Occidentale et Orientale Marocaine. Ph.D. Thesis, Louis Pasteur, Strasbourg, France, 1983.
77. Kharbouch, F. Le volcanisme dévono-dinantien du Massif central et de la Meseta orientale. *Bull. Inst. Sci.* **1994**, *18*, 192–200.
78. Chalot-Prat, F. Mise en évidence d'une dépression volcanotectonique associée à d'épais épanchements ignimbritiques hercyniens dans le massif du Tazekka (Maroc oriental). *Rev. Géologie Dyn. Géographie Phys.* **1986**, *27*, 193–203.
79. Chalot-Prat, F. Pétrogénèse d'un Volcanisme Intracontinental Tardi-Orogénique Hercynien. Etude du Complexe Volcanique Carbonifère du Tazekka et de Zones Volcaniques Comparables dans le Mekam et la Région de Jerada (Maroc Oriental). Ph.D. Thesis, P. et M. Curie, Paris, France, 1990.
80. Chakib, A.; Aït Brahim, L.; Tahiri, A. Tectonique cassante et paléochamps de contraintes dans les régions de Bou Chahraine, d'el Jemaa Moud Bled et Merzaga du Permien à l'actuel (bordure méridionale du bassin de Rommani-Khémisset). In Proceedings of the 13 éme Colloque des Bassins Sédimentaires Marocains, Marrakech, Morocco, 19–22 March 1996; pp. 69–70.
81. Aït Brahim, L.; Tahiri, A.; Saidi, A. Proposition d'une nouvelle chronologie des événements tectoniques responsables de la structuration du bassin sur décrochement permien de Bou-Achouch (bordure septentrionale de la Meseta central). In Proceedings of the 13 éme Colloque des Bassins Sédimentaires Marocains, Marrakech, Morocco, 19–22 March 1996; pp. 9–10.
82. Hmich, D.; Schneider, J.W.; Saber, H.; Voigt, S.; El Wartiti, M. New continental Carboniferous and Permian faunas of Morocco: Implications for biostratigraphy, palaeobiogeography and palaeoclimate. *Geol. Soc. Spec. Publ.* **2006**, *265*, 297–324. [CrossRef]
83. Kerp, H.; Broutin, J.; Lausberg, S.; Aassoumi, H. Discovery of Latest Carboniferous–Early Permian radially symmetrical peltaspermaceous megasporophylls from Europe and North Africa. *C. R. Acad. Sci. Ser. IIA Earth Planet. Sci.* **2001**, *332*, 513–519. [CrossRef]
84. Jébrak, M. Contribution à l'histoire Naturelle des Filons (F, Ba) du Domaine Varisque Français et Marocain. Essai de Caractérisation Structurale et Géochimique des Filons en Extension et en Décrochement. Ph.D. Thesis, Univiversity of Orleans, Orleans, France, 1984.
85. Aassoumi, H.; Broutin, J.; Youbi, N. Découverte de Scleromedulloxylon cf. aveyronense, bois fossile de coniférophyte, dans le complexe volcanique permien du bassin de Khnifra (Maroc): Intérêt stratigraphique et floristique. *Ann. Paléontol.* **1995**, *81*, 1–16.
86. Youbi, N. Le complexe volcanique permien de Khenifra (SE du Maroc central): Cartographie, volcanologie, minéralogie, géochimie, considérations structurales et implications géodynamiques. Ph.D. Thesis, Cadi Ayyad University, Marrakesh, Morocco, 1990.
87. Youbi, N.; Gaggero, L.; Assafar, H.; Hadimi, I.; Boumehdi, M.A.; Bensalah, M.K.; Linnemann, U.; Gaertner, A.; Mata, J.; Doblas, M.; et al. U-Pb Zircon Geochronological and Petrologic Constraints on the PostCollisional Variscan Volcanism of the Khénifra Basin (Western Meseta, Morocco): Stratigraphic and Petrogenetic Implications. In Proceedings of the 2nd International Congress on Permian and Triassic, Casablanca, Morocco, 25–27 April 2018; pp. 53–54.
88. Van Houten, F. Late Variscan Nonmarine Basin Deposits, Northwest Africa: Record of Hercynotype Orogeny. In *The Continental Permain in Central, West, and South Europe*; Springer: Dordrecht, The Netherlands, 1976; pp. 215–224.
89. Hadimi, I.; Ait Lahna, A.; Assafar, H.; Tassinari, C.C.G.; Youbi, N.; Boumehdi, M.; Bensalah, M.K.; Gaggero, L.; Mata, J.; Doblas, M. The Post-Collisional Hercynian Volcanism of Rehamna, Western Meseta, Morocco. Mineral Chemistry, Petrology and U-Pb Dating. In Proceedings of the 2nd International Congress on Permian and Triassic, Casablanca, Morocco, 25–27 April 2018; pp. 33–35.
90. Baudin, T.; Chèvremont, P.; Razin, P.; Youbi, N.; Andries, D.; Hoepffner, C.; Tegyey, M. Carte géologique du Maroc au 1/50000, feuille de Skhour des Rehamna, Mémoire explicatif. *Notes Mém. Serv. Géol. Maroc* **2003**, *435*, 114.
91. Bensalah, M.K. De la fin du cycle orogénique hercynien à l'ouverture et l'évolution de l'Atlantique Central: Les marqueurs magmatiques de cette transition. Etude pétrologique et géochimique. Exemple d'application dans le Permien, le TriasJurassique et le Jurassique-Cré. Ph.D. Thesis, CAADi AYYAD, Marrakesh, Morocco, 2012.

92. Bensalah, M.K.; Youbi, N.; Ait Lahna, A.; Tassinari, C.C.G.; Boumehdi, M.A.; Basei, M.A.S.; Sato, K.; Aarab, E.M.; Mata, J. Geochemical and Geochronological Constraints on the Origin of the Intermediate and Felsic Dyke and Sill Swarms from the Western Meseta and the High Atlas (Morocco): Implications for Tectonic Evolution. In Proceedings of the 2nd International Congress on Permian and Triassic, Casablanca, Morocco, 25–27 April 2018; pp. 20–22.
93. Hoepffner, C. Le magmatisme pré- et post-orogénique hercynien dans le Paléozoïque des Rehamna. *Serv. Géologique du Maroc Rabat* **1982**, *303*, 150–163.
94. Charif, A. Etude Pétrographique et Géochimique du Volcanisme "Post-Collisionnel" du Permien Inférieur (Autunien) des Rahamna (Mechraâ Ben Abbou et Nazt Lararcha, Messta-Maroc): Implications Géochimiques, Pétrogénétiques et Géodynamiques. Ph.D. Thesis, University of Chouaib Doukkali, El Jadida, Morocco, 2001.
95. Razin, P.; Baudin, T.; Chèvremont, P.; Andries, D.; Youbi, N.; Hœpffner, C.; Thiéblemont, D.; Chihani, E. Carte géologique du Maroc au 1/50 000, feuille de Jebel Kharrou, Mémoire explicatif. *Notes Mém. Serv. Géol. Rabat* **2003**, *436*, 1–105.

minerals

MDPI

Article

Detrital-Zircon Age Spectra of Neoproterozoic-Paleozoic Sedimentary Rocks from the Ereendavaa Terrane in NE Mongolia: Implications for the Early-Stage Evolution of the Ereendavaa Terrane and the Mongol-Okhotsk Ocean

Laicheng Miao [1,2,*], Mingshuai Zhu [1,2], Chenghao Liu [1,3], Munkhtsengel Baatar [4], Chimidtseren Anaad [4], Shunhu Yang [1,2] and Xingbo Li [1,3]

1 Key Laboratory of Mineral Resources, Institute of Geology and Geophysics, Chinese Academy of Sciences, Beijing 100029, China; zhumingshuai@mail.iggcas.ac.cn (M.Z.); liuchenghao@mail.iggcas.ac.cn (C.L.); yangshunhu@mail.iggcas.ac.cn (S.Y.); lixingboaries@163.com (X.L.)
2 Institutions of Earth Science, Chinese Academy of Sciences, Beijing 100029, China
3 College of Earth and Planetary Sciences, University of Chinese Academy of Sciences, Beijing 100049, China
4 School of Geology and Mining, Mongolian University of Science and Technology, Ulaanbaatar 120646, Mongolia; tsengel@must.edu.mn (M.B.); chimidtseren@must.edu.mn (C.A.)
* Correspondence: miaolc@mail.iggcas.ac.cn; Tel.: +86-10-82998561

Received: 29 June 2020; Accepted: 17 August 2020; Published: 22 August 2020

Abstract: The Mongol-Okhotsk orogenic belt (MOB) is considered to be the youngest division of the huge Central Asian Orogenic Belt, but its origin and evolution are still enigmatic. To better understand the history of the MOB, we conducted U-Pb geochronological analyses of detrital-zircon grains from Neoproterozoic-Paleozoic sedimentary sequences as well as a volcanic suite in the Ereendavaa terrane, the southern framing unit of the MOB, in NE Mongolia. Our results show that the protoliths of the quartzite assemblage of the Ereendavaa terrane basement (or proto-Ereendavaa terrane) was deposited after ca. 1.15 Ga on a passive continental margin. The detrital-zircon age spectra of the Silurian and Devonian sedimentary sequences of the terrane demonstrate that the source areas were dominated by proximal Cambrian-Ordovician arc rocks, likely resulting from the northward subduction of the Kherlen Ocean lithosphere beneath the Ereendavaa terrane. Based on a combination of our new data with those published, we show that the Mongol-Okhotsk Ocean split from an early Paleozoic domain during, or after, the early Silurian by a mantle plume, and developed an Andean-type margin along its northern rim possibly at Middle Devonian times, and a bidirection subduction system in mid-Carboniferous at approximately 325 Ma. This bipolar subduction of the Mongol-Okhotsk Ocean might have lasted until the Triassic.

Keywords: detrital-zircon age spectrum; Ereendavaa terrane; Mongol-Okhotsk orogenic belt; opening of the Mongol-Okhotsk ocean; northeastern Mongolia

1. Introduction

The Mongol-Okhotsk Belt (MOB), extending over 3000 km from central Mongolia in the southwest to the Uda Gulf of the Sea of Okhotsk in the northeast (Figure 1), was formed by the closing of the Mongol-Okhotsk Ocean (MOO) between the Northern Asian (Siberia) continent and the Amur Superterrane [1–4]. The MOO is considered to be the youngest domain within the Central Asian Orogenic Belt (CAOB), a vast region that resulted from the evolution of the Paleo Asian Ocean (PAO) and is bounded by the Siberia and Baltica cratons to the north and by the Sino-Korean and Tarim cratons to the south [5–8]. However, this is inconsistent with the recognition that the final closure of the PAO, with a long-lasting evolution at least from the Neoproterozoic until the Permian, took place along the Solonker-Linxi suture in northern Inner Mongolia, China, in the southern CAOB [2,6,8–10]. Therefore, the MOB is essential for understanding the geotectonic history of the Central Asian Orogenic Belt and also entire NE Asia.

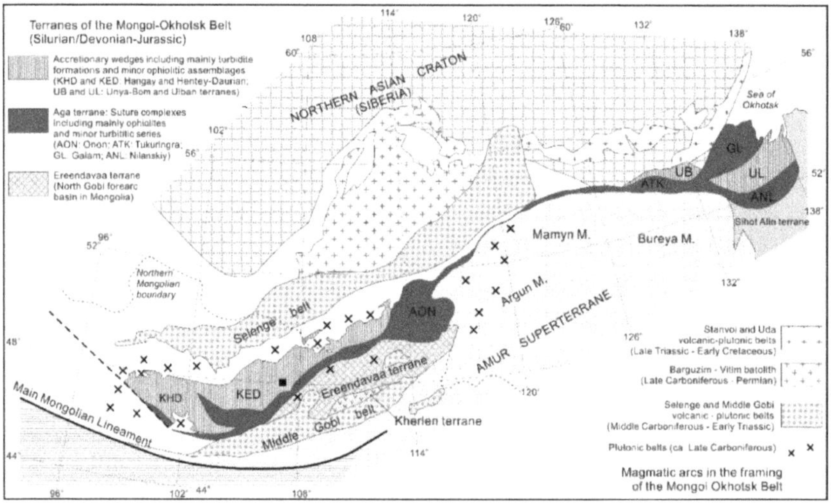

Figure 1. Sketch map showing the major tectonic units along the Mongol-Okhotsk belt (after [11]).

The evolution of the MOB is enigmatic, with contrasting models proposed concerning its development [6,11–13]. It appears that there is a general consensus regarding the timing of the final closure of the MOO despite the contrasting tectonic models. It is widely accepted that the closing of the MOO occurred in a scissor manner, with a progressive Jurassic to Early Cretaceous closing of the ocean and associated collisional orogenesis from the west towards the east [1,4,6,14–16]. Nevertheless, the time of opening is still a matter of controversy. Some authors have suggested that MOO has existed since Vendian (610–570 Ma) to Cambrian [5,6], whereas others have proposed that the ocean formed later in the Silurian [8,11,13,17] or even later [4,18].

Detrital zircon geochronology of sedimentary rocks is a powerful tool to study the history of sedimentary basins and it has been used worldwide [19–21]. Previous detrital-zircon geochronological studies for the sedimentary sequences within the MOB suggest that the MOO was opened during the early Paleozoic [11,22] or late Paleozoic [16,23]. These studies were mainly focused on the sedimentary sequences of the accretionary complexes inside the MOB, with little attention paid to those on the continental blocks fringing the MOO basin [11,22]. Although these studies do provide important constraints for the evolution of the MOO, some key information regarding the early-stage development of the ocean might have been undisclosed. This is because the accretionary complexes inside the MOB may not preserve the early-stage sedimentation records of the MOO due to the subduction consumption of the ocean lithosphere. In this regard, studies of the sedimentary sequences on the rifted margins of the MOO will be very useful, since the rifted margins witness the whole history of the ocean development. Despite some deformation of the rifted margins of the MOO, chiefly because of their involvement in continental collision following the closure of the ocean, the detrital zircon U-Pb geochronology for the sedimentary rocks deposited on the rifted margins is still capable of providing direct information of provenance of detritus and the ocean evolution, owing to the considerably high closure temperature of the zircon U-Pb isotope system, greater than 900 °C [24].

On the other hand, the age uncertainty of the basement on which an ocean opened will hamper the interpretation of provenance of detrital zircon in sedimentary sequences. This is particularly true for the Ereendavaa terrane that fringes the MOB to the south. Though it has long been interpreted as a cratonic terrane with basement rocks as old as the Neoarchean-Paleoproterozoic [13], there is no robust evidence so far to support the interpretation. Instead, recent reliable dating demonstrates that the metamorphic rocks of the terrane were mainly formed during the Paleozoic [25,26]. In other words, to ascertain whether early Precambrian basement rocks exist in the Ereendavaa terrane is also an issue to explore.

We assume if early Precambrian basement rocks exist in the Ereendavaa terrane, they would contribute much detritus to the Neoproterozoic-Paleozoic sedimentary sequences of the terrane, and thus generate detrital-zircon age spectra comprising major peaks of that age. In addition, it is also assumed that syn-sedimentary detrital zircon grains in the sedimentary rocks deposited at rifted margins depict the magmatic activity in subduction-related arcs at the framing margins of the MOO, and, therefore, provide key information concerning the subduction history of the ocean lithosphere. In this context, this paper presents results of zircon U-Pb geochronology of Neoproterozoic-Paleozoic (meta-) sedimentary rocks and andesitic-rhyolitic volcanic formations developed on the Ereendavaa terrane, which constitutes the southern margin of the Mongolian segment of the MOO, with aims to (1) verify whether early Precambrian rocks are present in the Ereendavaa terrane, and (2) constrain the early-stage evolution of both the MOO and the Ereendavaa terrane. Our results provide, for the first time, robust constraints not only for the depositional ages of the Neoproterozoic-Paleozoic sedimentary sequences but also for the development of magmatic arcs on the Ereendavaa terrane, and indicate that the MOO was opened after the early Silurian.

2. Geological Setting

The MOB, extending over 3000 km from central Mongolia in the southwest to the Uda Gulf of the Sea of Okhotsk in the northeast (Figure 1), formed by closing of the MOO between the Northern Asian (Siberia) continent and the Amur Superterrane [1–4]. Structurally, the MOB is sandwiched by the Northern Asian Caton (Siberia) to the north and the Amur Superterrane to the south, with an abrupt termination in the southwest (Figures 1 and 2). The core of the MOB is represented by a ribbon-like ophiolite-bearing suture zone and accretionary wedges [1,15,27]. The suture zone, which is termed the Aga terrane [2], is composed mainly of ophiolites and minor turbiditic series and is subdivided into the Onon, Tukuringra, Galam, and Nilanskiy terranes (Figure 1). Of these, the Onon terrane consists of the intra-oceanic Onon arc of assumed Devonian–Early Carboniferous age [4], which would be the oldest arc identified so far that was related to the southward subduction of the MOO lithosphere beneath the Amur Superterrane. The accretionary wedges are characterized by containing thick turbiditic sequences and minor ophiolitic assemblages, which include the Hangai, Kentei-Daurian, Unya-Bom, and Ulban terranes [2,28]. Spatially, the accretionary wedge terranes are mainly distributed on the northern side of the suture zone, although some also occur inside the suture zone at the eastern end of the MOB (Figure 1). Alternatively, some authors have considered all of these suture zone terranes and accretionary wedges together as one large accretionary complex resulting from subduction of the MOO lithosphere [6,22,23].

The territory of Mongolia has traditionally been divided into two tectonic domains separated by the Main Mongolian Lineament (MML), namely an early Paleozoic domain to the north and a late Paleozoic domain to the south [13,15]. The Mongolian segment of the MOB is located within the early Paleozoic domain (Figure 2) and here the MOO suture zone is represented by the Adaatsag and Dochgol terranes [11,13,15], corresponding to the Onon terrane in the northeastern prolongation in Russia [2,4]. The suture is bounded by the Hangay-Hentey turbiditic terrane to the north, by the Baydrag terrane to the southwest, and by the Ereendavaa terranes to the south (Figure 2).

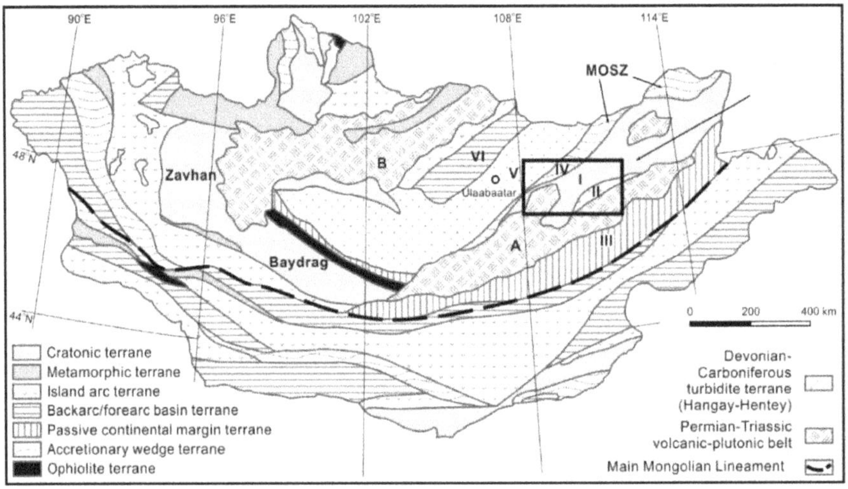

Figure 2. Sketch map showing tectonostratigraphic terranes of Mongolia and the approximate location of the study area of this paper (adapted from [13]). Terrane and belt names: I—Ereendavaa, II—Kherlen, III—Idermeg, IV—Adaatsag (Onon); V—Hangay-Hentey; VI—Haraa; A—Middle Gobi, B—Selenge; MOSZ—Mongol-Okhotsk suture zone.

The Hangay-Hentey terrane contains mainly Devonian-Carboniferous turbidites and Silurian-Devonian radiolarian cherts [11,13,17]. The authors of [13] interpreted the entire Hangay–Hentey sediment series to unconformably overlie an older basement of an unknown age. However, the authors of [17] and [11] suggested that the main body of the Hangay–Hentey belt, which is tectonically located between the early Paleozoic Haraa terrane in the northwest and the Mongol-Okhotsk suture zone (MOSZ) in the southeast, were formed in a trench-accretionary wedge environment. Previous detrital zircon geochronological studies show that the sedimentary rocks of the Hangay-Hentey terrane were mainly formed after the early Carboniferous [11,22,23] although the strata were previously believed to be Silurian, Devonian, or Carboniferous in age. In contrast, the sedimentary rocks of the Haraa terrane, which was considered Cambrian/Ordovician-Silurian in age, at the northern margin of the Hangay-Hentey basin display distinct detrital zircon age spectra, with the youngest age peaks (514–437 Ma) [11,23] being older than those of the Hangay-Hentey rocks, implying that the Haraa terrane formed earlier than the Hangay-Hentey terrane.

The Mongolia segment of the MOSZ, i.e., the Adaatsag and Dochgol terranes of [13], is a narrow belt of highly deformed rocks that were metamorphosed under greenschist and amphibolite facies conditions. It is interpreted as an accretionary wedge complex composed of schist, quartzite, metasandstone, phyllite, chert, metavolcanic rocks, coral limestone, and mélange containing fragments of ophiolite [13,15,28]. The MOSZ contains the 325-Ma Adaatsag ophiolite [15] at its western end and the 321-Ma Khuhu Davaa ophiolite [29] in its central part, and it is intruded by Triassic-Early-Jurassic granitic plutons [13,15]. These ophiolites have geochemical signatures of supra-subduction zone (SSZ)-type ophiolites and are interpreted to form during the subduction initiation of the MOO lithosphere [29]. In the central portion, the accretionary wedge complex of the suture was locally thrust southward over the Silurian-Devonian sedimentary sequence intruded by swarms of dolerite sills/dikes or gabbroic intrusions at the northern margin of the Ereendavaa terrane (Figures 2 and 3; [26]). In northeastern-most Mongolia and Russia, the Dochgol terrane consists predominately of late Permian-Triassic accretionary complexes containing ophiolite blocks of unknown age, metavolcanics with geochemical signatures of both MORBs and OIBs [30], and fossiliferous marine sediments [31]. Generally, the suture zone complexes form a fold-and-thrust belt with the top-to-SE structural polarity [15], in accordance with observations from the Ereendavaa terrane [26].

The area to the south of the MOSZ belongs to the eastern part of the early Paleozoic domain, which is composed, from north to south, of the Ereendavaa, Kherlen, and Idermeg terranes (Figures 2 and 3). The Ereendavaa terrane is considered as a cratonic massif and consists of Proterozoic gneiss, amphibolite, schist and marble, overlain by black schist, metasandstone, siltstone, limestone and minor conglomerate and volcanic rocks and intruded by granites of different age [13]. However, recent work seems to suggest that it predominately formed during the Paleozoic since the gneiss, amphibolite, and schist of the Ereendavaa terrane mainly have Paleozoic protolith ages, although some Precambrian zircon inheritance does exist in these rocks [26]. On the other hand, granitic rocks/gneisses with Proterozoic ages have been reported from the northeastern extension of the Ereendavaa terrane both in Russia and NE China. On the Russian side, biotite granite and leucogranite were dated at ca. 740 Ma by zircon U–Pb method and the Chinese portion of the terrane (also known as "Erguna" massif or block [32]) contains Paleoproterozoic granitic gneiss with ages of ca. 1840 Ma [33] and is intruded by Neoproterozoic mafic plutons with ages between 850 and 740 Ma [34]. Studies have shown that the Ereendavaa terrane, which constitutes the Ereendavaa Mountain Range, was exhumed during the Early Cretaceous at ca. 140 Ma [35].

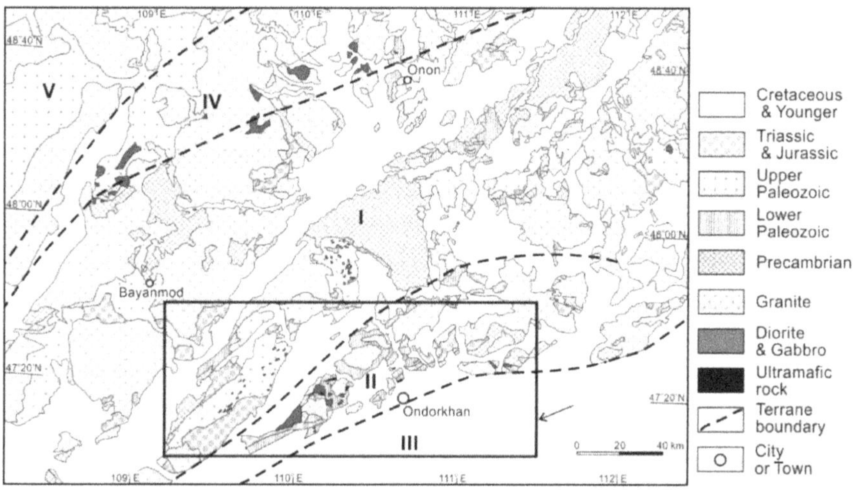

Figure 3. Simplified geological map of the Ondorkhan-Onon region in NE Mongolia (modified from [36]). Terranes: I—Ereendavaa, II—Kherlen, III—Idermeg, IV—Adaatsag (Onon), V—Hangay-Hentey.

The Idermeg terrane is located on the northern side of the MML and is mainly composed of marble, quartzite, conglomerate, sandstone, and limestone containing archeachyathes and stromatolites that are believed to be Neoproterozoic to Cambrian in age [13]. These rocks are assumed to overlie a crystalline basement composed of gneiss, amphibolite, schist, and phyllite [2,13]. The Idermeg terrane is classified by [13] as a passive continental margin terrane and is assumed to form in a Neoproterozoic–Cambrian continental shelf environment and represent the early Paleozoic southern margin of the Siberian amalgamated complex [8,37].

The Kherlen terrane is sandwiched between the two above-described metamorphic terranes (Figures 2 and 3). This terrane occurs as a narrow fault-bounded belt along the north bank of the Kherlen River. It consists of Neoproterozoic–early-Cambrian dismembered ophiolites in mélange, tholeiitic basalt, andesite, tuff, chert, volcaniclastic rocks, and minor archeacyathic limestone [13]. The Kherlen terrane was previously classified as an island arc terrane [13] but it is actually a Cambrian ophiolitic complex that was tectonically emplaced during the Silurian, and thus representing the suture between the Ereendavaa and Idermeg terranes [25]. This timing is in line with that of the amalgamation of the early Paleozoic domain of Mongolia [8,11,13].

The Ereendavaa terrane, possibly with the Idermeg and Kherlen ones, is considered to be the eastern prolongation of the Tuva-Mongolian Massif or the Central Mongolian Block (CMB) that frames the MOB in the northwest, west, and southwest. The CMB is considered either as an isolated microcontinent in the PAO during the late Proterozoic and Cambrian [4,38], a Precambrian continental ribbon connecting the Siberian craton [6,15,39], or a composite tectonic unit consisting of several smaller Precambrian blocks with uncertain tectonic relationships between each other [13]. Presently, the CMB forms a tight "V" or "horseshoe" in map view, opening towards the east [6,23]; the inside part of the "V" is considered to be the MOO realm and the outside part is the PAO one. Therefore, the Ereendavaa terrane, as well as the whole CMB, is the junction between the two orogenic systems.

3. Sample Descriptions

Samples of this study were collected from the southwestern segment of the Ereendavaa terrane, as well as the Kherlen terrane, which was likely an allochthonous unit thrust over the autochthonous Ereendavaa terrane in the Ondorkhan region. The sampling locations are shown in Figure 4, and the GPS coordinates and basic lithological features of the samples analyzed are listed in Table 1.

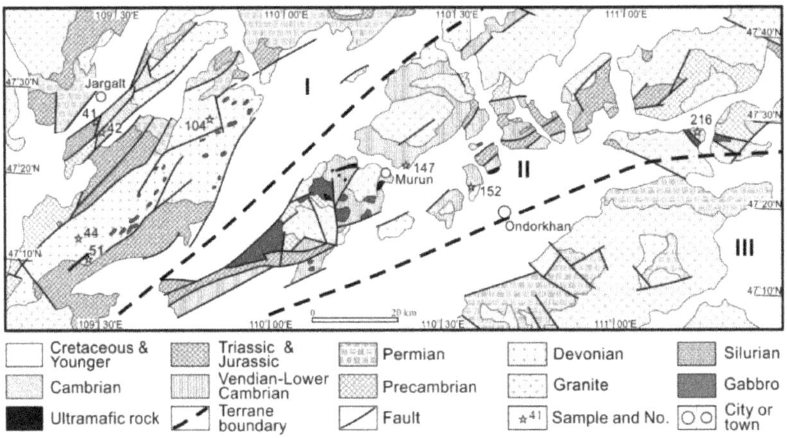

Figure 4. Geological map of the Ondorkhan area in northeastern Mongolia (modified from [36]). Terranes: I—Ereendavaa, II—Kherlen, III—Idermeg. Stars denote the localities where the dated samples were collected, and the numbers beside the stars are the last digitals of corresponding sample numbers, which have same prefix of "MOE-".

Table 1. Summary of basic petrographic features of analyzed samples from NE Mongolia.

Sample No.	GPS Coordinate	Lithology	Structure/Texture	Major Components	Stratum Age	Maximum Age [1] (Ma)
MOE-152	47.348° N 110.571° E	Quartzite	Massive to weakly foliated structure, coarse-grained,	Mainly quartz with minor muscovite and/or Fe-oxide minerals	Vendian-Cambrian [36]	1150
MOE-216	47.479° N 111.227° E	Quartzite	Foliated structure, coarse- to medium-grained texture	Mainly quartz with minor muscovite and Fe-oxide minerals	Riphean [36] or Vendian-Cambrian [40]	1200
MOE-147	47.389° N 109.376° E	Rhyolite	Magma flow structure, porphyritic and aphanitic or microcrystalline texture	Phenocrysts: quartz and K-feldspar; aphanitic or tiny felsic minerals in matrix	Vendian-Cambrian [36,40]	450 [2]
MOE-41	47.443° N 109.462° E	Sandstone	Foliated, bedding, kink and ripple fold structures porphyroclastic and medium-grained textures	Plagioclase, K-feldspar, quartz, and muscovite or sericite	Silurian [36,40]	447

Table 1. *Cont.*

Sample No.	GPS Coordinate	Lithology	Structure/Texture	Major Components	Stratum Age	Maximum Age [1] (Ma)
MOE-42	47.426° N 109.479° E	Siltstone	Foliated, kinking and ripple folding structures, fine-grained texture	Plagioclase, K-feldspar, quartz, and muscovite or sericite	Cambrian [36,40]	452
MOE-44	47.213° N 109.449° E	Arkose	Bedding structure, coarse-grained texture	Plagioclase, K-feldspar, quartz, debris, and/or sercite	Devonian [36,40]	428
MOE-51	47.178° N 109.460° E	Sandstone	Bedding structure, coarse-grained texture	Plagioclase, K-feldspar, quartz, debris, muscovite or sericite, and/or epidote	Devonian [36,40]	434
MOE-104	47.460° N 109.795° E	Sandstone	Foliated and mylonitic structures, porphyroclastic texture	Plagioclase, K-feldspar, quartz, debris, sericite, and/or epidote	Devonian [36,40]	436

[1] Maximum depositional ages inferred from the detrital zircon data of this study; [2] Crystallization age of rhyolite from an intermediate-felsic volcanic suite in the study area.

Sample MOE-152 is a quartzite sampled from an outcrop of the Precambrian basement rocks at approximately 8 km northwest of the Ondorkhan city (Figure 4). The geological setting of the quartzite is uncertain because of thick coverage of Cenozoic sediments. The quartzite is composed predominantly of quartz (>95 modal), with minor amounts of muscovite. Subcrystalline structure was locally observed (Figure 5A). This quartzite does not show clear effects of deformation, but the presence of muscovite suggests that it has experienced at least the upper greenschist facies metamorphism.

Sample MOE-216 was collected at approximately 55 km northeast of the Ondorkhan city, where Precambrian basement rocks crop out along both the northern and southern banks of the Kherlen River (Figure 4). These basement rocks were previously assigned to the Early-Middle Riphean [36] or Vendian-Early-Cambrian age [40] (note: Riphean and Vendian are two geological periods of the Geological Time Scale of Russia. The Riphean generally refers to the Mesoproterozoic-Early-Neoproterozoic period, i.e., from ca. 1800 to 680 Ma, and the Vendian to the Late Neoproterozoic from ca. 680 to 570 Ma; according this time scale, 1800 Ma and 570 Ma are the lower limit of the Mesoproterozoic and the upper limit of the Cambrian, respectively). This sample is a quartzite and has petrographic features similar to sample MOE-152 described above, except for the presence of a strong foliation defined by parallel bands of quartz, muscovite and Fe-oxide minerals (Figure 5B).

Sample MOE-147 is a rhyolite taken from a thick volcanic-volcanoclastic suite exposed to the northeast of the Murun town (Figure 4). This volcanic suite was previously considered to be Vendian-Early-Cambrian in age [36,40] and consists of mainly andesite, dacite, rhyolite, and corresponding volcanoclastic rocks. The sampled rhyolite shows magmatic foliation structure and porphyritic texture, with phenocrysts of quartz and K-feldspar in an aphanitic or microcrystalline matrix of similar compositions (Figure 5C). The purpose for selecting this rhyolite sample is to date the time of magmatic emplacement of the volcanic suite, which is crucial for understanding the evolution of both the Ereendavaa terrane and the MOO.

Samples MOE-41 and MOE-42 were taken from a metasedimentary sequence cropping out approximately 7 km south of the Jargalt Sum (town). This sequence has been strongly deformed and metamorphosed to some extent; it strikes in the northeast direction dipping to NNW (340°∠50°). Kink bands and folds were observed within this sequence in the field. This sequence overlies the Precambrian (Riphean) basement gneisses of the Ereendavaa terrane and was mapped as containing Cambrian and Silurian strata, but they are in a fault contact. Samples MOE-41 and MOE-42 show similar petrographic features, such as elongation of mineral grains, and a mineral assemblage of plagioclase, quartz, K-feldspar, and muscovite/sericite (Table 1), except for some difference in grain size, of which the latter sample is finer than the former one (Figure 5D,E). It is noted that the mineral grains in these two samples likely occur as porphyroclasts and that the debris clasts in the rocks are indiscernible due to modification by deformation and metamorphism.

Figure 5. Microphotographs (crossed polarizers) showing textures and constitutions of the dated rocks from the Ondorkhan area in NE Mongolia. Panels (**A–H**) are for samples MOE-152, -216, -147, -41, -42, -44, -51, and -104, respectively. Please refer to main text and Table 1 for their petrographic details. Abbreviation: Db—debris; Ep—epidote; Kfs—potassic feldspar; Mus—muscovite; Pl—plagioclase; Q—quartz; Ser—sericite.

Samples MOE-44, MOE-51, and MOE-104 were sampled from a thick Devonian sequence [36,40] cropping out approximately 35 km south of the Jargalt town and also extending in a northeasterly direction (Figure 4). The first two samples are coarse-grained arkose or sandstone (Figure 5F,G). Horizontal bedding and cross bedding structures were observed within the strata, which generally dip NNW, with relatively flat dip angles of 25–30°. Judging from the graded bedding, sample MOE-44 is stratigraphically higher than sample MOE-51. The last sample (MOE-104) was collected from the NE segment of the Devonian sequence, approximately 30 km east of the Jargalt town. This sample is a fine-grained sandstone and has experienced some extent of deformation as illustrated by slight elongation and orientation of the mineral grains (Figure 5H).

4. Analytical Results

Detrital or magmatic zircon grains from the above-described eight samples were analyzed using the LA-ICP-MS zircon U-Pb dating technique (Supplementary Material 1) [41–46]. The analytical data are presented in Table S1.

Sample MOE-152: This is a quartzite sampled from the basement rocks exposed at ca. 8 km northwest of the Ondorkhan city. Zircon grains from this sample are rounded or stubby in morphology, with or without oscillatory zoning (Figure 6A_1–A_6). Some zircon grains show a core-overgrowth internal structure, and the zircon cores mostly display oscillatory zoning whereas the overgrowths do not. These zircon grains have very smooth outlines that cut the magmatic oscillatory zoning, if there is any, or cut both the zircon cores and overgrowths at high angles, indicating that the detrital zircon grains have been subjected to long-distance transportation and/or long-time ablation. A total of 78 zircon grains were analyzed for U-Pb age determination and yielded apparent ages ranging from ca. 1021 to 2850 Ma (Table S1), with the most evident age peak at ca. 1426 Ma, two subordinate peaks at ca. 1115 and 1600 Ma, respectively, and three minor peaks at ca. 2090, 2660 and 2850 Ma (Figure 7A,B), respectively.

Sample MOE-216: This is a quartzite sample collected from the basement rocks exposed at ca. 55 km east of Ondorkhan city. Zircon grains of this sample display similar, if not identical, morphological and internal textural features to those from sample MOE-152 described above (Figure 6B_1–B_6). A total of 70 zircon grains of this sample were analyzed for U-Pb dating and gave an apparent age range from 1146 to 3043 Ma (Table S1) and generated an age spectrum also resembling that of sample MOE-152, except for slight differences in age range and peak age. This sample has the most predominant peak at ca. 1482 Ma, two subordinate peaks at ca. 1200 and 1722 Ma, respectively, and three minor peaks at ca. 2288, 2578 and 3040 Ma (Figure 7C; the 3040-Ma one is not shown in Figure 7C,D), respectively.

Sample MOE-147: Zircon grains from this rhyolite sample are mostly stubby prisms with well-developed oscillatory zoning (Figure 6C_1–C_5); rounded zircon grains with internal cracks were occasionally observed (Figure 6C_6). In total, 26 analyses on 26 zircon grains were selected to analyze for U-Pb age determination and gave apparent ages ranging from ca. 432 to 1452 Ma (Table S1). Of these, 24 analyses giving an age range of ca. 432–465 Ma formed a coherent group and yielded a weighted mean age of 450 ± 3 Ma (n = 24, mean square of weighted deviates (MSWD) = 1.9; Figure 8A,B). The remaining two analyses on two zircon grains that respectively have cracks inside and rounded morphology gave apparent ages of ca. 532 and 1452 Ma, respectively.

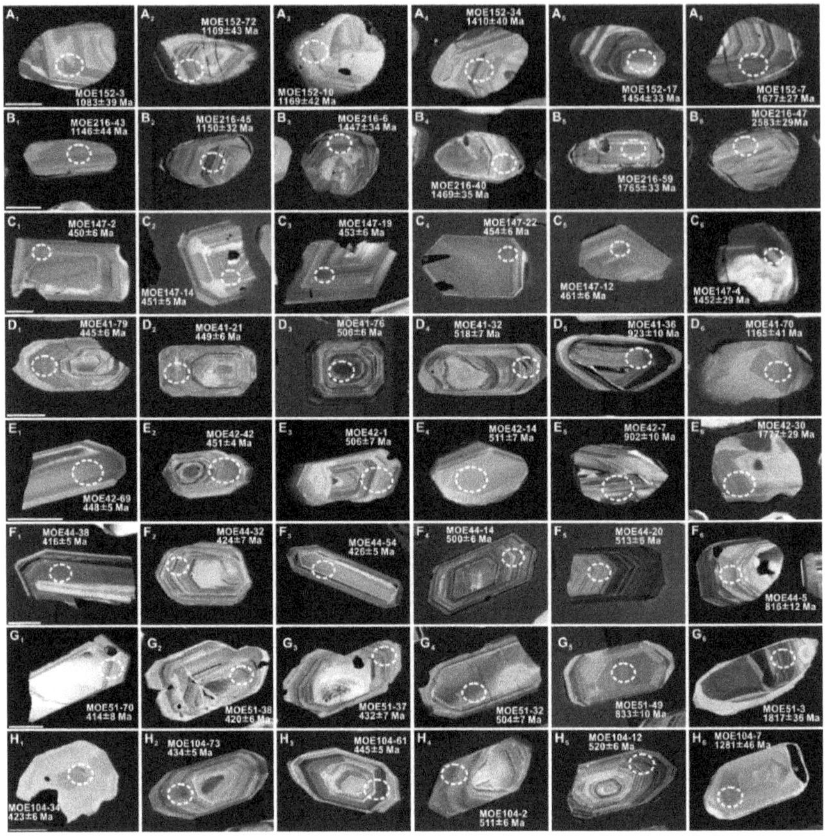

Figure 6. Representative cathode luminescence (CL) images of zircon grains of the dated samples from the Ereendavaa terrane in NE Mongolia. Also shown are approximate positions, numbers and apparent ages of the analytical spots (open circles of dashed line). Each raw image is for representative zircon grains from one sample; (**A–H**) are for samples MOE-152, MOE-216, MOE-147, MOE-41, MOE-42, MOE-44, MOE-51, and MOE-104, respectively. The analytical spot names are the same as those in Table S1. All scale bars are 50 μm long (please note that the scale bar for each panel in one raw image is the same and so it is just shown in the leftmost panel of each raw image).

Sample MOE-41: Zircon grains from this sandstone are mostly elongated to stubby prisms displaying well-developed oscillatory zoning despite their variable degrees of luminescence in the CL image, with several grains showing rounded morphology and core-overgrowth structures (Figure 6D_1–D_6). It is noteworthy that the rounded grains are essentially analogical in morphology and internal texture with those from quartzites MOE-152 and MOE-216. A total of 82 zircon grains or fragments were analyzed for U-Pb age determination and yielded apparent ages ranging from ~436 to 2924 Ma (Table S1), with the main peak at ca. 519 Ma and several minor peaks at ca. 447, 814 and 918 Ma (Figure 9A,B). In addition, this sample also contains Mesoproterozoic (ca. 1500 Ma) and Archean (ca. 2924 Ma) zircon grains. The youngest peak of ca. 447 Ma, composed of the nine youngest analyses, is identical to the weighted mean $^{206}Pb/^{238}U$ age of 447 ± 4 Ma (mean square of weighted deviates (MSWD) = 1.6; Figure 9B) of these 9 analyses.

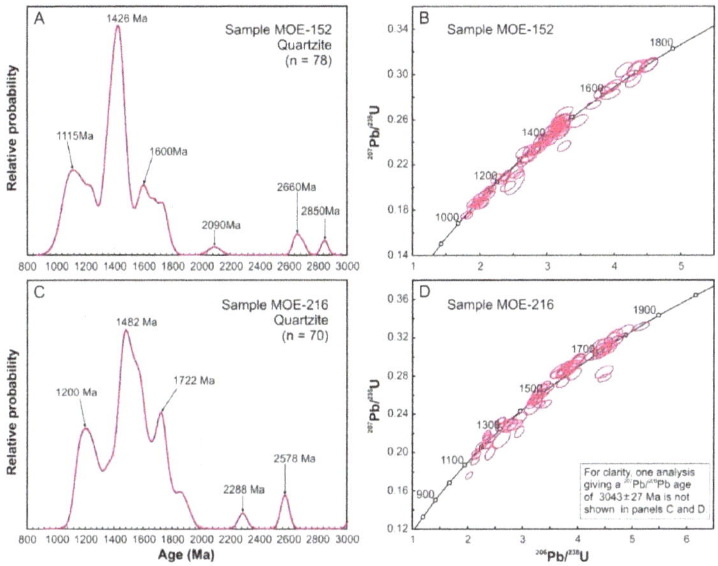

Figure 7. Detrital zircon age-relative probability curves (**left**) and concordia plots (**right**) for two quartzite samples MOE-152 (**A**,**B**) and MOE-216 (**C**,**D**) from the Ereendavaa terrane in NE Mongolia

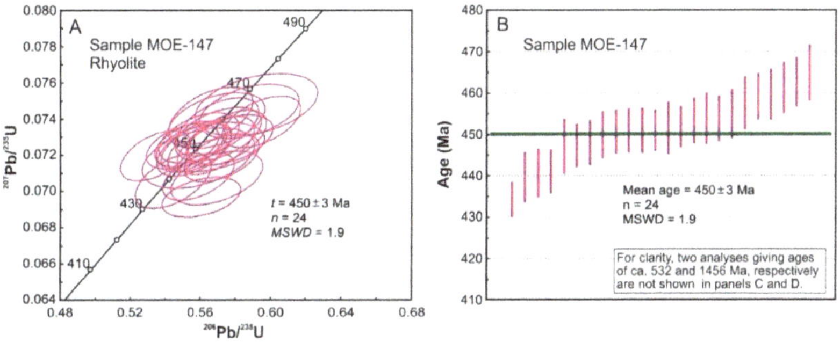

Figure 8. Zircon U-Pb data of a rhyolite sample MOE-147 from a 'Vendian-early Cambrian' volcanic suite on the Ereendavaa terrane in NE Mongolia. (**A**) Concordia plot and (**B**) Bar chart showing individual apparent ages and the weighted average (thick green line). Error ellipses in (**A**) and the error bars in (**B**) are at 1σ level.

Sample MOE-42: Zircon grains from this Cambrian siltstone are generally similar in morphology, grain size, and CL feature to those from sample MOE-41 described above (Figure 6E_1–E_6). In total, 68 zircon grains were analyzed for U-Pb age determination and yielded an age range from ~413 to 2504 Ma (Table S1), with the main peak at ca. 497 Ma (Figure 9C). Of these, there are seven zircon grains yielding a subordinate age peak at ca. 451 Ma, identical to their weighted mean ^{206}Pb/^{238}U age of 452 ± 4 Ma (n = 7, MSWD = 1.6; Figure 9D) if the analytical errors are considered. Similarly, this sample also contains Precambrian zircon grains with several small peaks of age at ca. 664, 893 and 1770 Ma (Figure 9C).

Sample MOE-44: Zircon grains of the arkose are mostly characterized by elongated to stubby prisms, with small amounts of grains showing rounded morphology (Figure 6F_1–F_6). In CL images, the former group of zircon grains displays well-developed oscillatory zoning (Figure 6F_1–F_5) whereas the rounded grains show relatively strong luminescence, likely a result of low concentrations of Th and U, with or without oscillatory zoning (Figure 6F$_6$). A total of 69 zircon grains were analyzed for U-Pb dating and yielded apparent ages ranging from ca. 412 to 2851 Ma (Table S1), with the most obvious age peak at ca. 496 Ma and several subordinate peaks at ca. 428, 591, 793, 960 and 1785 Ma (Figure 10A). Moreover, this sample also contains one Mesoproterozoic (ca. 1451 Ma) and one Archean (ca. 2851 Ma) zircon grains. The youngest peak age of ca. 428 Ma, which contains 15 youngest analyses, is consistent with the weighted mean $^{206}Pb/^{238}U$ age of 425 ± 5 Ma (n = 13, MSWD = 1.9; Figure 10B) of the 13 youngest analyses if the analytical errors are considered.

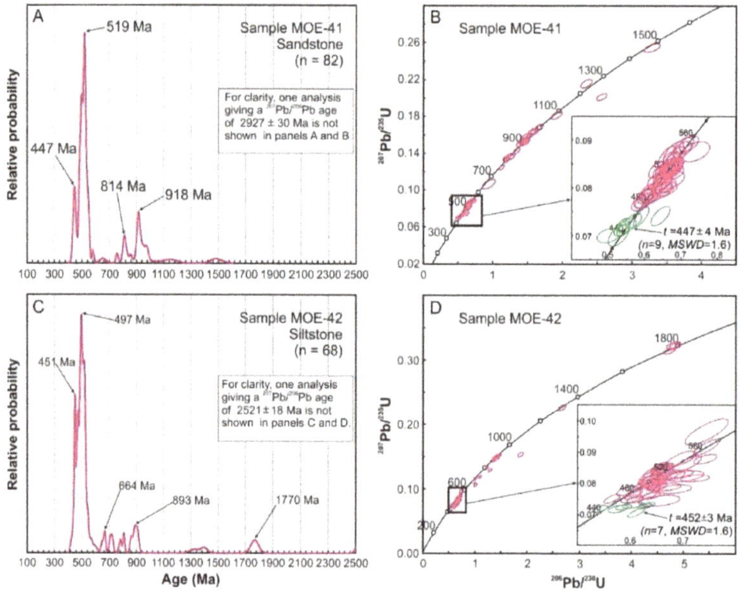

Figure 9. Detrital zircon age-relative probability diagrams (**left**) and concordia diagrams (**right**) for samples MOE-41 (**A**,**B**) and MOE-42 (**C**,**D**) from the Paleozoic sedimentary rocks on the Ereendavaa terrane in NE Mongolia.

Sample MOE-51: The sandstone contains zircon grains that are mostly characterized by elongated to stubby prisms with magmatic oscillatory zoning, and minor zircon grains are rounded in morphology without or with weak oscillatory zoning (Figure 6G$_1$–G$_6$). A total of 72 analyses were carried out on 72 zircons of this sample and generated an apparent age range from 408 to 2447 Ma (Table S1). They form an age spectrum resembling that of sample MOE-44, with the most evident peak at ca. 500 Ma and subordinate peaks at ca. 434, 830, 969 and 1825 Ma (Figure 10C). The youngest 12 analyses that constitute the youngest peak at ca. 434 Ma yielded a weighted mean $^{206}Pb/^{238}U$ age of 428 ± 6 Ma (n = 12, MSWD = 2.4; Figure 10D); these two ages, within analytical errors, overlap each other.

Sample MOE-104: Zircon grains from the sandstone are similar in morphology to those from samples MOE-44 and MOE-51 although the quantity of rounded zircon grains in this sample seem to be greater than in the other two samples (Figure 6H$_1$–H$_6$). A total of 76 analyses were done on 76 zircon grains and gave apparent ages ranging from 423 to 2415 Ma (Table S1), with the main age peak at ca, 510 Ma and subordinate peaks at ca. 436, 816, 897, 943, 1236 and 1830 Ma (Figure 10E). Apart from these, another three Neoproterozoic (588, 654 and 724 Ma) and three Paleoproterozoic (2087, 2107 and 2415 Ma) zircon grains were detected in this sample. The three analyses on the three youngest zircon grains of this sample gave a weighted mean $^{206}Pb/^{238}U$ age of 435±6 Ma (MSWD = 3.6; Figure 10F), in line with the youngest peak age of ca. 436 Ma.

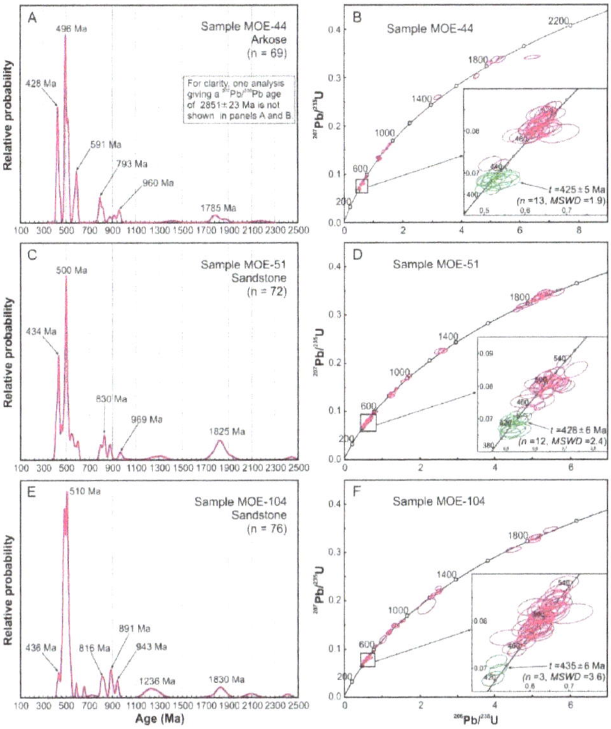

Figure 10. Detrital zircon age-relative probability curves (**left**) and concordia diagrams (**right**) for samples MOE-44 (**A**,**B**), MOE-51 (**C**,**D**), and MOE-104 (**E**,**F**) from Devonian sedimentary sequence on the Ereendavaa terrane in NE Mongolia.

5. Discussion

5.1. Deposition/Formation Time of the Dated Strata

The two quartzite samples (MOE-152 and -216) were collected from metamorphic strata in the Ereendavaa terrane exposed at two different places. Due to the lack of reliable paleontological and isotopic dating data, these metamorphic strata/rocks were previously assigned to a wide range of the protolith ages from the Early Riphean to the Middle Riphean [36], or from the Vendian to the early Cambrian [40], with the metamorphic age(s) unknown. According to our results, these two quartzite samples show similar age data of ca. 1021–2850 Ma and 1146–3043 Ma, with the youngest peak ages of ca. 1115 and 1200 Ma, respectively. From the CL images, it is certain that these ages represent the crystallization or metamorphic ages of the zircon grains or zircon domains themselves, rather than the metamorphic ages of the dated rocks. This is because, on the one hand, that the zircon grains or zircon domains giving these youngest ages mostly display very fine oscillatory zoning (Figure 6A_1,A_2), clearly suggesting a magmatic, not metamorphic, origin of the zircon grains. On the other hand, even if few analyzed zircon domains occur as metamorphic overgrowths without distinct zoning (Figure 6A_3), the smooth outline boundaries of the zircon grains cut the overgrowths and/or the zircon cores at high angles, indicating that the metamorphic overgrowths had already formed before the protolith sedimentation of the quartzites. Therefore, we conclude that the youngest peak ages of ca. 1115 and 1200 Ma of the two quartzite samples represent the maximum deposition ages of the protoliths of the sedimentary sequences from which the two samples came.

Though these two samples were taken from two localities, they are not far from each other, approximately 60 km along the strike of regional tectonic units (Figures 3 and 4). Considering the similarities of the lithology and detrital zircon age spectra, we suggest that the two quartzites belong to a single sedimentary sequence that was deposited after ca. 1200–1115 Ma. In this case, it is rational to deduce that the sequence was deposited during the latest Middle Riphean to early Late Riphean, rather than during the Vendian to early Cambrian as suggested by [40]. The reason for this is that the samples would contain some early Neoproterozoic (1000–680 Ma) detrital zircon grains if the sequence was formed during the Vendian to early Cambrian. Instead, zircon grains with early Neoproterozoic ages are relatively abundant in the samples from the Paleozoic sedimentary sequences (Figures 9 and 10).

Sample MOE-147 was taken from a thick volcanic sequence cropping approximately 20 km north of the Ondorkhan city (Figure 4), which is composed of basaltic andesites, andesites, dacites, rhyolites, and their volcanic agglomerates, breccias and pyroclastic rocks. This volcanic sequence was previously designated a Vendian to early Cambrian age [36,40]. However, the rhyolite sample yielded a zircon U-Pb age of 450 ± 3 Ma (Figure 8). This age is interpreted to date the magmatic emplacement of the volcanic rocks because it was obtained from zircon crystals that show typical magmatic oscillatory zoning (Figure 6C_1–C_5). This demonstrates that the volcanic sequence was formed in the Late Ordovician.

The two samples of MOE-41 and MOE-42 were collected from an early Paleozoic sedimentary sequence. The former sample (MOE-41), taken from the Silurian stratum, gave the youngest age peak at ca. 447, consistent with the weighted mean age of 447 ± 4 Ma of the youngest analyses (Figure 9B). These ages were obtained from the elongated zircon crystals with the well-developed oscillatory zoning (Figure 6D_1,D_2), and therefore the 447-Ma age constrains the maximum depositional age of the sedimentary strata from which this sample was collected. Consequently, deposition of the sampled rocks must be later than the Ordovician. The latter sample (MOE-42), from the Cambrian strata, contains 7 zircon grains that yielded the youngest age peak of ca. 451 Ma and a weighted mean age of 452 ± 3 Ma (Figure 9D), suggesting that this rock was formed after ca. 451 Ma. This proves that the sedimentary protolith is not Cambrian, but must be younger than the Late Ordovician. The age of 452 ± 3 of sample MOE-42 overlaps, within the analytical errors, with the age 447 ± 4 Ma of sample MOE-41, and thus this whole early Paleozoic sedimentary sequence was probably deposited after the Ordovician, most likely during the Silurian.

The three sandstone samples (MOE-44, -51 and -104), which were all collected from the Devonian sequence, gave the three youngest age peaks of ca. 428, 434, and 436 Ma and three weighted mean $^{206}Pb/^{238}U$ ages of 425 ± 5, 428 ± 6, and 435 ± 6 Ma (Figure 10), respectively. These ages overlap each other if the analytical errors are considered. Similarly, these youngest zircon grains are all euhedral prisms showing well-developed oscillatory zoning (Figure 6F_1–F_3, G_1–G_3 and H_1–H_3), and, therefore, we interpret them to represent the maximum ages of the sandstones. Although a maximum depositional age of late Silurian could be inferred from the dating data (e.g., the youngest age peak of ca. 437 Ma), we are inclined to interpret the youngest age of ca. 425 Ma as the lower limit of sedimentation time, which means that the deposition of the sequence occurred after the Gorstian, because this interpretation is in line with the fossil data of the sequence that suggests an Early–Middle Devonian stratigraphic age [36,40].

In summary, our new dating data provide direct evidence for the deposition/formation ages of Neoproterozoic metasedimentary rocks and Paleozoic sedimentary and volcanic successions overlying the Ereendavaa terrane in NE Mongolia. This provides robust constraints on the evolution of both the Ereendavaa itself and the MOO.

5.2. Constraints on the Evolution of the Ereendavaa Terrane

The Ereendavaa terrane, which forms the southern foreland of the MOB, witnessed the whole evolution of the MOO. This terrane was considered a cratonic block with basement rocks as old as the Paleoproterozoic or even the Archean [13,36,40]. However, recent SHRIMP zircon U-Pb dating showed that the so-called oldest basement rocks, such as orthogneisses, amphibolites, and schists, are actually Paleozoic and/or even Early Mesozoic in age [26]. The quartzite assemblage, the protolith deposition of which is constrained by this study to occur after 1200–1115 Ma, most likely during the latest Mesoproterozoic to early Neoproterozoic period, seems to be the oldest rock identified so far by radiogenic isotope dating. Importantly, the lithology of quartzite, derived from a quartz sandstone, suggests deposition in a passive continental margin setting. This corollary is supported by the morphology and internal texture of detrital zircon grains from the quartzites, which indicate a long-distance transport and long-term sorting of detritus. The main age peaks of ca. 1426 and 1482 Ma, within an age range of ca. 1.1–1.8 Ga, of the two quartzite samples suggest that the source area is mainly Mesoproterozoic in age. Several Paleoproterozoic and even Archean zircon grains detected in the quartzite indicate that either rocks with these ages also exist in the source area or that these zircon grains are recycled during the sedimentation. In terms of the paleogeography, the (proto) Ereendavaa terrane was probably located at a passive margin of Rodinia during the early Neoproterozoic.

All five samples from the Paleozoic sedimentary rocks have similar age spectra that show predominant age peaks between ca. 496 Ma and 520 Ma, with each exhibiting several subordinate Neoproterozoic age peaks between ca. 591 Ma and 970 Ma. This illustrates that these sedimentary rocks have similar provenance from a source area that supplied detritus of the middle Cambrian and Neoproterozoic age. The zircon grains contributing to the main peaks are euhedral prisms with well-developed oscillatory zoning (Figure 6), suggesting a proximal provenance. Cambrian intrusive rocks ranging in composition from gabbro to granite are widespread within the Ereendavaa terrane, which most likely represent products of Cambrian arc magmatism [13,25,26]. Consequently, it is possible that the zircon grains with ages belonging to the main peaks in the age spectrum were derived from the Cambrian magmatic arc. Similarly, the zircon grains making up several minor Neoproterozoic age peaks might have been derived from older igneous rocks although no such arc rocks have been proven so far. The volcanic sequence that was previously assigned to the Vendian–early-Cambrian was actually emplaced during the Ordovician, as evidenced by the crystallization ages of ca. 450 Ma (Figure 8) and 460–455 Ma [47] of the rhyolites, suggesting the presence of Ordovician magmatism on the Ereendavaa terrane. The lithological assemblage of andesite–dacite–rhyolite as well as their equivalent volcanoclastic rocks, the geochemical features, and zircon εHf(t) values (+0.1 to +4.07) [47] demonstrate that the Ordovician igneous rocks formed at an Andean-type margin at that time. Therefore, we suggest that the Ereendavaa terrane became an active continental margin during the Cambrian-Ordovician. There are some occurrences of early Silurian intrusions (ca. 430–440 Ma [25] and our unpublished data) within the Ereendavaa terrane, which are characterized by the presence of K-rich granites and by rounded shapes in map view. These features and the observation that these early Silurian granites intrude the Cambrian Kherlen ophiolite jointly indicate a postcollisional origin for these early Silurian granites, related to the collision between the Ereendavaa and Idermeg terranes [25]. Consequently, arc magmatism might have been active within the Ereendavaa terrane at least from the Cambrian to Late Ordovician and the postcollisional magmatism mainly occurred during the early Silurian. The zircon grains belonging to the youngest age peaks of the three Devonian samples (ca. 428, 434 and 436 Ma; Figure 10) were probably derived from the early Silurian postcollisional granites.

It is evident from the detrital-zircon age spectra of the Paleozoic sedimentary rocks (Figures 9 and 10) that a gap exists between ca. 600 Ma and 780 Ma, despite the subordinate Neoproterozoic age peaks dispersed between ca. 591 Ma and 970 Ma. The scarcity of zircon grains with ages of ca. 600–780 Ma in the source of detritus demonstrates that the middle to early late-Neoproterozoic magmatism was inactive, in contrast to the latest Neoproterozoic (ca. 600–540 Ma) and early Neoproterozoic (780–970 Ma). We interpret the ca. 600–540 Ma magmatism as an early phase of a Neoproterozoic-Cambrian-Ordovician magmatic arc. There are two possible explanations for the ca. 780–970 Ma magmatism. One possibility is a supra-subduction setting and the alternative scenario assumes a rifted continental margin. We tentatively prefer the second explanation because of the possible relationship to the separation of the Idermeg terrane from the Ereendavaa one. Summing up, our data suggest that the Ereendavaa terrane evolved into an active continental margin at least since the latest Neoproterozoic (ca. 600 Ma) until the Late Ordovician (ca. 450 Ma), and a collision to postcollisional tectonic setting was established during the Late Ordovician-early Silurian.

If the Ereendavaa terrane alone or together with the Idermeg terrane was part of the ribbon CMB, a question arises concerning the polarity of the Neoproterozoic-Ordovician subduction beneath the Ereendavaa terrane. Based on available data, we believe that the solution is northward subduction of the PAO lithosphere at the southern side of the Ereendavaa terrane (present coordinates). The justification for this is two-fold. On the one hand, the Cambrian Kherlen ophiolite, which was previously regarded as a volcanic arc terrane [13,15], occurs to the south of the Ereendavaa terrane and is interpreted to represent the suture between the Ereendavaa and Idermeg terranes [25]. The suturing time was constrained before ca. 442 Ma [25], slightly younger than the early Paleozoic arc magmatism. On the other hand, the Adaatsag-Dochgol terrane, the suture zone of the MOO, occurs to the north of the Ereendavaa terrane but the oldest ophiolite identified so far in this suture is Upper Mississippian in age, ca. 325 Ma [15,29], i.e., considerably younger than the early Paleozoic arc magmatism. Previous studies have documented that SSZ-type ophiolites in the Neo-Tethys belts normally formed during the subduction initiation [48–52]. The ca. 325-Ma ophiolites in the MOO suture zone display geochemical features of SSZ-type ophiolites, and thus the southward subduction of the MOO lithosphere probably started at ca. 325 Ma [29]. The Onon arc within the Russian segment of the suture zone, which was considered as an intra-ocean arc related to the southward subduction of the MOO, was proposed to be the oldest arc within the MOO and was presumed to be Late Devonian to Early Carboniferous in age [13]. Therefore, we propose that the Neoproterozoic to early Paleozoic arc magmatism was generated by the northward subduction of the PAO lithosphere beneath the Ereendavaa terrane.

The event that was associated with the tectonic emplacement of the Kherlen ophiolite at the end of the Ordovician (ca. 450–442 Ma) is likely a record of the amalgamation of the Siberian southern margin, resulting in the early Paleozoic domain of northern Mongolia.

5.3. Implications for the Evolution of the MOO

With respect to the evolution of the MOO, two contrasting models have been suggested. One model is that the MOO was a long-lived feature between ~620 Ma and 200 Ma, as a new ocean basin [6,53] or one evolving into a remnant ocean basin in the late Early-Paleozoic [4]. The other scenario is that the MOO opened later as a back-arc basin during the Silurian [11,13,54] or the Carboniferous [23], or as a new basin related to a mantle plume/hot-spot [22]. According to our new data and those previously published, we favor the second possibility, that is, the MOO was opened through break-up of the early Paleozoic basement. Supporting evidence for this is summarized below.

First, the detrital zircon age spectra of our samples from the Silurian-Devonian sequences (some were erroneously assigned to be Cambrian in age as evidenced by sample MOE-42) at the southern margin of the MOO, namely the Ereendavaa terrane, resemble those of the second group of samples of [23], which were taken from the "Cambrian-Silurian" sequences of the Haraa terrane at the northern margin of the MOO. That is, they all comprise Cambrian main age peaks centered at ca. 510 Ma, except for one sample of [23] with the main age peak at ca. 605 Ma, and several subordinate Neoproterozoic age peaks as well as early Paleoproterozoic and even Archean ones. This similarity implies that these sedimentary sequences located on the opposite margins might have been connected prior to the MOO opening. This corollary is in line with the observation that the source areas of the overlying Devonian-Carboniferous sequence and the underlying "Cambrian-Silurian" one at the northern margin were two different terranes, rather than one single terrane, the age distribution of which changed with time [23].

Second, the Silurian-Devonian samples from the Ereendavaa terrane contain no detrital-zircon grains of similar stratigraphic age, which is contrasting to the previously dated post-Devonian sedimentary samples that show youngest age peaks corresponding to their stratigraphic ages [11,22]. This is because of occurrence of contemporaneous volcanic arcs on both margins during the

Carboniferous and Triassic periods [13,15]. The dominance of the Cambrian-Ordovician zircon grains in the Silurian-Devonian samples suggests that their age spectra are inherited from reworked, older Neoproterozoic to early Paleozoic magmatic arc complexes [13,47]. It is noted that the authors of [11,22,55] reported a dominance of slightly negative εHf(t) values for Cambrian-Ordovician detrital zircons from the late Paleozoic and younger sedimentary rocks on the Ereendavaa terrane, which are different from the slightly positive εHf(t) values (+0.1–+4.07) of zircons from the Ordovician arc rhyolites [47]. This discrepancy could be explained in terms of different extents of mixing by crust- and mantle-derived melts for the magmas that formed the source rock of the detrital zircons and the arc rhyolites, respectively. This means that the magma forming the source rock of the detrital zircons contains more ancient crustal material than that of the rhyolites. Anyway, the age spectra of the Silurian-Devonian sedimentary rocks indicate no contemporaneous volcanic input during the Silurian-Devonian deposition on the southern margin of the MOO. This implies that extension and basement erosion likely prevailed at the southern margin of the MOO at that time. This is corroborated by the fact that swarms of doleritic sills intruding the Devonian sedimentary sequence in the Ereendavaa terrane (Figures 3 and 4) are indicative of an extensional regime. We presume that the extensional event was linked to the opening of the MOO, probably occurring at its rifted passive margin.

Third, the MOO suture physically dissects the Paleozoic domain that was amalgamated during the early Silurian [8,26,56] in central-NE Mongolia, as well as the Russian Transbaikal and northern Chinese Greater Xing'an regions (Figure 1). The Ereendavaa and Idermeg terrane basement blocks are welded by the early Paleozoic Kherlen suture in the southeast and the Precambrian continental blocks and orogenic collage comprising Neoproterozoic-early Paleozoic island arcs, ophiolites, and accretionary complexes extending in the northwest (Figure 2) [13]. These features demonstrate that the MOO developed later on the amalgamated early Paleozoic domain, which is further supported by the observation that the MOO lithosphere subduction-related igneous rocks penetrate and/or cover the adjacent blocks in both the north and south of the MOO suture [4,11,13,15]. These igneous rocks, respectively represented by the Selenge and Middle Gobi volcanic–plutonic belts in the north and south of the suture (Figure 2), are predominantly Permian-Triassic age [13,57,58]. Additionally, porphyry-type Cu-Mo mineralization related to the subduction of the MOO lithosphere was mainly formed during Permian-Triassic times [2,31,59–61].

Finally, no ophiolite that is older than the Carboniferous has been identified so far in the MOO suture zone (the Adaatsag and Dochgol terranes), with the oldest one having an age of ca. 325 Ma [15,29], similarly implying that the MOO opened relatively late. The authors of [17] reported deep-water radiolarian cherts of upper Silurian-Devonian age at the southeastern edge of the Hangay-Hentey belt, but the meta-sandstones from the same sequence at nearly the same location as that of the radiolarian chert samples have the maximum deposition age of ca. 437–340 Ma [11,22,23]. We interpret the cherts to be coeval with the ocean formation and later to be integrated into an accretionary complex.

Regarding the geodynamic setting of the MOO opening, we prefer the mantle plume origin [22] to the back-arc extensional one [11,13]. This is mainly owing to the recent recognition by [22] of rock packages characteristic of a drowned seamount within the accretionary complex of the MOB, which is robust evidence for the MOO opening above a mantle plume or hot-spot.

Based on the above discussion, we synthesize the tectonic evolution of the Ereendavaa terrane and the MOO as follows (Figure 11):

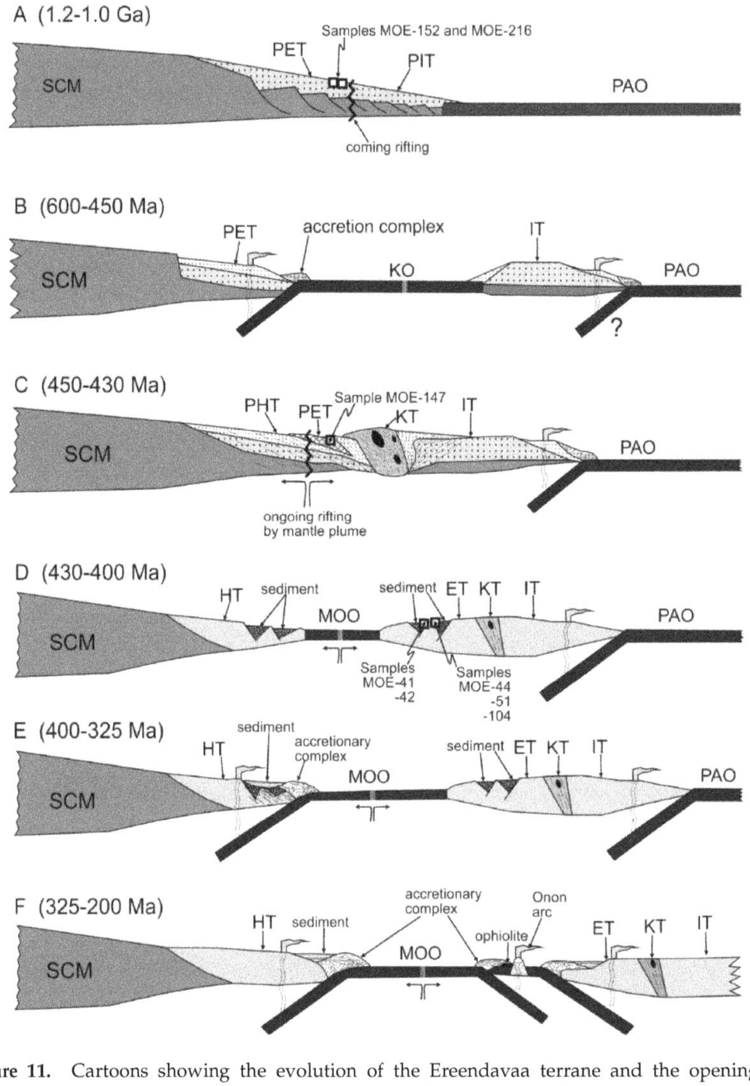

Figure 11. Cartoons showing the evolution of the Ereendavaa terrane and the opening and subduction stages of the Mongol-Okhotsk Ocean. The Mongol–Okhotsk ocean is suggested to have opened within the earlier amalgamated early Paleozoic domain in Mongolia (see text for details). ET—Ereendavaa terrane; HT—Haraa terrane; IT—Idermeg terrane; KT—Kherlen terrane; KO—Kherlen Ocean; MOO—Mongol-Okhotsk Ocean; PAO—Paleo Asian Ocean; PET—proto Ereendavaa terrane; PHT—proto Haraa terrane; PIT—proto Idermeg terrane; SCM—Siberian continental margin.

During the late Mesoproterozoic, mostly likely between ca. 1.2–1.0 Ga, the protolith of the quartzite sequence, belonging to the Ereendavaa basement (proto Ereendavaa terrane (PET) in Figure 11) was deposited on a passive continental margin, with a detritus source of mainly Mesoproterozoic age, though minor contributions from Paleoproterozoic and Archean rocks were also possible. We presume that the passive continental margin corresponded to the southern margin (present coordinates) of the early Precambrian Siberian block, one of the outer blocks of the Rodinian supercontinent at that time. The proto-Idermeg terrane (PIT in Figure 11), which has a passive continental margin origin [13], is presumed to form simultaneously, and link, with the PET (Figure 11A). During the early Neoproterozoic period (ca. 978–780 Ma), the passive continental margin seemed to experience extension and rifting, with the PIT drifting away from the margin of the PET, leading to the formation of the Kherlen Ocean (KO), and then during the late Neoproterozoic-Ordovician period (ca. 600–450 Ma) the northward subduction of the KO lithosphere generated the Andean-type continental margin on the PET (Figure 11B). At the same time, the northward subduction the PAO lithosphere beneath the Idermeg terrane (IT) was likely ongoing.

During the Late Ordovician to early Silurian (ca. 450–440 Ma), the KO was closed, leading to the collision of the IT with the PET, with the Kherlen suture between them (Figure 11C). This collisional event likely recorded the final amalgamation of the early Paleozoic domain of Mongolia, which was later split by the MOO. Postcollisional extension and magmatism dominated the Ereendavaa terrane during the early Silurian (ca. 440–430 Ma). Meanwhile, extension and rifting triggered by a mantle plume were likely ongoing during this period [22] (Figure 11C).

During the late Silurian to Early Devonian time (ca. 430–400 Ma), the amalgamated early Paleozoic domain was rifted, and break-up of the MOO occurred. Accompanying the extension and the MOO opening, the Silurian-Devonian sedimentary sequences formed on both the southern and northern margins of the MOO (Figure 11D). They presently exist on the northern edge of the Ereendavaa terrane and the southern edge of the Haraa terrane.

During the Middle Devonian-early Carboniferous (ca. 400–325 Ma), the northern margin of the MOO evolved from the rifted passive margin into an active one, whereas the southern margin remained passive (Figure 11E). This is based on the fact that the Devonian-Carboniferous sedimentary rocks to the north of the MOO suture zone contain numerous detrital zircon grains of Middle Devonian age [11,22,23] but contemporaneous sedimentary rocks within and to the south of the suture zone do not [11], indicating Devonian arc volcanism occurring only at the northern margin of the MOO.

At approximately the end of the early Carboniferous (ca. 325 Ma), southward subduction of the MOO lithosphere was initiated, forming the Adaatsag and Huhu Davaa ophiolites [15,29], and the MOO had evolved into a bidirection convergent system since then (Figure 11F). Two subduction zones might have existed at the southern margin of the MOO during the early stage of this period, which correspondingly generated two arcs: one intra-ocean arc (the Onon arc) [4,13] and one continental arc. This bidirectional convergent system might had continued until the end of the Triassic or even the Early Jurassic, resulting in the Selenge and Middle Gobi volcano–plutonic belts of mainly the Permian-Triassic age, which occur to the north and south of the Mongol-Okhotsk belt, respectively.

6. Conclusions

Based on our detrital zircon U-Pb results combined with published data, the following conclusions can be reached:

(1) Protoliths of the quartzite assemblage as part of the basement of the Ereendavaa terrane was deposited after ca. 1.2–1.15 Ga, most likely during the Late Mesoproterozoic (1.2–1.0 Ga), in a passive continental margin setting.
(2) The thick intermediate-felsic volcanic sequence to the north of the Ondorkhan was formed during the Late Ordovician (ca. 450 Ma), rather than during the Vendian-early Cambrian as previously suggested. This volcanic sequence is interpreted as part of a Late-Neoproterozoic-Ordovician arc formation resulting from the northward (present coordinates) subduction of the Kherlen Ocean lithosphere beneath the proto Ereendavaa terrane.
(3) The so-called Cambrian strata dated by this study, which yielded the youngest detrital zircon age peak at ca. 451 Ma that is similar to that of the nearby Silurian sequence, was deposited during the Silurian. All these strata and the nearby Devonian sedimentary sequence have a similar detrital zircon age spectrum with a maximum ca. 497–519 Ma, suggesting early Paleozoic arc provenance.
(4) The Mongol-Okhotsk Ocean split the amalgamated early Paleozoic domain above a mantle plume after the early Silurian and developed an Andean-type continental margin along its

northern margin during the Devonian and a bidirectional subduction system at ca. 325 Ma. This bidirectional subduction system might have lasted at least until the Triassic.

Supplementary Materials: The following are available online at http://www.mdpi.com/2075-163X/10/9/742/s1, Table S1: U-Pb analytical results of detrital zircon from the Neoproterozoic-Paleozoic sedimentary sequences in the Ereendavaa terrane in NE Mongolia, and Supplementary Material 1: LA-ICP-MS zircon U-Pb dating method.

Author Contributions: Conceptualization, M.B.; Data curation, C.L.; Formal analysis, X.L.; Investigation, M.Z., C.A. and S.Y.; Writing—original draft, L.M. All authors have read and agreed to the published version of the manuscript.

Funding: This research was funded by the National Natural Science Foundation of China (grant number 41772230), the Chinese Academy of Sciences (grant number GJHZ1805), and the National Key R&D Program of China (grant number 2017YFC0601306).

Acknowledgments: We thank Sanchir Dorjgochoo at the Institute of Geology, Mongolian Academy of Sciences, for providing some new geological material concerning the work area in Mongolia. K. Togtokh is thanked for her logistic assistance and sample treatment during the field expedition in Mongolia. The Gaonianlinghang company is acknowledged for the quick and high-quality preparation of zircon mounts analyzed. Help and technical support from Bei Xu and Yanjie Zhang during the LA-ICP-MS zircon U-Pb analyses at the Key Laboratory of Regional Geology and Mineralization, Hebei GEO University is highly appreciated. We give our special thanks to the two reviewers, and Wilfried Winkler, the Academic Editor, for their constructive suggestions and valuable comments.

Conflicts of Interest: The authors declare no conflict of interest. The funders had no role in the design of the study; in the collection, analyses, or interpretation of data; in the writing of the manuscript, or in the decision to publish the results.

References

1. Natal'in, B. History and modes of mesozoic accretion in southeastern russia. *Isl. Arc* **1993**, *2*, 15–34. [CrossRef]
2. Parfenov, L.M.; Popeko, L.I.; Tomurtogoo, O. Problems of tectonics of the Mongol–Okhotsk orogenic belt. *Geol. Pac. Ocean* **2001**, *16*, 797–830.
3. Zonenshain, L.P.; Kuzmin, M.I.; Natapov, L.M. Geology of the USSR: A plate tectonic synthesis. *Geodynamics* **1990**, *21*, 97–120.
4. Zorin, Y.A. Geodynamics of the western part of the Mongol–Okhotsk collisional belt, Trans-Baikal region (Russia) and Mongolia. *Tectonophysics* **1999**, *306*, 33–56. [CrossRef]
5. Sengör, A.M.C.; Natal'in, B.A.; Burtman, V.S. Evolution of the Altaid tectonic collage and Palaeozoic crustal growth in Eurasia. *Nature* **1993**, *364*, 299–306. [CrossRef]
6. Sengör, A.M.C.; Natal'in, B.A. Paleotectonics of Asia: Fragments of synthesis. In *The Tectonic Evolution of Asia*; Yin, A., Harrison, T.M., Eds.; Cambridge University Press: New York, NY, USA, 1996; pp. 486–640.
7. Jahn, B.-M.; Wu, F.; Chen, B. Granitoids of the central Asian orogen and continental growth in the phanerozoic. *Trans. R. Soc. Edinb. Earth Sci.* **2000**, *91*, 181–193.
8. Windley, B.F.; Alexeiev, D.; Xiao, W.; Kröner, A.; Badarch, G. Tectonic models for accretion of the central Asian orogenic belt. *J. Geol. Soc. Lond.* **2007**, *164*, 31–47. [CrossRef]
9. Khain, E.V.; Bibikova, E.V.; Kröner, A.; Zhuravlev, D.Z.; Sklyarov, E.V.; Fedotova, A.A.; Kravchenko-Bereznoy, I.R. The most ancient ophiolites of the central Asian fold belt: U–Pb, and Pb–Pb zircon ages for the dunzhugur complex, Eastern Sayan, Siberia, and geodynamic implications. *Earth Planet. Sci. Lett.* **2002**, *199*, 311–325. [CrossRef]
10. Xiao, W.; Windley, B.F.; Hao, J.; Zhai, M. Accretion leading to collision and the Permian suture, Inner Mongolia, China: Termination of the central Asian orogenic belt. *Tectonics* **2003**, *22*, 1069. [CrossRef]
11. Bussien, D.; Gombojav, N.; Winkler, W.; Quadt, A. The Mongol–Okhotsk belt in Mongolia—An appraisal of the geodynamic development by the study of sandstone provenance and detrital zircons. *Tectonophysics* **2011**, *510*, 132–150. [CrossRef]
12. Zorin, Y.A.; Belichenko, V.G.; Turutanov, E.K.; Kozhevnikov, V.M.; Ruzhentsev, S.V.; Dergunov, A.B.; Filippova, I.B.; Tomurtogoo, O.; Arvisbaatar, N.; Bayasgalan, T.; et al. The South Siberia–Central Mongolian transect. *Tectonophysics* **1993**, *225*, 361–378. [CrossRef]
13. Badarch, G.; Cunningham, W.D.; Windley, B.F. A new terrane subdivision for Mongolia: Implications for the Phanerozoic crustal growth of central Asia. *J. Asian Earth Sci.* **2002**, *21*, 87–110. [CrossRef]

14. Kravchinsky, V.A.; Cogné, J.P.; Harbert, W.P.; Kuzmin, M.I. Evolution of the Mongol-Okhotsk ocean as constrained by new paleomagnetic data from the Mongol-Okhotsk suture zone, Siberia. *Geophys. J. Int.* **2002**, *148*, 34–57. [CrossRef]
15. Tomurtogoo, O.; Windley, B.F.; Kröner, A.; Badarch, G.; Liu, D.Y. Zircon age and occurrence of the Adaatsag ophiolite and Muron shear zone, central Mongolia: Constraints on the evolution of the Mongol–Okhotsk ocean, suture and orogen. *J. Geol. Soc. Lond.* **2005**, *162*, 125–134. [CrossRef]
16. Sorokin, A.A.; Zaika, V.A.; Kovach, V.P.; Kotov, A.B.; Xu, E.; Yang, H. Timing of closure of the eastern Mongol–Okhotsk Ocean: Constraints from U–Pb and Hf isotopic data of detrital zircons from metasediments along the Dzhagdy Transect. *Gondwana Res.* **2020**, *81*, 58–78. [CrossRef]
17. Kurihara, T.; Tsukada, K.; Otoh, S.; Kashiwagi, K.; Chuluun, M.; Byambadash, D.; Boijir, B.; Gonchigdorj, S.; Nuramkhan, M.; Niwa, M.; et al. Upper Silurian and Devonian pelagic deep-water radiolarian chert from the Khangai–Khentei belt of central Mongolia: Evidence for middle Paleozoic subduction accretion activity in the central Asian orogenic belt. *J. Asian Earth Sci.* **2008**, *34*, 209–225. [CrossRef]
18. Zhao, X.; Coe, R.S.; Zhou, Y.; Wu, H.; Wang, J. New palaeomagnetic results from northern China: Collision and suturing with Siberia and Kazakhstan. *Tectonophysics* **1990**, *14*, 43–81.
19. Carrapa, B. Resolving tectonic problems by dating detrital minerals. *Geology* **2010**, *38*, 191–192. [CrossRef]
20. Lawton, T.F.; Hunt, G.J.; Gehrels, G.E. Detrital zircon record of thrust belt unroofing in Lower Cretaceous synorogenic conglomerates, central Utah. *Geology* **2010**, *38*, 463–466. [CrossRef]
21. Gehrels, G. Detrital zircon U-Pb geochronology applied to tectonics. *Annu. Rev. Earth Planet. Sci.* **2014**, *42*, 127–149. [CrossRef]
22. Ruppen, D.; Knaf, A.; Bussien, D.; Winkler, W.; Chimedtseren, A.; von Quadt, A. Restoring the Silurian to Carboniferous northern active continental margin of the Mongol–Okhotsk ocean in Mongolia: Hangay–Hentey accretionary wedge and seamount collision. *Gondwana Res.* **2014**, *25*, 1517–1534. [CrossRef]
23. Kelty, T.K.; Yin, A.; Dash, B.; Gehrels, G.E.; Ribeiro, A.E. Detrital-zircon geochronology of Paleozoic sedimentary rocks in the Hangay–Hentey basin, north-central Mongolia: Implications for the tectonic evolution of the Mongol–Okhotsk Ocean in central Asia. *Tectonophysics* **2008**, *451*, 290–311. [CrossRef]
24. Lee, J.; Williams, I.; Ellis, D. Pb, U and Th diffusion in natural zircon. *Nature* **1997**, *390*, 159–162. [CrossRef]
25. Miao, L.C.; Baatar, M.; Zhang, F.Q.; Anaad, C.; Zhu, M.S.; Yang, S.H. Cambrian kherlen ophiolite in northeastern Mongolia and its tectonic implications: SHRIMP zircon dating and geochemical constraints. *Lithos* **2016**, *261*, 128–143. [CrossRef]
26. Miao, L.C.; Zhang, F.Q.; Baatar, M.; Zhu, M.S.; Anaad, C. SHRIMP zircon U-Pb ages and tectonic implications of igneous events in the Ereendavaa metamorphic terrane in NE Mongolia. *J. Asian Earth Sci.* **2017**, *144*, 243–260. [CrossRef]
27. Yin, A.; Nie, S. A.; Nie, S. A Phanerozoic palinspastic reconstruction of China and its neighboring regions. In *The Tectonics of Asia*; Yin, A., Harrison, T.M., Eds.; Cambridge University Press: New York, NY, USA, 1996; pp. 442–485.
28. Parfenov, L.M.; Bulgatov, A.N.; Gordienko, I.V. Terranes and accretionary history of the Transbaikal orogenic belts. *Int. Geol. Rev.* **1995**, *37*, 736–751. [CrossRef]
29. Zhu, M.; Zhang, F.; Miao, L.; Baatar, M.; Anaad, C.; Yang, S.; Li, X. The late Carboniferous khuhu davaa ophiolite in northeastern Mongolia: Implications for the tectonic evolution of the Mongol–Okhotsk ocean. *Geol. J.* **2018**, *53*, 1263–1278. [CrossRef]
30. Gusev, G.S.; Peskov, A.I. Geochemistry and conditions of ophiolite formations of eastern Transbaikalia. *Geochem.* **1996**, *8*, 723–737.
31. Kovalenko, V.; Yarmolyuk, V.; Bogatikov, O. *Magmatism, Geodynamics, and Metallogeny of Central Asia*; MIKO—Commercial Herald Publishers: Moscow, Russia, 1995.
32. Wu, F.Y.; Sun, D.Y.; Ge, W.C.; Zhang, Y.B.; Grant, M.L.; Wilde, S.A.; Jahn, B.-M. Geochronology of the Phanerozoic granitoids in northeastern China. *J. Asian Earth Sci.* **2011**, *41*, 1–30. [CrossRef]
33. Sun, L.X.; Ren, B.F.; Zhao, F.Q.; Ji, S.P.; Geng, J.Z. Late Paleoproterozoic magmatic records in Eerguna massif: Evidence from the zircon U-Pb dating of granitic gneisses (in Chinese with English abstract). *Geol. Bull. China* **2013**, *32*, 341–352.
34. Tang, J.; Xu, W.; Wang, F.; Wang, W.; Xu, M.; Zhang, Y. Geochronology of Neoproterozoic magmatism in the Erguna Massif, NE China: Petrogenesis and implications for the breakup of the Rodinia supercontinent. *Precambrian. Res.* **2013**, *224*, 597–611. [CrossRef]

35. Daoudene, Y.; Ruffet, G.; Cocherie, A.; Ledru, P. Timing of exhumation of the Ereendavaa metamorphic core complex (north-eastern Mongolia)—U-Pb and $^{40}Ar/^{39}Ar$ constraints. *J. Asian Earth Sci.* **2011**, *62*, 98–116. [CrossRef]
36. Jamyandorj, U.; Tungalag, F.; Boishenko, A.F. *Geological Map of the Central and Eastern Mongolia, Scale 1: 500,000*; Institute of Geological Research Regional Geological Sector, Ministry of Heavy Industries: Ulaanbaatar, Mongolia, 1990.
37. Kröner, A.; Windley, B.F.; Badarch, G.; Tomurtogoo, O.; Hegner, E.; Jahn, B.M.; Gruschka, S.; Khain, E.V.; Demoux, A.; Wingate, M.T.D. Accretionary growth and crust-formation in the central Asian orogenic belt and comparison with the Arabian–Nubian shield. *Geol. Soc. Am. Mem.* **2007**, *200*, 181–209.
38. Mossakovsky, A.A.; Ruzhentsev, S.V.; Samygin, S.G.; Kheraskova, T.N. Central Asian fold belt: Geodynamic evolution and formation history. *Geotektonika* **1994**, *27*, 445–474.
39. Xiao, W.; Windley, B.F.; Han, H.; Liu, W.; Wan, B.; Zhang, J.; Ao, S.; Zhang, Z.; Song, D. Late Paleozoic to early Triassic multiple roll-back and oroclinal bending of the Mongolia collage in central Asia. *Earth Sci. Rev.* **2018**, *186*, 94–128. [CrossRef]
40. Tomurtogoo, O.; Badarch, G.; Makhbadar, T.S.; Orolmaa, D.; Khosbayar, P. *Geological Map of Mongolia, Scale 1: 1,000,000*; General Directorate of Mineral Research & Exploration of Turkey: Ankara, Turkey, 1999.
41. Jacobsen, Y.N.; Scherer, E.E.; Munker, C.; Mezger, K. Separation of U, Pb, Lu, and Hf from single zircons for combined U-Pb dating and Hf isotope measurements by TIMS and MC-ICPMS. *Chem. Geol.* **2005**, *220*, 105–120. [CrossRef]
42. Gerdes, A.; Zen, A. Combined U-Pb and Hf isotope LA-(MC-)ICP-MS analyses of detrital zircons: Comparison with SHRIMP and new constraints for the provenance and age of an Armorican metasediment in central Germany. *Earth Planet. Sci. Lett.* **2006**, *249*, 47–61. [CrossRef]
43. Liu, Y.S.; Hu, Z.C.; Zong, K.Q.; Gao, C.G.; Gao, S.; Xu, J.; Chen, H.H. Reappraisement and refinement of zircon U-Pb isotope and trace element analyses by LA-ICP-MS. *Chinese Sci. Bull.* **2010**, *55*, 1535–1546. [CrossRef]
44. Andersen, T. Correction of common lead in U-Pb analyses that do not report ^{204}Pb. *Chem. Geol.* **2002**, *192*, 59–79. [CrossRef]
45. Steiger, R.H.; Jager, E. Subcommission on geochronology: Convention on the use of decay constant in geo-and cosmochronology. *Earth Planet. Sci. Lett.* **1977**, *36*, 359–362. [CrossRef]
46. Ludwig, K.R. *User's Manual for Isoplot 3.00: A Geochronological Toolkit for Microsoft Excel*; Berkeley Geochronology Center: Berkeley, CA, USA, 2003; Volume 4, pp. 1–71.
47. Narantsetseg, T.; Orolmaa, D.; Yuan, C.; Wang, T.; Guo, L.; Tong, Y.; Wang, X.; Orshikh, O.E.; Oyunchimeg, T.; Delgerzaya, P.; et al. Early-middle Paleozoic volcanic rocks from the Ereendavaa terrane (Tsarigiin gol area, NE Mongolia) with implications for tectonic evolution of the Kherlen massif. *J. Asian Earth Sci.* **2019**, *175*, 138–157. [CrossRef]
48. Whattam, S.A.; Stern, R.J. The 'subduction initiation rule': A key for linking ophiolites, intra-oceanic forearcs, and subduction initiation. *Contrib. Miner. Petr.* **2011**, *162*, 1031–1045. [CrossRef]
49. Stern, R.J.; Reagan, M.; Ishizuka, O.; Ohara, Y.; Whattam, S. To understand subduction initiation, study forearc crust; to understand forearc crust, study ophiolites. *Lithosphere* **2012**, *4*, 469–483. [CrossRef]
50. Reagan, M.K.; Pearce, J.A.; Petronotis, K.; Almeev, R.R.; Avery, A.J.; Carvallo, C.; Chapman, T.; Christeson, G.L.; Ferré, E.C.; Godard, M.; et al. Subduction initiation and ophiolite crust: New insights from IODP drilling. *Int. Geol. Rev.* **2017**, *59*, 1439–1450. [CrossRef]
51. Stern, R.J.; Gerya, T. Subduction initiation in nature and models: A review. *Tectonophysics* **2018**, *746*, 173–198. [CrossRef]
52. Whattam, S.A.; Montes, C.; Stern, R.J. Early central American forearc follows the subduction initiation rule. *Gondwana Res.* **2020**, *79*, 283–300. [CrossRef]
53. Khain, E.V.; Bibikova, E.V.; Salnikova, E.E.; Kröner, A.; Gibsher, A.S.; Didenko, A.N.; Degtyarev, K.E.; Fedotova, A.A. The Palaeo-Asian Ocean in the Neoproterozoic and early Palaeozoic: New geochronologic data and palaeotectonic reconstructions. *Precambrian Res.* **2003**, *122*, 329–358. [CrossRef]
54. Gordienko, I.V. Paleozoic geodynamic evolution of the Mongol-Okhotsk fold belt. *J. Southeast Asian Earth Sci.* **1994**, *9*, 429–433. [CrossRef]
55. Winkler, W.; Bussien, D.; Baatar, M.; Anaad, C.; von Quadt, A. Detrital zircon provenance analysis in the central Asian orogenic belt of central and southeastern Mongolia—A Palaeotectonic model for the Mongolian Collage. *Minerals* **2020**, accepted.

56. Miao, L.; Zhang, F.; Jiao, S. Age, protoliths and tectonic implications of the Toudaoqiao blueschist, Inner Mongolia, China. *J. Asian Earth Sci.* **2015**, *105*, 360–373. [CrossRef]
57. Zhao, P.; Xu, B.; Jahn, B. The Mongol-Okhotsk Ocean subduction-related Permian peraluminous granites in northeastern Mongolia: Constraints from zircon U-Pb ages, whole-rock elemental and Sr-Nd-Hf isotopic compositions. *J. Asian Earth Sci.* **2017**, *144*, 225–242. [CrossRef]
58. Koval, P.V.; Grebenshchikova, V.I.; Lustenberg, E.E.; Henney, P.J. Database of granites in the Mongol-Okhotsk zone, Mongolia-Siberia, and its use in mineral exploration. *J. Geochem. Explor.* **1999**, *66*, 199–210. [CrossRef]
59. Berzina, A.P.; Sotnikov, V.I. Character of formation of the Erdenet-Ovoo porphyry Cu-Mo magmatic center (northern Mongolia) in the zone of influence of a Permo-Triassic plume. *Russ. Geol. Geophys.* **2007**, *48*, 141–156. [CrossRef]
60. Kang, Y.; She, H.; Lai, Y.; Wang, Z.; Li, J.; Zhang, Z.; Xiang, A.; Jiang, Z. Evolution of middle-late Triassic granitic intrusions from the Badaguan Cu-Mo deposit, Inner Mongolia: Constraints from zircon U-Pb dating, geochemistry and Hf isotopes. *Ore Geol. Rev.* **2018**, *95*, 195–215. [CrossRef]
61. Mi, K.; Liu, Z.; Li, C.; Liu, R.; Wang, J.; Peng, R. Origin of the badaguan porphyry Cu-Mo deposit, Inner Mongolia, northeast China: Constraints from geology, isotope geochemistry and geochronology. *Ore Geol. Rev.* **2017**, *81*, 154–172. [CrossRef]

minerals MDPI

Article

Detrital Zircon Provenance Analysis in the Central Asian Orogenic Belt of Central and Southeastern Mongolia—A Palaeotectonic Model for the Mongolian Collage

Wilfried Winkler [1,*], Denise Bussien [1,2], Munktsengel Baatar [3], Chimedtseren Anaad [3] and Albrecht von Quadt [1]

[1] Department of Earth Sciences, ETH Zurich, 8092 Zurich, Switzerland; denise.bussien.grosjean@gmail.com (D.B.); albrecht.vonquadt@erdw.ethz.ch (A.v.Q.)
[2] Aba Géol SA, 1530 Payerne, Switzerland
[3] School of Geology and Mining, Mongolian University of Science and Technology, Ulaanbaatar 120646, Mongolia; tsengel@must.edu.mn (M.B.); chimedtserena@yahoo.com (C.A.)
* Correspondence: wilfried.winkler@erdw.ethz.ch

Received: 24 June 2020; Accepted: 21 September 2020; Published: 2 October 2020

Abstract: Our study is aimed at reconstructing the Palaeozoic–early Mesozoic plate tectonic development of the Central Asian Orogenic Belt in central and southeast Mongolia (Gobi). We use sandstone provenance signatures including laser ablation U-Pb ages of detrital zircons, their epsilon hafnium isotope signatures, and detrital framework grain analyses. We adopt a well-established terran subdivision of central and southeastern Mongolia. However, according to their affinity and tectonic assemblage we group them into three larger units consisting of continental basement, rift-passive continental margin and arc elements, respectively. These are in today's coordinates: (i) in the north the late Cambrian collage from which the later Mongol-Okhotsk and the Central Mongolia-Erguna mountain ranges resulted, (ii) in the south a heterogeneous block from which the South Mongolia-Xin'gan and Inner Mongolia-Xilin belts developed, and (iii) in between we still distinguish the intra-oceanic volcanic arc of the Gurvansayhan terrane. We present a model for paleotectonic development for the period from Cambrian to Jurassic, which also integrates findings from the Central Asian Orogenic Belt in China and Russia. This mobilistic model implies an interplay of rift and drift processes, ocean formation, oceanic subduction, basin inversion, collision and suture formation in space and time. The final assemblage of the Central Asian Orogenic Belt occurred in Early Jurassic.

Keywords: Mongolia; detrital zircon; laser ablation U-Pb dating; Hafnium isotope-ratio; terrane definition; paleotectonic reconstruction

---

1. Introduction

The Central Asian Orogenic Belt or Altaid Belt records a long history of collage of different tectono-stratigraphic units (terranes), which amalgamated against the Siberian Craton from Early Neoproterozoic until Permian-Triassic (Figure 1; e.g., [1–9]). Accordingly, on Mongolian territory three main collisional events are recognized (Figure 1C): (i) a Neoproterzoic-Cambrian one in the NW and center (Tuva-Badrag terrane; Figure 1B), (ii) an arly Palaeozoic one in central Mongolia north of the Main Mongolian Lineament (MML, Figure 1B) (Central Mongolian terranes *sensu* [9]), and (iii) a Late Palaeozoic (Permian-Triassic) one in south-southeastern Mongolia, south of the MML (including the Gobi Altai, Mandalovoo, Gurvansayhan, Nuhetdavaa-Enshoo and Hutag Uul terranes *sensu* [10]; Figure 1C).

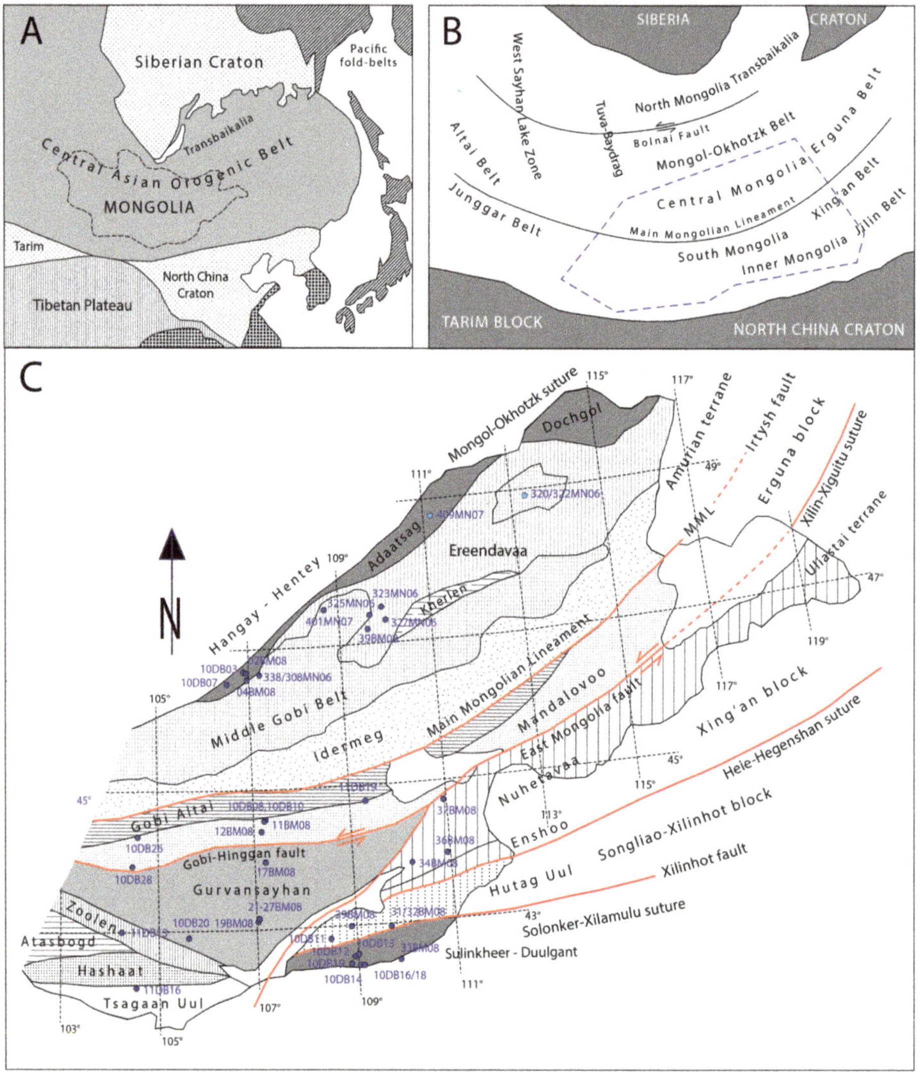

Figure 1. Index maps of orogens, terranes and sample locations in Mongolia. (**A**) Position of Central Asian Orogenic Belt. (**B**) Correlation of orogenic belts from Mongolia into adjacent countries. Adapted from [11–16]. Stippled frame corresponds to the approximate extent of subfigure (**C**). (**C**) Location of samples presented in this study based on the terrane subdivision of [10].

In South Mongolia, the colliding tectono-stratigraphic units comprised continental basement and oceanic terranes, seamounts, and continental and oceanic volcanic arcs nowadays preserved in continental slivers, sutures, arcs and accretionary complexes (e.g., [1,6,8,17–20]). Syn-tectonic Permian sediments were deposited in residual marine and continental (intermontane) basins [21,22]. Hence, the

Permo-Triassic sediments suffered post-depositional tectonic deformation indicating the final collision occurred during the Jurassic (e.g., [23–26]).

The situation is further complicated by two other factors: (i) the Mongol-Okhotsk Ocean opened in Silurian within the central Mongolian collage and subsequently closed scissor-like from Triassic on eastward (e.g., [24,27–31]) and, (ii) during the entire Palaeozoic Siberia experienced steady (>40°) northward drift and an ≈75° clockwise rotation from Middle Silurian to Permian [9,32]. In this respect it is a matter of debate, which terranes were when accreted, and as an integrated part of peri-Siberia they may have exercised rotations as well (e.g., [4,10,32]). With regard of point 1, the duration of the Mongol-Okhotsk Ocean, there exist earlier models which suggest its existence since the late Neoproterozoic [33]. Recent studies, however, have demonstrated the independence of this ocean from the larger Paleo-Asian Ocean (e.g., [4,10,11,27–31]), because it opened later.

The present work is a study of the central Mongolian and southeastern Mongolian terranes by analysing detrital modes and detrital zircons from sandstones comprised in the various terranes (Figure 1B,C). We suggest that ages and Hf-isotope signatures of the detrital zircons, gained by laser-ablation ICP-MS measurements, will allow to define the basement and volcanic sources of the terranes and discuss their plate-tectonic character [34–37]. Hafnium isotope ratios ($^{176}Hf/^{177}Hf$) in dated zircon grains may allow identification of the nature of the magmatic source rocks with respect to their derivation from depleted mantle, recycled continental crust or mixtures (e.g., [36]). Thus, giving arguments for plate tectonic processes working during the magmatic activity like rifting, passive continental margin and subduction settings.

Parts of the data from the Hangay-Hentey, Adaatsag and Ereendavaa terranes within the Mongol-Okhotsk Belt were discussed in earlier papers [24,31]. With the present study we can summerize the Palaeozoic tectonic history of central and southeastern Mongolia, and we attempt to sketch a palaeotectonic model for the central and southeastern Mongolian transect of the Central Asian Orogenic Belt. We expect that our hypothetical model will be tested and probably modified by further research in this area.

2. Geological Framework of Central and Southeastern Mongolia

Depending on scopes, scales and methods applied, numerous studies of the area exist, which use different terrane names, but mostly rely on Badarch et al. [10]. We also aim for the most accurate possible correlation of the Mongolian terranes with the established tectonic-stratigraphic units in China and Russia [6,8,14,37–42] with our results. The doubled Kipchak-arc model ([33,42]), and references therein) was modified during the last decades towards a model inferring the time-stepped closure of oceanic basins within the larger Palaeo-Asian Ocean driven by the collision of continental basement with arc complexes and suturing [1,6,9,43–46]. A main structure is the MML, which is assumed to coincide with the former Early Palaeozoic southern margin (in modern coordinates) of the late Cambrian-Ordovician peri-Siberian collage [1,6,9]. However, across the northeastern boundary of Mongolia within the Chinese and Russian territory this structure was not yet described.

2.1. *Terranes North of the MML*

2.1.1. Adaatsag Terrane

It comprises a highly deformed and metamorphosed ophiolite-bearing shaley association representing the suture of the Mongol-Okhotsk Belt with Ordovician shales, quartzites, sandstones, cherts and metavolcanic rocks. Mélanges contain serpentinites, fragments of serpentinized dunite and hartzburgite, gabbro, metabasalts and tholeiitic basalts [1,29]. From our detrital zircon dating results it becomes evident that the Addaatsag terrane comprises also deformed Permian maximum age sandstones (see Figure 2A and [24]).

Table 1. Samples listed by terranes, lithology, mapped age, maximum age according to detrital zircon dating and methods applied in the present study.

Terrane	Sample No.	Group/Formation	Colour Code on 1:1.5 Million Scale Geological Map of Erdenechimeg et al. [47]	Outcrop Lithology/Facies	Mapped Age	Maximum Depositional Age (Inferred According to Detrital Zircon Ages)	Methods: Detrital Zircons DZ, Sandstone Modal Analysis MA
Adaatsag	02BM08	Bayantsagaan Fm	C1-2*bc*	Greenish-grey shales, splintery breaking	Carboniferous	Ordovician-Silurian	DZ
	04BM08	Bayantsagaan Fm	C1-2*bc*	Very coarse sandstones with large-scale cross-bedding (slightly metamorphosed)	Carboniferous	Permian	DZ
	10DB03	Nomgon Fm	S-D*ad* or S-D*nn*	Greenish-brown massive shales with bryozoans, brachiopods and echinoderms	Carboniferous	Permian	DZ, MA
	10DB07	Bayantsagaan Fm	C1-2-*bc*	Medium-bedded graded sandstones and shales	Devonian	Permian	DZ, MA
	409MN07	Urtiin-Gol Fm	J2-3-*ug*	Alluvial coarse sanstones and conglomerates	Jurassic	Confirmed	DZ, MA
Ereendavaa	320MN06	Gazar Fm	P1-2*gz*	Sandstones, yellowish-grey	Permian	Permian - E-Triassic	DZ, MA
	322MN06	Sharil Fm	J3-*sr*	Coarse-grained immature sandstones, greenish-grey	Jurassic	Confirmed	DZ
	323MN06	Kavchuu Fm	D1-*hb*	Coarse-grained sandstones, greenish-grey	Devonian	Ordovician-Silurian	DZ, MA
	325MN06	Mungut-Uul Fm	T2-3-*mu*	Blueish-gray, coarse grained sandstone	Triassic	Permian	DZ, MA
	327MN06	Nariintai Fm	D1-*nr*	Light grey sandstones, slightly metamorphosed	Devonian	Ordovician	DZ, MA
	308MN06	Bayantsagaan Fm	C1-2*bc*	Greenish-grey, medium grained sandstone	Permian	Confirmed	DZ
	338MN06	Bayantsagaan Fm	C1-2-*bc*	Turbiditic sandstone in spotted (bioturbated) green shales	Carboniferous	L.Carb. - Permian	DZ, MA
	39BM08	Zamttolgoi Fm	NO2-*zm*	Coarsening upward quartz-rich turbiditic sandstones with prasinite sill	Devonian	Confirmed	DZ, MA
	401MN07	Ulz Fm	P2-*ul*	Siliceous shales and turbiditic sandstones	Permian	Confirmed	DZ
Goti Altai	11DB19	Dushovoo Fm	C2-*do*	Greenish-grey and minor reddish shales with coarse (turbiditic) sandstone intercalations	Devonian	Carboniferous	DZ, MA
Mandalovoo	10DB08	Khutul-Us-Khudag Fm	D1-*hu*	Volcaniclastic sandstone intercalated between quartz-bearing conglomerates, overlying bioturbated fossiliferous limestones and topped by bioturbated crinoidal limestones (*Thalassinoides*)	Early Devonian	Confirmed	Confirmed
	10DB10	Khutul-Us-Khudag Fm	D1-*hu*	Quartz-rich conglomerate with volcanic pebbles	Devonian	No data	MA
	10DB26	Mangal-Ovoo Fm	O3-S-*mo*	Greenish-grey, highly concentrated turbitity flow beds, partly slumped in finer facies	Ordovician-Silurian	Silurian	DZ, MA
	10DB28	Djadokhta Fm	K2-*bg*	Red shales and sandstones with saurian bones	Cretaceous	E. Cretaceous	DZ
	11BM08	Khutal-Us-Khudag Fm	D1-*hu*	Sandstones and conglomerates rich in vein-quartz pebbles with coarse cross-bedding; fan delta association	Devonian	Silurian-E. Devonian	DZ
	12BM08	Tal Fm	C1-*tl*	Dark greenish-brown turbiditic sandstones (30–50 cm) rich in red volcanic lithoclasts and plagioclase	Carboniferous	Confirmed	DZ, MA

Table 1. *Cont.*

Terrane	Sample No.	Group/Formation	Colour Code on 1:1.5 Million Scale Geological Map of Erdenechimeg et al. [47]	Outcrop Lithology/Facies	Mapped Age	Maximum Depositional Age (Inferred According to Detrital Zircon Ages)	Methods: Detrital Zircons DZ, Sandstone Modal Analysis MA
Gurvansayhan	17BM08	Manlai Fm	K1-*mn*	Conglomerates, sandstones and red shales	Cretaceous	Early Cretaceous	DZ
	19BM08	Alagbayan Gr	DA4b/c *	Andesltic sandstone	Late Devonian	No data	MA
	21/22BM08	Sainshanhuadag Fm	CS2a *	Fining-upward series of volcaniclastic (andesitic) conglomerates and sandstones, overlying ignimbrite	Tournaisian-Visean	Carboniferous	DZ, MA
	26BM08	Alagbayan Gr	DA2b *	Volcaniclastic sandstone	Devonian	Middle-Late Devonian	DZ
	27BM08	Alagbayan Gr	DA2b *	Volcaniclastic sandstone	Devonian	Late Devonian-Mississippian	DZ
Atasbogd-Zoolen	11DB13	Talynmeltes Fm	D1-2-*tm*	Green volcaniclastic sandstone	L. Silurian -E. Devonian	Late Devonian	DZ
Tsagaan Uul	11DB16	Tavankhudag Fm	D1-*th*	Massive dark green arkosic sandstones	Devonian	Late Devonian	DZ, MA
Nuhetdavaa (-Enshoo)	34BM08	Sangiindalai Fm	O3-*sd*	Massive brownish coarse sandstones (turbidites) assosiated with green siliceous shales	M.-I. Carboniferous	Permian-E. Triassiac	DZ, MA
	36BM08	Ulziit-Uul Fm	C2-*uu*	Green and minor red shales intercalated by dark greenish-grey graded sandstones (presumaby turbiditic), few breccias	Carboniferous	Ordovician	DZ, MA
	37BM08	Gunbajan Fm	C1-*gb*	Red and green spotted shales associated with lenticular poorly sorted red sand- and siltstones (presumably low energy meandring fluvial)	Carboniferous	No data	MA
Hutag Uul	10DB11	Lugiingol Fm	P2-*lg*	Greenish-grey, generally fine-grained turbiditic sandstones and shales	Devonian	Permian-E. Triassiac	DZ, MA
	29BM08	Uvgontolgoi Fm	P1-*ut*	Thin-bedded, pyrite-bearing black shales and thin » 5–10 cm thick sandstones	Permian	Confirmed	DZ
	31BM08	Toli-Uul Fm	NP-E1-*tu*	Greenish, coarse-grained volcanic litharenite, turbiditic facies, alternating with sandstones of 32BM08 type	Devonian	No data	MA
	32BM08	Toli-Uul Fm	NP-E1-*tu*	Light grey, coarse-grained arkosic sandstone, turbiditic facies, overthrusted by Devonian volcanics	Devonian	Permian	DZ, MA
Sulinkheer -Duulgant	10DB12	Hetsuul Fm	P1-2*hu*	Graded quartz-sandstones	U. Permian	Confirmed	DZ, MA
	10DB13	Hetsuul Fm	P1-2*hu*	Fine-grained quartz-sandstones, red shales, and limestone intercalations	Carboniferous	Silurian-E. Devonian	DZ
	10DB14	Agui-Uul Fm	C2-P1-*au*	Red shales intercalated by coarse sandstones and sand-supported conglomerates	Permian	Permian	DZ, MA
	10DB18	Agui-Uul Fm	C2-P1-*au*	Quartz-rich sandstones associated with green siliceous shales	Permian	No data	MA
	10DB19	P1-2	Unclassified	Arkosic sanstones and grain-supported conglomerates (cm-size pebbles, volcanics, vein-quartz, radiolarite)	Permian	Late Permian-Early Triassic	DZ, MA

* Denotes Drillhole Levels in the Oyu Dolgoi Mining District described by Dolgopolova et al. [48].

Figure 2. (**A**) Outcrop in Adaatsag terrane and locality of sandstone 04BM08 (Table 1, Figure 3), vertically bedded coarse sandstones and conglomerates with large scale cross-bedding, mapped "Carboniferous"; according to detrital zircon U-Pb results of Permian max. age; (**B**) Devonian series of Ereendavaa terrane showing a coarsening-upward series of quartz-rich turbiditic sandstones and shales at Bayantsogt Uul Mountain; (**C**) Detail of B with location of sample 39BM08 (Table 1) above a prasinitic sill (see hammer); (**D**) Devonian series of Mandalovoo terrane unconformably overlying Silurian shaley beds, from base to top volcaniclastic sandstones/conglomerates, bioturbated (Thalassinoides) cherty limestones (brachiopods, bryozoans), quartz sandstones and conglomerates in fan-delta facies (sample11BM08) with volcanoclastic intercalations (sample 10DB08); (**E**) Devonian-Carboniferous pebbly sandstones of the Gurvansayhan terrane bearing mainly clasts of basic volcanics, sandstones and shales, and spathic limestones; similar rocks we analysed from the Oyu Tolgoi mining district; (**F**) Cretaceous red beds onlapping the Gurvansayhan terrane in the Baron Gurvan area, view to SE, see

results from sample 17BM08; (**G**) Slightly southward dipping turbiditic sandstones and shales of Hutag Uul terrane, locality of sample 10DB11, mapped as "Devonian", detrital zircon U-Pb infer a Permian-Early Triassic max. age; (**H**) Permian, vertically bedded, E-W striking red shales intercalated by coarse sandstones and conglomerates; younging to the N, locality of sample 10DB14.

2.1.2. Ereendavaa-Kherlen-Idermeg Terrane Group and Middle Gobi Belt

In the Ereendavaa terrane, Silurian and Devonian marine shaley, sandy sediments and volcanics unconformably overly Palaeoproterozoic gneisses, Neoproterozoic schists, sandstones and marbles [10]. Recent results of Miao et al. ([11], this special volume) attest the presence of Proterozoic source rocks in the basement of the Ereendavaa terrane by the dating of detrital zircons in a quartzite formation deposited after 1.15 Ga. In a Devonian turbiditic sandstone succession (Bayantsogt Uul mountain, locality of sample 38BM08, Table 1) we recognised a prasinitic sill (Figure 2B,C), unfortunately not datable due to lack of zircons.

The Idermeg terrane consists of fossiliferous Neoproterozoic-Cambrian marbles, quartz-sandstones and conglomerates unconformably overlying gneisses, amphibolites and phyllites, which were intruded by Cambrian granites. The presence of Devonian-Permian and Triassic-Jurassic volcanic and sedimentary rocks we could not confirm. The Idermeg terrane is interpreted as a so-called microcontinent ([6], and references therein). In our view, this terrane represents a deeply eroded continental sliver like in the Baydrag terrane to the west as proposed by [16].

A former tectonic assembling of the Ereendavaa and Idermeg terranes (Figure 1C) at the Ordovician-Silurian transition is documented by the composition and ages derived from the Kherlen ophiolitic complex (eventually called Herlen). Miao et al. [16] dated gabbro and plagiogranite at ~500 Ma (SHRIMP zircon U-Pb) and post-collisional granites at ~440 Ma (Llandovery, early Silurian). These authors suggest that the Kherlen suture formed coevally and similarly with the Bayanhongor belt. Thus, this finding implies that the Baydrag and the Idermeg terranes during the Palaeozoic represented a common continental passive margin block related to the opening of the Mongol-Okhotsk Ocean [24,41].

In turn, in the Middle Gobi volcanic-plutonic belt Permian volcanic and sedimentary series unconformably top the abovementioned Palaeozoic turbiditic and pelagic sediments, and the Kherlen terrane (Figure 1C). Finally, flat-lying Lower Jurassic continental sandstones and conglomerates mark the youngest formations known from that terrane group [24].

In Russia, the above described Ereendavaa and Idermeg terranes continue in the form of the Precambrian Amurian superterrane bounded to the south by the MML [14,46] (Figure 1C). However, in contrast to Donskaya et al. [14] we suggest that these terranes represented the northern margin of the Mongol-Okhotsk Ocean and were subducted from Devonian on by the Mongol-Okhotsk Ocean plate.

2.2. Terranes South of the MML

2.2.1. Gobi-Altai and Mandalovoo Terranes

We genetically link here these terranes described by Badarch et al. [10] because we found similar detrital zircon age patterns and detrital modes in both (Figures 3 and 4). The Gobi Altai consists of Cambrian greenschist facies metamorphosed sandstones, mudstones, tuffs, and minor volcanic rocks, Ordovician to Silurian sandstones, Devonian to Mississippian conglomerates, sandstones, siltstones, fossiliferous limestone, andesite and tuff, overlain by Pennsylvanian, Permian to Triassic volcanic and sedimentary rocks [11,17]. The Mandalovoo terrane comprises Silurian to Carboniferous pelagic shales and turbiditic sandstones and conglomerates either rich in volcanic lithoclasts or quartz grains/pebbles. The Devonian series of the Mandalovoo terrane is also marked by the presence of shallow marine fossiliferous and bioturbated (*Thalassinoides*) crinoidal limestones. Devonian quartz-rich sandstones and conglomerates clearly show sand-dominated fluvio-deltaic facies (Figure 2D). In addition, Lamb and Badarch [17] and Badarch et al. [10] mention the presence of pillow basalts and Cretaceous post-accretionary sediments.

The Gobi Altai and Mandalovoo terranes may have their extension in the Erguna block in adjacent northeastern China (Figure 1). This contains Neoproterozoic (multi-) metamorphic basement (peak age ≈500 Ma; [39]), granitoids, Neoproterozoic meta-sediments, pre-Devonian sediments and Carboniferous strata [40]. However, Wu et al. [49] found that the great majority of the granitic intrusions in the Erguna block are Mesozoic in age related to the later subduction of the Palaeo-Pacific Ocean. Where the Gobi Altai and Mandalovoo terranes are not in contact with the Gurvansayhan terrane in southern Mongolia, Mandalovoo is separated from the Nuhetdavaa terrane by the blueschists and ophiolite-bearing Xilin-Xiguito suture [40]. Due to the completely different tectono-stratigraphic history of the Mandalovoo and Gurvansayhan terranes, their contact is the major Zumbayan (East Gobi) fault zone [15].

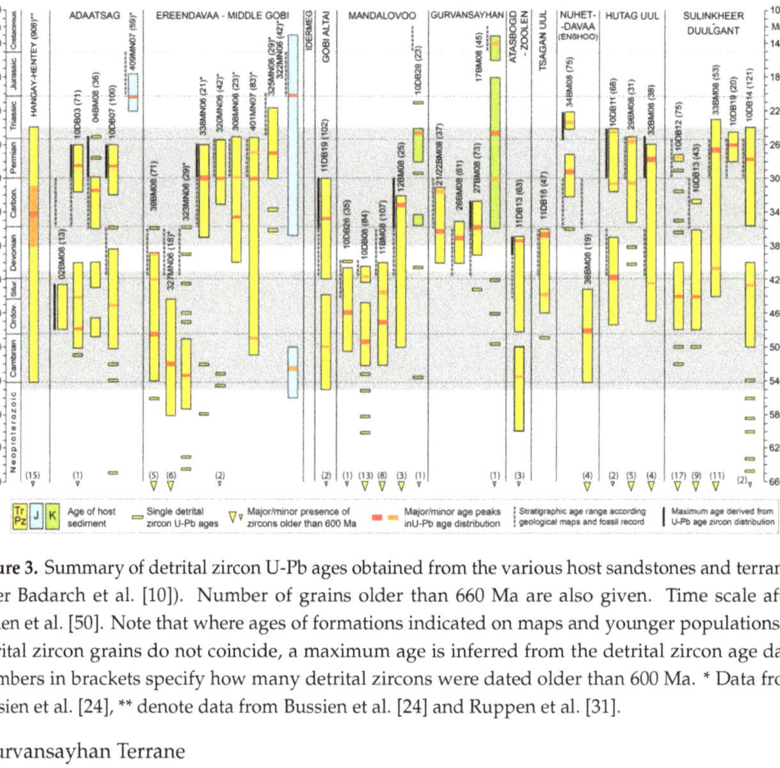

Figure 3. Summary of detrital zircon U-Pb ages obtained from the various host sandstones and terranes (after Badarch et al. [10]). Number of grains older than 660 Ma are also given. Time scale after Cohen et al. [50]. Note that where ages of formations indicated on maps and younger populations of detrital zircon grains do not coincide, a maximum age is inferred from the detrital zircon age data. Numbers in brackets specify how many detrital zircons were dated older than 600 Ma. * Data from Bussien et al. [24], ** denote data from Bussien et al. [24] and Ruppen et al. [31].

2.2.2. Gurvansayhan Terrane

This is a unique element in SE Mongolia. It comprises a varied oceanic series with pillow basalts, mélanges/olisthostromes, metamorphosed shales and sandstones, lower Silurian-lower Devonian radiolarites and Middle Devonian-Mississippian volcaniclastic sandstones and conglomerates (Figure 2E; [1,51,52]). The rock units are strongly dismembered. With Tsagaan Suvarga and Oyu Tolgoi the terrane comprises important porphyry copper mineralisation, the latter dated by Lamb and Lamb and Cox [52] and Perellò et al. [53] as Devonian. Dolgopolova et al. [48] summarised a U-Pb mean zircon peak age of ≈362 Ma (Late Devonian, Famennian) from basalts, quartz monzodiorites, dacites and granodiorites. However, granitoids did dot contributed material to the presently analyzed ± coeval sandstones because they are almost entirely free of quartz (see also below). The Carboniferous sediments unconformably overlie the folded Devonian series [54,55], and radiometric ages of post-accretionary dykes and intrusions suggest a collision of the oceanic arc latest in the Permian [56]. Cretaceous red beds unconformably overlap the terrane (Figures 2F and 3).

On Chinese territory, the Gurvansayhan terrane does not appear to have a counterpart. In Mongolia, that terrane disappears along the left-lateral Zuunboyan fault or Xilin-Xiguito suture, respectively (Figure 1C) (e.g., [15]; eventually called East Gobi or Eastern Mongolian fault zone [57]), which further to the NE directly separates the Mandalovoo from the Nuhetdavaa and the Uliastai terrane, respectively.

2.2.3. Nuhetdavaa and Enshoo Terranes

Badarch et al. [10] classified these units as back/forearc and arc terranes, respectively. The Nuhetdavaa terrane comprises Neoproterozoic basement (gneiss, schists, marbles), Cambrian-Silurian sandstones, phyllites and shales, and conglomerates. Devonian basalts, andesites and volcaniclastic sandstones are mentioned by Badarch et al. [10]. Various Silurian to Triassic granitoids intrude the terrane.

2.2.4. Hutag Uul Terrane

This cratonal terrane comprises Precambrian gneiss, schists, migmatites, marble, quartz-sandstones and stromatolithic limestones. Basalts, andesites, turbiditic volcaniclastic rocks and coral-bearing limestones represent the Devonian and Carboniferous series. Badarch et al. [10] interpreted the Devonian-Carboniferous granitoid intrusion as to be related to subduction processes. Permian marine sedimentary and volcanic rocks complete the stratigraphic section (e.g., [22]).

Some terrane comparisons linked the Hutag Uul unit either with the Inner Mongolia-Jilin belt in China [1,40] or the Baolidao arc-accretion complex [8], respectively. In Figure 1C, we have adapted the correlation proposed by Liu et al. [25], i.d., the Sangliao-Xilinhot block. In its Chinese occurrence, the Sangliao-Xilinhot block with Precambrian basement and Palaeozoic sandstones (≈Silurian) was correlated with the Hutag Uul terrane by Jian et al. [58] and Xu et al. [59]. The Hutag Uul terrane is bordered in the south by the Sulinkheer-Duulgant terrane and the Solonker-Xilamalu suture.

2.2.5. Sulinkheer-Duulgant Terrane

It is commonly classified as an accretionary wedge unit having its eastern prolongation in the Chinese Erdaojing accretion complex (Solonker suture; [3,60]) or, respectively, with the Xar Moron-Jilin Yanji suture (e.g., [40,61], Figure 1C). According to Badarch et al. [10] and own observations the unit shows a very composite tectonic configuration, i.e., represents a large mélange body containing dismembered ophiolitic and metamorphic slivers, Carboniferous and Permian volcaniclastic and carbonate rocks. Tholeiitic pillow basalts, radiolarian cherts and limestones point to the presence of oceanic basement. A bi-modal age distribution is evident from dating a diorite at 438 ± 4 Ma and a gabbro at 248 ± 2 [12]. Some highly sheared continental basement rocks (mylonitised gneiss, amphibolite, quartzite, marble, metasandstone and argilite) occur along the northern margin of the terrane [10]. Notably, Badarch et al. [10]. mention a personal communication of L.E. Webb (1999) that the muscovite-biotite schists and mylonites revealed Ar-Ar mica ages of 271–208 Ma, which may indicate a major (Permian-Triassic) deformation event in the accretionary wedge.

3. Materials and Methods

3.1. Sample Material

We sampled and partly completed our earlier data set [24,31] in the various terranes in a large section from central Mongolia to the Chinese border (Figure 1C). The location and lithologies of the samples are listed in Table 1. With respect to our data presented in Bussien et al. [24] we added four more samples in the Adaatsag and one more in the Ereendavaa terrane. All other samples from the southern terranes were collected later and represent new data sets. The collected sandstones and shales were used for petrographic analysis under the petrographic microscope and the analysis of detrital zircons (specifications are given in Supplementary Materials Table S1). Three samples from the Gurvansayhan terrane were derived from boreholes of the Oyu Tolgoi mining prospect [51,54,55].

3.2. Methods

The chronostratigraphic correlation of the samples primarily follows the ages inferred on the geological map of Mongolia 1:1.5 million [47]. If the detrital zircon U/Pb age results suggest a clearly younger age range, we suggest a "maximum sedimentation age" (Table 1, Figure 3), otherwise the detrital zircons confirmed the mapped ages.

An approximated rock composition of the detrital sources of the sandstones is evaluated by standard modal point-counts (200–400) on feldspar and carbonate stained thin-sections following grain types suggested by Folk [62] and Dickinson [63]. Because of missing age information on for example volcanic lithoclasts, this method cannot accurately provide the final plate-tectonic framework of the sampled basins. On contrary, at hands of the detrital zircon dating we partly must assume reworking of older deposits or volcanic sources into younger basins.

Zircon grains were hence isolated from crushed whole rock by hydrochloric acid (HCl, 25%) added to hydrogen peroxide (H_2O_2). Panning, heavy liquids (bromoform), and hand picking completed the separation and allowed to obtain grains ranging from ca. 200 μm to 70 μm. The grains were then mounted in epoxy and polished approximately at their half. Prior to isotopic analyses (U-Pb dating and Hf isotopes), all zircon grains were inspected by cathodoluminescence imaging to prove their homogeneous composition and recognize the magmatic growth by its oscillatory patterns. Only magmatic domains were analyzed and inherited cores were avoided.

U-Pb dating was performed at the Institute of Geochemistry and Petrology ETH Zurich with a laser-ablation inductively coupled plasma mass spectrometer (Elan 6100 instrument coupled to an in-house-built 193 nm Excimer laser). The complex growth domains and inherited cores of the detrital zircons were avoided. A focused laser beam of 30–40 μm (depending on grain size) in diameter was applied at a pulse rate of 10 Hz with 0.5 mJ/pulse. The ablated material was carried out in helium gas (1.1 L/min) to the spectrometer where the analytical signal was measured for 90 to 120 s after 30 s of background collection. In each analyzes, Hg^{202}, Pb^{204}, Pb^{206}, Pb^{207}, $Pb^{208,}$ Th^{232}, U^{235}, U^{238} masses were quantified sequentially. Prior and after each 15 zircons analysed we measured a set of primary and secondary standards represented by the GJ-1 zircon and the 91500Plesovice AUSZ7-5 zircon standard, respectively. Each run consisted of 3 measurement of GJ-1 to control the mass fractionation/downhole correction and one of the secondary reference materials to evaluate external reproducibility. Raw data were processed using Glitter software package (van Achtenberg et al. [64]). The Isoplot/Ex v. 4.15 [65] generated the Concordia diagrams and probability density distribution plots (divided into 25 my age bins). Repeated measurements on the Plesovice [66] give a Concordia age of 337.4 ± 1.3 Ma, (95% of confidence), which is identical within error to the reference value of 337.13 ± 0.37 Ma [67].

The in-situ Hf isotopic analyses were performed at the IGP (ETHZ, Zurich, Switzerland) using a Nu plasma MC-ICP-MS (Nu instrument Ltd.) coupled to an Excimer laser (GeoLas ArF, wavelength of 193 nm). A spot size of 60 μm in diameter was chosen to analyze already dated domains of the detrital zircon grains. The laser repetition rate was 4 Hz with an energy range of 10–20 J/cm^2. Again, helium was used as carrier gas (0.95 L/min), and the baseline was measured within 30 s and analytical signal within 60 s. Measurement of ^{171}Yb (^{176}Yb/^{171}Yb = 0.897145) and ^{175}Lu (^{176}Lu/^{175}Lu = 0.026549) allows the correction of the isobaric interference on ^{176}Hf of ^{176}Yb and ^{176}Lu, respectively. Age correction to calculate initial ^{176}Hf/^{177}Hf ratio (εHf (t)) were obtained using a ^{176}Lu decay constant of 1.867×10^{-11} year^{-1}, the measured ^{176}Lu/^{177}Hf ratio and the chondritic values of ^{176}Hf/^{177}Hf and ^{176}Lu/^{177}Hf (0.0332 and 0.282772, respectively [68]. Mud Tank and Temora zircon were used as a reference standard. The complete analytical results are provided in Supplementary Materials Tables S2–S4.

4. Results and Discussion

A summary of our results is illustrated in Figure 3. Accordingly, the general pattern of major and minor peaks of detrital zircon ages in Palaeozoic sandstones records two age groups: (i) Cambrian-Silurian, and (ii) Carboniferous-Permian. Except for the Gurvansayhan, Atasbogd-Zoolen and Tsagaan Uul terranes, which show Late Devonian peaks, the basins related to the other terranes were impoverished of Devonian detrital zircon influx. Few minor Triassic peaks are observed but are more evident in Jurassic and Cretaceous post-orogenic deposits. Figure 3 also informs about the variable presence of Proterozoic basement rocks (detrital zircon ages >600 Ma) in the source terranes. The obtained data sets are provided in Supplementary Materials Tables S1 and S2.

4.1. Adaatsag Terrane

Detrital zircons were analysed in four new samples (Table 1, Figure 3) and are combined with three samples presented in [24]. Detrital zircon ages show that the inferred maximum ages of the sandstones and mapped ages do not concur. Although, we can refer to one Ordovician-Silurian shale (02BM08, low zircon yield because of very fine grain size) and three Permian sandstones. The modal analyses of Permian sandstones reveal an arkosic to lithic arkosic composition suggesting basement uplift detrital sources with frequent volcanic lithoclasts (Figure 4a,b,d).

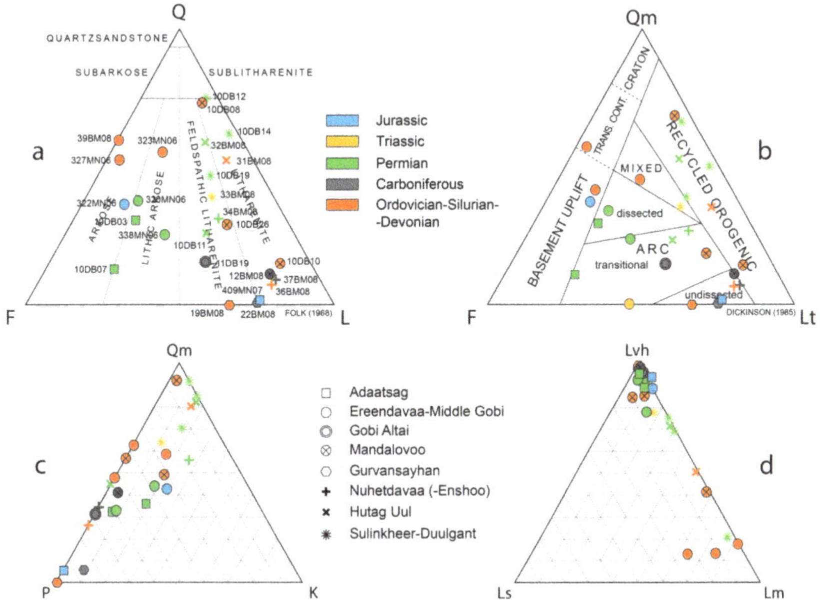

Figure 4. Detrital modes of analysed sandstones according to point-counting in ternary plots of Folk [62], (**a**) and Dickinson [63], (**b**–**d**) interpreting the composition and hypothetical plate tectonic position of the detrital sources. Abbreviation code: Q, total quartz; F, feldspar; L, lithoclasts; Qm, monocrystalline quartz; Lt, total lithoclasts (including quartzeous aphanitic ones); P, plagioclase; K, K-feldspar; Lvh, volcanic-hypabyssal lithoclasts; Ls, sedimentary lithoclasts; Lm, metamorphic lithoclasts.

As already observed by Bussien et al. [24], zircons older than 660 Ma (Cryogenian and older) are extremely rare (Figure 5). Regarding the detrital zircon age distribution, we observe that the mostly Permian sandstones also provide insight to the earlier history of the involved source areas (due to cannibalism). Therefore, we observe two main age clusters spanning the Ordovician-Silurian and Carboniferous-Permian (Figure 5a). The new analyses here available emphasise the strong Ordovician-Silurian population. Time corrected Epsilon Hf(t) ratios ($\varepsilon Hf_{(t)}$) show most detrital zircons to have been derived from moderately juvenile to juvenile melts. Including the negative ones and that along the CHUR, we discern a slight positive $\varepsilon Hf_{(t)}$ trend in time.

It is to note, the Adaatsag terrane is an ophiolitic suture zone comprising mélanges. By definition, suture zones may contain blocks and clastic formations, which were derived from both margins of a closing oceanic domain (polygenic mélanges *sensu* Raymond, [69]). The detrital zircon age and Hf-isotope results of Bussien et al. [24] and the present study stem from Permian-Jurassic maximum age sandstones deposited during early stages and during collision of the Hangay-Hentey and Ereendavaa margins of the Mongol-Okhotsk Belt, and therefore, we are not able to discriminate in detail the detrital input from the involved margins. However, the analysed Permian sandstones represent early collisional sandstones, because they are tectonically deformed in east-west striking belts (Figure 2A). Considering the well pronounced Ordovician-Silurian detrital zircon age population and related positive $\varepsilon Hf_{(t)}$ values, a strong similarity with results from the Hangay-Hentey terrane appears evident (Figures 4a and 5h). Therefore, a predominant detrital supply from the northern convergent margin (in modern coordinates) of the Mongol-Okhotsk Ocean is probable.

4.2. *Ereendavaa-Kherlen-Idermeg Terrane Group and Middle Gobi Belt*

For these terranes (called Ereendavaa-Middle Gobi for simplicity) we use mostly the data set of Bussien et al. [24] and an additional new one (39BM08) (Table 1). According to detrital zircon dating the analysed sandstones are of Ordovician to Jurassic maximum age (Figure 3). The Ordovician, Devonian and Permian-Early Triassic sandstones reveal a quartz-rich (≥50%) arkosic composition, which correlates approximately with continental basement sources (Figure 4).

The Silurian-Ordovician and Devonian sandstones likewise reveal scattered occurrence of detrital zircons older than 660 Ma. The Ereendavaa-Middle Gobi terrane records two main detrital zircon age clusters spanning the Late Neoproterozoic-Silurian, and the Carboniferous-Permian. While the former is characterised mostly by negative and transitional $\varepsilon Hf_{(t)}$ numbers, the latter shows positive ones in the range of +5 to +15 (Figure 5b). The Jurassic sandstone, unconformably topping the terrane, matches a similar trend, i.e., further reworking of similar source rocks into the Jurassic continental basin. Comparably to the Adaatsag samples, Devonian detrital zircons are rare again.

The trend depicted by the $\varepsilon Hf_{(t)}$ ratios infers a transition from evolved to moderately juvenile/juvenile magma sources, in which the zircons crystallised. This can be correlated with the change from extension-related (comprising a prasinitic sill; Figure 2C) to subduction-related volcano-plutonic activity in the Carboniferous (as suggested by Bussien et al. [24]) along the southern margin of the Mongol-Okhotsk Ocean. Some exceeding $\varepsilon Hf_{(t)}$ points of +10 to +20 in the age range of 500–380 Ma may suggest a connection with and transport from the adjacent Mandalovoo terrane to the south, where subduction occurred at that time (see Figures 4b and 5c).

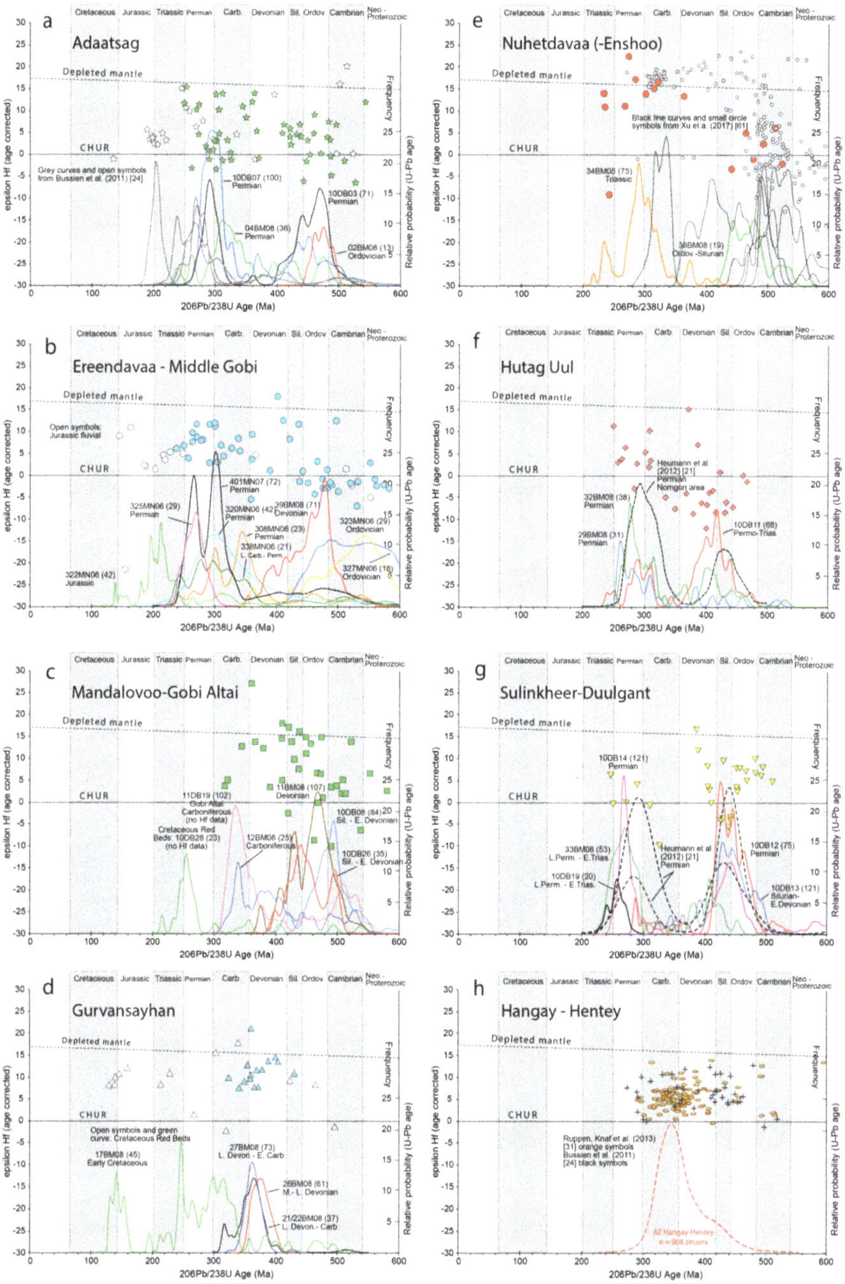

Figure 5. U-Pb detrital zircon age populations represented by relative probability curves plotted against age corrected epsilon Hafnium-isotope ratios as derived from sandstones in various terranes. The stratigraphic sample ages are indicated according to mapped ages or sample maximum age from detrital zircons presently analysed (see Table 1). Numbers in brackets after the sample numbers indicate the the number of detrital zircons analysed.

4.3. Mandalovoo-Gobi Altai Terranes

According to the obtained maximum detrital zircon ages, Silurian-Devonian and Carboniferous sandstones were analysed (Figures 3 and 5). They show a feldspatic litharenite and litharenite composition rich in volcanic clasts (Figure 4d). A sourcing in volcanic arc and recycled orogenic rock units can be inferred (Figure 4b). In the detrital zircon age distribution two major clusters are identified overlapping with the Cambrian to Silurian and Carboniferous periods. Again, Devonian ages appear subordinate (Figures 3 and 4c). Except for a few negative ones, $\varepsilon Hf_{(t)}$ values mainly plot in the moderately juvenile to depleted mantle source fields (Figure 5c).

The hafnium-isotope characteristics of the Mandalovoo terrane detrital zircons suggest a subduction-related volcano-plutonic source of the sandstone deposits active from at least later Cambrian until the Carboniferous. The positive trend in time suggests a successive depletion trend of the zircon source magmas. Younger sediments, except the discordantly overlying Cretaceous red beds of the Flaming Cliffs, we have not identified. These record reworking of Permian/Triassic detrital zircons as commonly observed (Figure 5c). According to Badarch et al. [10] the deformed rock series of the Mandalovoo terrane is unconformably overlain by Permian and Cretaceous clastic formations. Our finding of Silurian to Carboniferous maximum ages of the sandstones confirms this and suggests that the Mandalovoo depositional site was tectonically eliminated at the Carboniferous/Permian transition. The facies association comprising pelagic/turbiditic, shallow water limestones and continental alluvial fan deposits (Figure 2D) suggests that the terrane formed in a shallowing-upward forearc-accretionary wedge environment.

4.4. Gurvansayhan Terrane

Two representative sandstone samples were point-counted for their modal composition. They are very rich in volcanic lithoclasts and fall in the feldspatic litharenite and litharenite fields, virtually free of detrital quartz (Figure 4). An undissected volcanic arc source [63] is implied. In spite of the basic to intermediate chemistry of the supplying source, we were successful to obtain a reasonable number of detrital zircons which ages span the Devonian and Carboniferous (Figures 3 and 5d). Peak ages lie concurrently on the Devonian-Carboniferous transition. This compares well with the U-Pb average detrital zircon peak age of ≈362 Ma obtained by Dolgopolova et al. [48]. Older zircons from continental basement completely lack. The volcaniclastic facies and facies associations we studied in the Oyu Tolgoi mining district, the absence of continental influx to the basin and the confined detrital zircon age distributions corroborate an intra-oceanic arc environment [56], which soon after its construction was eliminated by tectonic processes.

The $\varepsilon Hf_{(t)}$ results from the discordantly overlying, post-collisional continental red beds document that moderate juvenile and juvenile volcano-plutonic sources continued activity during Mesozoic until Early Cretaceous [70,71]. The ~160–120 Ma detrital zircon age data we obtained match the regionally occurring surficial volcanic activity, which was sourced in a metasomatized depleted lithospheric mantle with minor crustal contamination [70,71]. Other older detrital zircon ages presumably were reworked from the short-lived Gurvansayhan intra-oceanic arc.

4.5. Nuhetdavaa (-Enshoo) Terrane

The area occupied by the Enshoo terrane is very narrow and characterised by poor outcrop conditions in Mongolia. According to Badarch et al. [10], the Nuhetdavaa terrane is classified as back or forearc basin and an island arc (Enshoo), respectively. We assume their common relationship and consider our results from Nuhetdavaa as also applying to the Enshoo unit. Following the maximum age attributions by our detrital zircon dating, the mapped ages are quite different (Table 1). We present results from an Ordovician, a Permian-Triassic, and a not age-confirmed (detrital zircons) Carboniferous sandstone (Figures 4 and 5e).

Petrographically, the two older sandstones (Ordovician and Carboniferous) are litharenites poor in quartz (≤10%), the Permian-Triassic one contains ≈40% of that grains. The former two plot in the undissected volcanic arc, and the latter almost in the recycled orogenic field (Figure 4). The Ordovician sandstone reveals $\varepsilon Hf_{(t)}$ values transitional along the CHUR line (ca. −3 to +8), which may indicate a crustal contaminated continental volcanic arc source input. The Permian–Triassic sandstone shows a main influx of later Carboniferous to Permian zircons derived from juvenile to depleted mantle volcano-plutonic sources. No older grains than Cambrian became obvious from our data.

Another analysis of detrital zircons in the north eastern extension of the Nuhetdavaa terrane on Chinese territory (Uliastai continental margin, Figure 1) is available from Xu et al. [42]. Therein, Ordovician-Devonian sandstones record Neo-Proterozoic to Cambrian sources with a large spread of $\varepsilon Hf_{(t)}$ from −10 to +20. This is interpreted by the authors to show derivation of the sands from mixed sources in their Northern accretionary orogen (Hutag Uul terrane in our terminology). The detrital zircon age data and Hafnium-isotope signatures of Xu et al. [42], and our results firstly corroborate the correlation of the Nuhetdavaa terrane with the Uliastai terrane in the Xing'an block. Secondly, the time trend depicted by the $\varepsilon Hf_{(t)}$ signatures suggests a Neo-Proterozoic rift/passive margin tectonic environment which inverted from about late Cambrian to a subduction related continental margin.

4.6. Hutag Uul Terrane

From this terrane we can present detrital zircon results from three samples correlatable with Permian and Permo-Triassic maximum ages (Table 1, Figure 3). Modal analysis of two sandstones classify them as feldspathic litharenites and litharenites and plot in the transitional arc and recycled orogenic fields (Figure 4). The zircon yield showed to be moderate but two age clusters appear consistent, Silurian-Middle Devonian and Pennsylvanian-Permian (Figure 5f). These two populations reveal very distinct hafnium-isotope ratios. The older one mostly has negative $\varepsilon Hf_{(t)}$ values down to −15, the younger one positive up to +10. Consequently, the Silurian-Middle Devonian zircons crystallised in crust-derived magmas, the Pennsylvanian-Permian ones in moderately juvenile and juvenile magmas, significant for crustal contamination of rising mantle magmas.

In plate-tectonic terms, the Hutag Uul continental (block) terrane [10] first experienced extension (rifting) during Silurian-Middle Devonian. It was inverted to a subduction regime environment with the establishment of a continental margin arc from about Pennsylvanian (late Carboniferous) on. Due to similar Permian-Triassic detrital zircon age distributions sandstones (Figure 5e,f), a common detrital source in the Nuhetdavaa continental margin arc may have existed.

Because of possible affinity of the Atasbogd/Zoolen and Tsagaan Uul terranes with such in south eastern Mongolia like Hutag Uul or Nuhetdavaa-Enshoo, some preliminary detrital zircon age dating on two Late Devonian sandstones are provided (Figure 3). Both, by revealing major Devonian peaks most likely cannot be correlated with the Hutag Uul and other terranes in the South Mongolian belt south of the MML. A common genesis with sandstones of the same age in Gurvansayhan terrane is impossible according to our sandstone framework data (Figure 4).

4.7. Sulinkheer-Duulgant Terrane

From this terrane (elsewhere also called Solonker multiple suture; e.g., [3,9]) a large data set is available, however, mostly from Permian sandstones (Figures 3 and 5). The point-counted Permian-Early Triassic sandstones have a feldspathic litharenite and litharenite composition quite rich in quartz grains (≈50–75%). The detrital source from a recycled orogen is inferred (Figure 4). With regard to the detrital zircon ages and $\varepsilon Hf_{(t)}$ measurements (Figure 5g), again a distinct bi-modal age distribution is observed: (i) an older population overlapping with later Cambrian and Early Devonian with major peaks in Ordovician and Silurian; the $\varepsilon Hf_{(t)}$ values are mostly positive up to +20, but a certain crustal contamination of the magmas seems also displayed, (ii) the younger one correlates with the Permian to Early Triassic. Less supported by the number of $\varepsilon Hf_{(t)}$ data points, a magma source of the detrital zircon with crustal contamination may be assumed. An identical bi-modal age distribution was reported by Heumann et al. [21] in Permian deposits in the Nomgon area of that terrane (peaks at ≈300 and 450 Ma, respectively). All Permian sample spots showed tectonic tilting to vertical bedding documenting post-depositional deformation.

4.8. Hangay-Hentey Terrane

In earlier project work we investigated the Mongol-Okhotsk Belt on the Mongolian territory [24,31]. The Hangay-Hentey accretionary complex comprises lithologic elements as oceanic basalts, seamount capped by crinoidal limestones, pelagic radiolarites, shales and turbiditic fan and forearc deposits [24,31,72,73]. Litho-, chrono- and biostratigraphic results clearly indicate that the Mongol-Okhotsk Ocean opened in Silurian. In summary, the Hangay-Hentey terrane represents the accretionary wedge and forearc that formed along the northern margin of the Mongol-Okhotsk Ocean. Detrital zircon age and hafnium-isotope signatures (Figure 5h) document continuous clastic influx from mainly moderately evolved volcano-plutonic continental arc source rocks [23,30], from Silurian until Permian.

With regard to our results, there are two points to emphasise: (i) also the analysed Permian sandstones give a precise insight into the earlier history of the terrane, i.e., no earlier collision has modified the supplying source rock composition, (ii) the rifting and opening of the Mongol-Okhotsk Ocean in Silurian is obviously not characterized by negative $\varepsilon Hf_{(t)}$ values of Silurian zircons as it would be expected. In view of the large number of zircons that have been reliably dated, we do not assume that the Hf values were biased by incorrect age attributions [74]. Presumably, the chemical signature of the newly produced magmatism was inherited from repeated earlier orogenic processes. In other words, the magmatic melts formed in a formerly (Cambrian) homogenised crust.

5. Palaeotectonic Evolution

5.1. General Trends

To distinguish between extensive and compressive plate tectonic processes we base our interpretations mainly on the age populations and related $\varepsilon Hf_{(t)}$ signatures of the detrital zircons in sandstones. Interestingly, it also turns out that the measured age populations in younger formations (e.g., Permian) provide insight into earlier paleotectonic events through the recycling of older formations.

From the overview of the detrital zircon ages (Figure 3) in the different terranes in the Central and South Mongolian belts it becomes evident that major magmatic (volcanic) activity took place in two periods, from Cambrian to Silurian and from Late Devonian to Permian, respectively. We identify two exceptions: (i) in the Gurvansayhan of Oyu Tolgoi detrital zircon ages are limited in majority to Late Devonian and less to Carboniferous, and (ii) in the Hangay-Heney basin Late Devonian to Carboniferous detrital zircon influx prevailed. Hence, this points to their distinctive geological development compared to the other terranes, because in the latter deficits of Devonian zircons are discerned (Figures 3 and 5). Another point to note is that our few data from the Zoolen and Tsaagan Uul terranes indicate that they seemingly cannot be correlated with any formations east of the Gurvansayhan terrane.

5.2. Development in Time

5.2.1. Late Cambrian-Silurian

According to our results, in the late Cambrian we are able to distinguish two composite continental blocks along and within the Palaeo-Asian Ocean (Figure 6) which were separated by the Ondor Sum Ocean as described in NW China (e.g., Miao et al. [12]; Wilhem et al. [6]): (i) to the north the Late Cambrian collage (A in Figure 6) comprising the amalgamated Baikalids (e.g., [14,75]) with the later Hangay-Hentey, Ereendavaa, Idermeg, Gobi Altai and Mandalovoo terranes, and the Kherlen oceanic suture (Miao et al. 2016); (ii) in the south a composite continental block including the later Nuhetdavaa-Enshoo, Hutag Uul and Sulinkheer-Duulgant terranes bordered by the Solonker.

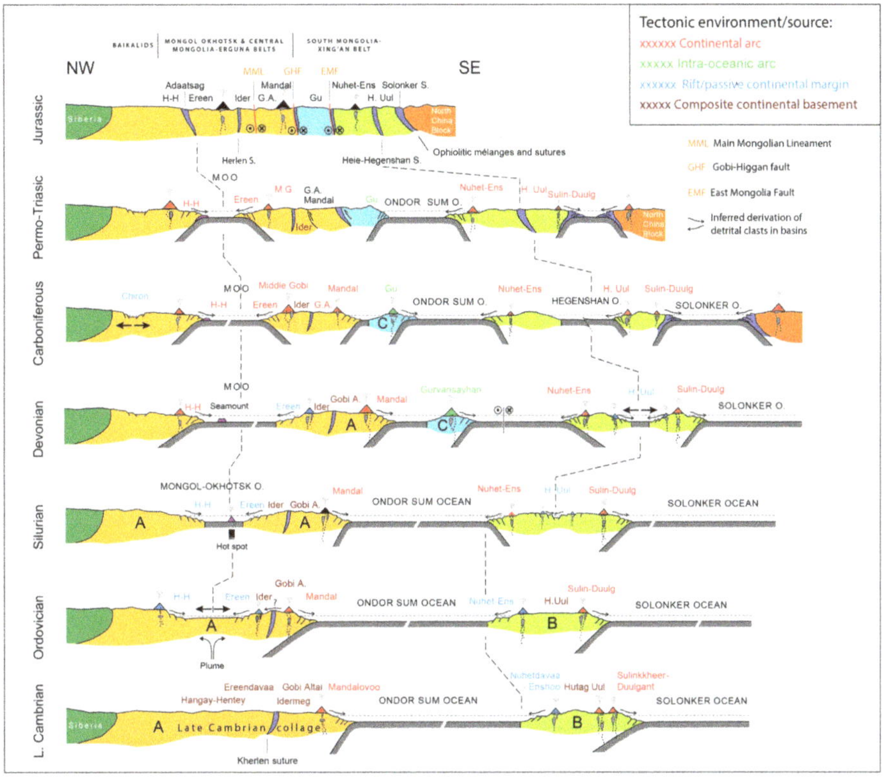

Figure 6. Palaeotectonic development of the Central Asian Orogenic Belt in Mongolia (oriented ≈N–S in modern coordinates) in geologic time from late Cambrian to Jurassic. The inferred plate tectonic environment of terranes and sources are highlighted in colour. Central Mongolian Superterrane, A; Southeastern Mongolian Superterrane, B; Gurvansayhan Terrane, C.

Ocean in the south (B in Figure 6). The Solonker Ocean belongs to the Palaeo-Asian Ocean situated between our block B and the North China Block.

Subduction took place at the southern edges of both blocks as the ages of the detrital zircon S and the mostly positive $\varepsilon Hf_{(t)}$ signatures (Figure 5c,g) in the Mandalovoo-Gobi Altai and Sulinkheer-Duulgant terranes infer. The two continental margins developed through the northward subduction of the Ondor Sum and Solonker oceans, respectively. Mixed mantle and crust generated magmas were produced in the sourcing continental volcanic arcs. Both units have in common that the volcanic influx decreased during Devonian most likely due to decelerated subduction. This coincides with the Gurvansayhan intra-oceanic arc generation and the initiation of subduction along the northern Mongol-Okhotsk Ocean (Figure 6; see also below). The mostly recycled orogenic source signature in Permian-Triassic sandstones in the Sulinkheer-Duulgant terrane (Figure 4b) reveal important presence of crustal rocks in the arc at that time. Mandalovoo-Gobi Altai Palaeozoic sandstones reflect a higher contribution of contemporaneous influx from the volcanic continental arc.

In the Nuhetdavaa (-Enshoo) terrane our sparse data are well completed by that of Xu et al. [43] in the comparable Uliastai passive margin terrane in Inner Mongolia. Negative and intermediate

$\varepsilon Hf_{(t)}$ values (Figure 5e) suggest crustal derived magmatic-volcanic activity in a passive margin setting. However, it will be inverted later (see below).

Epsilon Hf ratios suggest that in the late Cambrian collage, during Ordovician-Silurian the composite continental block, including the Hangay-Hentey, Ereendavaa and Idermeg terranes experienced rifting, extension and generation of oceanic lithosphere of the Mongol-Okhotsk Ocean in Silurian [24,27]. In the later Hangay-Hentey accretionary wedge terrane, the timing is well supported by biostratigraphic dating and seamount stratigraphy [31,72,73]. Along the southern margin of the Mongol-Okhotsk Ocean, in the Ereendava terrane, extension in a passive margin situation is suggested by a monotonous turbiditic facies association (including prasinitic dykes) and negative $\varepsilon Hf_{(t)}$ values. The spreading may have been driven by a mantle plume (Figure 6) rising from the mantle transition zone. Such process commonly is associated with hot-spot generation [47].

During Silurian the Mongol-Okhotsk Ocean oceanic lithosphere was created bearing a prominent volcanic seamount capped by crinoidal limestones which became progressively drowned and covered by pelagic and turbiditic deposits [31]. The highly negative $\varepsilon Hf_{(t)}$ signatures in detrital zircons detected by Xu et al. [43] and our measurements suggests the Nuhetdavaa (-Enshoo) and Uliastai passive margins to have been inverted to a continental arc regime during Silurian by the southward subduction of the Ondor Sum Ocean (Figure 6).

5.2.2. Devonian-Carboniferous

The Devonian period brought about great changes in the paleotectonic development of the Palaeo-Asian Ocean. Viewed from north to south these were: (i) start of northward subduction of the Mongol-Okhotsk Ocean under the Baikalids (see also [14,24,31]), (ii) the establishment of the intra-oceanic Gurvansayhan island arc in the Ondor Sum Ocean by north-dipping intra-oceanic subduction [6,53], and (iii) the opening of the Hegenshan Ocean [12]. The latter was preceded by the Silurian rifting in the Hutag Uul area.

In the Carboniferous period the southern passive margin of the Mongol-Okhotsk Ocean (Ereendavaa terrane) switched to subduction of the Mongol-Okhotsk Ocean with the establishment of a continental arc [24]. Medium positive $\varepsilon Hf_{(t)}$ rates (Figure 5b) suggest a mixed mantle-crust source of the melts. Similar $\varepsilon Hf_{(t)}$ signatures (0 to 15) of detrital zircons are observed in Hangay-Hentey sandstones (Figure 5h) which corroborates their common beginning in the late Cambrian collage. In the Hutag Uul terrane the $\varepsilon Hf_{(t)}$ data show a trend to positive values during the transition from Devonian to Carboniferous. This suggests that the former passive continental margin has been inverted into a subduction-linked one. The newly emerging continental volcanic arc obviously is continued in the Sunid-Baolidao volcanic arc in Inner Mongolia [2,9,12].

5.2.3. Permian-Triassic

The Permo-Triassic time was characterized by the elimination of several oceanic domains, and the collision of volcanic arcs as the youngest ages of the detrital zircons suggest (Figures 5 and 6). Due to the obtained minimum Carboniferous ages of the detrital zircons in both terranes, the Gurvansayhan intra-oceanic volcanic arc and the Gobi Altai-Mandalovoo continental island arc appear to have collided with the Idermeg continental sliver in the Permian. Cretaceous continental red beds overlay these terranes with unconformity. They partly show the reworking of older Palaeozoic material (Figure 5c,d) and distinct peaks at the Permian-Triassic transition. Probably this material was supplied from the still active Ereendavaa-Middle Gobi continental volcanic arc to the north. In the Baikalid realm, since the Devonian the northward subduction of the Mongol-Okhotsk Ocean has driven back-arc extension and mixed continental-marine sedimentation (e.g., Chiron basin; Popeko et al. [75]). Minor occurrence of Triassic detrital zircons in sandstones of the Hangay-Hentey, Ereendavaa-Middle Gobi, Nuhetdavaa (-Enshoo), Hutag Uul and Sulinkheer-Duulgant domains (Figure 3) attest their continued magmatic and sedimentary activity during the Permian and Triassic period.

5.2.4. Jurassic

The closure of the Mongol-Okhotsk Ocean and collision of the Hangay-Hentey accretionary complex with the Ereendavaa margin is timely constrained by generally flat-lying Jurassic continental sandstones and conglomerates, which unconformably cover deformed older series [22,24]. Both analysed Jurassic sandstones show peak detrital zircon ages at the Triassic-Jurassic transition (Figure 3). According to our variable age and Hf-isotope results (Figure 5a) from the Adaatsag suture, we infer that detrital debris from both margins was reworked into the analysed, tectonically deformed Permian-Triassic sediment series. Consequently, the final closure of the Mongol-Okhotsk Ocean was acquired at the Triassic-Jurassic transition (see also [44]). However, in other places (Noyon Uul area) Dimitru and Hendrix [76] documented continued collisional deformation and exhumation of Jurassic deposits. Our results from the southern realms (Nuhetdavaa-Enshoo, Hutag Uul and Sulinkheer-Duulgant) record youngest detrital zircons in majority of Permian and minor Triassic age. This explains the Late Triassic elimination of sedimentation sites associated with the consumption of the Ondor Sum and Solonker oceans as earlier suggested by e.g., [5,9,11]. Finally, the Jurassic closure of the Palaeo-Asian Ocean and collision of the terranes produced a peneplain surface, which stayed passive for about 150 My [77].

5.2.5. Terrane Accretion in Relationship with Siberian Rotation

In spite of the well documented clock-wise rotation of the Siberian cratonal core of the Central Asian Orogenic Belt [9,10], it is difficult to perceive the movements of the individual terranes. According to our results, the Central Mongolia-Erguna belt terranes (Gobi Altai, Mandalovoo and the Gurvansayhan) were welded to Siberia at the transition of late Carboniferous to Early Permian (Figure 6). They should have experienced similar rotation with Siberia from the late Carboniferous until end-Permian (e.g., [9,10]). For constraining the end of rotation, the results of Lehmann et al. [19] may be useful. According to the structural analysis of these authors, in southwestern Mongolia all terranes including the Baydrag block underwent S-N oriented tectonic shortening from ca. mid-Permian to Late Triassic. This means the southern South Mongolia-Xing'an and Inner Mongolia-Xilin Belts could have reached their position in the Jurassic without experiencing rotation. Later, during Jurassic and Cretaceous important coulissage of terranes along preferentially left-lateral fault zones and sutures in the range of tens to hundred kilometers occurred (e.g., [17,76,78]).

6. Conclusions

The present data set proves that detrital zircon age and Hf-isotope signatures strongly support provenance analysis aimed at reconstructing palaeogeographic and -tectonic processes in orogenic belts, and particularly in the Central Asian Orogenic Belt. Notably, the detrital zircon results from pre-collisional sandstone formations (Permian-Triassic) partly open a window on palaeotectonic processes even far in the past, because most larger blocks remained isolated during their history. Based on earlier terrane subdivisions in central and southern Mongolia we infer a new, clearer terrane subdivision by assembling smaller, genetically related terranes in larger superterranes.

A key element for the plate-tectonic reconstruction is the Gurvansayhan terrane. By its clear intra-oceanic setting, the division into a northern superterrane analogous with the Mongol-Okhotsk Belt and the Central Mongolia-Erguna Belt and a southern superterrane analogous with the South Mongolia-Xing'an and Inner Mongolia-Jilin belts) is acceptable (Figure 6). However, it persists an open question if and where the Gurvansayhan terrane can be traced into NW China.

These superterranes depict different tectonic and sedimentary evolutions in time and space. Each superterrane represents a plate tectonic assemblage of continental basement slivers, continental arc, forearc and backarc elements. However, there is no age or facies parallelism or repetition across the entire belt obvious, which would support doubling of terranes by oroclinal bending as supposed earlier (e.g., [33]).

Over-all terranes and time (late Cambrian-Cretaceous), the $\varepsilon Hf_{(t)}$ values show a trend towards positive numbers (see Supplementary Materials file). This suggests progressive removal of old crust and lithospheric mantle with increasing contribution of juvenile crust to magma reservoirs [79].

Supplementary Materials: The following are available online at http://www.mdpi.com/2075-163X/10/10/880/s1, Table S1: Table 1 extended providing informations on DGM sample positions, Table S2: All Terranes U-Pb and eHf LA-ICP-MS results, Table S3: LA-ICP-MS U-Pb detrital zircon ages of Bussien et al. 2011, Table S4: LA-ICP-MS eHf data of Bussien et al. 2011.

Author Contributions: W.W. and A.v.Q. were responsible for the project and participated with D.B., M.B. and C.A. in the fieldwork and the selection of the sample material. C.A. and M.B. contributed by their experience and knowledge of the geology of Mongolia. D.B. and M.B. carried out the various analyses in the laboratories of the Department of Earth Sciences ETH Zurich under the advice of A.v.Q. and W.W. All authors contributed to the design of the submitted manuscript. All authors have read and agreed to the published version of the manuscript.

Funding: This research was supported by Swiss Science Foundation grants 2-77927-06 and 2-77047-11 and ETH Zurich internal support.

Acknowledgments: We appreciate David Crane and Imants Kavalieris (Ivanhoe Corp.) for introduction and for providing borehole data and samples from the Oyu Tolgoi mining project. This project was fully supported by Swiss Science Foundation grants 2-77927-6 and 2-77047-11. We would like to thank three anonymous reviewers for constructive suggestions to improve the earlier versions of the manuscript.

Conflicts of Interest: The funders had no role in the design of the study; in the collection, analyses, or interpretation of data; in the writing of the manuscript, or in the decision to publish the results.

References

1. Zorin, Y.A.; Zorina, L.D.; Spiridonov, A.M.; Rutshtein, I.G. Geodynamic Setting of Gold Deposits in Eastern and Central Trans-Baikal (Chita Region, Russia). *Ore Geol. Rev.* **2001**, *17*, 215–232. [CrossRef]
2. Xiao, W.; Windley, B.F.; Hao, J.; Zhai, M. Accretion Leading to Collision and the Permian Solonker Suture, Inner Mongolia, China: Termination of the Central Asian Orogenic Belt. *Tectonics* **2003**, *22*. [CrossRef]
3. Jahn, B.-M.; Capdevila, R.; Liu, D.; Vernon, A.; Badarch, G. Sources of Phanerozoic granitoids in the transect Bayanhongor-Ulaan Baatar, Mongolia: Geochemical and Nd isotopic evidence, and implications for Phanerozoic crustal growth. *J. Asian Earth Sci.* **2004**, *23*, 629–653. [CrossRef]
4. Windley, B.F.; Alexeiev, D.; Xiao, W.; Kröner, A.; Badarch, G. Tectonic Models for Accretion of the Central Asian Orogenic Belt. *J. Geol. Soc.* **2007**, *164*, 31–47. [CrossRef]
5. Kröner, A.; Windley, B.F.; Badarch, G.; Tomurtogoo, O.; Hegner, E.; Jahn, B.M.; Gruschka, S.; Khain, E.V.; Demoux, A.; Wingate, M.T.D. Accretionary Growth and Crust Formation in the Central Asian Orogenic Belt and Comparison with the Arabian-Nubian Shield. *Mem. Geol. Soc. Am.* **2007**, *200*, 181–209. [CrossRef]
6. Wilhem, C.; Windley, B.F.; Stampfli, G.M. The Altaids of Central Asia: A Tectonic and Evolutionary Innovative Review. *Earth Sci. Rev.* **2012**, *113*, 303–341. [CrossRef]
7. Li, S.; Wang, T.; Wilde, S.A.; Tong, Y. Evolution, Source and Tectonic Significance of Early Mesozoic Granitoid Magmatism in the Central Asian Orogenic Belt (Central Segment). *Earth Sci. Rev.* **2013**, *126*, 206–234. [CrossRef]
8. Xiao, W.; Windley, B.F.; Sun, S.; Li, J.; Huang, B.; Han, C.; Yuan, C.; Sun, M.; Chen, H. A Tale of Amalgamation of Three Permo-Triassic Collage Systems in Central Asia: Oroclines, Sutures, and Terminal Accretion. *Annu. Rev. Earth Planet. Sci.* **2015**, *43*, 477–507. [CrossRef]
9. Cocks, L.R.M.; Torsvik, T.H. Siberia, the Wandering Northern Terrane, and Its Changing Geography through the Palaeozoic. *Earth Sci. Rev.* **2007**, *82*, 29–74. [CrossRef]
10. Badarch, G.; Cunningham, W.D.; Windley, B.F. A New Terrane Subdivision for Mongolia: Implications for the Phanerozoic Crustal Growth of Central Asia. *J. Asian Earth Sci.* **2002**, *21*, 87–110. [CrossRef]
11. Miao, L.; Zhu, M.; Liu, C.; Baatar, M.; Anaad, C.; Yang, S.; Li, X. Detrital-zircon age spectra of Neoproterozoic-Paleozoic sedimentary rocks on the Ereendavaa terrane in NE Mongolia: Implications for the early-stage evolution of the Ereendavaa terrane and the Mongol-Okhotsk ocean. *Minerals* **2020**, *10*, 742. [CrossRef]
12. Miao, L.; Zhang, F.; Fan, W.M.; Liu, D. Phanerozoic Evolution of the Inner Mongolia-Daxinganling Orogenic Belt in North China: Constraints from Geochronology of Ophiolites and Associated Formations. *Geol. Soc. Am. Spec. Publ.* **2007**, *280*, 223–237. [CrossRef]

13. Donskaya, T.V.; Gladkochub, D.P.; Mazukabzov, A.M.; Ivanov, A.V. Late Paleozoic-Mesozoic Subduction-Related Magmatism at the Southern Margin of the Siberian Continent and the 150 Million-Year History of the Mongol-Okhotsk Ocean. *J. Asian Earth Sci.* **2013**, *62*, 79–97. [CrossRef]
14. Porter, T.M. The Geology, Structure and Mineralisation of the Oyu Tolgoi Porphyry Copper-Gold-Molybdenum Deposits, Mongolia: A Review. *Geosci. Front.* **2016**, *7*, 375–407. [CrossRef]
15. Miao, L.; Baatar, M.; Zhang, F.; Anaad, C.; Zhu, M.; Yang, S. Cambrian Kherlen Ophiolite in Northeastern Mongolia and Its Tectonic Implications: SHRIMP Zircon Dating and Geochemical Constraints. *Lithos* **2016**, *261*, 128–143. [CrossRef]
16. Lamb, M.A.; Badarch, G. Paleozoic Sedimentary Basins and Volcanic-Arc Systems of Southern Mongolia: New Stratigraphic and Sedimentologic Constraints. *Int. Geol. Rev.* **1997**, *39*, 542–576. [CrossRef]
17. Blight, J.H.S.; Cunningham, D.; Petterson, M.G. Crustal Evolution of the Saykhandulaan Inlier, Mongolia: Implications for Palaeozoic Arc Magmatism, Polyphase Deformation and Terrane Accretion in the Southeast Gobi Mineral Belt. *J. Asian Earth Sci.* **2008**, *32*, 142–164. [CrossRef]
18. Kröner, A.; Lehmann, J.; Schulmann, K.; Demoux, A.; Lexa, O.; Tomurhuu, D.; Štípská, P.; Liu, D.; Wingat, M.T.D. Lithostratigraphic and Geochronological Constraints on the Evolution of the Central Asian Orogenic Belt in SW Mongolia: Early Paleozoic Rifting Followed by Late Paleozoic Accretion. *Am. J. Sci.* **2010**, *310*, 523–574. [CrossRef]
19. Lehmann, J.; Schulmann, K.; Lexa, O.; Corsini, M.; Kröner, A.; Štípská, P.; Tomurhuu, D.; Otgonbator, D. Structural Constraints on the Evolution of the Central Asian Orogenic Belt in SW Mongolia. *Am. J. Sci.* **2010**, *310*, 575–628. [CrossRef]
20. Heumann, M.J.; Johnson, C.L.; Webb, L.E.; Taylor, J.P.; Jalbaa, U.; Minjin, C. Paleogeographic Reconstruction of a Late Paleozoic Arc Collision Zone, Southern Mongolia. *Bull. Geol. Soc. Am.* **2012**, *124*, 1514–1534. [CrossRef]
21. Wu, G.Y.; Liu, M.; Wang, Y. Palinspastic Reconstruction and Geological Evolution of Permian Residual Marine Basins Bordering China and Mongolia. *J. Palaeogeogr.* **2014**, *3*, 219–232. [CrossRef]
22. Yakubchuk, A. Architecture and Mineral Deposit Settings of the Altaid Orogenic Collage: A Revised Model. *J. Asian Earth Sci.* **2004**, *23*, 761–779. [CrossRef]
23. Bussien, D.; Gombojav, N.; Winkler, W.; von Quadt, A. The Mongol-Okhotsk Belt in Mongolia-An Appraisal of the Geodynamic Development by the Study of Sandstone Provenance and Detrital Zircons. *Tectonophysics* **2011**, *510*, 132–150. [CrossRef]
24. Wang, W.; Tang, J.; Xu, W.L.; Wang, F. Geochronology and Geochemistry of Early Jurassic Volcanic Rocks in the Erguna Massif, Northeast China: Petrogenesis and Implications for the Tectonic Evolution of the Mongol-Okhotsk Suture Belt. *Lithos* **2015**, *218–219*, 73–86. [CrossRef]
25. Liu, Y.; Li, W.; Feng, Z.; Wen, Q.; Neubauer, F.; Liang, C. A Review of the Paleozoic Tectonics in the Eastern Part of Central Asian Orogenic Belt. *Gondwana Res.* **2017**, *43*, 123–148. [CrossRef]
26. Zorin, Y.A. Geodynamics of the Western Part of the Mongolia-Okhotsk Collisional Belt, Trans-Baikal Region (Russia) and Mongolia. *Tectonophysics* **1999**, *306*, 33–56. [CrossRef]
27. Parfenov, L.M. Problems of Tectonies of the Mongol-Okhotsk. *Geol. Pac. Ocean.* **2001**, *16*, 797–830.
28. Tomurtogoo, O.; Windley, B.F.; Kröner, A.; Badarch, G.; Liu, D.Y. Zircon Age and Occurrence of the Adaatsag Ophiolite and Muron Shear Zone, Central Mongolia: Constraints on the Evolution of the Mongol-Okhotsk Ocean, Suture and Orogen. *J. Geol. Soc. Lond.* **2005**, *162*, 125–134. [CrossRef]
29. Kelty, T.K.; Yin, A.; Dash, B.; Gehrels, G.E.; Ribeiro, A.E. Detrital-Zircon Geochronology of Paleozoic Sedimentary Rocks in the Hangay-Hentey Basin, North-Central Mongolia: Implications for the Tectonic Evolution of the Mongol-Okhotsk Ocean in Central Asia. *Tectonophysics* **2008**, *451*, 290–311. [CrossRef]
30. Ruppen, D.; Knaf, A.; Bussien, D.; Winkler, W.; Chimedtseren, A.; von Quadt, A. Restoring the Silurian to Carboniferous Northern Active Continental Margin of the Mongol-Okhotsk Ocean in Mongolia: Hangay-Hentey Accretionary Wedge and Seamount Collision. *Gondwana Res.* **2014**, *25*, 1517–1534. [CrossRef]
31. Van der Voo, R.; van Hinsbergen, D.J.J.; Domeier, M.; Spakman, W.; Torsvik, T.H. Latest Jurassic–Earliest Cretaceous Closure of the Mongol-Okhotsk Ocean: A Paleomagnetic and Seismological-Tomographic Analysis. *Geol. Soc. Am. Spec. Publ.* **2015**, *513*, 589–606. [CrossRef]
32. Sengör, A.M.C.; Natal'in, B.A.; Burtman, V.S. Evolution of the Altaid tectonic collage and Palaeozoic crustal growth in Eurasia. *Nature* **1993**, *364*, 299–306. [CrossRef]

33. Amelin, Y.; Lee, D.-C.; Halliday, A.N.; Pidgeon, R.T.; Hills, J. Nature of the Earth's earliest crust from hafnium isotopes in single detrital zircons. *Nature* **1999**, *399*, 252–255. [CrossRef]
34. Košler, J.; Fonneland, H.; Sylvester, P.; Tubrett, M.; Pedersen, R.B. U-Pb Dating of Detrital Zircons for Sediment Provenance Studies—A Comparison of Laser Ablation ICPMS and SIMS Techniques. *Chem. Geol.* **2002**, *182*, 605–618. [CrossRef]
35. Kemp, A.S.; Hawkesworth, C.J.; Foster, G.L.; Paterson, B.A.; Woodhead, J.D.; Hergt, J.M.; Gray, C.M.; Whitehouse, M.J. Magmatic and Crustal Differentiation History of Granitic Rocks from Hf-O Isotopes. *Science* **2007**, *315*, 980–983. [CrossRef] [PubMed]
36. Fryer, J.; Jackson, S.E.; Longerich, H.P. The Application of Laser Ablation Microprobe-Inductively Coupled Plasma-Mass Spectrometry (LA-ICP-MS) to in Situ (U)Pb Geochronology. *Chem. Geol.* **1993**, *109*, 1–8. [CrossRef]
37. Griffin, W.L.; Belousova, E.A.; Shee, S.R.; Pearson, N.J.; O'Reilly, S.Y. Archean crustal evolution in the northern Yilgarn Craton: U-Pb and Hf-isotope evidence from detrital zircons. *Precambrian Res.* **2004**, *131*, 231–282. [CrossRef]
38. Zhou, J.B.; Wilde, S.A.; Zhang, X.Z.; Zhao, G.C.; Liu, F.L.; Qiao, D.W.; Ren, S.M.; Liu, J.H. A >1300 km Late Pan-African Metamorphic Belt in NE China: New Evidence from the Xing'an Block and Its Tectonic Implications. *Tectonophysics* **2011**, *509*, 280–292. [CrossRef]
39. Li, S.; Wilde, S.A.; He, Z.; Jiang, X.; Liu, R.; Zhao, L. Triassic Sedimentation and Postaccretionary Crustal Evolution along the Solonker Suture Zone in Inner Mongolia, China. *Tectonics* **2014**, *33*, 960–981. [CrossRef]
40. Xu, B.; Zhao, G.; Li, J.; Liu, D.; Wang, B.; Han, Y.; Eizenhöfer, P.R.; Zhang, X.; Hou, W.; Liu, Q. Ages and Hf Isotopes of Detrital Zircons from Paleozoic Strata in the Chagan Obo Temple Area, Inner Mongolia: Implications for the Evolution of the Central Asian Orogenic Belt. *Gondwana Res.* **2017**, *43*, 149–163. [CrossRef]
41. Sorokin, A.A.; Zaika, V.A.; Kovach, V.P.; Kotov, A.B.; Xu, W.; Yang, H. Timing of Closure of the Eastern Mongol–Okhotsk Ocean: Constraints from U–Pb and Hf Isotopic Data of Detrital Zircons from Metasediments along the Dzhagdy Transect. *Gondwana Res.* **2020**, *81*, 58–78. [CrossRef]
42. Şengör, A.M.C.; Natal-in, B.A.; Sunal, G.; Van Der Voo, R. A New Look at the Altaids: A Superorogenic Complex in Northern and Central Asia as a Factory of Continental Crust Part I: Geological Data Compilation (Exclusive of Palaeomagnetic Observations). *Austrian J. Earth Sci.* **2014**, *107*, 169–232.
43. Filippova, I.B.; Bush, V.A.; Didenko, A.N. Middle Paleozoic Subduction Belts: The Leading Factor in the Formation of the Central Asian Fold-and-Thrust Belt. *Russ. J. Earth Sci.* **2001**, *3*, 405–426. [CrossRef]
44. Zorin, Y.A.; Sklyarov, E.V.; Belichenko, V.G.; Mazukabzov, A.M. Evolution of Island Arcs and Geodynamics of the Eastern Central Asian Foldbelt in the Neogea. *Dokl. Earth Sci.* **2007**, *412*, 39–42. [CrossRef]
45. Safonova, I.; Seltmann, R.; Kröner, A.; Gladkochub, D.; Schulmann, K.; Xiao, W.; Kim, J.; Komiya, T.; Sun, M. A New Concept of Continental Construction in the Central Asian Orogenic Belt: (Compared to Actualistic Examples from the Western Pacific). *Episodes* **2011**, *34*, 186–196. [CrossRef]
46. Tang, J.; Xu, W.L.; Wang, F.; Wang, W.; Xu, M.J.; Zhang, Y.H. Geochronology and Geochemistry of Early-Middle Triassic Magmatism in the Erguna Massif, NE China: Constraints on the Tectonic Evolution of the Mongol-Okhotsk Ocean. *Lithos* **2014**, *184–187*, 1–16. [CrossRef]
47. Erdenechimeg, D.; Damdinjav, B.; Enkhbayar, B.; Boldbaatar, G. *National Geological Map of Mongolia at the Scale of 1:500,000 (MUGZ-500) (37 Lists)*; Report#8480; Tumurtogoo, O., Orolmaa, D., Eds.; Geological Investigation Center: Lanzhou, China, 2017.
48. Dolgopolova, A.; Seltmann, R.; Armstrong, R.; Belousova, E.; Pankhurst, R.J.; Kavalieris, I. Sr-Nd-Pb-Hf Isotope Systematics of the Hugo Dummett Cu-Au Porphyry Deposit (Oyu Tolgoi, Mongolia). *Lithos* **2013**, *164–167*, 47–64. [CrossRef]
49. Wu, F.Y.; Sun, D.Y.; Ge, W.C.; Zhang, Y.B.; Grant, M.L.; Wilde, S.A.; Jahn, B.M. Geochronology of the Phanerozoic Granitoids in Northeastern China. *J. Asian Earth Sci.* **2011**, *41*, 1–30. [CrossRef]
50. Cohen, K.M.; Finney, S.C.; Fan, J.-X. The ICS International Chronostratigraphic Chart. *Episodes* **2013**, *36*, 199–204. [CrossRef]
51. Crane, D.; Kavalieris, I. Geologic overview of the Oyu Tolgoi porphyry Cu-Au-Mo deposits, Mongolia. In *Geology and Genesis of Major Copper Deposits and Districts of the World-A tribute to Richard H Sillitoe*; Hedenquist, J.W., Harris, M., Camus, F., Eds.; Society of Economic Geologists Special Publication: Littleton, CO, USA, 2012; Volume 16, pp. 187–213.

52. Lamb, M.A.; Cox, D. New ^{40} Ar/^{39} Ar age data and implications for porphyry copper deposits of Mongolia. *Econ. Geol.* **1998**, *93*, 524–529. [CrossRef]
53. Perelló, J.; Cox, D.; Garamjav, D.; Sanjdorj, S.; Diakov, S.; Schissel, D.; Munkhbat, T.-O.; Oyun, G. Oyu Tolgoi, Mongolia—Silurian-Devonian porphyry Cu-Au-(Mo) and high-sulfidation Cu mineralization with a Cretaceous chalcocite blanket. *Econ. Geol.* **2001**, *96*, 1407–1428. [CrossRef]
54. Kavalieris, I. *Notes on 21 samples for zircon U–Pb geochronology from the Oyu Tolgoi Project, South Gobi Desert, Mongolia*; Technical Report for Ivanhoe Mines Mongolia; Ivanhoe Mines: Khanbogd, Mongolia, 2005; p. 48.
55. Tolgoi, O. *Oyu Tolgoi Geological Summary Report*; Internal Technical Report for Ivanhoe Mines Mongolia; Ivanhoe Mines: Khanbogd, Mongolia, 2008; p. 129.
56. Johnson, C.L.; Amory, J.A.; Zinniker, D.; Lamb, M.A.; Graham, S.A.; Affolter, M.; Badarch, G. Sedimentary response to arc-continent collision, Permian, southern Mongolia. In *Formation and Applications of the Sedimentary Record in Arc Collision Zones*; Draut, A., Clift, P.D., Scholl, D.W., Eds.; Geological Society of America: Boulder, CO, USA, 2007; Volume 436, pp. 363–390.
57. Webb, L.E.; Johnson, C.L. Tertiary Strike-Slip Faulting in Southeastern Mongolia and Implications for Asian Tectonics. *Earth Planet. Sci. Lett.* **2006**, *241*, 323–335. [CrossRef]
58. Jian, P.; Liu, D.; Kröner, A.; Windley, B.F.; Shi, Y.; Zhang, W.; Zhang, F.; Miao, L.; Zhang, L.; Tomurhuu, D. Evolution of a Permian intraoceanic arc–trench system in the Solonker suture zone, Central Asian Orogenic Belt, China and Mongolia. *Lithos* **2010**, *118*, 169–190. [CrossRef]
59. Xu, B.; Zhao, P.; Wang, Y.; Liao, W.; Luo, Z.; Bao, Q.; Zhou, Y. The Pre-Devonian Tectonic Framework of Xing'an-Mongolia Orogenic Belt (XMOB) in North China. *J. Asian Earth Sci.* **2015**, *97*, 183–196. [CrossRef]
60. Zhao, P.; Fang, J.; Xu, B.; Chen, Y.; Faure, M. Early Paleozoic Tectonic Evolution of the Xing-Meng Orogenic Belt: Constraints from Detrital Zircon Geochronology of Western Erguna-Xing'an Block, North China. *J. Asian Earth Sci.* **2014**, *95*, 136–146. [CrossRef]
61. Li, S.; Chung, S.L.; Wilde, S.A.; Wang, T.; Xiao, W.J.; Guo, Q.Q. Linking Magmatism with Collision in an Accretionary Orogen. *Sci. Rep.* **2016**, *6*, 25751. [CrossRef]
62. Folk, R.L. *Petrology of Sedimentary Rocks*; Hemphill Publishing Company: Austin, TX, USA, 1974; p. 182.
63. Dickinson, W.R. Interpreting detrital modes of greywacke and arkose. *J. Sediment. Petrol.* **1970**, *40*, 695–707. [CrossRef]
64. van Achterberg, E.; Ryan, C.G.; Jackson, S.E.; Griffin, W. Data reduction software for LA652 ICP-MS. In *Laser Ablation-ICPMS in the Earth Science: Principles and 653 Applications*; Sylvester, P., Ed.; Short Course Series; Mineralogical Association of Canada: Ottawa, ON, Canada, 2001; Volume 29, pp. 239–243.
65. Jackson, S.E.; Pearson, N.J.; Griffin, W.L.; Belousova, E.A. The application of laser ablation-inductively coupled plasma-mass spectrometry to in situ U–Pb zircon geochronology. *Chem. Geol.* **2004**, *211*, 47–69. [CrossRef]
66. Sláma, J.; Košler, J.; Condon, D.J.; Crowley, J.L.; Gerdes, A.; Hanchar, J.M.; Horstwood, M.S.A.; Morris, G.A.; Nasdala, L.; Norberg, N.; et al. Plešovice zircon—A new natural reference material for U–Pb and Hf isotopic microanalysis. *Chem. Geol.* **2008**, *249*, 1–35. [CrossRef]
67. Ludwig, K.R. ISOPLOT. Free Software. 2009. Available online: http://sourceforge.net/projects/isoplot/ (accessed on 22 September 2020).
68. Blichert-Toft, J.; Albarède, F. The Lu-Hf geochemistry of the chondrites and the evolution of the mantle crust system. *Earth Planet. Sci. Lett.* **1997**, *148*, 243–258. [CrossRef]
69. Raymond, L.A. Classification of melanges. In *Melanges: Their Origin and Significance*; Raymond, L.A., Ed.; Geological Society of America: Boulder, CO, USA, 1984; Volume 198, pp. 7–22.
70. Bars, A.; Miao, L.; Fochin, Z.; Baatar, M.; Aanad, C.; Togtock, K. Petrogenesis and tectonic implication of the Late Mesozoic volcanic rocks in East Mongolia. *Geol. J.* **2018**, *53*, 2449–2470. [CrossRef]
71. Peretyazhko, I.S.; Savina, S.A.; Dril, S.I. Early Cretaceous trachybasalt–trachyte–trachyrhyolitic volcanism in the Nyalga basin (Central Mongolia): Sources and evolution of continental rift magmas. *Russ. Geol. Geophys.* **2018**, *59*, 1679–1701. [CrossRef]
72. Kurihara, T.; Tsukada, K.; Otoh, S.; Kashiwagi, K.; Chuluun, M.; Byambadash, D.; Boijir, B.; Gonchigdorj, S.; Nuramkhan, M.; Niwa, M.; et al. Upper Silurian and Devonian pelagic deep-water radiolarian chert from the Khangai-Khentei belt of Central Mongolia: Evidence for Middle Paleozoic subduction accretion activity in the Central Asian Orogenic Belt. *J. Asian Earth Sci.* **2009**, *34*, 209–225. [CrossRef]

73. Hara, H.; Kurihara, T.; Tsukada, K.; Kon, Y.; Uchino, T.; Suzuki, T.; Takeuchi, M.; Nakane, Y.; Nuramkhaan, M.; Chuluun, M. Provenance and origins of a Late Paleozoic accretionary complex within the Khangai–Khentei belt in the Central Asian Orogenic Belt, central Mongolia. *J. Asian Earth Sci.* **2013**, *75*, 141–157. [CrossRef]
74. Fisher, C.M.; Vervoort, J.D.; Hancharb, J.M. Guidelines for reporting zircon Hf isotopic data by LA-MC-ICPMS and potential pitfalls in the interpretation of these data. *Chem. Geol.* **2014**, *363*, 125–133. [CrossRef]
75. Popeko, L.I.; Smirnova, Y.N.; Zaika, V.A.; Sorokin, A.A.; Dril, S.I. Provenance and Tectonic Implications of Sedimentary Rocks of the Paleozoic Chiron Basin, Eastern Transbaikalia, Russia, Based on Whole-Rock Geochemistry and Detrital Zircon U–Pb Age and Hf Isotopic Data. *Minerals* **2020**, *10*, 279. [CrossRef]
76. Dumitru, T.A.; Hendrix, M.S. Fission-Track Constraints Constraints on on Jurassic Jurassic Folding Folding and and Thrusting Thrusting in Southern Southern Mongolia Mongolia and and Their Their Relationship Relationship to to the the in Beishan Thrust Thrust Belt Belt of of Northern. *Geol. Soc. Am. Mem.* **2001**, *194*, 215–229.
77. Jolivet, M.; Ritz, J.F.; Vassallo, R.; Larroque, C.; Braucher, R.; Todbileg, M.; Chauvet, A. Mongolian Sumits: An Uplifted, Flat, Old but Still Preserved Erosion Surface. *Geology* **2007**, *35*, 871–874. [CrossRef]
78. Cunningham, W.D.; Windley, B.F.; Dorjnamjaa, D.; Saandar, M. Late Cenozoic Transpression in Southwestern Mongolia and the Gobi Altai-Tien Shan Connection. *Earth Planet. Sci. Lett.* **1996**, *140*, 67–81. [CrossRef]
79. Dobretsov, N.L.; Buslov, M.M.; De Grave, J.; Sklyarov, E.V. Interplay of Magmatism, Sedimentation, and Collision Processes in the Siberian Craton and the Flanking Orogens. *Russ. Geol. Geophys.* **2013**, *54*, 1135–1149. [CrossRef]

minerals

MDPI

Article

Tectonic Setting of the Eastern Margin of the Sino-Korean Block in the Pennsylvanian: Constraints from Detrital Zircon Ages

Mun Gi Kim [1], Yong Il Lee [2,*] and Taejin Choi [3]

[1] Global Ocean Research Center, Korea Institute of Ocean Science & Technology, Busan 49111, Korea; mgkim@kiost.ac.kr
[2] School of Earth and Environmental Sciences, Seoul National University, Seoul 08826, Korea
[3] Department of Energy Resources Engineering, Chosun University, Gwangju 61542, Korea; tchoi@chosun.ac.kr
* Correspondence: lee2602@plaza.snu.ac.kr

Received: 20 April 2020; Accepted: 8 June 2020; Published: 9 June 2020

Abstract: To test the previous hypothesis that upper Paleozoic sediments in the eastern Sino-Korean Block were mostly derived from the paleo-orogen located to the east, we compared published and new U–Pb age data of detrital zircons from Pennsylvanian strata distributed in the Sino-Korean Block (SKB). The age distributions of detrital zircons from different localities of Pennsylvanian strata in North China reflect varying contributions from the Inner Mongolia Paleo-uplift in the north and the Central China Orogenic Belt in the south. The supply of detritus from the northern source to distant areas, however, appears to have been limited during the Pennsylvanian times. The age distributions of detrital zircons from Korean Pennsylvanian strata located in the east of the SKB are characterized by a dense cluster of 1.84–1.90 Ga and differ from those of North China. The Korean age characteristic is best explained by strong influences of the detritus derived from the Paleoproterozoic Yeongnam Massif in southeastern Korea. Along with the significant number of zircons that record syn- to near-depositional magmatic activities, this observation supports the hypothesis of the existence of an active continental margin setting in the east of the SKB.

Keywords: Pennsylvanian; Sino-Korean Block; provenance; detrital zircons; U–Pb ages

1. Introduction

Detrital zircon U–Pb geochronology has greatly improved our understanding on the late Paleozoic tectonic evolution of the Sino-Korean Block (SKB; also called the North China Block). The upper Paleozoic succession in North China, particularly its lowermost strata deposited upon the Late Ordovician–Pennsylvanian unconformity, has been rigorously analyzed in many different localities [1–6]. Results pointed to two major source terrains, each carrying the "northern" and "southern" signatures in the U–Pb age and Hf isotopic composition of detrital zircons: the uplifted northern margin of the SKB, commonly referred to as the Inner Mongolian Paleo-uplift (IMPU), and the Central China Orogenic Belt (CCOB) including the North Qinling Block (NQB) in the south, respectively (Figure 1). It is increasingly accepted that both northern and southern terrains had contributed detritus significantly to the Pennsylvanian deposits [7], while the former had dominant influence on the overlying Permian deposits [8,9]. Wang et al. [7] described the Pennsylvanian depositional setting as a "walled continental basin", which was set before the Permian southward tilt of the basin due to further uplift in the north [10].

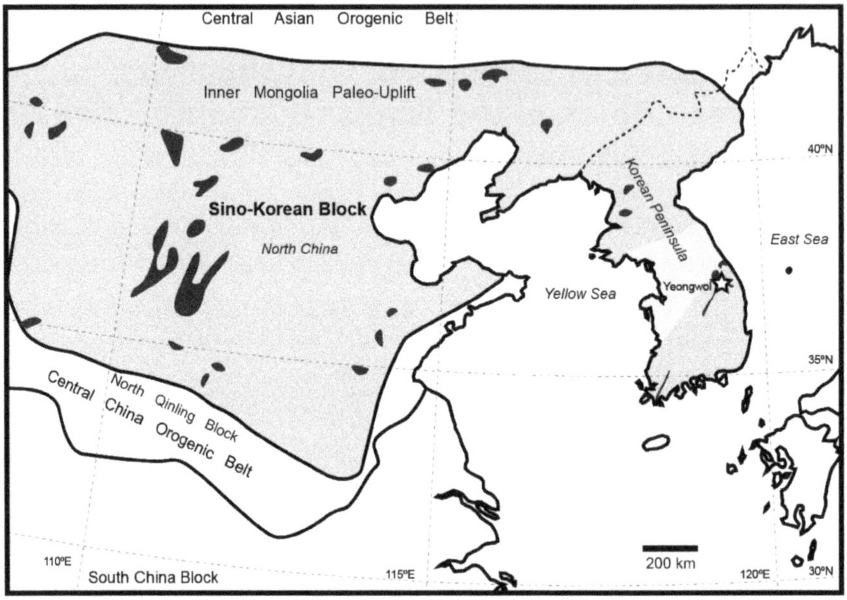

Figure 1. Simplified tectonic map of the Sino-Korean Block (gray color). Light gray color in the Korean Peninsula represents the tectonic provinces under debate. Dark gray color outlines the distribution of Pennsylvanian strata in the Sino-Korean Block (modified from [11,12]). Yellow star indicates the Yeongwol site where the newly analyzed sample was obtained.

Paleozoic sedimentary successions in Korea are well correlatable to the equivalent strata in North China [13–20], and thus they are widely recognized as being deposited on the SKB in various tectonic models [21–24]. For the upper Paleozoic sediments in the southeastern Korean Peninsula, the eastern derivation of detritus in a foreland basin setting has long been assumed based on paleocurrent directions and sandstone composition [25–28]. Recent detrital zircon provenance studies have also supported the presence of an active continental margin setting that had formed in the east [29–31]. However, as many previous detrital zircon provenance studies conducted in North China interpreted their results simply upon north-south cross-section, the possible eastern source has been commonly overlooked in the large-scale discussions on the SKB.

This study proposes the existence of a 'third wall' in the east of the SKB during the Pennsylvanian, probably formed by the subduction of the paleo-Pacific plate beneath the eastern margin of the SKB [32,33]. The general framework of this setting has been put forward earlier by Kim and Lee [20]. In this study, we elaborate the idea further by comprehensively comparing the detrital zircon age populations reported from Pennsylvanian deposits spread over the SKB, with particular focus on their localities in the eastern side of the block. The Pennsylvanian sediments in the SKB, the oldest upper Paleozoic sedimentary strata deposited before the dominance of northern signature in the Permian, provide the most suitable time record that can distinguish the potential eastern contribution.

2. Geological Background and Methods

The SKB is bounded to the north by the Central Asian Orogenic Belt (CAOB) and to the south by the CCOB (Figure 1). Tectonic evolution of the southern CAOB is not fully understood, but there is a broad consensus that the Paleo-Asian Ocean subducted southward beneath the northern margin of the SKB during the Pennsylvanian–Permian times [34–36], which led to the uplifting and exhumation of the IMPU [8,37]. The timing of complicated accretionary and collisional processes in the CCOB is also controversial, but it is widely accepted that the NQB, the northern part of CCOB, had formed highlands along the southern margin of the SKB, at least by the Carboniferous [38–40]. Taking these into account, many previous studies on the provenance of upper Paleozoic sediments in North China considered the IMPU and the CCOB as two primary potential source regions [8,9,41]. Archean–Paleoproterozoic basement rocks intruded by Devonian–Triassic plutons are widely exposed in the IMPU [35,42], whereas multiple orogenies in the Meso- to Neoproterozoic and early Paleozoic times are recorded in the highly metamorphosed rock assemblages in the CCOB [43,44].

The Paleozoic of the SKB is mostly represented by the Cambrian–Middle Ordovician marine sediments and the Pennsylvanian–Triassic paralic to non-marine sediments. A regional unconformity spanning more than 100 Ma separates the lower and upper Paleozoic sequences (Figure 2). In North China, the basal stratigraphic unit of the upper Paleozoic succession is named the Benxi Formation. In South Korea, it is called the Manhang Formation (and locally, the Yobong Formation). Both the Benxi and Manhang formations are composed of siliciclastic sediments with some intercalated limestones, and are unconformably underlain by Middle Ordovician limestones. The average thicknesses of the Benxi and Manhang formations are known to be 10–150 m and 100–300 m, respectively, although they reach more than 1 km in a few localities [10,20]. Correlatable fusulinid zones and floral assemblages have been established for both formations, and indicate that they are Bashkirian–Moscovian (early Pennsylvanian) in age [45–48]. However, in some localities in northern North China, several detrital zircon grains that were dated as young as the earliest Permian times have been reported from the Benxi Formation [2,37]. This might be due to the fact that the lowermost strata of the upper Paleozoic succession are partly diachronous across the SKB [49] (Figure 2).

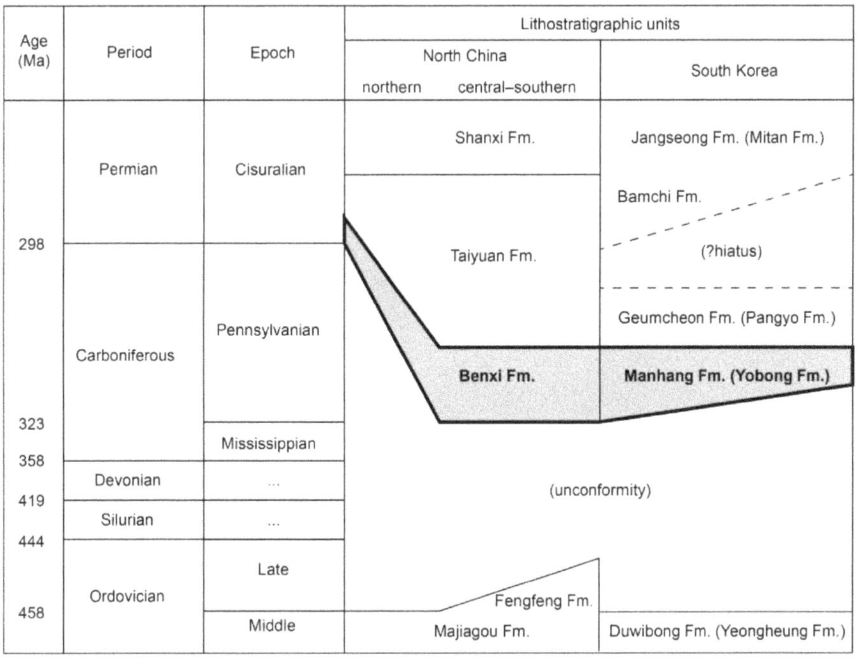

Figure 2. Middle Ordovician–Early Permian (Cisuralian) stratigraphy of North China and South Korea (modified from references [20,49,50]). The gray color represents the strata from which the detrital zircon U–Pb age data were obtained.

From the literature, we obtained data on the detrital zircon U–Pb age of the Pennsylvanian deposits from two localities in South Korea and 19 localities in North China. Data collection was limited to the lowermost units of the upper Paleozoic succession (Figure 2). Three localities of the Benxi Formation of possible earliest Permian age are included; however, caution needs to be exercised when interpreting data derived from strata known to have such young ages. In addition to the literature data, we analyzed detrital zircon U–Pb ages of one sandstone sample collected from the basal part of the Yobong Formation in the Yeongwol area of central eastern Korea (Figure 1). Analytical methods for detrital zircon U–Pb dating of the Yobong Formation sample are presented in Appendix A.

The collected literature and newly obtained age data were processed under the same measure for accurate comparison of multiple dataset. For detrital zircon grains younger and older than 1000 Ma, ^{238}U–^{206}Pb dates and ^{207}Pb–^{206}Pb dates, respectively, were used for interpretation. Consistent discordance cut (exclusion of ages with normal or reverse discordance >10%) was applied. Discordance was calculated based on the difference between ^{238}U–^{206}Pb and ^{235}U–^{207}Pb dates (1-(^{238}U–^{206}Pb date/^{235}U–^{207}Pb date)) for detrital zircon grains younger than 1000 Ma, and that between ^{238}U–^{206}Pb and ^{206}Pb–^{207}Pb dates (1-(^{238}U–^{206}Pb date/^{206}Pb–^{207}Pb date)) for those older than 1000 Ma. Few grains with apparent ages that passed the discordance cut, but were undeniably younger (e.g., 288 ± 2 Ma at the Taebaek locality and 182 ± 21 Ma at the Pingquan locality) than the upper limit of depositional age (Figure 2), or had excessively large analytic errors (e.g., 1594 ± 105 Ma) were removed from further consideration. Only age data meeting all the listed criteria were used for plotting and interpretation.

3. Results

A total number of 1669 concordant or slightly discordant zircon U–Pb ages, including 49 newly analyzed ages (Table S1), were obtained from the Pennsylvanian strata at 22 localities in the SKB. A summary of the combined detrital zircon ages is presented in Figure 3. Six major age populations were identified: (1) Carboniferous (ca. 310 Ma); (2) Devonian (ca. 390 Ma); (3) Ordovician–Silurian (ca. 440 Ma); (4) Neo- to Mesoproterozoic (a notable peak at ca. 960 Ma with several subordinate peaks); (5) Late Paleoproterozoic (a wide peak centering at ca. 1870 Ma with a subsidiary peak at ca. 1930 Ma), and (6) Early Paleoproterozoic to Neoarchean (a broad peak centering at ca. 2500 Ma).

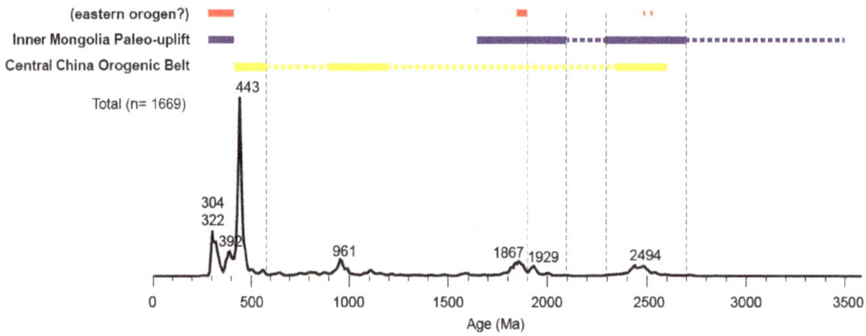

Figure 3. Composite detrital zircon age spectrum (Kernel Density Estimate; Vermeesch [50]) from the Pennsylvanian strata distributed in the Sino-Korean Block. Thick colored lines represent the characteristic age ranges of zircons that are distributed in the respective source terrains, while dotted lines represent that occurring in subordinate amount (modified from Zhu et al. [9]).

Figure 4 summarizes the distributions of detrital zircon ages of the Pennsylvanian strata deposited in sedimentary basins on the SKB, represented by a pie diagram in each studied area. The full age data are available in the online supplementary material (Table S2).

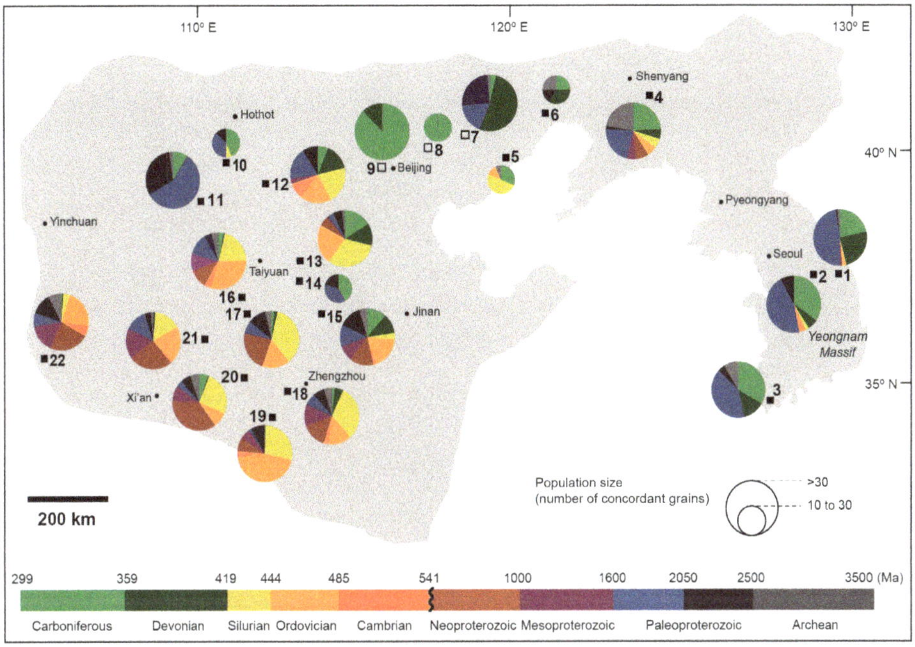

Figure 4. Pie diagrams of detrital zircon age distribution over the Sino-Korean Block. Locations: 1. Taebaek; 2. Yeongwol; 3. Gangjin; 4. Benxi; 5. Qinhuangdao; 6. Nanpiao; 7. Pingquan; 8. Yingshouyingzi; 9. Western Beijing; 10. Qingshuihe; 11. Baode; 12. Shuozhou; 13. Yangquan; 14. southern Yangquan; 15. Wu'an; 16. Jinzhong; 17. Changzhi; 18. Gongyi; 19. Lushan; 20. Sanmenxia; 21. Hancheng; 22. Pingliang (see Table S2 for references). The empty rectangles indicate where the Benxi Formation is supposed to be younger (see text for details).

4. Discussion

4.1. Sediment Dispersal over the SKB

4.1.1. Northern vs. Southern Signature

It is generally agreed that the Carboniferous, Devonian, and Late Paleoproterozoic zircon grains represent the northern signature derived from the IMPU, whereas the Ordovician–Silurian (with negative $\varepsilon_{Hf}(t)$) and Neo- to Mesoproterozoic components represent the southern signature derived from the CCOB [7–9,40], although such occurrences are not completely exclusive to each other. In this regard, grains with the northern signature dominantly occur in areas close to the northern margin (sites 6–11 in Figure 4). The Silurian to Mesoproterozoic grains mainly occupy the zircon population in central to southern North China (sites 13–22), suggesting a sediment supply from the south. Such distinct distributions of detrital zircon ages clearly show that competing northern and southern signatures are reflected in Pennsylvanian deposits in North China.

Notably, the southern signature is visible at several northern sites, including the Benxi and Qinhuangdao sites in northeastern China (sites 4 and 5), as well as the Shuozhou site in north-central China (site 12). The occurrence of ca. 440 Ma zircon grains characterized by negative $\varepsilon_{Hf}(t)$ in northern localities was deemed enigmatic by Liu et al. [2], because they only considered the source rocks in the north. According to Wang et al. [7], these grains may be reinterpreted to have been originated from the NQB. Significant southern signature in these three sites suggests that sediment supply from the north was not dominant. In the frontal region of the eastern IMPU, it has been interpreted that the Benxi Formation was deposited in coast–offshore settings, including fan deltas at the basin margin [49].

Such environmental settings explain both the mixing of northern and southern influences in the distal part (sites 4 and 5) and highly variable detrital zircon age distributions among sites in the proximal part (sites 6–9), suggestive of the northern provenance being of local importance. Besides, some of the zircon age records with the dominant northern signature (sites 7–9) may represent the earliest Permian time instead of the Pennsylvanian (empty rectangles in Figure 4). It is therefore likely that the supply of detritus from the northern margin of the SKB to the distant areas was rather limited during the Pennsylvanian times, particularly in the northeastern side of the block.

4.1.2. Potential Eastern Signature

Based on the interpretation above, the northern signature-like detrital zircons from the Pennsylvanian strata in southern Korea (sites 1–3) located in the eastern margin of the SKB are difficult to interpret as being derived from the north. Consistent age distribution patterns are observed in Taebaek, Yeongwol, and Gangjin localities (Figure 5a): major age groups at ca. 320 Ma and ca. 1.85 Ga, and minor groups at ca. 390 Ma and ca. 2.5 Ga. The lack of southern signature is understandable considering that these sites were located relatively far from the NQB. Given the presumed setting in northern North China during the Pennsylvanian times, however, major sediment supply from the northern margin of the SKB is not plausible, since it has to account for the large amount of sediments reaching several hundred meters thick in these distant sites. An alternative explanation for the source terrain is then required.

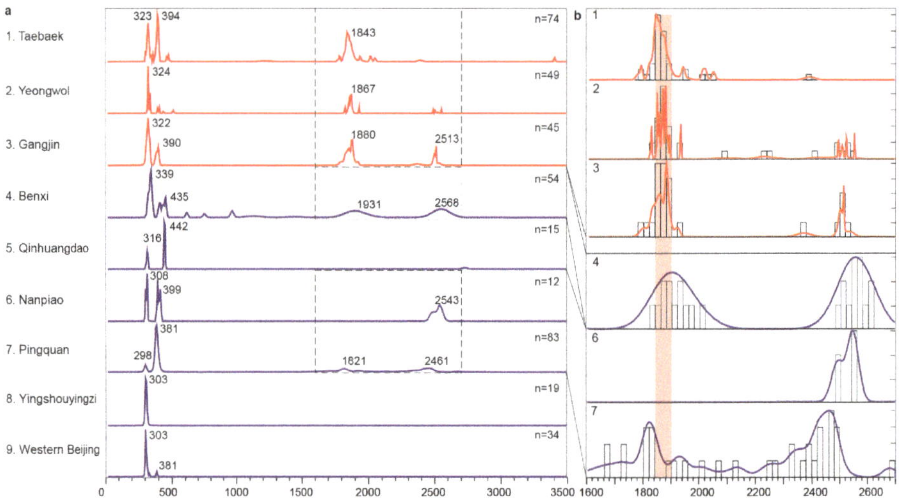

Figure 5. Probability density plot and age histograms of detrital zircons from the Pennsylvanian strata in the eastern part of the Sino-Korean Block: (**a**) full age spectra, and (**b**) enlarged views of the Neoarchean–Paleoproterozoic portion for selected localities. See Figure 4 for site locations.

Close looking into the Paleoproterozoic–Neoarchean age groups provides an important clue (Figure 5b). Records from northern North China exhibit wide variance of Paleoproterozoic–Neoarchean ages. The Pennsylvanian strata in Korea, in contrast, are characterized by a vast majority of Precambrian grains belonging to narrow time spans of 1.84–1.90 Ga and 2.48–2.54 Ga, with the former being more abundant than the latter. This observation is in line with a previous report on composite detrital zircon ages of Upper Paleozoic sediments in southern Korean Peninsula [18]. The consistency of 1.84–1.90 Ga peak in all three localities in Korea suggests a regional characteristic, a signature that is best explained by derivative sediments from the Precambrian massifs in the Korean Peninsula, particularly the Yeongnam Massif located in the southeast. Besides the widespread distribution of basement rocks with these ages [51,52], sediments from the Nakdong River that drains the Yeongnam Massif prove that the Yeongnam Massif indeed provides detritus that is dominantly composed of 1.84–1.90 Ga detrital zircons [53]. This feature is distinct from the age characteristics of zircons from river sediments of the Yongding and Luan rivers in eastern North China. Both latter rivers have a drainage area similar to the Nakdong River, but their sediments contain much heterogeneous Paleoproterozoic detrital zircon population [54]. In summary, the detrital zircon ages of the Pennsylvanian sediments in Korea may, at first glance, be thought to have been derived from the IMPU, but are much likely to have been derived from another source terrain.

4.2. Tectonic Implications

In general, the significant proportion of zircon grains that were young at the time of deposition is considered as the characteristic of a basin close to the convergent plate margin [55]. In a situation where major supplies from the northern and southern sources are not likely, another orogenic belt for Pennsylvanian strata in the Korean Peninsula could be set to the east (Figure 6). This interpretation is in agreement with the paleocurrent directions and inferred basin topography [25–27,56], and better explains the generally euhedral shape of syndepositional zircons (Figure A1). Some of these Carboniferous zircons also exhibit high Th/U values (Figure 7), which is typical of the zircons derived from intermediate to mafic igneous rocks [57–59]. Previous detrital zircon provenance studies in different localities in Korea [29–31] also favored the same interpretation. Our interpretation of the eastern source terrains is supported by the hypothesis of previous studies that the subduction of the paleo-Pacific plate took place beneath the eastern margin of the SKB during the Late Paleozoic [32,33]. Wider acceptance of this hypothesis is currently hindered by the limited occurrence of Pennsylvanian plutons in the east, but some evidence for Pennsylvanian magmatism can be found in the Hida Belt [60–62], a continental fragment in Japan [63]. Even if the subduction had taken place in the eastern margin of the SKB, corresponding plutons might have been largely removed by later rapid uplift of widespread Jurassic plutons (i.e., [64]), Cretaceous sedimentation in southeastern Korea, and Upper Cretaceous regional exhumation of the Korean Peninsula [65] and the Japanese islands [66]. Recent U–Pb zircon age analysis on paragneiss in the Hida Belt reported inherited/detrital core ages that cluster at ca. 300 and 330 Ma, which also accompany Paleoproterozoic zircon grains that almost entirely fall within 1.81–1.90 Ga [67]. This finding is highly significant considering the pre-Cenozoic configuration of the Hida–Oki Belt in Japan, which is believed to have been attached to the east of the Korean Peninsula before the opening of the East Sea (Sea of Japan) [66,68,69].

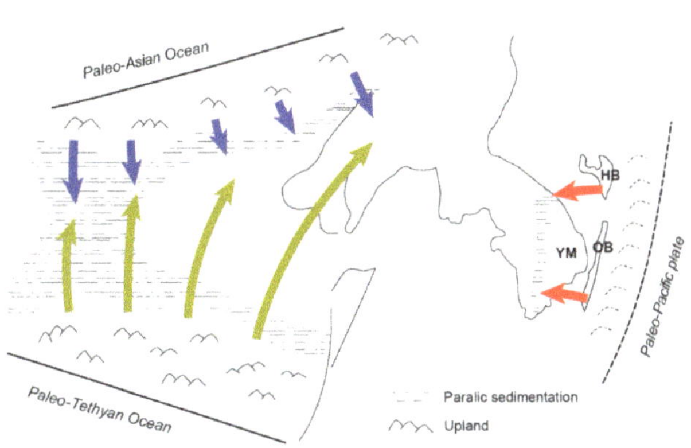

Figure 6. Sediment dispersal patterns and inferred paleogeography and tectonic configuration of the Sino-Korean Block (SKB) during the Pennsylvanian (modified from Liu [10]). YM = Yeongnam Massif, HB = Hida Belt, OB = Oki Belt. Tectonic blocks are reconstructed to their inferred original position before the Tanlu fault-movement and the opening of the East Sea (Sea of Japan).

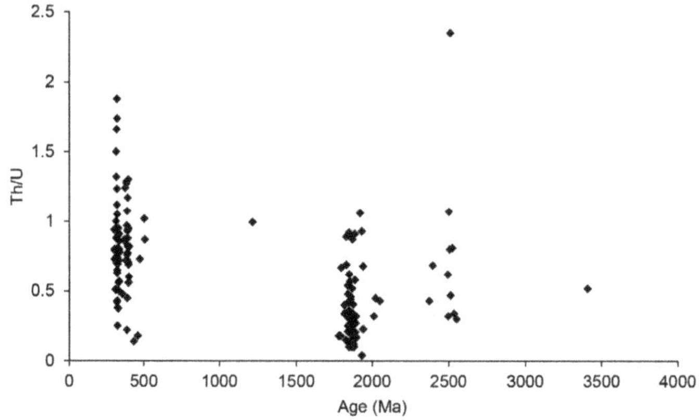

Figure 7. Th/U vs. Age plot for detrital zircon grains from the Pennsylvanian Yobong Formation in the Yeongwol area, central eastern Korea.

Based on our new perspective, it is also noteworthy that the same characteristic 1.84–1.90 Ga-aged zircons consistently dominate the detrital zircon populations throughout the overlying Permian–Lower Triassic strata in Korea [31,70,71], whereas detrital zircons from the coeval successions in North China generally show much wider spread of ca. 1.9 Ga peak and larger portion of ca. 2.5 Ga peak [9,37,72]. This comparison suggests that (1) high concentration in the 1.84–1.90 Ga range is the unique signature of the eastern source terrain and (2) its influence on the proximal areas remained constant throughout late Paleozoic times. How far this sedimentation regime in the east had expanded further to the inner part of the SKB is unknown at present, partly due to the lack of data in areas adjacent to the western Korean Peninsula. Overlap in U–Pb ages and $\varepsilon_{Hf}(t)$ values of the Yeongnam Massif zircons (i.e., as reported previously [52]) with those of the IMPU makes it difficult to evaluate how much contribution was from the eastern sources, but our findings may prove useful in case where the eastern signature is suspected to occur. Reinterpretation of the existing detrital zircon U–Pb age data and future studies on the upper Paleozoic strata in the SKB, especially for the eastern North China, are recommended to take this potential contribution from the east into account.

Supplementary Materials: The following are available online at http://www.mdpi.com/2075-163X/10/6/527/s1, Table S1: LA-ICP-MS U–Pb dating data of detrital zircons from Yobong Formation sample; Table S2: Detrital zircon U–Pb ages of Pennsylvanian sediments in the Sino-Korean block.

Author Contributions: Conceptualization, M.G.K. and Y.I.L.; formal analysis, M.G.K.; investigation, M.G.K. and T.C.; resources, T.C.; writing—original draft preparation, M.G.K.; writing—review and editing, Y.I.L. and T.C.; visualization, M.G.K.; supervision, Y.I.L.; funding acquisition, Y.I.L. All authors have read and agreed to the published version of the manuscript.

Funding: This research was funded by the National Research Foundation of Korea, grant number 2014R1A2A2A01005404 to YIL and grant number 2012H1A2A1049032 to MGK. MGK also acknowledges support from the Korea Institute of Ocean Science & Technology (PE99824).

Acknowledgments: We thank the Editors, Wilfried Winkler and Albrecht von Quadt, and two anonymous journal reviewers for their constructive comments on our manuscript.

Conflicts of Interest: The authors declare no conflict of interest.

Appendix A

Zircon grains were extracted from a sandstone sample, by following the conventional heavy mineral separation methods. After hand picking under binocular microscope, 100 randomly selected zircon grains were embedded in a PFA teflon sheet. Cathodoluminescence (CL) images of each grain were obtained by using a scanning electron microscope (JEOL JSM-5400, JEOL, Akishima, Japan). U–Pb dating of zircons was conducted by using laser ablation inductively coupled plasma mass spectrometry

(LA-ICP-MS) housed at the Korea Basic Science Institute (Cheongju, Korea, Nu Plasma II MC ICP-MS). Detailed instrumental settings and analytical procedures are similar to those described by Lee et al. [73]. Laser ablation was conducted at a spot size of 15 μm, energy density of 2–3 J/cm^3, and pulse repetition rate of 5 Hz. The reference zircon 91,500 was used as the primary standard [74] and Plesovice as the secondary standard. A single spot per grain was analyzed. Magmatic rim with oscillatory zoning was preferred to core when selecting a spot for the age dating of zircon grains, unless the mixing of different age domain was concerned. Metamorphic overgrowths were not usually present, and were not analyzed (Figure A1).

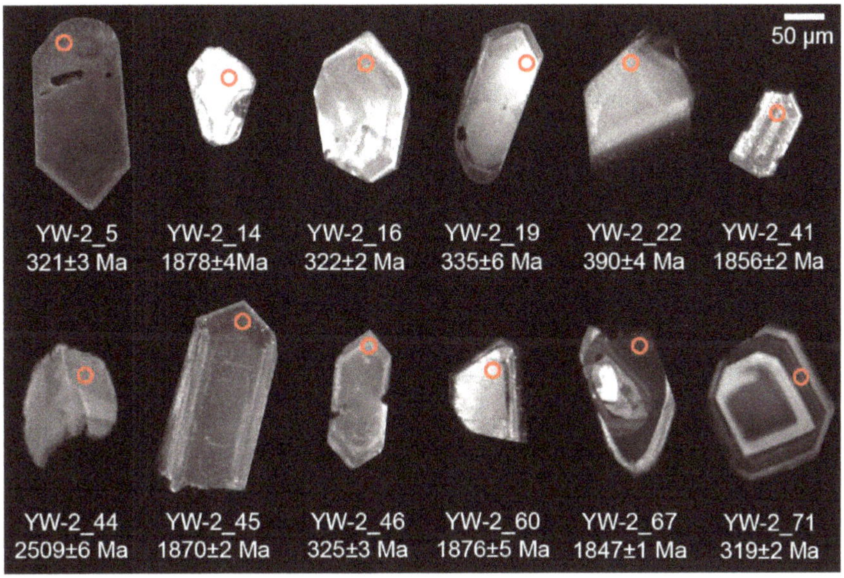

Figure A1. Representative SEM-CL images of detrital zircon grains from the Yobong Formation in the Yeongwol area, central eastern Korea.

References

1. Wang, Y.; Zhou, L.; Zhao, L.; Ji, M.; Gao, H. Palaeozoic uplands and unconformity in the North China Block: Constraints from zircon LA-ICP-MS dating and geochemical analysis of Bauxite. *Terra Nova* **2010**, *22*, 264–273. [CrossRef]
2. Liu, J.; Zhao, Y.; Liu, A.; Zhang, S.; Yang, Z.; Zhuo, S. Origin of Late Palaeozoic bauxites in the North China Craton: Constraints from zircon U–Pb geochronology and in situ Hf isotopes. *J. Geol. Soc.* **2014**, *171*, 695–707. [CrossRef]
3. Cai, S.; Wang, Q.; Liu, X.; Feng, Y.; Zhang, Y. Petrography and detrital zircon study of late Carboniferous sequences in the southwestern North China Craton: Implications for the regional tectonic evolution and bauxite genesis. *J. Asian Earth Sci.* **2015**, *98*, 421–435. [CrossRef]
4. Zhao, L.; Liu, X. Metallogenic and tectonic implications of detrital zircon U–Pb, Hf isotopes, and detrital rutile geochemistry of late carboniferous karstic bauxite on the southern margin of the North China Craton. *Lithos* **2019**, *350–351*, 105222. [CrossRef]
5. Jia, L.; Zhong, D.; Sun, H.; Yan, R.; Zhang, C.; Mo, W.; Qiu, C.; Dong, Y.; Li, B.; Liao, G. Sediment provenance analysis and tectonic implication of the Benxi Formation, Ordos Basin. *Acta Sediment. Sin.* **2019**, *37*, 1087–1103. (In Chinese)

6. Wang, A.-Q.; Yang, D.-B.; Yang, H.-T.; Mu, M.-S.; Quan, Y.-K.; Hao, L.-R.; Xu, W.-L. Late Palaeozoic tectonic evolution of the southern North China Craton: Constraints from detrital zircon dating and Hf–O isotopic compositions of the Benxi Formation, Sanmenxia area, North China Craton. *Geol. J.* **2020**, *55*, 1320–1331. [CrossRef]
7. Wang, Q.; Deng, J.; Liu, X.; Zhao, R.; Cai, S. Provenance of Late Carboniferous bauxite deposits in the North China Craton: New constraints on marginal arc construction and accretion processes. *Gondwana Res.* **2016**, *38*, 86–98. [CrossRef]
8. Li, H.-Y.; He, B.; Xu, Y.-G.; Huang, X.-L. U–Pb and Hf isotope analyses of detrital zircons from Late Paleozoic sediments: Insights into interactions of the North China Craton with surrounding plates. *J. Asian Earth Sci.* **2010**, *39*, 335–346. [CrossRef]
9. Zhu, X.-Q.; Zhu, W.-B.; Ge, R.-F.; Wang, X. Late Paleozoic provenance shift in the south-central North China Craton: Implications for tectonic evolution and crustal growth. *Gondwana Res.* **2014**, *25*, 383–400. [CrossRef]
10. Liu, G. Permo-Carboniferous paleogeography and coal accumulation and their tectonic control in the North and South China continental plates. *Int. J. Coal. Geol.* **1990**, *16*, 73–117. [CrossRef]
11. Yang, Z. The Carboniferous system. In *The geology of China*; Yang, Z., Cheng, Y., Wang, H., Eds.; Oxford University Press: New York, NY, USA, 1986; pp. 102–112.
12. Lee, C.Z. The Late Paleozoic strata. In *Geology of Korea*; Cheng, C.H., Ed.; Sigma Press: Seoul, Korea, 1999; pp. 129–203.
13. Doh, S.-J.; Piper, J.D.A. Palaeomagnetism of the (Upper Palaeozoic-Lower Mesozoic) Pyongan Supergroup, Korea: A Phanerozoic link with the North China Block. *Geophys. J. Int.* **1994**, *117*, 850–863. [CrossRef]
14. Jeong, H.; Lee, Y.I. Late Cambrian biogeography: Conodont bioprovinces from Korea. *Palaeogeogr. Palaeoclimatol. Palaeoecol.* **2000**, *162*, 119–136. [CrossRef]
15. Choi, D.K.; Kim, D.H.; Sohn, J.W. Ordovician trilobite faunas and depositional history of the Taebaeksan Basin, Korea: Implications for palaeogeography. *Alcheringa* **2001**, *25*, 53–68. [CrossRef]
16. Kim, J.H.; Lee, Y.I.; Li, M.; Bai, Z. Comparison of the Ordovician-Carboniferous boundary between Korea and NE China: Implications for correlation and tectonic evolution. *Gondwana Res.* **2001**, *4*, 39–53. [CrossRef]
17. Jeong, H.; Lee, Y.I. Nd isotopic study of Upper Cambrian conodonts from Korea and implications for early Paleozoic paleogeography. *Palaeogeogr. Palaeoclimatol. Palaeoecol.* **2004**, *212*, 77–94. [CrossRef]
18. Choi, D.K.; Kim, E.Y. Occurrence of Changshania (Trilobita, Cambrian) in the Taebaeksan Basin, Korea and its stratigraphic and paleogeographic significance. *Palaeogeogr. Palaeoclimatol. Palaeoecol.* **2006**, *242*, 343–354. [CrossRef]
19. Li, Z.; Ni, L.; Xu, J. The Upper Proterozoic-Paleozoic records of sedimentary sequences and detrital zircon geochronology in Korean Peninsula and North China: Implications for tectonic attributes and division. *Acta Petrol. Sin.* **2016**, *32*, 3139–3154.
20. Kim, M.G.; Lee, Y.I. The Pyeongan Supergroup (upper Paleozoic–Lower Triassic) in the Okcheon Belt, Korea: A review of stratigraphy and detrital zircon provenance, and its implications for the tectonic setting of the eastern Sino-Korean Block. *Earth-Sci. Rev.* **2018**, *185*, 1170–1186. [CrossRef]
21. Chough, S.K.; Kwon, S.-T.; Ree, J.-H.; Choi, D.K. Tectonic and sedimentary evolution of the Korean peninsula: A review and new view. *Earth-Sci. Rev.* **2000**, *52*, 175–235. [CrossRef]
22. Ishiwatari, A.; Tsujimori, T. Paleozoic ophiolites and blueschists in Japan and Russian Primorye in the tectonic framework of East Asia: A synthesis. *Isl. Arc* **2003**, *12*, 190–206. [CrossRef]
23. Chang, K.-H.; Zhao, X. North and South China suturing in the east end: What happened in Korean Peninsula? *Gondwana Res.* **2012**, *22*, 493–506. [CrossRef]
24. Cho, M.; Kim, T.; Yang, S.-y.; Yi, K. Paleoproterozoic to Triassic crustal evolution of the Gyeonggi Massif, Korea: Tectonic correlation with the North China craton. In *Linkages and Feedbacks in Orogenic Systems*; Law, R.D., Thigpen, J.R., Merschat, A.J., Stowell, H.H., Eds.; Geological Society of America: Boulder, CO, USA, 2017; pp. 165–198.
25. Kim, H.M. Paleozoic and Mesozoic paleocurrents of the Danyang coalfield district, Korea. *J. Geol. Soc. Korea* **1971**, *7*, 257–276.
26. Kim, H.M. Neritic paleocurrent analysis of Pennsylvanian Tethyan sea at Samcheog coalfield, Korea. *J. Korea Inst. Min. Geol.* **1978**, *11*, 21–38. (In Korean)
27. Lee, Y.I.; Lim, C. Provenance and compositional variance of the Carboniferous Manhang sandstones, central eastern Korea. *J. Geol. Soc. Korea* **1995**, *31*, 637–652.

28. Lee, Y.I.; Sheen, D.-H. Detrital modes of the Pyeongan Supergroup (Late Carboniferous–Early Triassic) sandstones in the Samcheog coalfield, Korea: Implications for provenance and tectonic setting. *Sediment. Geol.* **1998**, *119*, 219–238. [CrossRef]
29. Kim, H.S.; Ree, J.-H.; Kim, J. Tectonometamorphic evolution of the Permo-Triassic Songrim (Indosinian) orogeny: Evidence from the late Paleozoic Pyeongan Supergroup in the northeastern Taebaeksan Basin, South Korea. *Int. J. Earth Sci.* **2012**, *101*, 483–498. [CrossRef]
30. Kim, H.S.; Seo, B.; Yi, K. Medium Temperature and Lower Pressure Metamorphism and Tectonic Setting of the Pyeongan Supergroup in the Munkyeong Area. *J. Petrol. Soc. Korea* **2014**, *23*, 311–324. (In Korean) [CrossRef]
31. Kim, M.G.; Lee, Y.I.; Choi, T.; Orihashi, Y. The tectonic setting of the eastern margin of the Sino-Korean Block inferred from detrital zircon U–Pb age and Nd isotope composition of the Pyeongan Supergroup (late Paleozoic–Early Triassic), Korea. *Geol. Mag.* **2019**, *156*, 471–484. [CrossRef]
32. Windley, B.F.; Maruyama, S.; Xiao, W.J. Delamination/thinning of sub-continental lithospheric mantle under Eastern China: The role of water and multiple subduction. *Am. J. Sci.* **2011**, *310*, 1250–1293. [CrossRef]
33. Kusky, T.M.; Windley, B.F.; Wang, L.; Wang, Z.; Li, X.; Zhu, P. Flat slab subduction, trench suction, and craton destruction: Comparison of the North China, Wyoming, and Brazilian cratons. *Tectonophysics* **2014**, *630*, 208–221. [CrossRef]
34. Zhang, S.-H.; Zhao, Y.; Song, B.; Yang, Z.-Y.; Hu, J.-M.; Wu, H. Carboniferous granitic plutons from the northern margin of the North China block: Implications for a late Palaeozoic active continental margin. *J. Geol. Soc.* **2007**, *164*, 451–463. [CrossRef]
35. Zhang, S.-H.; Zhao, Y.; Liu, J.-M.; Hu, Z.-C. Different sources involved in generation of continental arc volcanism: The Carboniferous–Permian volcanic rocks in the northern margin of the North China block. *Lithos* **2016**, *240–243*, 382–401. [CrossRef]
36. Eizenhöfer, P.R.; Zhao, G.; Zhang, J.; Sun, M. Final closure of the Paleo-Asian Ocean along the Solonker Suture Zone: Constraints from geochronological and geochemical data of Permian volcanic and sedimentary rocks. *Tectonics* **2014**, *33*, 441–463. [CrossRef]
37. Ma, S.; Meng, Q.; Duan, L.; Wu, G. Reconstructing Late Paleozoic exhumation history of the Inner Mongolia Highland along the northern edge of the North China Craton. *J. Asian Earth Sci.* **2014**, *87*, 89–101. [CrossRef]
38. Dong, Y.; Zhang, X.; Liu, X.; Li, W.; Chen, Q.; Zhang, G.; Zhang, H.; Yang, Z.; Sun, S.; Zhnag, F. Propagation tectonics and multiple accretionary processes of the Qinling Orogen. *J. Asian Earth Sci.* **2015**, *104*, 84–98. [CrossRef]
39. Liu, L.; Liao, X.; Wang, Y.; Wang, C.; Santosh, M.; Yang, M.; Zhang, C.; Chen, D. Early Paleozoic tectonic evolution of the North Qinling Orogenic Belt in Central China: Insights on continental deep subduction and multiphase exhumation. *Earth-Sci. Rev.* **2016**, *159*, 58–81. [CrossRef]
40. Yang, D.-B.; Yang, H.-T.; Shi, J.-P.; Xu, W.-L.; Wang, F. Sedimentary response to the paleogeographic and tectonic evolution of the southern North China Craton during the late Paleozoic and Mesozoic. *Gondwana Res.* **2017**, *49*, 278–295. [CrossRef]
41. Wang, Y.; Yang, W.; Zheng, D.; Zuo, P.; Qi, S. Detrital zircon U–Pb ages from the Middle to Late Permian strata of the Yiyang area, southern North China Craton: Implications for the Mianlue oceanic crust subduction. *Geol. J.* **2019**, *54*, 3527–3541. [CrossRef]
42. Zhang, S.-H.; Zhao, Y.; Song, B.; Hu, J.-M.; Liu, S.-W.; Yang, Y.-H.; Chen, F.-K.; Liu, X.-M.; Liu, J. Contrasting Late Carboniferous and Late Permian–Middle Triassic intrusive suites from the northern margin of the North China craton: Geochronology, petrogenesis, and tectonic implications. *Geol. Soc. Am. Bull.* **2009**, *121*, 181–200. [CrossRef]
43. Wang, X.; Wang, T.; Zhang, C. Neoproterozoic, Paleozoic, and Mesozoic granitoid magmatism in the Qinling Orogen, China: Constraints on orogenic process. *J. Asian Earth Sci.* **2013**, *72*, 129–151. [CrossRef]
44. Dong, Y.; Santosh, M. Tectonic architecture and multiple orogeny of the Qinling Orogenic Belt, Central China. *Gondwana Res.* **2016**, *29*, 1–40. [CrossRef]
45. Cheong, C.H. Stratigraphy and paleontology of the Late Paleozoic of Korea (1). *J. Korea Earth Sci. Edu. Soc.* **1979**, *1*, 3–16. (In Korean)
46. Chun, H.Y. Permo-Carboniferous plant fossils from the Samcheok coalfield, Gangweondo, Korea Part 1. *J. Paleontol. Soc. Korea* **1985**, *1*, 95–122.

47. Cleal, C.J.; Ziqiang, W. A new and diverse plant fossil assemblage from the upper Westphalian Benxi Formation, Shanxi, China, and its palaeofloristic significance. *Geol. Mag.* **2002**, *139*, 107–130. [CrossRef]
48. Wang, J. Late Paleozoic macrofloral assemblages from Weibei Coalfield, with reference to vegetational change through the Late Paleozoic Ice-age in the North China Block. *Int. J. Coal Geol.* **2010**, *83*, 292–317. [CrossRef]
49. Ma, S.; Meng, Q.; Wu, G.; Duan, L. Late Paleozoic exhumation of the Inner Mongolia Paleo-Uplift: Evidence from sedimentary records. *Acta Geol. Sin.* **2014**, *88*, 1771–1789. (In Chinese)
50. Vermeesch, P. On the visualisation of detrital age distributions. *Chem. Geol.* **2012**, *312–313*, 190–194. [CrossRef]
51. Lee, S.-R.; Cho, K.-O. Precambrian crustal evolution of the Korean Peninsula. *J. Petrol. Soc. Korea* **2012**, *21*, 89–112. (In Korean) [CrossRef]
52. Lee, Y.; Cho, M.; Yi, K. In situ U–Pb and Lu–Hf isotopic studies of zircons from the Sancheong–Hadong AMCG suite, Yeongnam Massif, Korea: Implications for the petrogenesis of ~1.86 Ga massif-type anorthosite. *J. Asian Earth Sci.* **2017**, *138*, 629–646. [CrossRef]
53. Choi, T.; Lee, Y.I.; Orihashi, Y. Crustal growth history of the Korean Peninsula: Constraints from detrital zircon ages in modern river sediments. *Geosci. Front.* **2016**, *7*, 707–714. [CrossRef]
54. Yang, J.; Gao, S.; Chen, C.; Tang, Y.; Yuan, H.; Gong, H.; Xie, S.; Wang, J. Episodic crustal growth of North China as revealed by U–Pb age and Hf isotopes of detrital zircons from modern rivers. *Geochim. Cosmochim. Acta* **2009**, *73*, 2660–2673. [CrossRef]
55. Cawood, P.A.; Hawkesworth, C.J.; Dhuime, B. Detrital zircon record and tectonic setting. *Geology* **2012**, *40*, 875–878. [CrossRef]
56. Kim, M.G.; Lee, Y.I. The stratigraphy and correlation of the upper Paleozoic Pyeongan Supergroup of southern Korean Peninsula—A review. *J. Geol. Soc. Korea* **2017**, *53*, 321–338, (in Korean with English abstract). [CrossRef]
57. Kaczmarek, M.A.; Müntener, O.; Rubatto, D. Trace element chemistry and U–Pb dating of zircons from oceanic gabbros and their relationship with whole rock composition (Lanzo, Italian Alps). *Contrib. Mineral. Petr.* **2008**, *155*, 295–312. [CrossRef]
58. Linnemann, U.; Ouzegane, K.; Drareni, A.; Hofmann, M.; Becker, S.; Gärtner, A.; Sagawe, A. Sands of West Gondwana: An archive of secular magmatism and plate interactions—A case study from the Cambro-Ordovician section of the Tassili Ouan Ahaggar (Algerian Sahara) using U–Pb-LA-ICP-MS detrital zircon ages. *Lithos* **2011**, *123*, 188–203. [CrossRef]
59. Wang, X.; Griffin, W.L.; Chen, J.; Huang, P.; Li, X. U and Th contents and Th/U ratios of zircon in felsic and mafic magmatic rocks: Improved zircon-melt distribution coefficients. *Acta Geol. Sin. Engl.* **2011**, *85*, 164–174.
60. Arakawa, Y. Rb–Sr ages of the gneiss and metamorphosed intrusive rocks of the Hida metamorphic belt in the Urushiyama area, Gifu Prefecture, central Japan. *J. Jap. Ass. Min. Petr. Econ. Geol.* **1984**, *79*, 431–442. [CrossRef]
61. Arakawa, Y.; Saito, Y.; Amakawa, H. Crustal development of the Hida belt, Japan: Evidence from Nd–Sr isotopic and chemical characteristics of igneous and metamorphic rocks. *Tectonophysics* **2000**, *328*, 183–204. [CrossRef]
62. Wakita, K. Geology and tectonics of Japanese islands: A review—The key to understanding the geology of Asia. *J. Asian Earth Sci.* **2013**, *72*, 75–87. [CrossRef]
63. Ehiro, M.; Tsujimori, T.; Tsukada, K.; Nuramkhaan, M. Palaeozoic basement and associated cover. In *The Geology of Japan*; Moreno, T., Wallis, S., Kojima, T., Gibbons, W., Eds.; Geological Society of London: London, UK, 2016; pp. 25–60.
64. Kim, S.W.; Oh, C.W.; Choi, S.G.; Ryu, I.C.; Itaya, T. Ridge Subduction-Related Jurassic Plutonism in and around the Okcheon Metamorphic Belt, South Korea, and Implications for Northeast Asian Tectonics. *Int. Geol. Rev.* **2005**, *47*, 248–269. [CrossRef]
65. Choi, T.; Lee, Y.I. Thermal histories of Cretaceous basins in Korea: Implications for response of the East Asian continental margin to subduction of the Paleo-Pacific Plate. *Isl. Arc* **2011**, *20*, 371–385. [CrossRef]
66. Maruyama, S.; Isozaki, Y.; Kimura, G.; Terabayashi, M. Paleogeographic maps of the Japanese Islands: Plate tectonic synthesis from 750 Ma to the present. *Isl. Arc* **1997**, *6*, 121–142. [CrossRef]
67. Takahashi, Y.; Cho, D.-L.; Mao, J.; Zhao, X.; Yi, K. SHRIMP U–Pb zircon ages of the Hida metamorphic and plutonic rocks, Japan: Implications for late Paleozoic to Mesozoic tectonics around the Korean Peninsula. *Isl. Arc* **2018**, *27*, e12220. [CrossRef]

68. Isozaki, Y. Jurassic accretion tectonics of Japan. *Isl. Arc* **1997**, *6*, 25–51. [CrossRef]
69. Stampfli, G.M.; Hochard, C.; Vérard, C.; Wilhem, C. The formation of Pangea. *Tectonophysics* **2016**, *593*, 1–19. [CrossRef]
70. Lee, Y.I.; Lim, H.S.; Choi, T.; Orihashi, Y. Detrital zircon U–Pb ages of the late Paleozoic Sadong Formation in the Pyeongchang coalfield, Gangweon-do Province, Korea: Implications for depositional age and provenance. *J. Geol. Soc. Korea* **2010**, *46*, 73–81. (In Korean)
71. Kim, S.W.; Park, S.-I.; Jang, Y.; Kwon, S.; Kim, S.J.; Santosh, M. Tracking Paleozoic evolution of the South Korean Peninsula from detrital zircon records: Implications for the tectonic history of East Asia. *Gondwana Res.* **2017**, *50*, 195–215. [CrossRef]
72. Cope, T.; Ritts, B.D.; Darby, B.J.; Fildani, A.; Graham, S.A. Late Paleozoic Sedimentation on the Northern Margin of the North China Block: Implications for Regional Tectonics and Climate Change. *Int. Geol. Rev.* **2005**, *47*, 270–296. [CrossRef]
73. Lee, T.-H.; Park, K.-H.; Yi, K. Nature and evolution of the Cretaceous basins in the eastern margin of Eurasia: A case study of the Gyeongsang Basin, SE Korea. *J. Asian Earth Sci.* **2018**, *166*, 19–31. [CrossRef]
74. Wiedenbeck, M.; Hanchar, J.M.; Peck, W.H.; Sylvester, P.; Valley, J.; Whitehouse, M.; Kronz, A.; Morishita, Y.; Nasdala, L.; Fiebig, J.; et al. Further characterisation of the 91500 zircon crystal. *Geostand. Geoanal. Res.* **2004**, *28*, 9–39. [CrossRef]

MDPI
St. Alban-Anlage 66
4052 Basel
Switzerland
Tel. +41 61 683 77 34
Fax +41 61 302 89 18
www.mdpi.com

Minerals Editorial Office
E-mail: minerals@mdpi.com
www.mdpi.com/journal/minerals

www.ingramcontent.com/pod-product-compliance
Ingram Content Group UK Ltd.
Pitfield, Milton Keynes, MK11 3LW, UK
UKHW060122010826
14089UKWH00006B/107

9 783036 549545